Offshore Renewable Energy

Offshore Renewable Energy

Ocean Waves, Tides and Offshore Wind

Special Issue Editors

Eugen Rusu
Vengatesan Venugopal

MDPI • Basel • Beijing • Wuhan • Barcelona • Belgrade

MDPI

Special Issue Editors
Eugen Rusu
"Dunarea de Jos" University of Galati
Romania

Vengatesan Venugopal
The University of Edinburgh
UK

Editorial Office
MDPI
St. Alban-Anlage 66
4052 Basel, Switzerland

This is a reprint of articles from the Special Issue published online in the open access journal *Energies* (ISSN 1996-1073) from 2018 to 2019 (available at: https://www.mdpi.com/journal/energies/special_issues/offshore)

For citation purposes, cite each article independently as indicated on the article page online and as indicated below:

LastName, A.A.; LastName, B.B.; LastName, C.C. Article Title. *Journal Name* **Year**, *Article Number*, Page Range.

ISBN 978-3-03897-592-2 (Pbk)
ISBN 978-3-03897-593-9 (PDF)

Cover image courtesy of Eugen Rusu.

Contents

About the Special Issue Editors

Eugen Rusu received a diploma in Naval Architecture (1982) and a PhD in Mechanical Engineering (1997). In 1999–2004, he worked as a post doc fellow at the at Hydrographical Institute of the Portuguese Navy, where he was responsible for the wave modelling and participated in providing environmental support in some major situations, such as: the accident of the M/V Prestige (2002) and the NATO exercises 'Unified Odyssey' (2002) and 'Swordfish' (2003). Eugen Rusu also worked as a consulting Scientist at the NATO Centre for Maritime Research and Experimentation, La Spezia, Italy (2005), having as his main tasks: Modelling coastal waves and surf zone processes.

Starting in 2006, in parallel with his activity as professor at University Dunarea de Jos of Galati, he currently works as a professor collaborator at CENTEC—Centre for Marine Technology and Ocean Engineering, University of Lisbon, Portugal. Furthermore, since 2012, he has also acted as an expert for the European Commission. Eugen Rusu has published more than 150 works in the fields of renewable energy and marine engineering and has received the awards of Doctor Honoris Causa (2015), at the Maritime University of Constanta, Romania, Outstanding Contribution in Reviewing for the *Renewable Energy* (2015) and *Ocean Engineering* (2016) journals and Top 1% World Reviewers in the field of Engineering (2018). He is also the President of the Council of the Doctoral Schools in Galati University and of the Romanian National Commission of Mechanical Engineering. In 2018, he became corresponding member of the Romanian Academy, the highest scientific and cultural forum in Romania.

Vengatesan Venugopal holds a personal chair in Ocean Engineering at the School of Engineering, University of Edinburgh, United Kingdom. He graduated with a Bachelor of Civil Engineering degree (1991), Master of Technology in Ocean Engineering (1994), and PhD in Ocean Engineering (2003). Since 2000, his research activity has focused on wave and tidal power resource modelling, marine energy device array modelling and its interactions with the environment, numerical and physical modelling of offshore/coastal structures, and wave–current loadings on fixed and floating offshore structures. He has led and co-led several research projects, including UK EPSRC FloWTurb (EP/N021487/1), TeraWatt (EP/J010170/1), EcoWatt2050 (EP/K012851/1) and Adaptation and Resilience in Energy Systems (EP/I035773/1), and EU-funded ('EQUIMAR Protocols', 'PolyWEC') research consortia. He has authored over 140 peer reviewed journal and conference articles, and numerous research reports. He is a Chartered Engineer and Fellow of the Institution of Mechanical Engineers (IMechE).

Preface to "Offshore Renewable Energy"

Among the many forms of energy that can be extracted from the World Oceans, the technologies that are developed to harvest commercial scale electricity productions have been well proven for Offshore wind, wave, and tidal energy sources. Numerous research activities which have been undertaken worldwide to understand how to characterise these ocean resources, convert them into useful electricity using machines, store them, transport them to where they are needed, and distribute them by demand have all been well played and understood. However, as with any other technologies, there has been always the need to fill gaps in research which will improve various elements in each type of energy conversion technologies, leading up to cost reduction and increase of reliability and safety. These cannot be achieved without further research, learning, and communication of the findings relevant to offshore energy conversion. The purpose of this book is to provide further updates and knowledge on the above three ocean sources to the readers.

Technical articles describing various aspects of the offshore wind, wave, and tidal energies, such as resource prediction, shape optimisation of energy converters, optimal design of rotors for cost reductions, numerical modelling of large scale array energy converters, numerical simulation of electricity converting machines, hybrid energy converters, control system for generators, farm interactions, assessing economic benefits, and energy production benefits and so on have been included. This book comprises seventeen original research articles, one review paper, and one editorial. All have been written in easily readable language, but with enriched technical materials addressing some of the current challenges and solutions useful to researchers and industries working in offshore renewables.

The editors of the book would like to record their sincere thanks and acknowledgements to all the contributors of the articles and the continuous support they received from the *Energies* journal editorial staff team, without whose dedication it would have not been possible to publish this book.

Eugen Rusu, Vengatesan Venugopal
Special Issue Editors

energies

MDPI

Editorial

Special Issue "Offshore Renewable Energy: Ocean Waves, Tides and Offshore Wind"

Eugen Rusu [1] and **Vengatesan Venugopal** [2,*]

1   Department of Mechanical Engineering, University Dunarea de Jos of Galati, Galati 800008, Romania; Eugen.Rusu@ugal.ro
2   Institute for Energy Systems, The University of Edinburgh, Edinburgh EH8 9YL, UK
*   Correspondence: V.Venugopal@ed.ac.uk; Tel.: +44-(0)131-650-5652

Received: 18 December 2018; Accepted: 4 January 2019; Published: 7 January 2019

Offshore renewable energy includes several forms of energy extraction from oceans and seas, and the most common and successful offshore technologies developed so far are based on wind, wave and tides. In addition to other resources, wind, waves and tides are considered to be abundant, inexhaustible, and harvestable zero-carbon resources which benefit the human race in tackling energy-related problems, mitigating climate change, and other environmental issues.

Energy production from offshore wind turbines is leading other ocean renewable energy technologies with significant growth since the first installation in Denmark in 1991. According to the Global Wind Energy Council [1], the installed offshore wind capacity at the end of 2017 in 17 countries across the globe (UK, Germany, PR China, Denmark, Netherlands, Belgium, Sweden, Vietnam, Finland, Japan, South Korea, United States, Ireland, Taiwan, Spain, Norway, and France) accounts for 18,814 MW. The UK leads the offshore wind market with over 36% of installed capacity, with Germany in second place with 28.5%. About 84% (15,780 MW) of all offshore installations at the end of 2017 were located in the waters off the coasts of the above-mentioned 11 European countries, and the remaining 16% is located largely in China, Vietnam, Japan, South Korea, the United States and Taiwan.

With wave and tidal energy technologies, although various studies report differing numbers in quantifying resources, the theoretical wave energy potential is estimated to be 32 PWh/year [2], within which the Asian region shares the highest resource of 6200 TWh/year. Only in Europe, a large number of technological advancements has been undertaken, including both research and prototype testing. Similar to wave resources, the quantification of a reliable estimate of global tidal stream energy potential also appears to have variable numbers which are estimated from numerical models; however, the estimated global resource of 3 TW, which includes both tidal ranges and tidal streams [3], indicates its significance. Nevertheless, only a fraction of this could be harvestable, due to several constraints. Unlike the offshore wind sector, only a handful of commercial wave and tidal energy projects have been undertaken globally, which demonstrates in many cases the industry's immaturity, the costs of energy production using these technologies, the lack of investor confidence, political and other market challenges within this particular sector.

The above information illustrates that the Earth is blessed with enormous resources of offshore wind, wave and tidal energy, and an expansion in technologies to harvest them. The research interest in harvesting marine energy is ever-growing, and hence the outcome of these research materials must be widely shared with the research community to increase awareness and enable knowledge transfer activities in relation to new methodologies, modelling techniques, software tools, optimization methods, and the laboratory testing of technologies etc. used in offshore renewables. The editors of this special issue on "Offshore Renewable Energy: Ocean Waves, Tides and Offshore Wind" have made an attempt to publish a book containing original research articles addressing various elements of wind, wave and tidal energies. This book contains research articles written by authors from various countries (Belgium, China, France, Greece, Japan, Malaysia, Netherlands, Romania, Portugal, Spain,

Sweden, Tunisia, United Kingdom) which elaborated several aspects of offshore renewable energy. It covers, through its 18 articles, a broad range of topics including the resource modeling of waves, tides and offshore wind, technologies for energy conversion, numerical and physical modelling of marine energy converters, hybrid energy converters, the shape optimization of energy converters, the modelling of arrays of energy converters; electrical power generation, the control of energy converters, and a macro-economic and cost–benefit analysis.

Regarding offshore wind, the articles discuss the evaluation of state-of-the-art wind technologies suitable for specific locations based on data analysis, the cost of energy evaluated, and longer-term resources estimated for specific areas. Nearshore wind resources in the Black Sea area produced from the European Centre for Medium Weather Forecast (ECMWF) ERA-Interim and AVISO (Archiving, Validation and Interpretation of Satellite Oceanographic data) satellite measurements were used to estimate what type of wind turbines and wind farm configurations would be more suitable for coastal environments [4]. The results indicated that the Crimea Peninsula has the best wind resources; however, considering the geopolitical situation, the western part of this basin (Romania and Bulgaria) was found to be a viable location for developing offshore wind projects. A method was proposed in [5] to minimize the cost of energy (COE) of offshore wind turbines, in which two design parameters, the rated wind speed and rotor radius, are optimally designed, and the relation between the COE and the two design parameters is explored. The recent-past and near-future wind power potential in the Black Sea basin was explored in [6]. An analysis of the wind climate was also undertaken, and the wind-power potential from the recent past was assessed based on two different sources each covering the 30-year period 1981–2010.

In coastal areas, seawater can be desalinated through reverse osmosis (RO) and transformed into freshwater for human use; however, this requires a large reliable electricity supply. An analysis of wave power resource availability in Kilifi-Kenya and an evaluation of the possible use of a wave power converter (WEC) to power desalination plants was described in [7]. Wave energy propagation patterns in the western side of the Iberian nearshore was evaluated in [8]. Several data assimilation techniques were implemented for the model validation. A novel hybrid wind–wave system that integrates an oscillating water column wave energy converter with an offshore wind turbine on a jacket-frame substructure was detailed in [9], in which a scale model of 1:50 was tested under regular and irregular waves to characterise the hydrodynamic response of the WEC sub-system. This study appeared to have led to a proof of concept of this novel hybrid system. Another novel method of estimating wave energy converter performance in variable bathymetry regions was presented in [10], which takes into the account of the interaction of the floating units with the bottom topography. The proposed method used a coupled model which was able to resolve the 3D wave field for the propagation of the waves over the general bottom topography, in combination with a boundary element method (BEM) for the treatment of the diffraction/radiation problems and the evaluation of the flow details on the local scale of the energy absorbers.

A numerical model was proposed in [11], considering not only the interference effect in the multiple floating structures, but also the controlling force of each linear electrical generator. The copper losses in the electrical generator are taken into account when the electrical power is computed. This paper established a relationship between the interference effect and electric powers from wave energy converters. A sliding mode control scheme aimed at oscillating water column (OWC) generation plants using Wells turbines and DFIGs (Doubly Fed Induction Generators) was proposed in [12]. The papers discussed an adaptive sliding mode control scheme that does not require calculating the bounds of the system uncertainties, a Lyapunov analysis of stability for the control algorithm against system uncertainties and disturbances, and a validation of the proposed control scheme through numerical simulations. A generic coupling methodology which allows the modelling of both near-field and far-field effects was presented in [13]. The methodology was exemplified using the mild slope wave propagation model MILDwave and the open source boundary-element method (BEM) code called NEMOH. This paper [14] focused on one of the point absorber wave energy converters (PAWs) of the hybrid platform W2POWER. Two of the model predictive controllers (MPCs) have been designed with the addition of an embedded integrator. In order to analyze and compare the MPCs with a conventional PI type control, a study was carried out to assess the performance and robustness through computer simulations, in which uncertainties in the WEC dynamics were discussed.

A coupled techno–macro-economic model which was used to assess the macro-economic benefit of installing a 5.25 MW farm of oscillating water column wave energy devices at two locations, Orkney in Scotland and Leixoes in Portugal, was presented in [15]. Through an input–output analysis, the wide-reaching macro-economic benefit of the prospective projects was highlighted. The results presented in this paper demonstrated the merit of macro-economic analysis for understanding the wider economic benefit of wave energy projects, while providing an understanding of key physical factors which will dominate the estimated effects. A shape optimization method of a truncated conical point absorber wave energy converter is presented in [16]. This method converts the wave energy absorption efficiency into the matching problem between the wave spectrum of the South China Sea and the buoy's absorption power spectrum. An objective function which combines these two spectra is established to reflect the energy absorbing efficiency. Through a frequency domain hydrodynamic analysis and the response surface method (RSM), the radius, cone angle and draft of the buoy are optimized.

An electrical model of a vertical axis tidal current turbine in Simulink is coupled with a hydrodynamic vortex-model, and its validation is carried out by a comparison with experimental data in [17]. The current turbine was connected to a permanent magnet synchronous generator in a direct drive configuration. The fuzzy gain scheduling (FGS) technique was used in [18] to control the blade pitch angle of a tidal turbine, to protect it from a strong tidal range. Rotational speed control was investigated by means of back-to-back power converters. The optimal speed was provided by using the maximum power point tracking (MPPT) strategy to harness maximum power from the tidal speed. A methodology was presented in [19] to implement an actuator disc approach to model tidal turbines using the Reynolds-averaged Navier–Stokes (RANS) momentum source term for a 20-m diameter turbine in an idealized channel. The model was tuned to match the known coefficient of thrust and operational profiles for a set of validation cases based on published experimental data. Predictions of velocity deficit and turbulent intensity as a function of grid size/mesh resolution used in modelling the turbine were discussed. The results demonstrated that the accuracy of the actuator disc method was highly influenced by the vertical resolutions, as well as the grid density of the disc enclosure.

An up-to-date review of hybrid systems based on marine renewable energies is proposed in [20]. Main characteristics of the different sources, such as solar, wind, tidal, and wave energies, which can provide electrical energy in remote maritime areas are included in the review. A review of multi-source systems based on marine energies was also presented. Offshore locations at the west of Crete shows a wind availability of about 80%; combining this with the installation of large-scale modern wind turbines is expected to result in higher annual benefits. The spatio-temporal correlation of wind and wave energy production shows that wind and wave hybrid stations can contribute significant amounts of clean energy, while at the same time reducing spatial constraints and public acceptance issues. The analysis reported in [21] discussed the benefits of co-located wind–wave technology for Crete.

The above-mentioned articles which constitute this book critically reviewed various technologies of marine energy, investigated the theoretical, numerical and experimental methodologies of modelling various energy converters and their control systems and provided systematic solutions for the readers to easily understand the concepts used and outcomes produced. The editors believe that this book will be useful to many researchers and industries working on offshore renewable energy.

Conflicts of Interest: The authors declare no conflict of interest.

References

1. GWEC—Global Wind Energy Council. Available online: http://gwec.net/policy-research/reports/ (accessed on 17 December 2018).
2. World Energy Council. World Energy Resources. 2016. Available online: https://www.worldenergy.org/data/resources/resource/marine/ (accessed on 17 December 2018).
3. Charlier, R.H.; Justus, J.R. *Ocean Energies: Environmental, Economic and Technological Aspects of Alternative Power Sources*; Elsevier: Amsterdam, The Netherlands, 1993.
4. Onea, F.; Rusu, L. Estimation of the Near Future Wind Power Potential in the Black Sea. *Energies* **2018**, *11*, 2452. [CrossRef]
5. Luo, L.; Zhang, X.; Song, D.; Tang, W.; Yang, J.; Li, L.; Tian, X.; Wen, W. Optimal Design of Rated Wind Speed and Rotor Radius to Minimizing the Cost of Energy for Offshore Wind Turbines. *Energies* **2018**, *11*, 2728. [CrossRef]
6. Ganea, D.; Mereuta, E.; Rusu, L. Evaluation of Some State-Of-The-Art Wind Technologies in the Nearshore of the Black Sea. *Energies* **2018**, *11*, 3198. [CrossRef]
7. Francisco, F.; Leijon, J.; Boström, C.; Engström, J.; Sundberg, J. Wave Power as Solution for Off-Grid Water Desalination Systems: Resource Characterization for Kilifi-Kenya. *Energies* **2018**, *11*, 1004. [CrossRef]
8. Rusu, E. Numerical Modeling of the Wave Energy Propagation in the Iberian Nearshore. *Energies* **2018**, *11*, 980. [CrossRef]
9. Perez-Collazo, C.; Greaves, D.; Iglesias, G. A Novel Hybrid Wind-Wave Energy Converter for Jacket-Frame Substructures. *Energies* **2018**, *11*, 637. [CrossRef]
10. Belibassakis, K.; Bonovas, M.; Rusu, E. A Novel Method for Estimating Wave Energy Converter Performance in Variable Bathymetry Regions and Applications. *Energies* **2018**, *11*, 2092. [CrossRef]
11. Li, Q.; Murai, M.; Kuwada, S. A Study on Electrical Power for Multiple Linear Wave Energy Converter Considering the Interaction Effect. *Energies* **2018**, *11*, 2964. [CrossRef]
12. Barambones, O.; Gonzalez de Durana, J.M.; Calvo, I. Adaptive Sliding Mode Control for a Double Fed Induction Generator Used in an Oscillating Water Column System. *Energies* **2018**, *11*, 2939. [CrossRef]
13. Fernandez, G.V.; Balitsky, P.; Stratigaki, V.; Troch, P. Coupling Methodology for Studying the Far Field Effects of Wave Energy Converter Arrays over a Varying Bathymetry. *Energies* **2018**, *11*, 2899. [CrossRef]
14. Guardeño, R.; Consegliere, A.; López, M.J. A Study about Performance and Robustness of Model Predictive Controllers in a WEC System. *Energies* **2018**, *11*, 2857. [CrossRef]
15. Draycott, S.; Szadkowska, I.; Silva, M.; Ingram, D.M. Assessing the Macro-Economic Benefit of Installing a Farm of Oscillating Water Columns in Scotland and Portugal. *Energies* **2018**, *11*, 2824. [CrossRef]
16. Wen, Y.; Wang, W.; Liu, H.; Mao, L.; Mi, H.; Wang, W.; Zhang, G. A Shape Optimization Method of a Specified Point Absorber Wave Energy Converter for the South China Sea. *Energies* **2018**, *11*, 2645. [CrossRef]
17. Forslund, J.; Goude, A.; Thomas, K. Validation of a Coupled Electrical and Hydrodynamic Simulation Model for a Vertical Axis Marine Current Energy Converter. *Energies* **2018**, *11*, 3067. [CrossRef]
18. Ghefiri, K.; Garrido, A.J.; Rusu, E.; Bouallègue, S.; Haggège, J.; Garrido, I. Fuzzy Supervision Based-Pitch Angle Control of a Tidal Stream Generator for a Disturbed Tidal Input. *Energies* **2018**, *11*, 2989. [CrossRef]
19. Rahman, A.; Venugopal, V.; Thiebot, J. On the Accuracy of Three-Dimensional Actuator Disc Approach in Modelling a Large-Scale Tidal Turbine in a Simple Channel. *Energies* **2018**, *11*, 2151. [CrossRef]
20. Roy, A.; Auger, F.; Dupriez-Robin, F.; Bourguet, S.; Tran, Q.T. Electrical Power Supply of Remote Maritime Areas: A Review of Hybrid Systems Based on Marine Renewable Energies. *Energies* **2018**, *11*, 1904. [CrossRef]
21. Lavidas, G.; Venugopal, V. Energy Production Benefits by Wind and Wave Energies for the Autonomous System of Crete. *Energies* **2018**, *11*, 2741. [CrossRef]

energies

MDPI

Article

A Novel Hybrid Wind-Wave Energy Converter for Jacket-Frame Substructures

Carlos Perez-Collazo * , Deborah Greaves and Gregorio Iglesias

School of Engineering, University of Plymouth, Reynolds Building, PL4 8AA Plymouth, UK;
deborah.greaves@plymouth.ac.uk (D.G.); gregorio.iglesias@plymouth.ac.uk (G.I.)
* Correspondence: carlos.perezcollazo@Plymouth.ac.uk; Tel.: +44-1752-586151

Received: 27 February 2018; Accepted: 11 March 2018; Published: 13 March 2018

Abstract: The growth of the offshore wind industry in the last couple of decades has made this technology a key player in the maritime sector. The sustainable development of the offshore wind sector is crucial for this to consolidate within a global scenario of climate change and increasing threats to the marine environment. In this context, multipurpose platforms have been proposed as a sustainable approach to harnessing different marine resources and combining their use under the same platform. Hybrid wind-wave systems are a type of multipurpose platform where a single platform combines the exploitation of offshore wind and wave energy. In particular, this paper deals with a novel hybrid wind-wave system that integrates an oscillating water column wave energy converter with an offshore wind turbine on a jacket-frame substructure. The main objective of this paper is to characterise the hydrodynamic response of the WEC sub-system of this hybrid energy converter. A 1:50 scale model was tested under regular and irregular waves to characterise the hydrodynamic response of the WEC sub-system. The results from this analysis lead to the proof of concept of this novel hybrid system; but additionally, to characterising its behaviour and interaction with the wave field, which is a requirement for fully understanding the benefits of hybrid systems.

Keywords: wave energy; hybrid wind-wave; concept development; oscillating water column (OWC); physical modelling; hydrodynamic response

---

1. Introduction

In the last couple of decades, offshore wind energy has become a major player in the world's renewable energy sector, with 15.8 GW of installed capacity in Europe at the end of 2017 [1]. This exceptional development has been, to a large extent, driven by the relatively shallow waters and good wind resources of the North Sea, which washes the shores of one of the most industrialised regions of the planet [2]. The great potential for development of offshore wind has raised the expectations that this will play a leading role in Europe's future energy supply, pushing its industry to establish a target of 460 GW of installed capacity by 2050 [3]. It is clear that, for this target to be realised, a significant increase must be achieved, especially by developing deep water and floating substructure systems.

In a global scenario of climate change and amid mounting threats to the marine environment [4–7], the sustainable development of offshore wind is not only crucial for the consolidation of the industry, but also to providing a reliable and accessible source of renewable energy. In this context, multipurpose platforms have been suggested as a sustainable means of exploitation of certain maritime resources, which are usually in the same area [8–11]—e.g., marine renewable energies (MREs), food resources (fisheries and aquaculture), maritime transport and leisure, among others. On the basis of the strong synergies between offshore wind and wave energy [12–14], hybrid wind-wave systems have been proposed as one of the most promising types of multipurpose platforms [15].

Previous works on hybrid systems have mostly been grouped around some EU-funded projects, whose aim was to develop some conceptual ideas and set the basis for future developments, defining guidelines and recommended practices for the wider group of multipurpose platforms [16–20]. This work has been complemented with some concepts proposed by the industry, e.g., [21–24]. At the moment of writing, there are only a few scientific publications dealing with hybrid systems [25–28], with most of the previous work around the wider group of combined wind-wave systems [29]. The characterisation of the combined resource together with the study of the potential combination of both technologies has been studied by [30–32]—e.g., through the co-location feasibility index [33,34]. The effects of the temporal correlation of both wind and wave resources on the combined power output and its grid integration have been studied by [35–43]. The study of the shadow-effect of co-located wind-wave farms on the operation and cost of the overall farm was carried out by [44–46].

In particular, this research deals with the development of a novel hybrid wind-wave energy converter for jacket-frame offshore wind substructures. The proposed hybrid system integrates an oscillating water column (OWC) wave energy converter (WEC) sub-system with a jacket-frame type of offshore wind substructure. An intensive test campaign was carried out using a 1:50 scale model of the hybrid device to characterise the hydrodynamic response of the WEC sub-system. This was carried out following a three-step methodology: (i) the interaction between the device and its surrounding wave field was studied by means of an incident and reflected wave analysis (IRWA); (ii) the performance of the OWC was studied using the capture width ratio; and (iii) the response of the main parameters influencing the performance of the OWC—i.e., the free surface elevation and the pneumatic pressure inside the OWC chamber—was studied by means of the response amplitude operator (RAO).

The content of this article is structured as follows. Section 2 defines the hybrid device's WEC sub-system. Section 3 tackles the materials and methods for the experimental campaign, including the physical model, the experimental set-up and programme, and the data analysis. The results are presented in Section 4 and discussed in Section 5. Finally, conclusions are drawn in Section 6.

2. The OWC WEC Sub-System

The hybrid wind-wave energy converter concept considered for this work builds on that presented in [47] (Figure 1a). An OWC WEC sub-system prototype (Figure 1c) was outlined in the framework of a new patent [48], with number WO2016185189A1. A novel hybrid wind-wave energy converter is defined, where the OWC chamber forming the WEC sub-system has the capability to self-adapt to different wave heights and tidal ranges as well as to the direction of the incident waves. The adaptability of the OWC chamber is achieved by means of a self-adaptable skirt and the change of the relative position between the chamber and the substructure.

Figure 1b shows a schematic representation of one of the possible configurations of the prototype. The figure shows frontal and top views of the device, where some of its components and parts are indicated. The proposed device is formed by a chamber (1); a substructure system (2) to link the device to the seabed (i.e., usually the substructure system will be shared with a wind turbine); a ballast tank (3), defined as part of the hull of the chamber between the inner (7) and external walls of the chamber; a skirt (4) or extension at the bottom of the chamber; one or more air turbines (5), which act as the OWC power take-off, driving the electric generator to produce electricity; a security and control system including pressure relief valves (6); and a set of bulkheads (8) that provide structural strength and divide the internal part of the chamber into separate segments (9). Note that the numbers shown in brackets refer to those in the figure.

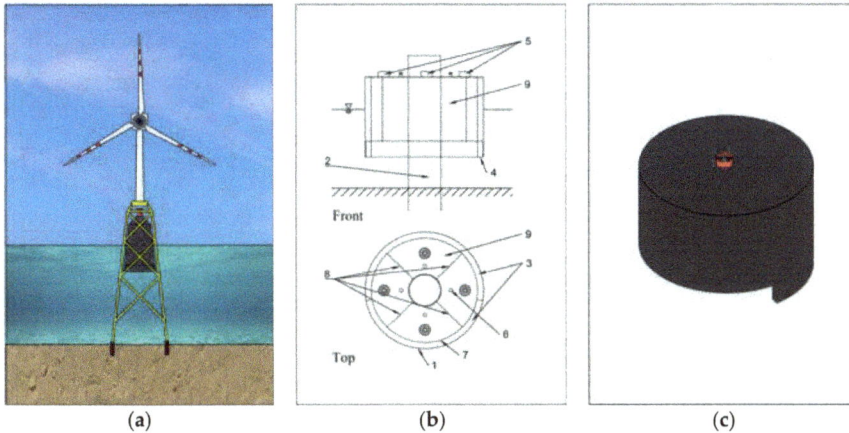

Figure 1. Proposed hybrid wind-wave energy converter for jacket-frame offshore wind substructures: (a) conceptual representation of the hybrid model; (b) front and top views of the prototype showing its different parts; and (c) a perspective view of the WEC sub-system.

The hybrid system proposed in the patent includes an OWC WEC sub-system that integrates a *skirt* of a certain length l_S (Figure 2a) over a certain angular sector α (Figure 2b). The device can be either designed for the skirt length and aperture angle to be constant, or equipped with a mechanism that enables the aperture angle and depth to be modified.

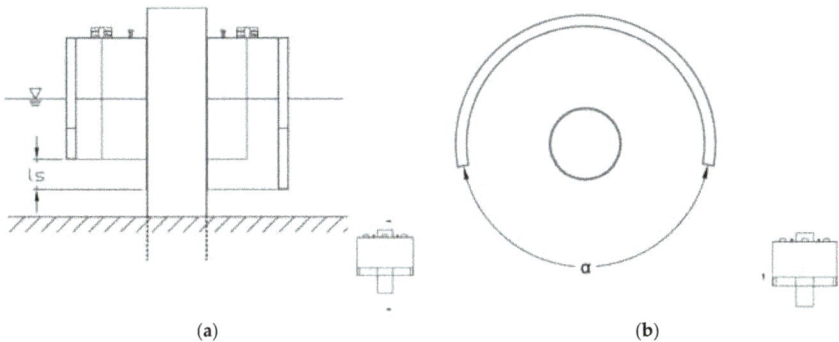

Figure 2. Schematic representation of the OWC skirt, its length (l_S) and aperture angle (α): (a) cut view of the device with a vertical lateral plane; and (b) cut view of the device with a horizontal plane at the skirt level (partially reproduced from [48]).

3. Materials and Methods

3.1. The Physical Model

A 1:50 scale model of the proposed hybrid wind-wave energy converter was built. The design of the model (Figure 3) considered the limitations of the experimental facility—e.g., the wave maker capabilities and main dimensions of the flume [49], and tank blockage effects [50]—together with various guidelines and recommended practices for physical modelling of WECs [51,52]. A jacket-frame substructure proposed by [53] was considered to define the model for a 50 m water depth site [54]. Froude similitude and geometrical similarity were considered to define the jacket frame and the section of the OWC chamber below the mean water level. However, the volume of the pneumatic section of the OWC—i.e., the OWC chamber above the mean water level—was scaled [55–57] using λ^2 as the scale ratio, rather than the λ^3 dictated by Froude similarity, to account for air compressibility [58,59]. The jacket-frame substructure was the limiting factor in designing the OWC subsystem, and in particular in defining its diameter, which was selected to fit within the jacket-frame, and so that the connection pipe between the OWC chamber and the air reservoir could pass through the top aperture of the jacket-frame. Table 1 shows the main characteristics and dimensions of the model.

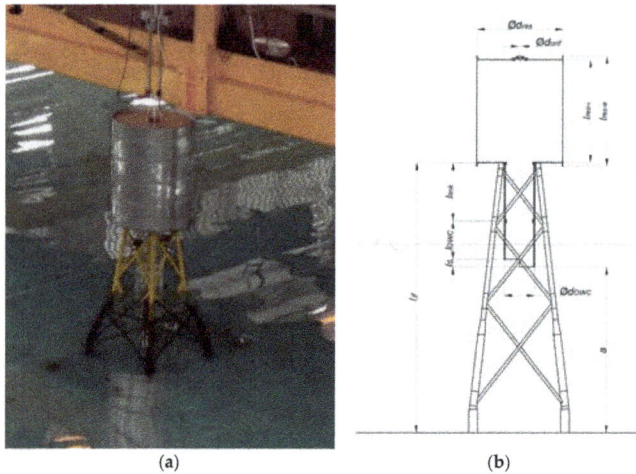

(a)   (b)

Figure 3. 1:50 model of the hybrid wind-wave energy converter: (**a**) during tests at the University of Plymouth's COAST Laboratory, and (**b**) cross-sectional view of the model.

Table 1. Model characteristics and dimensions.

Parameter	Symbol	Dimension
Air reservoir external diameter	d_{res}	0.450 m
Air reservoir external length	l_{res-e}	0.585 m
Air reservoir internal length	l_{res-i}	0.545 m
Air reservoir wall thickness	e_{res}	1.5×10^{-3} m
Chamber draught	c	8.0×10^{-2} m
Chamber external diameter	d_{OWC}	0.160 m
Chamber length	l_{OWC}	0.200 m
Chamber-reservoir link length	l_{link}	0.294 m
Chamber wall-thickness	e_{OWC}	4.0×10^{-3} m
Distance from the skirt to the floor	c	0.884 m
Jacket-frame length	l_{jf}	1.438 m
Skirt length	l_s	4.0×10^{-2} m
Skirt angle	α	180 deg
Water depth	h	1.0 m

The model was built out of four different parts. First, a lattice of welded carbon steel pipes was used for the jacket-frame. A clear acrylic pipe of 0.16 m diameter was used for the OWC chamber. The air reservoir—for the additional volume of air—was built using galvanised steel pipe of 0.45 m diameter and galvanised sheets. Finally, the section linking the OWC chamber and the air reservoir was built using the same acrylic pipe as for the OWC chamber.

The damping exerted on the OWC chamber by an impulse turbine can be modelled by means of an orifice [60,61]. But if further calculations concerning the efficiency of the turbine are required, the orifice may be replaced by an Actuator Disk Model in the case of a numerical simulation [62,63]. In order to study the effect of the turbine-chamber coupling in the model, three different orifice sizes (turbine damping) were considered [64]. The diameter of the orifices was selected for three values of the area coefficient—i.e., the area coefficient is defined as the ratio between the area of the orifice and the water plane area of the inner OWC chamber—of 0.5%, 1% and 1.5% [65].

3.2. Experimental Set-Up and Testing Programme

The ocean basin at the University of Plymouth's COAST Laboratory was the facility selected to conduct the experimental campaign. This has a total length of 35 m, a width of 15.5 m and a variable floor depth that, for the purpose of this study, was adjusted at 1.0 m to match the Wave Hub test site—i.e., a test centre of the North coast of Cornwall and in particular in selecting the wave conditions. Waves are generated from a flap-type wave-maker, from Edinburg Designs, Ltd. (EDL, Edinburgh, UK). The reference system adopted for the experimental set-up defines: the longitudinal axis (Ox), passing through the mid plane of the basin, with $x = 0$ at the wave-makers and positive towards the model; the vertical axis (Oz), with positive direction upwards and $z = 0$ at the still water level; and the transversal axis (Oy), perpendicular to the basin, with positive direction such that the trihedral $Oxyz$ has a positive orientation.

The free surface displacement along the basin was measured using four conductive wave gauges (WGs), and the displacement of the free surface inside the OWC chamber with an additional WG (Figure 4). The first group of WGs (WG1, WG2 and WG3) were positioned along the centreline of the basin, at $x_1 = 9.43$ m, $x_2 = 9.87$ m and $x_3 = 10.12$ m, to record the data for an IRWA. The fourth (WG4) was positioned in the lee of the model along the centreline of the basin, at $x_4 = 14.23$ m, to record the transmitted wave. The remaining wave gauge (WG5) was positioned inside the OWC chamber at $x_5 = 12.72$ m, to measure the free surface oscillation inside the chamber. In addition, a differential pressure transducer (PT), PX2300-0.5BDI, from Omega was used to measure the differential pneumatic pressure between inside and outside the OWC chamber. Data were acquired using Edinburg Designs hardware and a National Instruments acquisition system for the wave WG and the PT, respectively, both at a sampling frequency of 128 Hz.

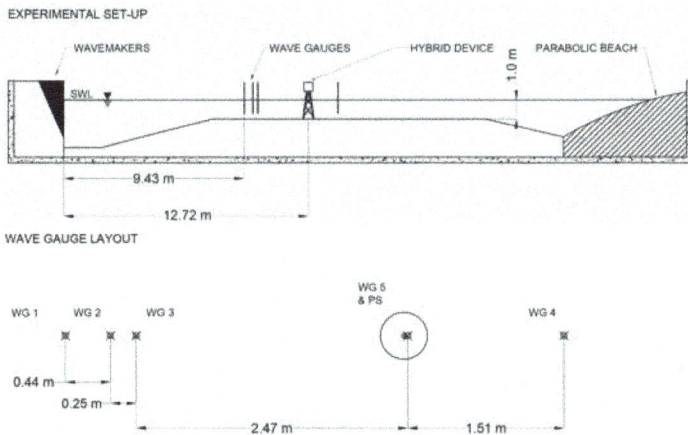

Figure 4. Side and top views of the experimental set-up.

The experimental programme was defined for a range of regular and irregular wave conditions and three different orifice sizes—note that for irregular waves, only the intermediate orifice size was used. Following [51,66], the tests were structured into three different series, one for regular waves (Series A) and two for irregular waves (Series B and Series C). Series A defines regular waves by combining five wave heights ($H = 1.5, 2.5, 3.5, 4.5$ and 5.5 m, in prototype values) and seven wave periods ($T = 7, 8, 9, 10, 11, 12$ and 13 s, also in prototype values). The duration of the tests was defined to cover at least 100 waves. Series B defines six sea states using a joint North Sea wave project (JONSWAP) spectrum [49], to study the hydrodynamic response of the device under irregular waves.

In addition, the effect of the wave period on the response of the device is studied in Series C, which is defined for seven JONSWAP sea states with the same significant wave height and different peak wave periods (Table 2). The duration of all the irregular tests was selected to match 60 min at prototype scale following [51], covering between 271 and 571 waves.

Table 2. Wave conditions for the two irregular wave series (data in prototype values).

Test Series	Test Number	H_S	T_E	T_Z	T_P
	B01	0.5 m	6.05 s	5.04 s	7.06 s
	B02	1.5 m	6.49 s	5.41 s	7.57 s
Series B	B03	2.5 m	6.98 s	5.82 s	8.14 s
	B04	3.5 m	8.00 s	6.67 s	9.33 s
	B05	4.5 m	8.46 s	7.05 s	9.87 s
	B06	5.5 m	9.10 s	7.58 s	10.62 s
	C01		5.40 s	4.50 s	6.30 s
	C02		6.60 s	5.50 s	7.70 s
	C03		7.20 s	6.00 s	8.40 s
Series C	C04	3.5 m	8.40 s	7.00 s	9.80 s
	C05		9.00 s	7.50 s	10.50 s
	C06		9.60 s	8.00 s	11.20 s
	C07		11.40 s	9.50 s	13.30 s

The time series of the free surface elevation recorded from the wave gauges at their different positions along the basin, the free surface recorded by the wave gauge inside the oscillating water column and the differential pneumatic pressure between inside and outside the OWC chamber recorded by the pressure transducer are presented in Figure 5 as an example of an irregular waves test.

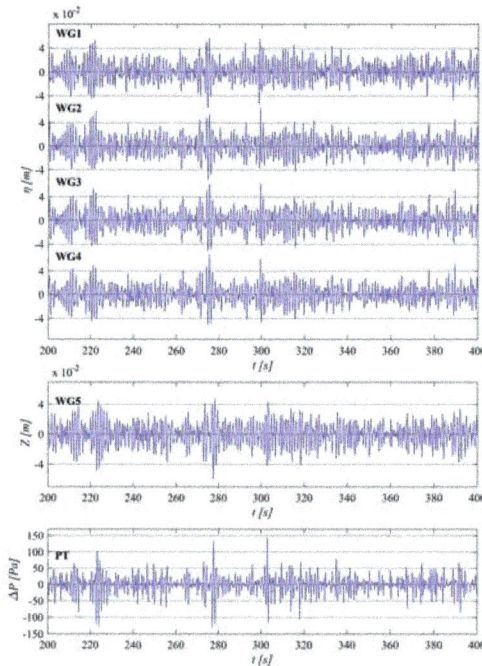

Figure 5. Recorded data, during part of a test, from the free surface elevations along the flume, the oscillation of the water column and the differential pressure between the inner OWC chamber and the atmosphere (H_S = 3.5 m, T_P = 9.33 s, B^* = 47.87).

3.3. Data Analysis

Incident and reflected waves were characterised following the incident and reflected wave analysis (IRWA) method proposed by [67,68]. Data from the frontal group of wave gauges (WG1, WG2 and WG3) were used as input for the method. Based on the incident and reflected wave heights and on the transmitted wave height obtained from the wave gauge in the lee of the model (WG4), the reflection and transmission coefficients (K_R and K_T) can be defined for regular waves [69] as

$$K_R = \frac{H_R}{H_I}, \tag{1}$$

$$K_T = \frac{H_T}{H_I}, \tag{2}$$

and for irregular waves as

$$K_R = \sqrt{\frac{m_{0R}}{m_{0I}}}, \tag{3}$$

$$K_T = \sqrt{\frac{m_{0T}}{m_{0I}}}, \tag{4}$$

where m_{0i} is the generic zero order moment,

$$m_{0i} = \int_{f_{min}}^{f_{max}} S_i(f)df, \tag{5}$$

S_i is a generic power spectral density, and the respective incident, reflected and transmitted zero-order moments (m_{0I}, m_{0R} and m_{0T}) can be obtained by replacing S_i with the power spectral density of the respective incident, reflected and transmitted waves.

The wave energy flux, or mean power for the incident waves per metre of wave front (J), can be calculated from the incident wave from the IRWA for regular waves

$$J = \frac{\rho_w g H_I^2 c_g}{8}, \tag{6}$$

and for irregular waves

$$J = \rho_w g \sum_{i=1}^{N} S_i(c_g)_i \Delta f, \tag{7}$$

where ρ_w is the water density; g the gravitational acceleration; H_I the incident wave height: N is the number of frequency components or bands (for each Δf), and S_i and $(c_g)_i$ are the spectral density and the group velocity for the i-th band, respectively. The group velocity is given by

$$(c_g)_i = n_i c_i, \tag{8}$$

$$n_i = \frac{1}{2}\left(1 + \frac{2k_i h}{\sinh(2k_i h)}\right), \tag{9}$$

where h is the water depth, k_i the wave number for the i-th frequency band and c_i is the phase celerity,

$$c_i = \frac{\omega_i}{k_i}, \tag{10}$$

where ω_i is the angular frequency of the i-th band, obtained from the dispersion relationship,

$$\omega^2 = g k_i \tanh(k_i h). \tag{11}$$

The mean pneumatic power of the OWC (P_m) during a test can be defined, following [64], by

$$P_m = \frac{1}{t_{max}} \int_0^{t_{max}} \Delta p \, q \, dt, \tag{12}$$

where Δp is the relative pressure, i.e., the pressure difference between the inner chamber and the atmosphere, q is the volumetric air flow rate through the chamber's orifice, t is time and t_{max} is the duration of the test. Here, Δp is directly obtained from the differential pressure transducer (PT) data, while q is approximated assuming incompressible flow and using the velocity of the free surface inside the inner chamber [60,70], obtained by numerical differentiation of the free surface elevation recorded by wave gauge WG5.S.

The capture width ratio (C_{WR}) is the parameter used to evaluate the performance of the WEC sub-system. C_{WR} is defined as the ratio between the power absorbed by the WEC—i.e., the mean pneumatic power (P_m)—and the wave power incident on the device per metre of wave front (J) times a relevant dimension of the device (b) in m—i.e., for this paper, this is the external diameter of the chamber (d_{OWC})—

$$C_{WR} = \frac{P_m}{J \, b}. \tag{13}$$

The response amplitude operator (RAO) is used to characterise the response of the two main parameters controlling the performance of the device—i.e., the amplitude of the free surface oscillation and the pneumatic pressure of the OWC chamber—against the incident wave. The RAO operator for the translation motion of the chamber's free surface oscillation in heave (RAO_C) can be rewritten as

$$RAO_C = \frac{H_C}{H_I}, \tag{14}$$

where H_I is the incident wave height and H_C is the chamber's free surface oscillation height. A similar approach is followed for the RAO of the pneumatic pressure (RAO_P), but divided by the water density (ρ_w) and the gravitational acceleration (g) to make the RAO non-dimensional

$$RAO_P = \frac{1}{\rho_w g} \frac{H_P}{H_I}, \tag{15}$$

where H_I is the incident wave height and H_P is the variation of the pneumatic pressure height.

The five parameters defined to characterise the hydrodynamic response of the hybrid device (K_R, K_T, C_{WR}, RAO_C and RAO_P) depend not only on the wave conditions (wave height and period) but also on the damping induced by the orifice on the OWC system [71]. To quantify its influence, the dimensionless damping coefficient (B^*) can be defined, following [49], as

$$B^* = \frac{\Delta p^{1/2}}{q \rho_a^{1/2}}, \tag{16}$$

where Δp is the pressure between the chamber and the atmosphere, q is the volumetric air-flow rate through the chamber's orifice, and ρ_a is the air density. For this work, the damping coefficients for the three orifice diameters tested (d_o = 11, 15 and 19 mm) are B^* = 64.10, 47.87 and 39.59, respectively.

4. Results

4.1. Incident and Reflected Wave Analysis (IRWA)

An IRWA was carried out with data from the experimental campaign to obtain the incident and reflected wave heights and to determine the reflection and transmission coefficients (K_R and K_T). The results for regular waves are presented in Figure 6 for the three damping coefficients versus the wave steepness (S). In addition, Figure 7 presents the results for irregular waves for the intermediate damping coefficient ($B^* = 47.87$) versus the significant wave steepness (S_S)—refer to Appendix A for the definition of both wave steepnesses (S and S_S).

K_R values range from 0.09 to 0.40, with an average value of 0.19, for regular waves; while it ranges from 0.40 to 0.64, with an average value of 0.46, for irregular waves. K_T values range from 0.28 to 0.42, with an average value of 0.35, for regular waves; and from 0.27 to 0.58, with an average value of 0.39, for irregular waves. It is clear that both coefficients (K_R and K_T) are, in general, driven by the wave period and, to a small extent, the wave steepness and turbine damping. Data are, in general, well grouped, except for the two smallest periods ($T = 7$ and 8 s) in regular waves, which show more scattered results.

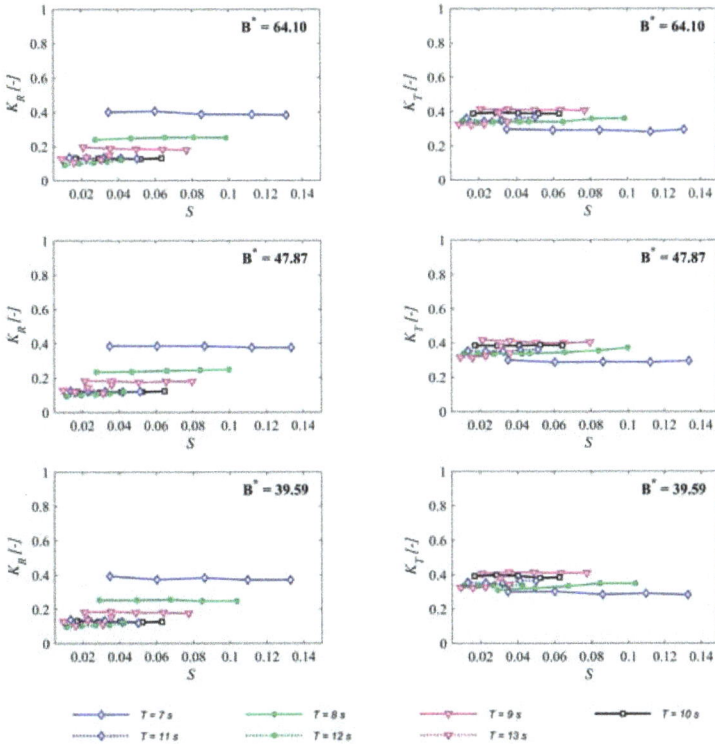

Figure 6. Variation of the reflection and transmission coefficients (K_R and K_T) with the wave steepness (S) for regular waves (Series A); and for different values of the wave period (T) and damping coefficient (B^*) (prototype data).

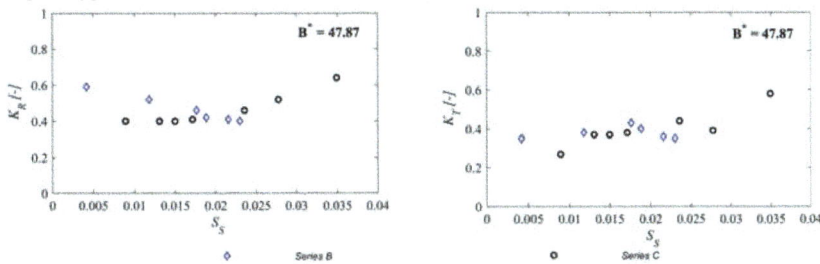

Figure 7. Variation of the reflection and transmission coefficients (K_R and K_T) with the significant wave steepness (S_S) for irregular waves (Series B and C); and for the intermediate damping coefficient (prototype data).

For a given value of wave height (Series C), the reflection coefficient (K_R) does not vary significantly with the wave steepness for wave steepness values of up to 0.02, beyond with the reflection coefficient increases with the wave steepness (Figure 6). This means that the influence of the wave steepness on the reflection coefficient is limited to large wave periods. Furthermore, it can also be noticed that K_R increases when the damping coefficient increases—i.e., when the orifice diameter size decreases. The transmission coefficient (K_T) shows, in general, well-grouped values around 0.35–0.40 for both regular and irregular waves. For regular waves, contrary to the reflection coefficient, K_T increases with the wave period and decreases with the damping coefficient; similarly, for irregular waves, K_T increases with the significant wave steepness.

4.2. OWC Performance

The capture width ratio (C_{WR}) was used to evaluate the performance of the hybrid wind-wave energy converter. The C_{WR} is represented for regular waves in Figure 8 for the three damping coefficients tested versus the wave steepness (S); and for irregular waves in Figure 9 for the intermediate damping coefficient ($B^* = 47.87$) versus the significant wave steepness (S_S). In addition, the C_{WR} and mean pneumatic power (P_m) are also represented as the capture width and power matrices in Figures 10 and 11, respectively.

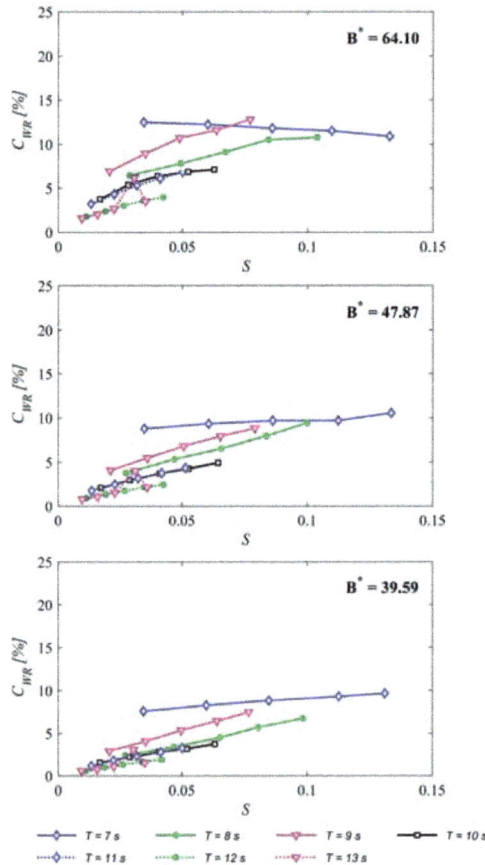

Figure 8. Variation of the capture width ratio (C_{WR}) with the wave steepness (S) for regular waves (Series A); and for different values of the wave period (T) and damping coefficient (B^*) (prototype data).

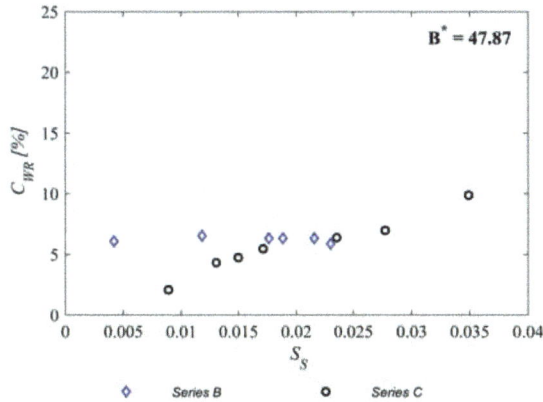

Figure 9. Variation of the capture width ratio (C_{WR}) with the significant wave steepness (S_S) for irregular waves (Series B and C); and for the intermediate damping coefficient (prototype data).

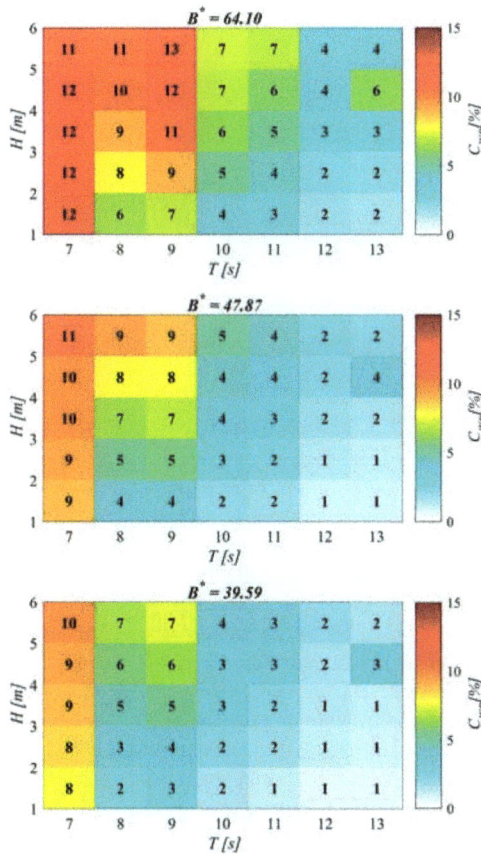

Figure 10. Variation of the capture width ratio (C_{WR}) with the wave height (H) and wave period (T), or C_{WR} matrix, for regular waves (Series A) and for different values of the damping coefficient (B^*) (prototype data).

15

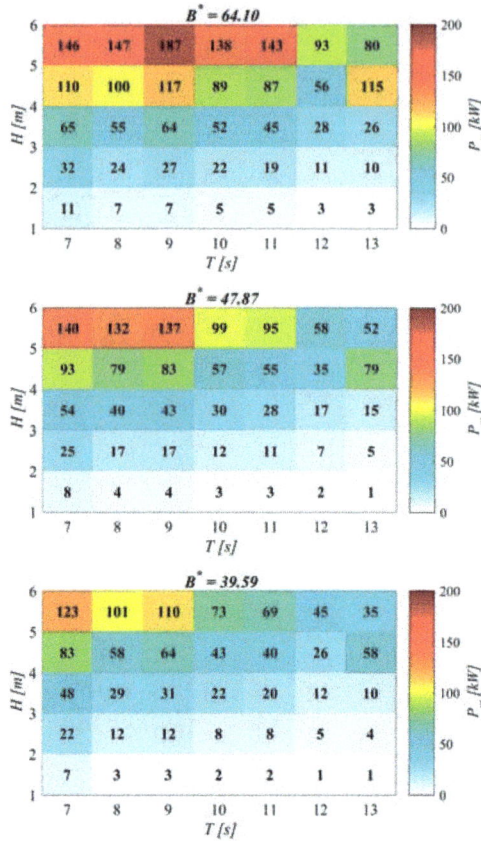

Figure 11. Power matrix showing the variation of the mean pneumatic power (P_m) with the wave height (H) and period (T) for regular waves (Series A) and for different values of the damping coefficient (B^*) (prototype data).

C_{WR} values range from 1% to 13%, with an average value of 5%, for regular waves; and from 2% to 10% for irregular waves, with an average value of 6%. It is clear that the C_{WR} is mostly driven by the turbine damping and the wave period, and to a small extent by the wave steepness. Data are, in general, well grouped and show a similar trend, except for the smallest period ($T = 7$ s) in regular waves, which shows more scattered results. Note that the regular wave condition with a wave height ($H = 4.5$ m) and a wave period ($T = 13$ s) seems to be an outlier (Figure 10).

From Figures 8 and 9, it can be observed that the capture width ratio (C_{WR}) is highly influenced by the turbine damping—with average values of the C_{WR} 7%, 5% and 4% for turbine damping values of $B^* = 64.10$, 47.87 and 39.59, respectively, for regular waves and 6% for irregular waves and $B^* = 47.87$. The largest turbine damping value—i.e., the smallest orifice diameter size ($B^* = 64.10$)—thus, not only the best performance—i.e., the maximum C_{WR} for this damping is 13% while the maximum value for the other two damping values is 11% and 10% for $B^* = 47.87$ and 39.59, respectively—but also a wider region of larger efficiency. Furthermore, the wave period does exert a strong influence on the C_{WR}, increasing, in general, when the wave period decreases; with the exception of the smallest wave period ($T = 7$ s) for the largest damping value ($B^* = 64.10$). This behaviour can also be clearly identified for irregular waves (Figure 9) by comparing the results of both Series. While C_{WR} remains, in general, constant for Series B, it increases with the significant wave steepness (S_S) for Series C, and so it does when the wave period decreases (Series C keeps a constant significant wave height).

To better understand the relevance of the different parameters influencing the performance of the device, the capture width matrix is represented, for regular waves, in terms of the wave height (H) and the wave period (T) for the three turbine damping values (B^*) (Figure 10). An area of best performance can be identified for the three damping values for the lower wave periods ($T < 10$ s). The intermediate and the smallest damping values ($B^* = 47.87$ and 39.59) have their peaks of maximum efficiency at $T = 7$ s; by contrast, the largest damping value ($B^* = 64.10$) has a primary efficiency peak at $T = 9$ s, followed closely by a secondary efficiency peak at $T = 7$ s. Complementarily, the mean pneumatic power (P_m) may be presented in terms of the wave height (H) and period (T) for the three turbine damping values (B^*) in the form of power matrices (Figure 11). An area of best performance can be identified for the larger wave heights ($H > 3$ m). Similarly to the capture width matrix, two peaks of power output are found for the three damping values at $T = 7$ s and $T = 9$ s.

4.3. Device Response

The response amplitude operator (RAO) was used to evaluate the response of the two main parameters influencing the performance of the OWC—i.e., the free surface elevation and the pneumatic pressure inside the OWC chamber (RAO_C and RAO_P respectively). The results for regular waves are presented in Figure 12 for the three damping coefficient values versus the wave steepness (S) and in Figure 13 versus the wave frequency.

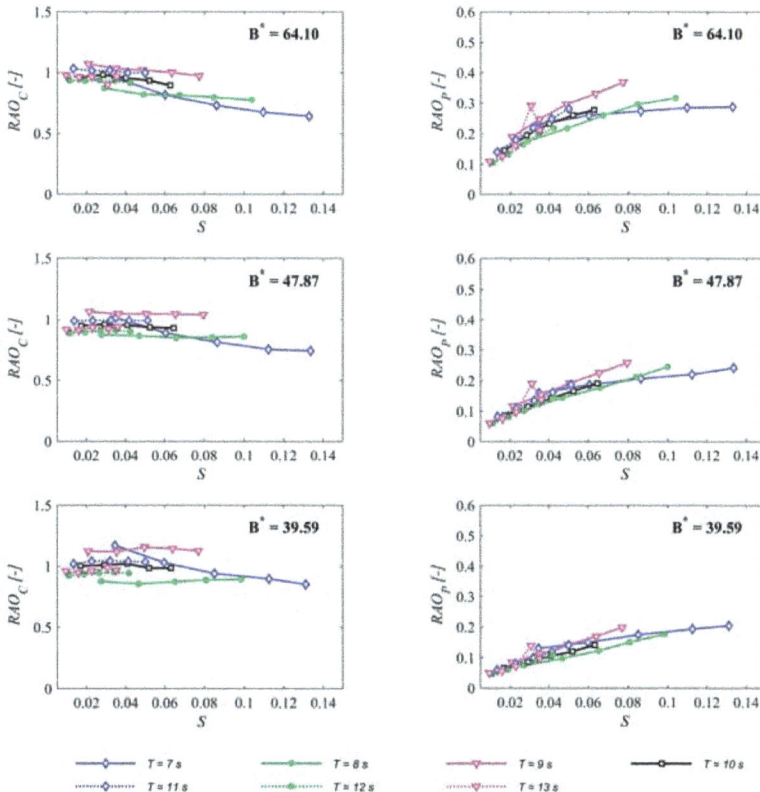

Figure 12. Variation of the response amplitude operator (RAO) for the free surface elevation and the differential pressure between inside the OWC chamber and the atmosphere (RAO_C and RAO_P respectively) with the wave steepness (S) for regular waves (Series A): and for different values of the wave period (T) and damping coefficient (B^*) (prototype data).

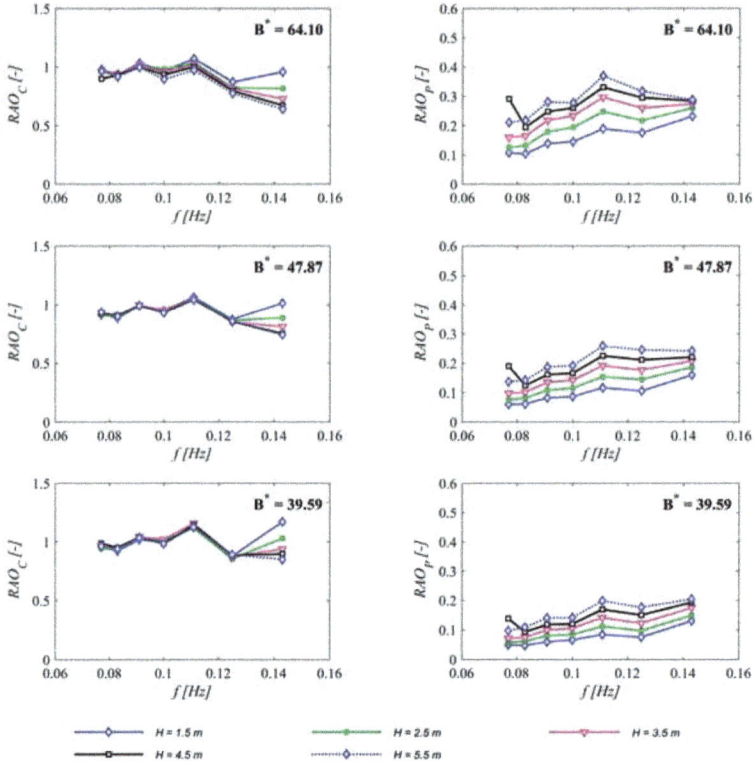

Figure 13. Variation of the response amplitude operator (RAO) for the free surface elevation and the differential pressure between inside the OWC chamber and the atmosphere (RAO_C and RAO_P respectively) with the wave frequency (f) for regular waves (Series A); and for different values of the wave height (H) and damping coefficient (B^*) (prototype data).

RAO_C values range from 0.64 to 1.17, with an average value of 0.95 while RAO_P values range from 0.05 to 0.37, with an average value of 0.16. Data are, in general, well grouped and show similar trends. Note that the wave condition with a wave height $H = 4.5$ m and a wave period $T = 13$ s seems to be an outlier, as identified in the previous section.

In Figure 12, it can be clearly observed that both RAOs are mainly driven by the damping coefficient (B^*) and the wave steepness (S). RAO_C and RAO_P show opposite behaviours, when RAO_C increases RAO_P decreases and vice versa. The higher the damping coefficient value and the higher the wave steepness, the lower the amplitude of the free surface oscillation (RAO_C). Therefore, the higher the damping coefficient and the higher the wave steepness, the lower the amplitude of the differential pneumatic pressure between the inner OWC chamber and the atmosphere (RAO_P).

Figure 13 shows the traditional representation for RAOs—i.e., versus wave frequency (f). From the analysis of this figure, the following observations can be made:

(1) RAO_C values converge, in general, to 1 for lower wave frequencies for all damping values.
(2) The effect of the wave height on RAO_C is, in general, very limited for medium to low wave frequency values; however, this influence becomes more relevant for larger wave frequency values ($f > 0.12$ Hz, in prototype values).
(3) Two peaks of maximum free surface oscillations (RAO_C) values are found at the frequency values corresponding to the wave periods $T = 7$ and 9 s—i.e., matching the two periods of maximum OWC efficiency identified in the previous section (Figure 10).

(4) RAO_C diverges for the larger wave frequencies.

(5) The parameter affecting the RAO_P values most is the damping coefficient, increasing with RAO_P value with the damping.

(6) The higher the wave height and the lower the wave period, the higher the RAO_P.

(7) Two peaks of maximum values of RAO_P can be observed for the largest damping value ($B^* = 64.10$) at the frequencies ($f = 0.111$ and 0.143 Hz) while only the second one is observed for the other two damping values—i.e., these correspond to the same peaks observed for the C_{WR} at $T = 9$ and 7 s, respectively.

5. Discussion

A comprehensive series of physical model tests were carried out as a first step in the development of this novel concept of a hybrid wind-wave energy converter for jacket-frame substructures. A simplified version of the WEC sub-system was defined and tested. Its hydrodynamic response was characterised to better understand the performance of the device and its interaction with the wave field; and to set the reference for future developments of the device.

Three parameters were selected to investigate the hydrodynamic response of the hybrid energy converter: the wave height, the wave period and the damping coefficient—i.e., different turbine damping values were modelled by considering different orifice plates, with three different orifice diameter sizes. In total, 118 tests were performed, considering regular and irregular waves, and these were structured into three test series. The methodology followed to characterise the hydrodynamic response of the device was carried out considering three different sets of analysis techniques: (i) an incident and reflected wave analysis (IRWA), to determine the reflection and transmission coefficients (K_R and K_T); (ii) the analysis of the capture width ratio (C_{WR}), to study the efficiency or ratio between the pneumatic power output of the OWC and the incident wave power; and (iii) the analysis of the RAOs of the free surface oscillation and the pneumatic pressure inside the OWC chamber (RAO_C and RAO_P respectively), to study the relationship between these key components of the OWC power output with the incident wave.

The IRWA identified the wave period as the parameter that influences the wave reflection and transmission coefficients most—i.e., the influence of the wave period is much more relevant than that of the turbine damping coefficient or the wave steepness. As the wave period increases, K_R decreases—note that this behaviour is the opposite to that observed for a coastal structure extending down to the seabed, but similar to the one observed for other WECs, e.g., [72,73]. As the wave period increases, K_T increases very slightly. The IRWA shows an interaction of the device with the wave field that reflects between 9% and 40% of the incident wave power and reflects between 28% and 42%. Note that this interaction is crucial to understanding the implications that the 'shadow effect' may have at a larger scale—at the wind farm scale or at the nearest coasts.

The analysis of the hybrid energy converter performance identified the damping and the wave period as the parameters influencing the C_{WR} the most. The accentuated influence of the turbine damping on the device performance highlights the importance of the appropriate selection of the turbine damping when designing an OWC device, as shown previously by, e.g., [49,74–78]. For the wave conditions and damping coefficient values tested, the largest damping coefficient ($B^* = 64.10$) is the one showing, generally, the highest values of C_{WR}. A peak of C_{WR} was found for the three damping values at $T = 7$ s, and a second peak was found for the larger damping coefficient value ($B^* = 64.10$) at $T = 9$ s. The capture width ratio matrix shows an area of maximum efficiency for the lower wave period ($T < 10$ s) and across most of the wave heights. The maximum value of C_{WR} is approx. 13%, with average values between 4% and 7%. These results are in line with the best-fit equation based on the statistical analysis of about 20 different OWC devices [79]. Indeed, following the best-fit equation, and considering the diameter of the OWC chamber (8 m, at prototype scale) as the width of the device, a C_{WR} value of about 15.5% is to be expected. Notwithstanding, previous work has shown that maximum values that are remarkably higher can be attained, e.g., 80% [60] or

87% [76]. In principle, these figures would appear to indicate that there is plenty of margin to optimise the performance of the OWC chamber. However, given the dependence of the C_{WR} on the chamber width that is apparent in the best-fit equation, the C_{WR} is limited by the restriction imposed on the chamber dimensions by the jacket-frame substructure within which it is to be mounted.

Comparing the shape of the capture width and power matrices with those from previous works, certain differences are apparent. For example, comparing the previous Figure 10 with the results in [49], it may be seen that maximum C_{WR} values correspond to greater wave heights in the case of the present OWC—and in both cases to lower wave periods ($T < 10$ s). The power matrix from Figure 11 bears some resemblance to those of point-absorbers in [80]—further research is needed in this respect.

Finally, the analysis of the response of the two main parameters that influence the power output of an OWC—the relative pneumatic pressure between inside the chamber and the atmosphere, and the free surface oscillation inside the chamber—was carried out by means of the RAO. Both RAOs (RAO_C and RAO_P) are strongly driven by the damping coefficient and wave steepness—when the damping coefficient or the wave steepness increases, RAO_C decreases and RAO_P increases. RAO_C converges to 1 for the lower wave frequencies and wave steepness—i.e., the amplitude of the free surface oscillation inside the OWC chamber equals the incident wave amplitude, reducing the efficiency of the device (Figure 8). On the one hand, the effect of the wave height on RAO_P is quite marked, increasing the RAO with the wave height. On the other hand, the effect of the wave height on RAO_C is of little significance with the exception of the longest wave frequency tested—corresponding to the shortest wave period, $T = 7$ s. Furthermore, both RAOs show peaks of maximum values for the wave frequencies corresponding to the wave periods $T = 7$ s and 9 s, matching the periods where the peaks of maximum efficiency are found.

6. Conclusions

In this work, a novel hybrid wind-wave energy converter for jacket-frame substructures was successfully studied by means of an intensive physical modelling test campaign. Based on the results of the model tests, two main objectives were achieved: (i) the proof of concept of the proposed WEC sub-system was successfully carried out for a jacket-frame substructure; and (ii) the hydrodynamic response of the OWC WEC sub-system was characterised following a comprehensive methodology that makes it possible to better understand the performance of the device and its interaction with the wave field.

Previous research on either hybrid or WEC devices was mostly focused on individual parameters, such as the efficiency or the RAOs. This work follows a comprehensive methodology to characterise the hydrodynamic response of a hybrid system's WEC sub-system; a methodology based on three main pillars: (i) the interaction between the device and the wave field; (ii) the OWC efficiency; and (iii) the response of the two main parameters driving the efficiency of the OWC—the free surface oscillation and the pressure drop inside the OWC chamber—to the incident wave field. This methodology makes it possible to fully characterise the behaviour of the WEC sub-systems and will constitute a starting point for future research to further evaluate the effects of this behaviour at a larger scale (e.g., the behaviour of a hybrid wind-wave farm).

Based on the results from this work, it may be concluded that the proposed hybrid wind-wave energy converter constitutes a viable solution for installation on jacket-frame substructures; notwithstanding, further work is required for its development and to tackle some fundamental issues—e.g., the increased loads on the substructure. Thus, the results of this work are but an initial step towards the development of the proposed prototype.

Supplementary Materials: The research materials supporting this publication may be accessed at http://hdl. handle.net/10026.1/11045. If you have any question regarding these research materials, please contact the corresponding author of this paper.

Energies **2018**, *11*, 637

Acknowledgments: This work was carried out with the financial support of the Higher Education Innovation Fund (HEIF) and the School of Engineering of the University of Plymouth. The Authors are grateful to Johan Skanberg-Tippen, Tim Hatton and James Harvey for their help during the experimental campaign.

Author Contributions: Carlos Perez-Collazo and Gregorio Iglesias conceived and designed the physical modelling; Carlos Perez-Collazo conducted the experimental campaign, analysed the data and wrote the paper; and Deborah Greaves and Gregorio Iglesias gave helpful comments and revised the paper.

Conflicts of Interest: The authors declare no conflict of interest.

Appendix A

The wave steepness parameter (S) used to analyse the results for regular waves is defined as

$$S = 2\pi\frac{H}{gT^2},\tag{A1}$$

where H is the wave height, g is the acceleration of gravity (9.81 m/s^2) and T is the wave period. For irregular waves, the significant steepness parameter (S_S) is defined as

$$S_S = 2\pi\frac{H_S}{gT_P^2},\tag{A2}$$

where H_s is the significant wave height, g is the acceleration of gravity and T_P is the peak wave period.

References

1. Remy, T.; Mbistrova, A. *Offshore Wind in Europe: Key Trends and Statistics 2017*; Wind Europe: Brussels, Belgium, 2018; p. 36.
2. Royal Belgiam Institute of Natural Sciences NORTH SEA FACTS Web Page. Available online: http://www.mumm.ac.be/EN/NorthSea/facts.php (accessed on 14 December 2012).
3. Jeffrey, H.; Sedgwick, J. *ORECCA. European Offshore Renewable Energy Roadmap*; ORECCA: Munchen, Germany, 2011; pp. 1–201.
4. Weisse, R.; von Storch, H.; Callies, U.; Chrastansky, A.; Feser, F.; Grabemann, I.; Günther, H.; Pluess, A.; Stoye, T.; Tellkamp, J.; et al. Regional Meteorological–Marine Reanalyses and Climate Change Projections. *Bull. Am. Meteorol. Soc.* **2009**, *90*, 849–860. [CrossRef]
5. Pryor, S.C.; Barthelmie, R.J. Climate change impacts on wind energy: A review. *Renew. Sustain. Energy Rev.* **2010**, *14*, 430–437. [CrossRef]
6. Azzellino, A.; Conley, D.; Vicinanza, D.; Kofoed, J.P. Marine Renewable Energies: Perspectives and Implications for Marine Ecosystems. *Sci. World J.* **2013**, *2013*, 547563. [CrossRef] [PubMed]
7. Astariz, S.; Iglesias, G. Wave energy vs. other energy sources: A reassessment of the economics. *Int. J. Green Energy* **2016**, *13*, 747–755. [CrossRef]
8. Quevedo, E.; Delory, M.; Castro, A.; Llinas, O.; de Lara, J.; Papandroulakis, N.; Anasrasiadis, P.; Bard, J.; Jeffrey, H.; Ingram, D.; et al. Multi-use offshore platform configurations in the scope of the FP7 TROPOS Project. In Proceedings of the OCEANS MTS/IEE Conference, Bergen, Norway, 10–13 June 2013; pp. 1–7.
9. Stuiver, M.; Soma, K.; Koundouri, P.; van den Burg, S.; Gerritsen, A.; Harkamp, T.; Dalsgaard, N.; Zagonari, F.; Guanche, R.; Schouten, J.J.; et al. The Governance of Multi-Use Platforms at Sea for Energy Production and Aquaculture: Challenges for Policy Makers in European Seas. *Sustainability* **2016**, *8*, 333. [CrossRef]
10. Van den Burg, S.; Stuiver, M.; Norrman, J.; Garção, R.; Söderqvist, T.; Röckmann, C.; Schouten, J.J.; Petersen, O.; García, R.; Diaz-Simal, P.; et al. Participatory Design of Multi-Use Platforms at Sea. *Sustainability* **2016**, *8*, 127. [CrossRef]
11. Astariz, S.; Iglesias, G. Enhancing Wave Energy Competitiveness through Co-Located Wind and Wave Energy Farms. A Review on the Shadow Effect. *Energies* **2015**, *8*, 7344–7366. [CrossRef]
12. Perez-Collazo, C.; Jakobsen, M.M.; Buckland, H.; Fernandez Chozas, J. Synergies for a wave-wind energy concept. In Proceedings of the European Offshore Wind Energy Conference, Frankfurt, Germany, 21 November 2013; EWEA: Frankfurt, Germany, 2013; pp. 1–10.

13. Casale, C.; Serri, L.; Stolk, N.; Yildiz, I.; Cantù, M. *Synergies, Innovative Designs and Concepts for Multipurpose Use of Conversion Platforms*; Results of ORECCA Project—WP4; ORECCA: Munchen, Germany, 2012; pp. 1–77.

14. Astariz, S.; Abanades, J.; Perez-Collazo, C.; Iglesias, G. Improving wind farm accessibility for operation & maintenance through a co-located wave farm: Influence of layout and wave climate. *Energy Convers. Manag.* **2015**, *95*, 229–241.

15. Perez-Collazo, C.; Greaves, D.; Iglesias, G. A review of combined wave and offshore wind energy. *Renew. Sustain. Energy Rev.* **2015**, *42*, 141–153. [CrossRef]

16. CORDIS MARINA Platform. Available online: http://cordis.europa.eu/project/rcn/93425_en.html (accessed on 24 February 2018).

17. CORDIS ORECCA. Available online: http://cordis.europa.eu/project/rcn/94058_en.html (accessed on 24 February 2018).

18. CORDIS TROPOS. Available online: http://cordis.europa.eu/project/rcn/101556_en.html (accessed on 24 February 2018).

19. CORDIS H2OCEAN. Available online: http://cordis.europa.eu/project/rcn/102016_en.html (accessed on 24 February 2018).

20. CORDIS MERMAID. Available online: http://cordis.europa.eu/project/rcn/101743_en.html (accessed on 24 February 2018).

21. Floating Power Plant AS Poseidon Floating Power Web Page. Available online: http://www.floatingpowerplant.com/ (accessed on 22 November 2016).

22. NEMOS GmBH NEMOS Web Page. Available online: http://www.nemos.org/english/technology/ (accessed on 10 October 2015).

23. Wave Star AS Wave Star Energy Web Page. Available online: http://wavestarenergy.com/ (accessed on 5 November 2015).

24. Pelagic Power AS W2Power Web Page. Available online: http://www.pelagicpower.no/ (accessed on 31 March 2016).

25. Hanssen, J.E.; Margheritini, L.; O'Sullivan, K.; Mayorga, P.; Martinez, I.; Arriaga, A.; Agos, I.; Steynor, J.; Ingram, D.; Hezari, R.; et al. Design and performance validation of a hybrid offshore renewable energy platform. In Proceedings of the 2015 Tenth International Conference on Ecological Vehicles and Renewable Energies (EVER), Monte-Carlo, Monaco, 31 March–2 April 2015; pp. 1–8.

26. O'Sullivan, K. Feasibility of Compbined Wind-Wave Energy Platform. Ph.D. Thesis, University College Cork, Cork, Ireland, 2014.

27. O'Sullivan, K.; Murphy, J. Deterministic Economic Model for Wind-Wave Energy Hybrid Energy Conversion Systems. In Proceedings of the Fourth International Conference on Ocean Energy (ICOE), Dublin, The Republic of Ireland, 17–19 October 2012; p. 7.

28. Chen, W.; Gao, F.; Meng, X.; Chen, B.; Ren, A. W2P: A high-power integrated generation unit for offshore wind power and ocean wave energy. *Ocean Eng.* **2016**, *128*, 41–47. [CrossRef]

29. Zanuttigh, B.; Angelelli, E.; Kortenhaus, A.; Koca, K.; Krontira, Y.; Koundouri, P. A methodology for multi-criteria design of multi-use offshore platforms for marine renewable energy harvesting. *Renew. Energy* **2016**, *85*, 1271–1289. [CrossRef]

30. Veigas, M.; Iglesias, G. A Hybrid Wave-Wind Offshore Farm for an Island. *Int. J. Green Energy* **2014**, *12*, 570–576. [CrossRef]

31. Fusco, F.; Nolan, G.; Ringwood, J.V. Variability reduction through optimal combination of wind/wave resources—An Irish case study. *Energy* **2010**, *35*, 314–325. [CrossRef]

32. Fernandez Chozas, J.; Kofoed, J.P.; Sørensen, H.C. *Predictability and Variability of Wave and Wind: Wave and Wind Forecasting and Diversified Energy Systems in the Danish North Sea*; No. 156; DCE Technical Reports; Aalborg University, Department of Civil Engineering: Aalborg, Denmark, 2013.

33. Astariz, S.; Iglesias, G. Selecting optimum locations for co-located wave and wind energy farms. Part I: The Co-Location Feasibility index. *Energy Convers. Manag.* **2016**, *122*, 589–598. [CrossRef]

34. Astariz, S.; Iglesias, G. Selecting optimum locations for co-located wave and wind energy farms. Part II: A case study. *Energy Convers. Manag.* **2016**, *122*, 599–608. [CrossRef]

35. Fernandez Chozas, J.; Helstrup Jensen, N.E.; Sørensen, H.C. Economic benefit of combining wave and wind power productions in day-ahead electricity markets. In Proceedings of the Fourth International Conference on Ocean Energy (ICOE), Dublin, Ireland, 17–19 October 2012.

36. Fernandez Chozas, J.; Kofoed, J.P.; Kramer, M.M.; Sørensen, H.C. Combined production of a full-scale wave converter and a full-scale wind turbine—A real case study. In Proceedings of the Fourth International Conference on Ocean Energy (ICOE), Dublin, Ireland, 17–19 October 2012.

37. Azzellino, A.; Ferrante, V.; Kofoed, J.P.; Lanfredi, C.; Vicinanza, D. Optimal siting of offshore wind-power combined with wave energy through a marine spatial planning approach. *Int. J. Mar. Energy* **2013**, *3–4*, e11–e25. [CrossRef]

38. Lund, H. Large-scale integration of optimal combinations of PV, wind and wave power into the electricity supply. *Renew. Energy* **2006**, *31*, 503–515. [CrossRef]

39. Stoutenburg, E.D.; Jenkins, N.; Jacobson, M.Z. Power output variations of co-located offshore wind turbines and wave energy converters in California. *Renew. Energy* **2010**, *35*, 2781–2791. [CrossRef]

40. Wang, L.; Jan, S.-R.; Li, C.-N.; Li, H.-W.; Huang, Y.-H.; Chen, Y.-T. Analysis of an integrated offshore wind farm and seashore wave farm fed to a power grid through a variable frequency transformer. In Proceedings of the 2011 IEEE Power and Energy Society General Meeting, Detroit, MI, USA, 24–29 July 2011; pp. 1–7.

41. Wang, L.; Jan, S.-R.; Li, C.-N.; Li, H.-W.; Huang, Y.-H.; Chen, Y.-T.; Wang, S.-W. Study of a hybrid offshore wind and seashore wave farm connected to a large power grid through a flywheel energy storage system. In Proceedings of the 2011 IEEE Power and Energy Society General Meeting, Detroit, MI, USA, 24–29 July 2011; pp. 1–7.

42. Cradden, L.; Mouslim, H.; Duperray, O.; Ingram, D. Joint exploitation of wave and offshore wind power. In Proceedings of the Nineth European Wave and Tidal Energy Conference (EWTEC), Southampton, UK, 5–9 September 2011; pp. 1–10.

43. Astariz, S.; Iglesias, G. Output power smoothing and reduced downtime period by combined wind and wave energy farms. *Energy* **2016**, *97*, 69–81. [CrossRef]

44. Astariz, S.; Iglesias, G. Co-located wind and wave energy farms: Uniformly distributed arrays. *Energy* **2016**, *113*, 497–508. [CrossRef]

45. Astariz, S.; Perez-Collazo, C.; Abanades, J.; Iglesias, G. Co-located wind-wave farm synergies (Operation & Maintenance): A case study. *Energy Convers. Manag.* **2015**, *91*, 63–75.

46. Astariz, S.; Perez-Collazo, C.; Abanades, J.; Iglesias, G. Towards the optimal design of a co-located wind wave farm. *Energy* **2015**, *84*, 15–24. [CrossRef]

47. Perez, C.; Iglesias, G. Integration of wave energy converters and offshore windmills. In Proceedings of the Fourth International Conference on Ocean Energy (ICOE), Dublin, Ireland, 30 October 2012; pp. 1–6.

48. Collazo, C.P.; Rodriguez, J.G.I.; Greaves, D. Wave Energy Capture Device. Patent WO2016185189 A1, 21 May 2015.

49. López, I.; Pereiras, B.; Castro, F.; Iglesias, G. Performance of OWC wave energy converters: Influence of turbine damping and tidal variability. *Int. J. Energy Res.* **2015**, *39*, 472–483. [CrossRef]

50. Chakrabarti, S.K. *Offshore Structure Modeling*; World Scientific: Singapore, 1994; Volume 9, p. 460.

51. Nielsen, K. *Annex II Report 2003. Development of Recommended Practices for Testing and Evaluating Ocean Energy Systems*; IEA—Ocean Energy Systems: Lisbon, Portugal, 2003; pp. 1–62.

52. Holmes, B. *Tank Testing of Wave Energy Conversion Systems*; The European Marine Energy Centre (EMEC): London, UK, 2009; pp. 1–82.

53. Jonkman, J.; Butterfield, S.; Musial, W.; DScott, G. *Definition of a 5-MW Reference Wind Turbine for Offshore System Development*; NREL/TP-500-38060; National Renewable Energy Laboratory: Golden, CO, USA, 2009.

54. Kenny, J.P. *SW Wave Hub—Metocean Design Basis*; SWRDA: Hants, UK, 2009; p. 64.

55. Falcão, A.F.O.; Henriques, J.C.C. Model-prototype similarity of oscillating-water-column wave energy converters. *Int. J. Mar. Energy* **2014**, *6*, 18–34. [CrossRef]

56. Elhanafi, A.; Macfarlane, G.; Fleming, A.; Leong, Z. Scaling and air compressibility effects on a three-dimensional offshore stationary OWC wave energy converter. *Appl. Energy* **2017**, *189*, 1–20. [CrossRef]

57. Simonetti, I.; Cappietti, L.; Elsafti, H.; Oumeraci, H. Evaluation of air compressibility effects on the performance of fixed OWC wave energy converters using CFD modelling. *Renew. Energy* **2018**, *119*, 741–753. [CrossRef]

58. Weber, J.W. Representation of non-linear aero-thermodynamic effects during small scale physical modelling if OWC WECs'. In Proceedings of the Seventh European Wave and Tidal Energy Conference (EWTEC), Porto, Portugal, 11–14 September 2007; pp. 11–14.

59. Sarmento, A.J.N.A.; Falcão, A.F.O. Wave generation by an oscillating surface-pressure and its applications in wave energy extraction. *J. Fluid Mech.* **1985**, *150*, 467–485. [CrossRef]

60. Morris-Thomas, M.T.; Irvin, R.J.; Thiagarajan, K.P. An investigation into the hydrodynamic efficiency of an oscillating water column. *J. Offshore Mech. Arct. Eng.* **2007**, *129*, 273–278. [CrossRef]

61. Dizadji, N.; Sajadian, S.E. Modeling and optimization of the chamber of OWC system. *Energy* **2011**, *36*, 2360–2366. [CrossRef]

62. Moñino, A.; Medina-López, E.; Clavero, M.; Benslimane, S. Numerical simulation of a simple OWC problem for turbine performance. *Int. J. Mar. Energy* **2017**, *20*, 17–32. [CrossRef]

63. Medina-López, E.; Moñino, A.; Borthwick, A.G.L.; Clavero, M. Thermodynamics of an OWC containing real gas. *Energy* **2017**, *135*, 709–717. [CrossRef]

64. López, I.; Pereiras, B.; Castro, F.; Iglesias, G. Optimisation of turbine-induced damping for an OWC wave energy converter using a RANS–VOF numerical model. *Appl. Energy* **2014**, *127*, 105–114. [CrossRef]

65. Sheng, W.; Lewis, A.; Alcorn, R. On wave energy extraction of oscillating water column device. In Proceedings of the Fourth International Conference on Ocean Energy (ICOE), Dublin, Ireland, 17–19 October 2012.

66. Hann, M.; Perez-Collazo, C. Chapter 7: Physical Modelling. In *Wave and Tidal Energy*; Greaves, D., Iglesias, G., Eds.; John Wiley & Sons Ltd.: West Sussex, UK, 2018; in press; pp. 233–288.

67. Mansard, E.P.; Funke, E.R. The Measurement of incident and reflected spectra using a least squares method. In Proceedings of the International Conference on Coastal Engineering (ICCE), Sydney, Australia, 23–28 March 1980.

68. Baquerizo, A.; Losada, M.A.; Smith, J.M. Wave reflection from beaches: A predictive model. *J. Coast. Res.* **1998**, *14*, 291–298.

69. Hughes, S.A. *Physical Models and Laboratory Techniques in Coastal Engineering*; World Scientic Publishing Co. Pte. Ltd.: Singapore, 1993; p. 568.

70. Wang, D.J.; Katory, M.; Li, Y.S. Analytical and experimental investigation on the hydrodynamic performance of onshore wave-power devices. *Ocean Eng.* **2002**, *29*, 871–885. [CrossRef]

71. Pereiras, B.; López, I.; Castro, F.; Iglesias, G. Non-dimensional analysis for matching an impulse turbine to an OWC (oscillating water column) with an optimum energy transfer. *Energy* **2015**, *87*, 481–489. [CrossRef]

72. Zanuttigh, B.; Angelelli, E.; Kofoed, J.P. Effects of mooring systems on the performance of a wave activated body energy converter. *Renew. Energy* **2013**, *57*, 422–431. [CrossRef]

73. Fernandez, H. WaveCat: Concepto y Desarrollo Mediante Modelización Física de un Nuevo Convertidor de Energía del Oleaje. Ph.D. Thesis, University of Santiago de Compostela, Santiago de Compostela, Spain, 2012.

74. Kamath, A.; Bihs, H.; Arntsen, Ø.A. Numerical investigations of the hydrodynamics of an oscillating water column device. *Ocean Eng.* **2015**, *102*, 40–50. [CrossRef]

75. Elhanafi, A.; Fleming, A.; Macfarlane, G.; Leong, Z. Underwater geometrical impact on the hydrodynamic performance of an offshore oscillating water column–wave energy converter. *Renew. Energy* **2017**, *105*, 209–231. [CrossRef]

76. Simonetti, I.; Cappietti, L.; Elsafti, H.; Oumeraci, H. Optimization of the geometry and the turbine induced damping for fixed detached and asymmetric OWC devices: A numerical study. *Energy* **2017**, *139*, 1197–1209. [CrossRef]

77. López, I.; Iglesias, G. Efficiency of OWC wave energy converters: A virtual laboratory. *Appl. Ocean Res.* **2014**, *44*, 63–70. [CrossRef]

78. López, I.; Castro, A.; Iglesias, G. Hydrodynamic performance of an oscillating water column wave energy converter by means of particle imaging velocimetry. *Energy* **2015**, *83*, 89–103. [CrossRef]

79. Babarit, A. A database of capture width ratio of wave energy converters. *Renew. Energy* **2015**, *80*, 610–628. [CrossRef]

80. Babarit, A.; Hals, J.; Muliawan, M.J.; Kurniawan, A.; Moan, T.; Krokstad, J. Numerical benchmarking study of a selection of wave energy converters. *Renew. Energy* **2012**, *41*, 44–63. [CrossRef]

energies

MDPI

Article

Numerical Modeling of the Wave Energy Propagation in the Iberian Nearshore

Eugen Rusu [ID]

Department of Mechanical Engineering, Faculty of Engineering, "Dunărea de Jos" University of Galati, 47 Domneasca Street, 800008 Galati, Romania; Eugen.Rusu@ugal.ro; Tel.: +40-740-205-534

Received: 30 March 2018; Accepted: 13 April 2018; Published: 18 April 2018

Abstract: In the present work the wave energy propagation patterns in the western side of the Iberian nearshore were evaluated. This assessment takes into account the results provided by a wave modelling system based on spectral phase averaged wave models, which considers subsequent computational domains with increasing resolution towards the coast. The system was previously validated against both in situ measurements and remotely sensed data. Moreover, several data assimilation techniques were implemented as well. In this way, the reliability of the wave predictions was significantly increased. Although extended wave hindcasts have already been carried out close to the Iberian coast of the Atlantic Ocean, including wave energy assessments, they might not be completely accurate because of recent changes in the dynamics of the ocean and coastal wave climate. Thus, the present work considers wave nowcasts that correspond to the most recent and relevant wave energy propagation patterns in the targeted coastal environment. In order to perform this analysis, four different computational levels were considered. The first level corresponds to the sub oceanic domain and it is linked directly to the oceanic wave model, which is implemented over the entire North Atlantic Ocean. The second is related to the coarser computational domains of the coastal areas, while the third relates to the high-resolution domains. These three levels are defined as spherical coordinates (longitude, latitude). Finally, the last computational level includes some coastal areas which have the highest spatial resolution, defined considering the Cartesian coordinates. Moreover, for each level several computational domains have been considered. This work illustrates the most recent and significant wave transformation and energy propagation patterns corresponding to 18 computational domains with various resolutions in the western Iberian coastal environment.

Keywords: Iberian nearshore; coastal areas; wave power; numerical modeling; Simulating waves nearshore (SWAN)

1. Introduction

All over the world, ambitious objectives for low carbon energy sources and climate change policy have been set. The European Union (EU) targets aim to increase the share of renewable energy to 20% by 2020 and at least to 27% by 2030 [1]. In order to achieve such targets, significant investments in the area of renewable energy, including the acceleration of research and innovation (R & I), are expected. Nowadays, coastal environments represent the most challenging areas for renewable energy extraction. This is because the nearshore has huge energy potential, especially with regard to offshore wind, waves and tides. The most spectacular evolutions can be seen in the area of offshore wind. Thus, both fixed wind turbines and floating platforms are being quickly implemented and they have become commercially efficient [2,3]. Most of the advances in offshore wind energy is due to the high performance and innovative design of large wind turbines (WT). The major challenge in offshore renewable energy is the achievement of a significant reduction in the levelized cost of energy (LCOE). From this perspective, a revolutionary idea in the wind industry is represented by the multi-rotor concept. This means having more rotors on a single support structure, which avoids the upscaling disadvantages of the unit turbine and at the same time facilitates the benefits of large unit capacities. An analysis using a cost model compared 20 MW multi-rotor systems to 10 MW reference turbines in a specified 500 MW wind farm, shows major benefits for this concept [3]. According to this analysis, the multi-rotor system achieved 7.2 ct€/kWh with a cost reduction of about 33% relative to the LCOE of the reference turbines.

The very high dynamics of offshore wind has also encouraged the development of other technologies for renewable energy extraction in the marine environment. In this context, the EU SET-Plan (European Union Strategic Energy Technology Plan) [4] sets the LCOE targets for tidal streams as 15 ct€/kWh by 2025 and 10 ct€/kWh by 2030, while for wave energy there is a five-year shift in the LCOE target, which means 15 ct€/kWh by 2030 and 10 ct€/kWh by 2035. Nevertheless, it is expected that the advances in the offshore wind industry in relation to wave energy extraction technologies will have a great impact. This is because it creates the possibility of collocation of wave devices in the vicinity of wind farms. Such an approach has three major advantages: (a) The grid connection and other infrastructure costs that have already been completed for the wind farm can also be used by the wave farm. Thus, a substantial capital expenditure (CAPEX) reduction is expected. (b) Many parts of the maintenance operations for WTs and wave energy converters (WECs) can be combined. In this way a substantial operational expenditure (OPEX) reduction is also expected. (c) The wave farm could be used as an obstacle to protect the wind farm, or alternatively the wave farm may provide coastal protection. Thus, although wave extraction technologies are not yet considered to be mature, significant advances are expected in the near future.

Wave energy is abundant, since there are large areas of the coastal environment with significant wave power potential. Furthermore, it has higher density and predictability than wind or solar energy. Thus, although the technologies for wave energy extraction are not yet fully evolved, significant advances are expected in the near future and various types of wave energy converters are being developed (see for example [5,6]). A very important issue in increasing the efficiency of wave energy conversion is related to the development of more advanced power take-off (PTO) systems, through which the wave power is transformed into electric power, and for this reason a lot of theoretical and experimental research is being carried out in this area [7,8].

Therefore, the objective of the present work is to provide an updated picture of the wave energy propagation patterns in the western side of the Iberian nearshore. There are two reasons for this choice. First, this coastal environment is very attractive for wave energy developers [9–11] and any new study related to the wave energy patterns here should be of considerable interest. In 2008, in the coastal environment north of Porto, the world's first multiple machine wave power project was installed, known as the Aguçadoura Wave Farm. This was based on three first generation Pelamis type wave energy converters, with a total capacity of 2.25 MW. The farm first generated electricity in July 2008 but was taken offline in November 2008.

The second reason is related to the fact that the coastal wave climate may change and as a result new propagation patterns, either global or/and local, may occur [12]. Thus, although there are many studies concerning the wave propagation patterns in this coastal environment, some of them are focused only on the evaluation of the wave energy patterns in the Spanish [13–16] or in the Portuguese [17–19] nearshores. Furthermore, evaluations related to the expected performances of the current state-of-the-art wave energy converters in the Iberian coastal environment have been carried out [20–22]. The results look promising and if we compare the west Iberian nearshore with some of the most energetic locations in the Mediterranean Sea where similar investigations are currently being performed, see for example [23–25], it can be considered as one of the most attractive European coastal environments for the implementation of marine energy farms in the future.

2. Materials and Methods

2.1. Theoretical Background of the Spectral Wave Models

The wave models based on the spectrum concept integrate an advection type equation that gives the propagation of the wave spectrum in a space defined by five dimensions: time, geographical and spectral spaces [26,27]:

$$\frac{DN}{Dt} = \frac{S}{\sigma}. \tag{1}$$

The spectrum that is considered in the great majority of spectral wave models is the action density spectrum (N), rather than the energy density spectrum. This is because in the presence of the currents action, density is conserved while energy density is not. The concept of wave action was introduced by Bretherton and Garret [28], and the action density is equal to the energy density (E) divided by the relative frequency (σ). S from the right-hand side of Equation (1) represents the source terms. The radian relative frequency is related to the wave number (k) by the dispersion relationship:

$$\sigma^2 = gk \tanh(kd), \tag{2}$$

where g is the acceleration of gravity and d the water depth. In the presence of currents, the absolute radian frequency (ω) is given by the usual Doppler shift:

$$\omega = \sigma + \vec{k}\vec{U}, \tag{3}$$

where $\vec{k}$ is the wave number vector and $\vec{U}$ is the current velocity.

For large scale applications the governing Equation (1) is expressed for the geographical space in spherical coordinates, longitude (λ) and latitude (φ), while the spectral space is defined by the relative frequency (σ) and the wave direction (θ):

$$\frac{\partial N}{\partial t} + \frac{\partial}{\partial \lambda}\dot{\lambda}N + \frac{1}{\cos \varphi}\frac{\partial}{\partial \varphi}\dot{\varphi}N \cos \varphi + \frac{\partial}{\partial \sigma}\dot{\sigma}N + \frac{\partial}{\partial \theta}\dot{\theta}N = \frac{S}{\sigma}, \tag{4}$$

For coastal applications the Cartesian coordinates (x) and (y) are mostly used and the action balance equation in this case becomes:

$$\frac{\partial N}{\partial t} + \frac{\partial}{\partial x}\dot{x}N + \frac{\partial}{\partial y}\dot{y}N + \frac{\partial}{\partial \sigma}\dot{\sigma}N + \frac{\partial}{\partial \theta}\dot{\theta}N = \frac{S}{\sigma} \tag{5}$$

The left side of the action balance Equations (1), (4) or (5) is the kinematic part which reflects the action propagation in the space with fifth dimensions accounting also for phenomena as wave diffraction or refraction. On the right-hand side the source (S) is expressed in terms of energy density. In deep water, three components are significant. They correspond to the atmospheric input, nonlinear quadruplet interactions and whitecapping dissipation. Besides these three terms, in shallow water additional terms corresponding to phenomena such as bottom friction, depth induced wave breaking or triad nonlinear wave–wave interactions may play an important role and the total source becomes:

$$S = S_{in} + S_{nl} + S_{dis} + S_{bf} + S_{br} + S_{tr} + \cdots. \tag{6}$$

In spectral wave models, the wave power components (expressed in W/m, i.e., energy transport per unit length of wave front), are computed with the relationships [29]:

$$P_x = \rho g \iint c_x\, E(\sigma, \theta) d\sigma\, d\theta \tag{7}$$

$$P_y = \rho g \iint c_y E(\sigma, \theta) d\sigma\, d\theta$$

In the above equation x, y are the problem coordinate system and c_x, c_y are the propagation velocities of wave energy in the geographical space (absolute group velocity components) defined as:

$$(c_x, \; c_y) = \frac{d\vec{x}}{dt}. \tag{8}$$

The absolute value of the energy transport (denoted also as wave power) will be:

$$P = \sqrt{P_x^2 + P_y^2}, \tag{9}$$

2.2. Description of the Wave Prediction System

A wave prediction system based on spectral phase averaged models has been considered in the present work. The ocean forcing is provided by the wave generation model Wave Modeling (WAM) Cycle 4 [30], in an improved version that allows for two-way nesting [31]. This model covers the entire North Atlantic basin. Alternatively, implementation of the Wave Watch 3 (WW3) model [32] over the same generation area has also been completed [33]. For the sub oceanic scale and the coastal domains, the Simulating Waves Nearshore (SWAN) model [34] was nested in the ocean models and various computational schemes including subsequent domains with increasing resolution towards the shore were defined [35]. Furthermore, based on these schemes an operational wave prediction system was applied to the major Portuguese harbors [36], providing nowcasts and 3-day forecasts. In this system the Fifth-Generation US National Center for Atmospheric Research (NCAR)/Penn State Mesoscale Model (MM5) model [37] is used for providing high resolution wind fields. The MM5 model is based on meteorological fields provided by the US National Oceanic and Atmospheric Administration (NOAA) Atmospheric Model Global Forecast System (GFS). The model allows for a broad set of meteorological fields at different scales and with the desired temporal resolution.

It should be noted at this point that the WAM model was mainly utilized in the wave modeling system used in the present work. WW3 was used only as a backup and it runs in parallel with WAM in some of the more sensitive cases, for example in the generation of the high energy storms in the North Atlantic Ocean and their impact in the Iberian nearshore. In general, the results provided by the two models are consistent [33], although the WAM version considered, was intensively tested and calibrated for this particular coastal environment and appears to provide slightly more reliable results.

In the present work, four different SWAN computational levels are considered. The first level is related to the sub-oceanic scale and represents the link between the ocean models and the coastal domains. Three different SWAN domains are considered at this level and their characteristics are presented in Table 1. In this table $\Delta\lambda$ and $\Delta\varphi$ represent the spatial resolution, Δt the time resolution, nf number of frequencies, $n\theta$ number of directions, $n\lambda$ number of grid points in longitude, $n\varphi$ number of grid points in latitude and np is total number of grid points. These domains are denoted as wave drivers (D), (D1) covers the northwestern part of the Iberian Peninsula, (D2) is related to the entire western Iberian coastal environment and (D3) covers the Portuguese continental nearshore. For all these driver domains the spatial resolution is 0.05° for longitude and also 0.05° for latitude for (D1) and (D3), while for (D2) the resolution in latitude is 0.1°. Figure 1 illustrates the geographical space of (D1) in a bathymetric map, Figure 2a shows the domain (D2) and Figure 2b shows the geographical space of (D3).

Table 1. Characteristics of the driver computational domains defined for the Simulating Waves Nearshore model (SWAN) simulations in the western Iberian nearshore.

Driver Domains	$\Delta\lambda \times \Delta\varphi$	Δt (min)	nf	$n\theta$	$ng\lambda \times ng\varphi = np$
D1-Northern	0.05° × 0.05°	10 non-stat	24	24	141 × 121 = 17,061
D2-Western	0.05° × 0.1°	10 non-stat	24	24	101 × 101 = 10,201
D3-Portuguese	0.05° × 0.05°	10 non-stat	24	24	81 × 121 = 9801

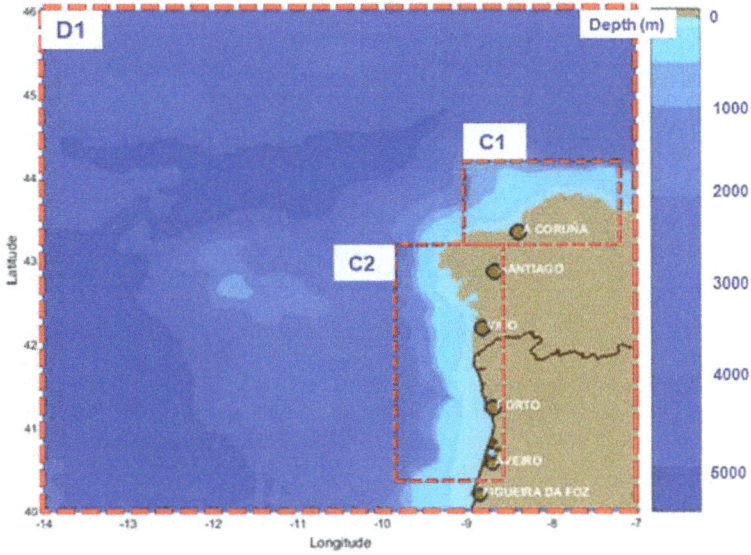

Figure 1. The northern wave driver and the subsequent coastal computational domains.

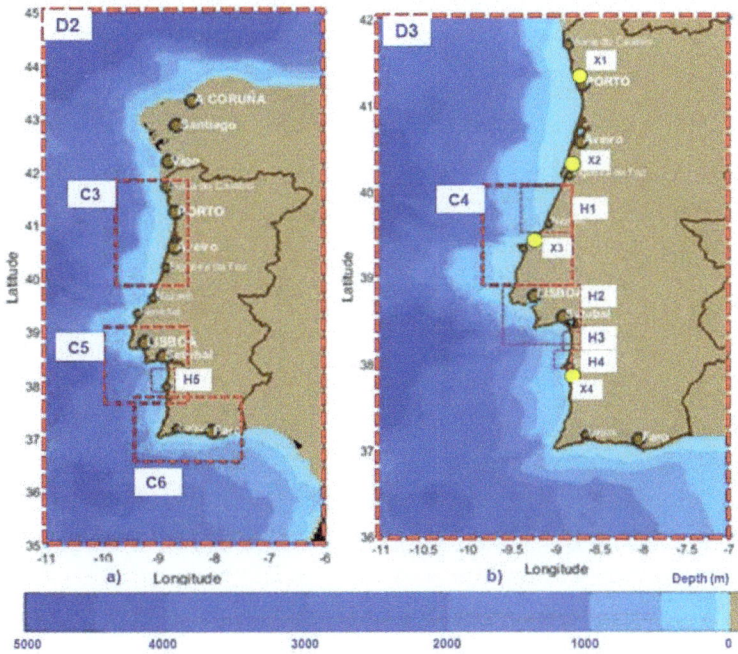

Figure 2. The Iberian (**a**) and the Portuguese (**b**) wave drivers and the subsequent computational domains corresponding to the computational levels: coastal, high resolution and Cartesian, respectively.

Six SWAN coastal domains (denoted as the C domains) are considered in the present work. These are (C1) in the north, (C2) and (C3) in the northwest, (C4) and (C5) in the center, and (C6) in the south. Most of these domains have a resolution in the geographical space of 0.02°. However, for (C1), (C2) and (C3) the resolution in the direction normal to the mean coastline direction is 0.01°. The characteristics of the coastal domains are presented in Table 2. It should be noted that the choice of defining the resolution in the geographical space is also linked with the computational effectiveness. The coastal SWAN domains (C1) and (C2) are illustrated in Figure 1, (C3), (C5) and (C6) in Figure 2a and (C4) in Figure 2b.

Table 2. Characteristics of the coastal coarse computational domains defined for the SWAN simulations in the Iberian nearshore.

Coastal Domains	$\Delta\lambda \times \Delta\varphi$	Δt (min)	nf	$n\theta$	$ng\lambda \times ng\varphi = np$
C1-Northern	$0.02° \times 0.01°$	10 non-stat	24	36	$101 \times 101 = 10,201$
C2-North Western 01	$0.01° \times 0.02°$	10 non-stat	24	36	$101 \times 141 = 14,241$
C3-North Western 02	$0.01° \times 0.02°$	10 non-stat	24	36	$101 \times 101 = 10,201$
C4-Central 01	$0.02° \times 0.02°$	10 non-stat	24	36	$76 \times 91 = 6916$
C5-Central 02	$0.02° \times 0.02°$	10 non-stat	24	36	$63 \times 76 = 4788$
C6-Southern	$0.02° \times 0.02°$	10 non-stat	24	36	$111 \times 76 = 8436$

Furthermore, five SWAN high-resolution domains (denoted as the H domains) are considered in the present work. They have spatial resolutions in the range 0.001°–0.01° and these are: (H1) corresponding to the nearshore area of Peniche, (H2) in the vicinity of Lisbon, (H3) is the Pinheiro da Cruz nearshore, south of the Portuguese city of Setubal, (H4) the nearshore area located in the north of the Sines port, and (H5) the nearshore area centered on the port of Sines. The characteristics of the high-resolution SWAN domains are presented in Table 3. In Figure 2a the (H5) domain is illustrated, while Figure 2b shows the geographical spaces corresponding to all the other high-resolution computational domains (H1, H2, H3 and H4).

Table 3. Characteristics of the high-resolution computational domains considered for the SWAN simulations in the western Iberian nearshore.

High Resolution Domains	$\Delta\lambda \times \Delta\varphi$	Δt (min)	nf	$n\theta$	$ng\lambda \times ng\varphi = np$
H1-Peniche	$0.01° \times 0.01°$	10 non-stat	30	36	$91 \times 81 = 7371$
H2-Lisbon	$0.005° \times 0.01°$	10 non-stat	30	36	$91 \times 81 = 7371$
H3-Pinheiro da Cruz	$0.001° \times 0.0025°$	10 non-stat	30	36	$91 \times 81 = 7371$
H4- Sines North	$0.002° \times 0.003°$	10 non-stat	30	36	$91 \times 81 = 7371$
H5-Sines	$0.005° \times 0.005°$	10 non-stat	30	36	$101 \times 101 = 10,201$

Finally, four Cartesian SWAN domains have also been considered. They have spatial resolutions in the range 25–100 m and they are denoted as: (X1) in the vicinity of the Leixoes harbor, close to the Portuguese city of Porto, (X2) related to the nearshore area of Figueira da Foz, (X3) the nearshore area of Obidos in the central part of continental Portugal, and (X4) in the vicinity of the Sines harbor, south of Lisbon [38]. The characteristics of the Cartesian SWAN domains are presented in Table 4, while the corresponding geographical spaces are illustrated in Figure 2b, being represented with yellow circles.

Table 4. Characteristics of the Cartesian computational domains considered for the SWAN simulations in the west Iberian nearshore.

Cartesian Domains	$\Delta x \times \Delta y$ (m)	Δt (min)	nf	$n\theta$	$ngx \times ngy = np$
X1-Leixoes harbor	25×25	60 stat	30	36	$236 \times 216 = 50,976$
X2-Figueira da Foz	25×50	60 stat	30	36	$65 \times 106 = 6890$
X3-Obidos	100×100	60 stat	30	36	$156 \times 328 = 51,168$
X4-Sines harbor	25×25	60 stat	30	36	$261 \times 201 = 52,461$

2.3. Model System Validations and Data Assimilation

Validations were carried out for the wave modeling system using both in situ and remotely sensed measurements. The results are discussed below, considering first the model system output for various computational levels versus in situ measurements. For this, data provided by two wave rider type directional buoys were considered. One buoy was located offshore the Leixoes port ($-9.0883°$ W/$41.2033°$ N) and operates at a water depth of about 83 m, close to the offshore boundary of the Cartesian domain (X1). The second buoy was located offshore the Sines port ($-8.9289°$ W/$37.9211°$ N) and operates at approximately 97 m, close to the offshore boundary of the Cartesian domain (X4). Both are maintained by the Hydrographic Institute of the Portuguese Navy [36]. Statistical results for the simulated wave parameters against the buoy measurements are presented in Table 5 for the time interval from 5 October 2010 to 31 May 2011.

The statistical parameters presented in Tables 5 and 6 are *Bias*, root mean square error (*RMSE*), scatter index (*SI*), correlation coefficient (*r*) and symmetric slope (*S*). If X_i represents the measured values, Y_i the simulated values and n the number of observations the aforementioned parameters can be defined with the relationships:

$$X_{med} = \widetilde{X} = \frac{\sum_{i=1}^{n} X_i}{n}, \ Bias = \frac{\sum_{i=1}^{n}(X_i - Y_i)}{n}, \ RMSE = \sqrt{\frac{\sum_{i=1}^{n}(X_i - Y_i)^2}{n}} \tag{10}$$

$$SI = \frac{RMSE}{\widetilde{X}}, \ r = \frac{\sum_{i=1}^{n}\left(X_i - \widetilde{X}\right)\left(Y_i - \widetilde{Y}\right)}{\left(\sum_{i=1}^{n}\left(X_i - \widetilde{X}\right)^2 \sum_{i=1}^{n}\left(Y_i - \widetilde{Y}\right)^2\right)^{\frac{1}{2}}}, \ S = \sqrt{\frac{\sum_{i=1}^{n} Y_i^2}{\sum_{i=1}^{n} X_i^2}} \tag{11}$$

Table 5. The statistical results of the wave parameters at the buoys for the time interval 5 October 2010 to 31 May 2011. N represents the number of data points [36].

Parameter	Grid	Bias	RMSE	SI	R	S	Buoy
H_s (m)	D2	0.09	0.47	0.22	0.92	1.02	
	C3	0.12	0.40	0.18	0.95	1.04	Leixoes
T_{m02} (s)	D2	−0.47	1.29	0.18	0.80	0.95	N = 1374
	C3	−0.55	1.06	0.15	0.86	0.93	
	D2	0.25	0.44	0.25	0.92	1.13	
	C5	0.15	0.35	0.20	0.93	1.07	
	H5	0.14	0.35	0.20	0.93	1.06	Sines
	D2	−0.37	1.26	0.18	0.81	0.96	N = 1087
T_{m02} (s)	C5	−0.55	1.09	0.15	0.86	0.93	
	H5	−0.57	1.07	0.15	0.87	0.93	

In this table *RMSE* represents the root mean square error, *SI* the scatter index, *R* the correlation coefficient and *S* the symmetric slope. The results presented in Table 5 show that in general, the wave model predictions have good accuracy in all the computational levels considered. Closer to the coast, better accuracy in the higher resolution domains would be expected. Furthermore, in order to improve the wave predictions, an assimilation scheme for the satellite data has also been implemented at the sub oceanic levels [39]. This scheme is based on the optimal interpolation method. The basic philosophy of data assimilation (DA) is to combine the information coming from measurements with the results of the numerical models to enable the optimal estimation of the field of interest [40]. The assimilation of wave data is usually performed in terms of significant wave height (H_s). Measurements of this wave parameter are available locally (generally coming from buoys) and widespread (satellite data). The sequential methods combine all the observations falling within a particular time window and update the model solution without reference to the model dynamics. The most widely adopted DA schemes are based on instantaneous sequential procedures, such as the optimal interpolation

(OI) [41] or the successive correction method (SCM) [9]. These methods are attractive, especially due to their lower computational demands, and consequently DA schemes based on OI are widely used in wave forecasting. In fact, nowadays most of the weather prediction centers with wave modelling capabilities assimilate altimeter measurements, using various assimilation procedures based either on OI or SCM techniques.

Table 6 presents some statistical results obtained for the significant wave height values simulated with SWAN without assimilation (H_{s-WDA}), and obtained after the assimilation of the altimeter data (H_{s-DA}), against the measurements from the buoys of Leixoes and Sines in the west Iberian nearshore, considering the time interval from 5 October 2010 to 31 May 2011 [39]. As the results presented in Table 6 show, the assimilation scheme slightly improves all parameters. Also, in order to illustrate how this data assimilation scheme influences the wave field, Figure 3 presents the significant wave height scalar fields (Hs) corresponding to the time frame of 8 February 2017 without data assimilation (a) and with data assimilation (b). The satellite track, corresponding to JASON2, is also represented in Figure 2a.

Table 6. Statistical results obtained for the H_s values simulated with SWAN without assimilation (H_{s-WDA}), and obtained after the assimilation of the altimeter data (H_{s-DA}), against buoy measurements in the west Iberian nearshore, time interval from 5 October 2010 to 31 May 2011 [39].

Parameter	*MeanMes* (m)	*MeanSim* (m)	*Bias* (m)	*RMSE* (m)	*SI*	*R*	*S*	Buoy
H_{s-WDA}	2.15	2.23	0.08	0.47	0.22	0.91	1.03	**Leixoes**
H_{s-DA}		2.21	0.06	0.42	0.20	0.92	1.01	(*N* = 1374)
H_{s-WDA}	1.72	1.98	0.26	0.44	0.25	0.92	1.13	**Sines**
H_{s-DA}		1.95	0.23	0.39	0.23	0.93	1.12	(*N* = 1087)

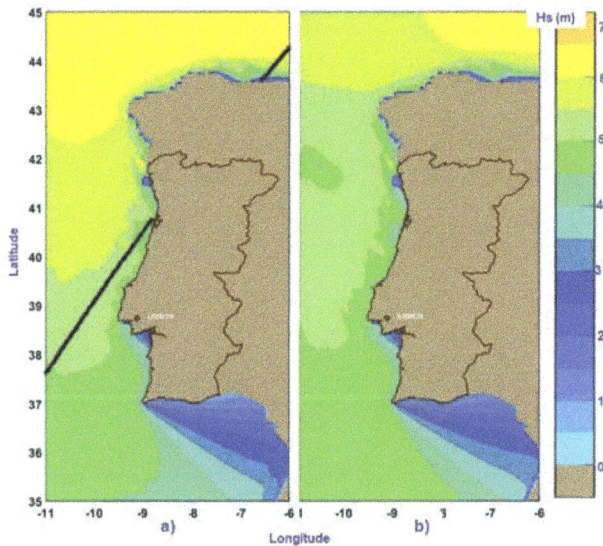

Figure 3. Significant wave height scalar fields corresponding to the time frame 8 February 2017 (a) results without data assimilation, the corresponding trajectory of JASON2 is also represented and (b) results with data assimilation.

3. Results and Discussion

Model system simulations were performed starting in October 2016 and continued until the end of 2017. Emphasis was given to the analysis of the average wintertime conditions. Since the nowcast predictions (corresponding to zero hours of each day) were found to be more accurate, and the reliability of the predictions decreased as they moved ahead into the forecast [24], the results presented next are related only to the nowcast. The physical processes activated in the SWAN simulations, corresponding to the four defined computational levels (Driver, Coastal, High Resolution and Cartesian) are presented in Table 7. In this table, *Wave* indicates the wave forcing, *Tide* the tide forcing, *Wind* the wind forcing, *Curr* the current field input, *Gen* generation by wind, *Wcap* the whitecapping process, *Quad* the quadruplet nonlinear interactions, *Triad* the triad nonlinear interactions, *Diff* diffraction process, *Bfric* bottom friction, *Set up* the wave induced set up and *Br* depth induced wave breaking. For each computational level the most relevant processes were considered. Thus, most of the deep-water processes (such as whitecapping or generation by wind) are also valid in shallow water, which is why they have been considered in all computational levels. However, for the first two levels (sub oceanic and coastal) the tide was not considered since for deep-water wave propagation the tide level does not influence the model results in any way, while building an accurate tide level matrix for a large-scale domain is a quite difficult task. In shallow water, corresponding to the high resolution and Cartesian computational domains, the tide might significantly influence the results, see for example [42] and thus the tide should be considered in the model simulations. A similar reason relates to the activation of diffraction in the high-resolution domains. Although triad nonlinear interaction is a process characteristic of shallow water, it was also activated in the coastal computational levels, because occasionally it may occur in intermediate water as well. Finally, the wave induced set up is a process that can be activated only in Cartesian coordinates.

Table 7. Physical processes activated in the SWAN simulations, corresponding to the four defined computational levels (Driver, Coastal, High Resolution and Cartesian). X—process activated, 0—process inactivated.

Input/Process Domains	*Wave*	*Wind*	*Tide*	*Curr*	*Gen*	*Wcap*	*Quad*	*Triad*	*Diff*	*Bfric*	*Set up*	*Br*
Driver	X	X	0	0	X	X	X	0	0	X	0	X
Coastal	X	X	0	0	X	X	X	X	0	X	0	X
High Resolution	X	X	X	0	X	X	X	X	X	X	0	X
Cartesian	X	X	X	0	X	X	X	X	X	X	X	X

The results of the 15-month period of model simulations, performed in the 18 SWAN domains corresponding to the four defined computational levels, were analyzed focusing on the average wintertime conditions. It has to be noted that in these analyses, wintertime is considered the 6-month period from October to March. Some case studies that were found to be representative will be presented and discussed below.

Figure 4 illustrates some relevant wave energy propagation patterns in the northwestern side of the Iberian nearshore. These model simulations correspond to 22 October 2016, which reflects an average winter time wave energy situation. The results corresponding to the driver domain (D1) are presented in Figure 4a, while those corresponding to the coastal domains (C1) and (C2) are illustrated in Figure 4b,c, respectively. The energy propagation patterns are represented through the normalized wave power and the energy transport vectors. The mean direction of the incoming waves is about 330° (nautical convention) in the offshore, while in the nearshore this might be drastically changed due to the refraction process. Thus, in the vicinity of the Spanish city of A Corunha (see Figure 4a) the mean wave direction is about 310°, while in the nearshore of the Portuguese city Aveiro the mean wave direction is around 290°. As regards the wave power, expressed in KW/m (kilowatts over a meter of wave front) this is normalized in the maps presented in Figure 4 and then by dividing its value by 100. For the sake of consistency, the same color bar is used for all the energy maps presented.

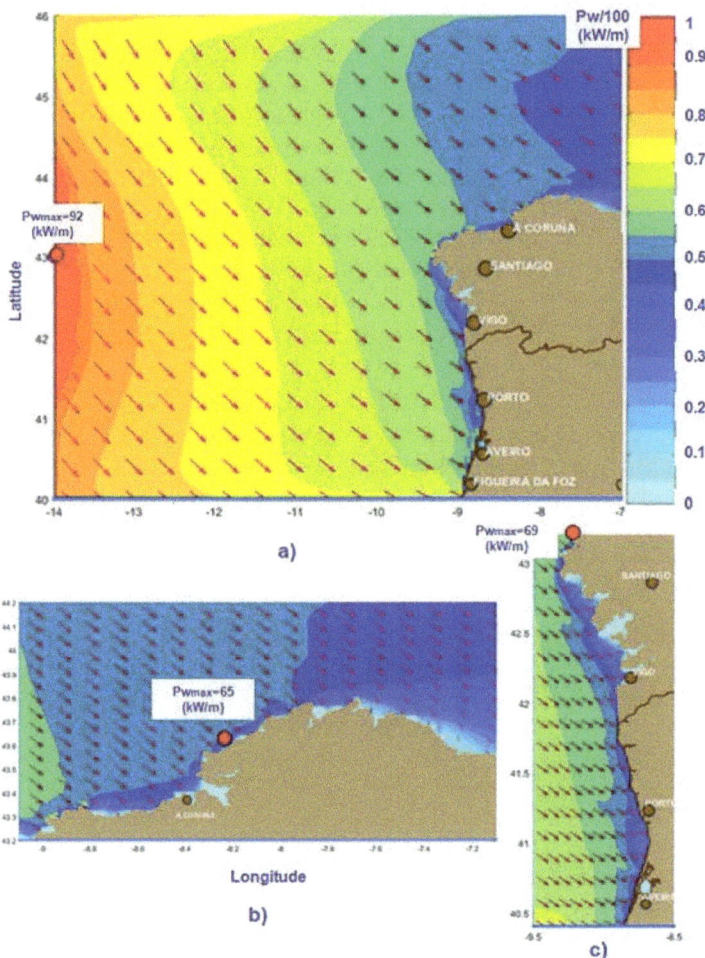

Figure 4. Normalized wave power and energy transport vectors illustrating an average winter time wave energy pattern in the northwestern Iberian nearshore. (**a**) Northern wave driver (D1); (**b**) Coastal domain (C1); (**c**) Coastal domain (C2). Model results correspond to the time frame 22 October 2016.

Some wave energy hot spots can be also identified in Figure 4b,c and they are marked with red circles in these figures. In this case study, they have estimated wave powers of 65 kW/m and 69 kW/m, respectively. These two hot spots are persistent in the sense that they occur as such in most of the simulations performed in the coastal computational domains (C1) and (C2). Figure 5 presents a relevant case study related to the wave energy propagation towards the Portuguese port of Sines, located in the central part of continental Portugal. The model results correspond to the time frame of 10 March 2017 and they show the model system focusing on wave power via the Iberian driver (D2) in Figure 5a, through the coastal central computational domain (C5) in Figure 5b and finishing with the SWAN high resolution domain (H5), in Figure 5c. The mean wave direction is this time was around 315° offshore and varies to about 300° in the vicinity of the Sines harbor. For this regular wave energy propagation pattern, when the waves are coming from the northwest, the coastal environment south of Lisbon, including the nearshore in the vicinity of the Sines port is relatively sheltered as illustrated very clearly in all subplots belonging to Figure 5.

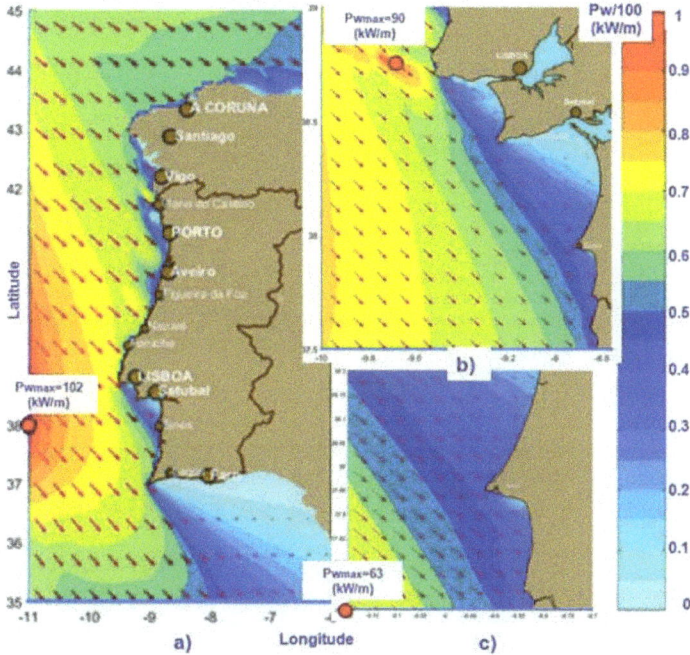

Figure 5. Wave energy propagation towards the Portuguese port of Sines. (**a**) Iberian wave driver (D2); (**b**) Coastal domain (C5); (**c**) High resolution area in the nearshore of Sines (H5). Model results correspond to the time frame 10 March 2017.

A hot spot as regards the wave energy can be also noticed in Figure 5b. This is marked with a red circle; it has an estimated wave power of 90 kW/m and is located in the vicinity of Capo DA Roca, which is the western point of continental Europe. This is also a persistent hot spot, since this coastal area is well known as being more energetic.

Figure 6 presents two different wave energy propagation patterns. Thus, Figure 6a illustrates the results for the Portuguese wave driver (D3) on 18 November 2017. This is a very common wave energy propagation pattern in the west Iberian nearshore with the waves coming from the northwest and the energy decreasing gradually from north to south. The maximum wave power is about 100 kW/m and is located offshore in the northwestern corner of the computational domain. An interesting result relates to the coastal area indicated with a red arrow. This was the location of the Aguçadora wave farm which operated in 2008 and is now one of two Portuguese pilot areas to demonstrate marine renewable energy extraction. As it can be noticed from Figure 6a, although it is not a global hot spot, there is a concentration of wave energy in this coastal environment. Figure 6b illustrates another wave energy propagation pattern in the coastal domain (C3), which corresponds to the time frame of 2 December 2017. Here, the wave energy is more focused in the central part of the Portuguese nearshore with a wave power peak of about 70 kW/m offshore the Portuguese city of Figueira da Foz. The location of the Aguçadora pilot area is also represented with a red arrow and even in such propagation conditions, enhanced local wave energy in the nearshore can be noticed.

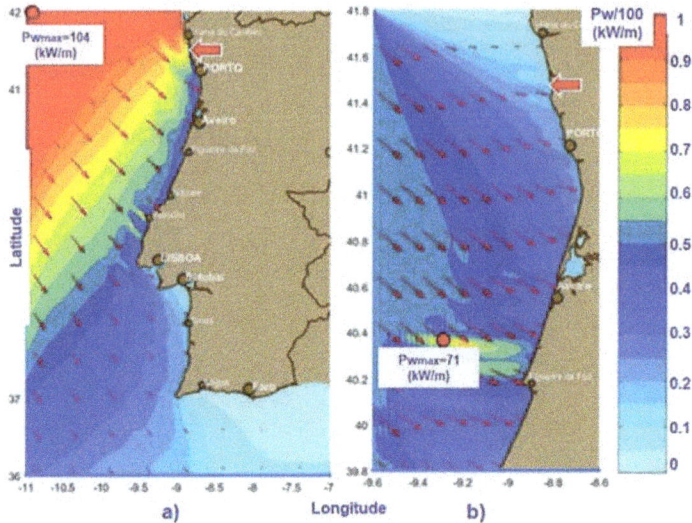

Figure 6. Wave energy propagation in the Portuguese nearshore. (**a**) Portuguese wave driver (D3), simulation corresponds to the time frame 18 November 2017; (**b**) Coastal domain (C3), simulation corresponds to the time frame 2 December 2017. The red arrow indicates one of the two Portuguese pilot areas.

Figure 7 illustrates another wave propagation pattern that corresponds to 14 October 2017, when the waves are coming from the west. Although in most cases the mean wave direction is from the northwest, there are still about 10% of situations when the dominant wave direction is from the west. Furthermore, the results of this analysis indicated a slight increase in this percentage over the last two years. Figure 7a presents the results for the coastal domain (C4). An energy peak of 78 kW/m can be noticed in the same location as in Figure 5b, which is marked with a red circle and is offshore Cabo da Roca. This demonstrates that it is a persistent hot spot. Another persistent hot spot can be seen in the nearshore of the Portuguese city of Peniche since the energy focused in this area can be also be seen in Figure 6a. Figure 7b illustrates the wave energy propagation in the southern coastal domain (C6) for the same time frame. Even though the wave propagation pattern corresponds to the waves coming from the west, the southwest of the Iberian Peninsula is sheltered and only in quite rare cases when the waves are coming from the southwest, can this nearshore be characterized by relatively high wave energy.

Figure 8 presents several wave energy propagation patterns in the high resolution computational domains. As in the previous cases presented above, they are also related to waves coming from the west. Thus, Figure 8a illustrates the wave energy propagation in the area of Peniche (H1), with the results corresponding to 18 September 2017. Again, the energy peak (89 kW/m) is located in front of Cabo da Roca. Another relevant energy peak can be noticed in the north of the computational domain in the area indicated with a red arrow, which corresponds to the second Portuguese pilot area, Sao Pedro de Moel. Figure 8b presents the wave energy propagation patterns in the nearshore of Lisbon (H2) for a simulation corresponding to the time frame 17 December 2017. The power peak (96 kW/m) also occurs in front of Cabo da Roca. Finally, Figure 8c,d present the results from the high-resolution areas Pinheiro da Cruz (H3) and North of Sines (H4), respectively. Both these results are from 19 November 2107. As it can be noticed from these figures, when the mean direction of the incoming waves is from the west, the coastal area north of Sines is no longer sheltered and an energy peak (76 kW/m for this case) occurs in the vicinity of the Sines harbor.

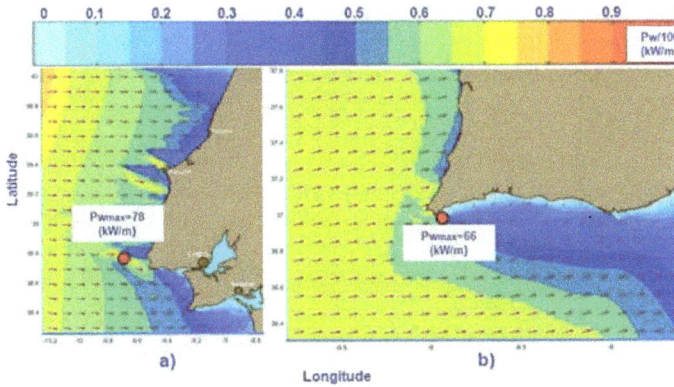

Figure 7. Wave energy propagation in the central and southern coastal domains, simulation corresponding to the time frame 14 October 2017. (**a**) Results for the central coastal domain (C4), (**b**) Results for the southern coastal domain (C6).

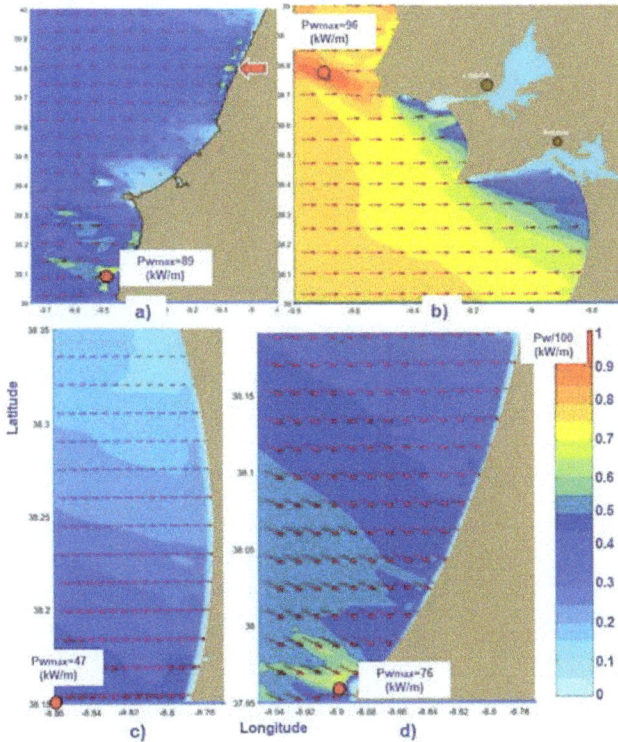

Figure 8. Wave energy propagation in the high resolution computational domains. (**a**) Peniche area (H1), simulation corresponding to the time frame 18 September 2017, the red arrow indicates one of the two Portuguese pilot areas; (**b**) Lisbon nearshore (H2), simulation corresponding to the time frame 17 December 2017; (**c**) Pinheiro da Cruz area (H3) and (**d**) North of Sines (H4), simulations corresponding to the time frame 19 November 2017.

Figures 9 and 10 illustrate case studies of wave energy propagation considering the Cartesian computational domains. Thus, Figure 9a presents a SWAN simulation corresponding to the time frame 17 September 2017 in the nearshore area Figueira da Foz (X2). The resolution of 25 m considered in the geographical space allows a good representation of the breaking process so that some very clear energy peaks that occur just before breaking are noticed in this figure, with a maximum wave power of 69 kW/m in the central part of the computational domain. Figure 9b presents a simulation in the coastal area of Obibos (X3) corresponding to the time frame 17 November 2017.

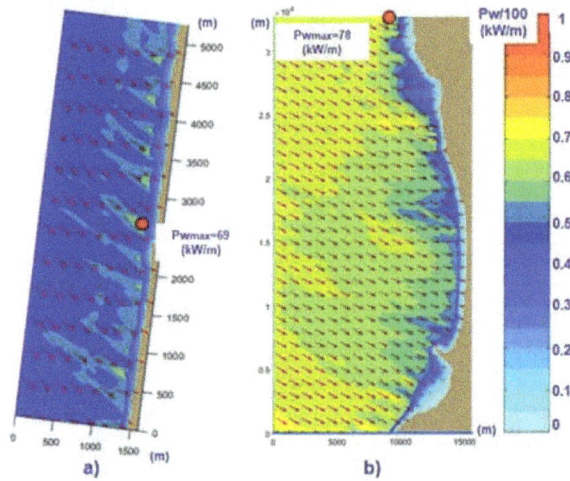

Figure 9. Wave energy propagation in the Cartesian computational domains. (**a**) Figueira da Foz (X2), simulation corresponding to the time frame 17 September 2017; (**b**) Obibos (X3), simulation corresponding to the time frame 17 November 2017.

Finally, Figure 10 presents the wave energy propagation in the vicinity of two important Portuguese harbors. These are Leixoes (X1), close to the city of Porto and Sines (X4), located south of Lisbon in the central part of the Portuguese nearshore. The results presented correspond to the time frames 17 November 2017 and 10 March 2017, respectively. No significant wave energy peaks can be seen in the vicinity of these harbors.

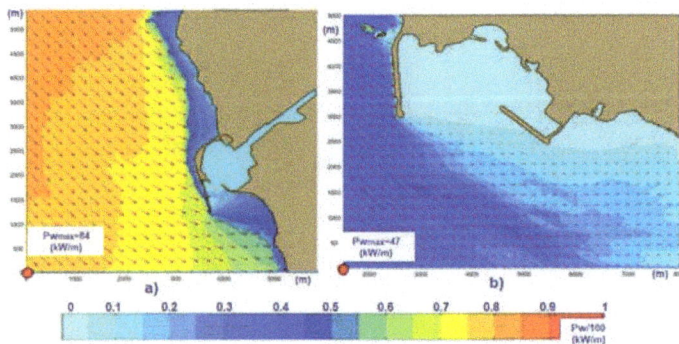

Figure 10. Wave energy propagation in the Cartesian computational domains defined close to the Portuguese harbors. (**a**) Leixoes (X1), simulation corresponding to the time frame 17 November 2017; (**b**) Sines (X4), simulation corresponding to the time frame 10 March 2017.

4. Conclusions

Starting from the fact that climate changes may induce mutations in nearshore wave energy propagation patterns, the objective of the present work is to present and analyze several recent relevant case studies that reflect the spatial distribution of the wave power in the Iberian coastal environment. For this purpose, a complex multi-level wave modelling system based on spectral phase averaged models was used. This system is focused towards the coast, with increasing spatial resolution. Furthermore, data assimilation techniques were also implemented to increase the reliability of the wave predictions, so that the results provided by this system can be considered credible. Simulations were performed for a 15-month period starting in October 2016 and finishing at the end of 2017. Some of the most relevant results are presented, considering 18 different SWAN computational domains that correspond to four geographic levels (sub oceanic, coastal, high resolution and Cartesian very high resolution).

A comparison with the results from similar studies that used the same modeling system to estimate the spatial distribution of wave energy in the west Iberian nearshore [43–46], and which were undertaken in the last 25 years, show many similar features, but also some differences. Thus, an important observation coming from the analysis of the case studies presented in this work, is that most of the wave energy hot spots already identified in the western Iberian nearshore appear to be persistent, since they were identified again in various computational domains and for various time frames. Thus, a concentration of energy waves has been very clearly identified in the nearshore in the vicinity of Cabo da Roca, the coastal environment of Peniche, the two Portuguese pilot areas and also some areas in the northern side of the Iberian Peninsula, as the nearshore gets close to the city of A Corunha. Some other general observations resulting from the analysis of the model results for the entire period considered are that, although the dominant wave direction in the western Iberian nearshore is from the northwest, the incidence of waves coming from the west appears to have increased slightly. Since the general wave energy distribution pattern in the Iberian nearshore is characterized by a gradual decrease from north to south, this new pattern may induce a global increase in the wave power in the central part of the western Iberian coastal environment. Another issue is related to the frequency and the intensity of the storms. Although the results of the model simulations performed in this 15-month period do not allow clear conclusions, they do indicate some changes in the storm dynamics. As regards the marine energy, this means careful attention should be paid to safety and survival issues.

Finally, it can be concluded that, although advances in the technologies of wave energy conversion are still needed, the high dynamics of offshore wind energy extraction will encourage momentum in the field of wave energy extraction. Furthermore, wave farms can provide protection for wind farms [47] and they can also influence coastal dynamics in various ways [48,49]. From this perspective, the present work provides a more comprehensive picture of the most recent wave energy propagation patterns in the western Iberian nearshore, which represents a very interesting coastal environment that has high potential for the future development of marine energy parks.

Acknowledgments: This work was carried out in the framework of the research project REMARC (Renewable Energy extraction in Marine environments and its Coastal impact), supported by the Romanian Executive Agency for Higher Education, Research, Development and Innovation Funding—UEFISCDI, grant number PN-III-P4-IDPCE-2016-0017. The author would like also to express his gratitude to the reviewers for their suggestions and observations that helped in improving the present work.

Conflicts of Interest: The author declares no conflict of interest.

Nomenclature

CAPEX	Capital Expenditure
DA	Data Assimilation
GFS	Global Forecast System
LCOE	Levelized Cost of Energy

EU	European Union
Hs	Significant Wave Height
MM5	Fifth-Generation Mesoscale Model
NCAR	US National Center for Atmospheric Research
NOAA	US National Oceanic and Atmospheric Administration Model
OI	Optimal Interpolation
OPEX	Operational Expenditure
Pw	Wave power
PTO	Power take-off
SCM	Successive Correction Method
SET	Strategic Energy Technology
SWAN	Simulating Waves Nearshore
WAM	Wave Modeling
WEC	Wave Energy Converter
WT	Wind Turbines
WW3	Wave Watch 3

References

1. European Commission. *Strategic Energy Technology Plan*; European Commission: Brussels, Belgium, 2014.
2. JRC. JRC Wind Energy Status Report 2016 Edition. 2016. Available online: http://publications.jrc.ec.europa.eu/repository/bitstream/JRC105720/kjna28530enn.pdf (accessed on 22 February 2018).
3. Jensen, P.H.; Chaviaropoulos, T.; Natarajan, A. LCOE Reduction for the Next Generation Offshore Wind Turbines; Outcomes from the INNWIND.EU Project. 2017. Available online: file:///C:/Users/user/Downloads/Innwind-final-printing-version.pdf (accessed on 24 February 2018).
4. SET Plan—Declaration of Intent on Strategic Targets in the Context of an Initiative for Global Leadership in Ocean Energy. 2016. Available online: https://setis.ec.europa.eu/system/files/integrated_set-plan/declaration_of_intent_ocean_0.pdf (accessed on 24 February 2018).
5. Zhang, X.; Yang, J. Power capture performance of an oscillating-body WEC with nonlinear snap through PTO systems in irregular waves. *Appl. Ocean Res.* **2015**, *52*, 261–273. [CrossRef]
6. Zhang, X.; Lu, D.; Guo, F.; Gao, Y.; Sun, Y. The maximum wave energy conversion by two interconnected floaters: Effects of structural flexibility. *Appl. Ocean Res.* **2018**, *71*, 34–47. [CrossRef]
7. Rezanejad, K.; Soares, C.G.; Lópezb, I.; Carballo, R. Experimental and numerical investigation of the hydrodynamic performance of an oscillating water column wave energy converter. *Renew. Energy* **2017**, *106*, 1–16. [CrossRef]
8. Liu, Z.; Han, Z.; Shi, H.; Yang, W. Experimental study on multi-level overtopping wave energy convertor under regular wave conditions. *Int. J. Nav. Archit. Ocean Eng.* **2017**, in press. [CrossRef]
9. Rusu, L.; Onea, F. The performance of some state-of-the-art wave energy converters in locations with the worldwide highest wave power. *Renew. Sustain. Energy Rev.* **2017**, *75*, 1348–1362. [CrossRef]
10. Veigas, M.; López, M.; Iglesias, G. Assessing the optimal location for a shoreline wave energy converter. *Appl. Energy* **2014**, *132*, 404–411. [CrossRef]
11. Bernardino, M.; Guedes Soares, C. Evaluating marine climate change in the Portuguese coast during the 20th century. In *Maritime Transportation and Harvesting of Sea Resources*; Guedes Soares, C., Teixeira, A.P., Eds.; Taylor & Francis Group: London, UK, 2018; pp. 1089–1095.
12. Iglesias, G.; Carballo, R. Wave energy and nearshore hot spots: The case of the SE Bay of Biscay. *Renew. Energy* **2010**, *35*, 2490–2500. [CrossRef]
13. Iglesias, G.; Carballo, R. Offshore and inshore wave energy assessment: Asturias (N Spain). *Energy* **2010**, *35*, 1964–1972. [CrossRef]
14. Iglesias, G.; Carballo, R. Wave energy potential along the death coast (Spain). *Energy* **2009**, *34*, 1963–1975. [CrossRef]
15. Iglesias, G.; Carballo, R. Wave energy resource in the Estaca de Bares area (Spain). *Renew. Energy* **2010**, *35*, 1574–8154. [CrossRef]

16. Rusu, E.; Soares, C.G. Wave energy assessments in the coastal environment of Portugal continental. In Proceedings of the 27th International Conference on Offshore Mechanics and Arctic Engineering, Estoril, Portugal, 15–20 June 2008; Volume 6, pp. 761–772.

17. Rusu, L.; Pilar, P.; Soares, C.G. Reanalysis of the wave conditions on the approaches to the Portuguese port of Sines. In *Maritime Transportation and Exploitation of Ocean and Coastal Resources*; Taylor & Francis Group: London, UK, 2005; Volume 1, pp. 1137–1142.

18. Silva, D.; Rusu, E.; Soares, C.G. Evaluation of Various Technologies for Wave Energy Conversion in the Portuguese Nearshore. *Energies* **2013**, *6*, 1344–1364. [CrossRef]

19. Silva, D.; Martinho, P.; Soares, C.G. Wave power resources at Portuguese test sites from 11-year hindcast data. In *Renewable Energies Offshore*; Taylor & Francis Group: London, UK, 2015; pp. 113–121.

20. Carballo, R.; Sánchez, M.; Ramos, V.; Taveira-Pinto, F.; Iglesias, G. A high resolution geospatial database for wave energy exploitation. *Energy* **2014**, *68*, 572–583. [CrossRef]

21. Rusu, E.; Silva, D.; Soares, C.G. Efficiency assessments for different WEC types operating in the Portuguese coastal environment. In *Developments in Maritime Transportation and Exploitation of Sea Resources*; Taylor & Francis Group: London, UK, 2014; pp. 961–972.

22. Silva, D.; Rusu, E.; Soares, C.G. Evaluation of the expected power output of wave energy converters in the north of the Portuguese nearshore. In *Progress in Renewable Energies Offshore*; Taylor & Francis Group: London, UK, 2016; pp. 875–882.

23. Iuppa, C.; Cavallaro, L.; Vicinanza, D.; Foti, E. Investigation of suitable sites for Wave Energy Converters around Sicily (Italy). *Ocean Sci.* **2015**, *11*, 543–557. [CrossRef]

24. Iuppa, C.; Cavallaro, L.; Foti, E.; Vicinanza, D. Potential wave energy production by different wave energy converters around sicily. *J. Renew. Sustain. Energy* **2015**, *7*. [CrossRef]

25. Vicinanza, D.; Contestabile, P.; Ferrante, V. Wave energy potential in the north-west of sardinia (Italy). *Renew. Energy* **2013**, *50*, 506–521. [CrossRef]

26. Holthuijsen, H. *Waves in Oceanic and Coastal Waters*; Cambridge University Press: Cambridge, UK, 2007; p. 387.

27. Rusu, E. Strategies in using numerical wave models in ocean/coastal applications. *J. Mar. Sci. Technol.* **2011**, *19*, 58–75.

28. Bretherton, F.P.; Garrett, C.J.R. Wave trains in inhomogeneous moving media. *Proc. R. Soc. Lond. Ser. A* **1968**, *302*, 529–554. [CrossRef]

29. SWAN Team. *Scientific and Technical Documentation*; SWAN Cycle III; Department of Civil Engineering, Delft University of Technology: Delft, The Netherlands, 2017.

30. WAMDI Group. The WAM model—A third generation ocean wave prediction model. *J. Phys. Oceanogr.* **1988**, *18*, 1775–1810.

31. Gómez Lahoz, M.; Carretero Albiach, J.C. A two-way nesting procedure for the WAM model: Application to the Spanish Coast. *J. Offshore Mech. Arctic Eng.* **1997**, *119*, 20–24. [CrossRef]

32. Tolman, H.L. A third-generation model for wind waves on slowly varying, unsteady and inhomogeneous depths and currents. *J. Phys. Oceanogr.* **1991**, *21*, 782–797. [CrossRef]

33. Silva, D.; Martinho, P.; Guedes Soares, C. Modeling wave energy for the Portuguese coast. In *Maritime Engineering and Technology*; Guedes Soares, C., Garbatov, Y., Sutulo, S., Santos, T.A., Eds.; Taylor & Francis Group: London, UK, 2012; pp. 647–653.

34. Booij, N.; Ris, R.C.; Holthuijsen, L.H. A third generation wave model for coastal regions. Part 1: Model description and validation. *J. Geophys. Res.* **1999**, *104*, 7649–7666. [CrossRef]

35. Rusu, E.; Soares, C.V.; Rusu, L. Computational strategies and visualisation techniques for the wave modelling in the Portuguese nearshore. In *Maritime Transportation and Exploitation of Ocean and Coastal Resources*; Taylor & Francis: London, UK, 2005; Volume 1, pp. 1129–1136.

36. Soares, C.G.; Rusu, L.; Bernardino, M. An operational wave forecasting system for the Portuguese continental coastal area. *J. Oper. Oceanogr.* **2011**, *4*, 17–27. [CrossRef]

37. Dudhia, J. A nonhydrostatic version of the Penn State–NCAR mesoscale model: Validation tests and simulation of an Atlantic cyclone and cold front. *Mon. Weather Rev.* **1993**, *121*, 1493–1513. [CrossRef]

38. Rusu, L.; Soares, C.G. Evaluation of a high-resolution wave forecasting system for the approaches to ports. *Ocean Eng.* **2013**, *58*, 224–238. [CrossRef]

39. Rusu, L.; Soares, C.G. Impact of assimilating altimeter data on wave predictions in the western Iberian coast. *Ocean Model.* **2015**, *96*, 126–135. [CrossRef]

40. Kalnay, E. *Atmospheric Modeling, Data Assimilation and Predictability*; Cambridge University Press: Cambridge, UK, 2003; p. 341.

41. Lionello, P.; Günther, H.; Janssen, P.A.E.M. Assimilation of altimeter data in a global third-generation wave model. *J. Geophys. Res.* **1992**, *97*, 14453–14474. [CrossRef]

42. Silva, D.; Rusu, E.; Soares, C.G. High-resolution wave energy assessment in shallow water accounting for tides. *Energies* **2016**, *9*, 761. [CrossRef]

43. Rusu, E.; Guedes Soares, C. Numerical modeling to estimate the spatial distribution of the wave energy in the Portuguese nearshore. *Renew. Energy* **2009**, *34*, 1501–1516. [CrossRef]

44. Soares, C.G.; Bento, A.R.; Goncalves, M.; Silva, D.; Martinho, P. Numerical evaluation of the wave energy resource along the Atlantic European coast. *Comput. Geosci.* **2014**, *71*, 37–49. [CrossRef]

45. Silva, D.; Bento, A.R.; Martinho, P.; Soares, C.G. High resolution local wave energy modelling for the Iberian Peninsula. *Energy* **2015**, *91*, 1099–1112. [CrossRef]

46. Bento, A.R.; Martinho, P.; Soares, C.G. Wave energy resource assessment for Northern Spain from a 13-year hindcast. In *Renewable Energies Offshore*; Taylor & Francis Group: London, UK, 2015; pp. 63–69.

47. Silva, D.; Martinho, P.; Soares, C.G. Trends in the available wave power at the Portuguese pilot zone. In *Progress in Renewable Energies Offshore*; Taylor & Francis Group: London, UK, 2016; pp. 53–59.

48. Onea, F.; Rusu, E. The expected efficiency and coastal impact of a hybrid energy farm operating in the Portuguese nearshore. *Energy* **2016**, *97*, 411–423. [CrossRef]

49. Rusu, E.; Onea, F. Study on the influence of the distance to shore for a wave energy farm operating in the central part of the Portuguese nearshore. *Energy Convers. Manag.* **2016**, *114*, 209–223. [CrossRef]

energies

MDPI

Article

Wave Power as Solution for Off-Grid Water Desalination Systems: Resource Characterization for Kilifi-Kenya

Francisco Francisco * [ID], Jennifer Leijon, Cecilia Boström, Jens Engström and Jan Sundberg

Division of Electricity, Uppsala University, 751 21, Uppsala, Sweden; jennifer.leijon@angstrom.uu.se (J.L.);
Cecilia.Bostrom@angstrom.uu.se (C.B.); Jens.Engstrom@angstrom.uu.se (J.E.);
Jan.Sundberg@angstrom.uu.se (J.S.)
* Correspondence: Francisco.Francisco@angstrom.uu.se; Tel.: +46-18-471-5843

Received: 16 February 2018; Accepted: 26 March 2018; Published: 20 April 2018

Abstract: Freshwater scarcity is one of humanity's reoccurring problems that hamper socio-economic development in many regions across the globe. In coastal areas, seawater can be desalinated through reverse osmosis (RO) and transformed into freshwater for human use. Desalination requires large amounts of energy, mostly in the form of a reliable electricity supply, which in many cases is supplied by diesel generators. The objective of this work is to analyze the wave power resource availability in Kilifi-Kenya and evaluate the possible use of wave power converter (WEC) to power desalination plants. A particular focus is given use of WECs developed by Uppsala University (UU-WEC). The results here presented were achieved using reanalysis—wave data revealed that the local wave climate has an approximate annual mean of 7 kW/m and mode of 5 kW/m. Significant wave height and wave mean period are within 0.8–2 m and 7–8 s respectively, with a predominant wave mean direction from southeast. The seasonal cycle appeared to be the most relevant for energy conversion, having the highest difference of 6 kW/m, in which April is the lowest (3.8 kW/m) and August is the peak (10.5 kW/m). In such mild wave climates, the UU–WEC and similar devices can be suitable for ocean energy harvesting for water desalination systems. Technically, with a capacity factor of 30% and energy consumption of 3 kWh/m^3, a coastal community of about five thousand inhabitants can be provided of freshwater by only ten WECs with installed capacity of 20 kW.

Keywords: wave power resource; desalination; freshwater; wave energy converter; Kilifi; Kenya

1. Introduction

The worldwide deficiency of clean freshwater causes sanitation problems, food shortage and sometimes even conflicts [1–3]. UN's sustainable development goal (SDG), number 6 aim to ensure access to water and sanitation for all [4]—a goal currently far from met. For example, in 2015, 58% of the Kenyan population had access to safe drinking water, 30% had access to safe sanitation and only 14% had access to a proper handwashing facility [5,6]. The energy-food-water nexus describes an entangled relationship for the need of fulfilling basic human needs; water shortage in the agricultural sector inevitably leads to a decrease in food production [7]. Therefore, a holistic and interdisciplinary view on this problem is necessary. As an example, recent studies reflect upon effects of water usage in rural regions of Kenya [8,9]; discussing the farmers' willingness to pay for water for agricultural purposes; payments for water services in rural communities; the overall wealth; the use of water saving technologies; and the possibility of handling different types of future changes in the climate in this region [8,9].

In many coastal places around the world, the inhabitants still live without reliable source of clean freshwater even though there are an overflow of seawater and ocean waves. One such example is small desalination plant located in an educational center in Kilifi-Kenya, which also was taken as an example for the present work. Instead of using conventional diesel generators [10], this desalination plant can be powered by renewable energy technologies such as wave energy converts (WECs).

Seawater desalination is used to produce freshwater around the world, using thermal processes or technologies involving membranes [11,12]. Three different desalination processes are reverse osmosis (RO), electrodialysis (ED) and multistage flash (MSF). To briefly describe these processes: With RO, the salt in the seawater is excluded by the use of a semipermeable membrane and an applied pressure, as described in [13]; with ED, an electric field is applied to the water, producing freshwater with the use of ion-selective membranes, with applications such as recently discussed in [14,15], and with MSF distillation, there are different evaporation stages in the steps towards production of freshwater [12,16]. RO technology is one of the most energy efficient for seawater desalination [17,18]. Several studies such as [18–22], have estimated that the energy consumption in RO desalination plants are mainly within 2–5 kWh/m^3. This is mostly due to substantial advances in RO technologies that may include energy recovery techniques and efficient membranes [18]. Energy consumption will decrease even more with gains in membrane and energy recovery systems performances [18,23].

Dependable electric grid is not always available where the freshwater is needed. Therefore, solutions to implement renewable energy systems (RES) via smaller autonomous grids connected to desalination plants are suggested [24]. An issue regarding RES powered desalination, for technologies such as RO, is the intermittency of the RES. Whereas, a variable power output from the RES (if no energy storage system is included) causes a pressure variation on the RO membranes which may damage or lower their lifetime and can affect the freshwater output. As such, a RES with lower variability/intermittency is preferable.

Among the RES, wave power has high potential in terms of energy density, resource availability and predictability [4,25–28]. Especially in locations where seas and swell waves occur permanently, suggesting that wave power can be a better choice to combine with RO than solar or wind power. Moreover, as the energy source (ocean waves), and the water resource to desalinate (seawater) are found at the same location, there is a clear opportunity in combining wave power and desalination, taking care of the natural resources at a certain location [29]. However, the technology for wave power conversion is still in its early stages. This infers that there are still constrains on the implementation of large-scale wave power farms that are able to deliver large amounts of power (in the order of GWh) in a reliable manner. However, with advances in information technologies, computational modeling and prototyping, the actual WECs have gained substantial improvements in power take-off as well as decrease in production costs, making them ready-enough to be used in small scale. Information on WECs concepts can be found in for example [28,30].

Previous studies such as [31,32], have estimated that the wave climate in Kenya is similar to other regions situated within the equatorial belt, with offshore peak values within 20–30 kW/m. However, there is a need to access the nearshore wave power resource. Therefore, the overall objective of this research is to access the feasibility of utilizing wave power in desalination plants. In particular this study aims to investigate the local wave power resource in Kilifi as a starting step of evaluating the possibility of using wave power converter developed by Uppsala University (UU-WEC) technology for desalination systems. The present work lies within the research domain of the authors and an existent collaboration between researchers at Uppsala University in Sweden and Strathmore Energy Research Center in Kenya. The results of the present work will provide insight information of a potential solution for freshwater shortage in coastal areas.

The UU–WEC Technology

The UU-WEC (Figure 1) is a point absorber that comprises a submerged linear generator connected to a heaving buoy e.g., [33]. The WECs are connected through an offshore marine substation before the electricity is transmitted to an onshore connection point [34]. The voltage output of UU-WECs is about 450 V, and the estimated operational capacity is between of 10 kW and 100 kW but can be scaled up. The power rating can be adjusted to match the local wave climate [33,35,36], and is mostly determined by the electrical configurations and by the size and shape of the buoy and generator. These three parameters can be adjusted to fit a local wave climate. This WEC technology is currently used for offshore experiments in Lysekil on the west coast of Sweden, and the experimental research has progressed for more than ten years [33], including studies on environmental aspects of wave energy conversion. These studies revealed very limited negative impacts of the UU-WECs to the marine environment e.g., [37,38]. Instead, positive effects, such as artificial reefs, were observed [37,38]. Other recent studies on the same WEC concept include control strategies, compensations for tides and discussions on the design of wave power farms with several WECs [39–41].

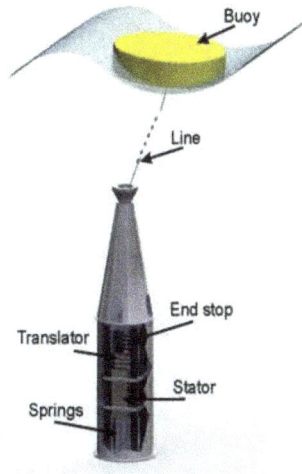

Figure 1. Comprehensive illustration of Uppsala University's main WEC concept, comprising of a linear generator on the ocean floor, with a movable part (the translator) connected via a line to a buoy.

2. Studied Region

The studied region covers the coastal area of Kilifi, located north of Mombasa in Kenya (Figure 2). The local bathymetry varies from 0 m to 30 m depth at close proximity to the beach. The seabed in this area is dominated by sedimentary rocks, loose clay, sand, mud flats, and coral reef that is normally distributed between 16 m and 45 m depth at distances from shore between 500 m to 2 km [42].

The marine environment is generally characterized by seagrass communities, mangrove forest and sandy subtracts [43]. The marine fauna is diverse and, includes sensitive species such as turtles and dugongs [42]. The wind pattern in this region is mostly dominated by monsoons, sea breeze, and occasionally cyclones. The tide in Kilifi is semi-diurnal with an amplitude of 3.5 m [27]. These tides contribute to intertidal platforms, mud flats and rather rocky communities that get exposed for several hours during low tides, and sub tidal platforms that are highly productive and populated by coral reefs, reef flats and susceptible to nearshore wave action.

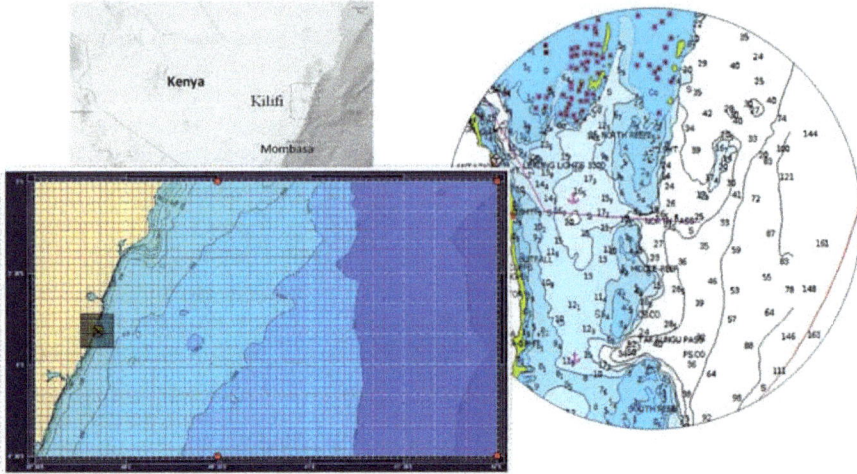

Figure 2. Studied location in Kilifi (3.815° S, 39.846° E), ca. 60 km north of Mombasa, Kenya. The Yellow dashed circle represents the data extraction point (*Fugro*-No. 162230-1-R0). Local bathymetry and costal features are shown within the circle.

3. Wave Data

The wave data used was provided by *Fugro* (document No. 162030-1-R0) (see Supplementary Materials). The data covers a period from January 1997 to December 2015 with a temporal resolution of 6 h, and was based on analysis of the world waves data source of the *European Centre for Medium-range Weather Forecasts (ECWMF)* wave model. Re-analyses were conducted by *Fugro Oceanor* combined with a global buoy at location with 34 m bottom depth, and multi-satellite altimeter database (*Topex, Jason, Geosat, GFO* and *Envisat*) by transforming offshore grid points to a nearshore point (3.815° S, 39.846° E) using *Simulating Wave Nearshore (SWAN)* model.

The data set contains wave field variables: significant wave height (H_s), mean wave energy period (T_e), and mean wave direction (*MDir*). The variables were estimated from a wave energy spectrum (f, θ) with moment m_0. The H_s is equivalent to the mean height of the highest one-third of the waves in a sea-state ($H_s = 4\sqrt{m_0}$). The T_e is equivalent to the spectral period and is given by $T_e = m_{-1}/m_0$. *MDir* is the direction of the most energetic spectral band.

The wave power resource (P_w) was estimated using Equation (1), which defines P_w as the average transport rate of energy per meter of wave front, it depends mostly on the wave height and wave period. By definition, P_w can be affected by wave dissipation, shoaling, reflection, refraction, trapping, diffraction, and other non-linear phenomena [44–46].

$$P_w = \rho g \int \int Cg\,(f,\,\theta)\,S(f,\,\theta)df\,d\theta \tag{1}$$

where ρ is the water density (1025 kg/m^3 for sea water), g is the acceleration due to gravity (9.8 m/s^2), Cg is the group velocity which depends of the water depth h, f is the frequency, and θ is the direction of propagation.

Assuming that the water depth is deep enough so that $h/L > 0.5$, then $Cg = g/4\pi f$; therefore the simplified form of Equation (1) (in (kW/m)) becomes:

$$P_w \cong 0.5\,H_s^2 T_e \tag{2}$$

The Wave Powered Desalination Approach—An Estimation

The validation of the possibility of utilizing wave energy for desalination through RO was conducted taking into account the following parameters: Available wave power resource (P_w), installed capacity per unit WEC (P_{wec}), energy per unit WEC per day at capacity factor of 30% ($W_{day.wec} = P_{wec} \times 24 \times 0.3$ h), energy needed per unit volume of treated water (E_{ro} = 3 kWh/m^3), taken from an interval between 2 kWh/m^3 and 5 kWh/m^3 [18–22]. Taking into account that the basic daily need of volume of water for one person (V_i), is around 0.02 m^3 [1]; however, this study uses $V_i \cong 0.1$ m^3. A fundamental question can be: How many WECs (Equation (3)) are needed to meet the water demand of the local inhabitants at daily basis?

$$X_{wec} = X_p / X_{p.wec} \tag{3}$$

where X_{wec} is the number of WECs, X_p is the number of inhabitants, and $X_{p.wec} = \frac{W_{day.wec}}{E_{ro}} / 1000$ L, is the number of inhabitants supplied by freshwater from a single WEC, recalling that a 1000 L is equivalent to 1 m^3.

4. Results

4.1. Local Wave Climate

The mean value of wave power, P_w, for this data set is 7 kW/m, the median values is 5 kW/m, the mode is 5 kW/m, and the minimum and maximum are 0.2 and 53 kW/m respectively. This data reveals a wide difference of wave power between the periods of slack versus periods of high seas. It is important is to analyse the frequency of occurrence and magnitude of sea states (Figures 3–5) in order to understand how the energy is distributed over time. The significant wave height, H_s, (Figures 3 and 4) shows that the predominant values are within 0.75 m to 2 m, in which the southeast sea state, with a frequency of occurrence of 56%, has values of H_s in the interval between 1 m and 1.5 m. The second most dominant sea state, with 28% of occurrence, has easterly waves within 1.5–2 m. 12% of the waves compose the east-north-easterly seas state with H_s within 0.4–0.7 m. Higher sea states account for approximately 4% of occurrence and have values of H_s within 2–3 m. In what regards the wave mean period, results in Figure 3 show that the most prevalent sea state, 79% of occurrence, has values of within 7–8 s, followed by 10% of sea states with T_e of 5 s, 10% with 9–10 s, and 1% for sea states with values of T_e of 4 s and above 11 s, respectively.

The distribution of H_s and T_e (Figures 3 and 4) in time also show sea states with different wave power values starting from 2 kW/m up to 50 kW/m with a higher density between 10 kW/m and 20 kW/m (Figure 5). Approximately 52% of the waves have power values within 2 kW/m–5 kW/m, 33% have P_w within 7–10 kW/m, and the remaining 15% has values of P_w greater than 12 kW/m (Figure 6).

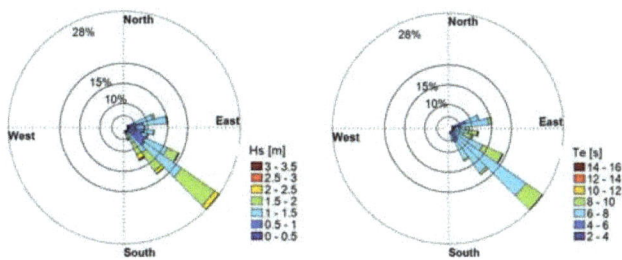

Figure 3. Total probability distribution of significant wave height H_s wave mean period T_e and wave mean direction versus mean wave direction MD_{ir} in Kilifi. The frequencies of occurrence of sea states are given in percentage.

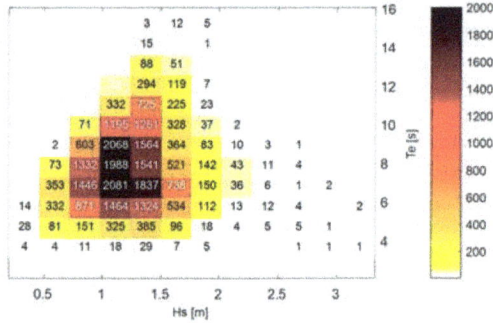

Figure 4. Total probability distribution of sea sates at six-hourly occurrence in Kilifi.

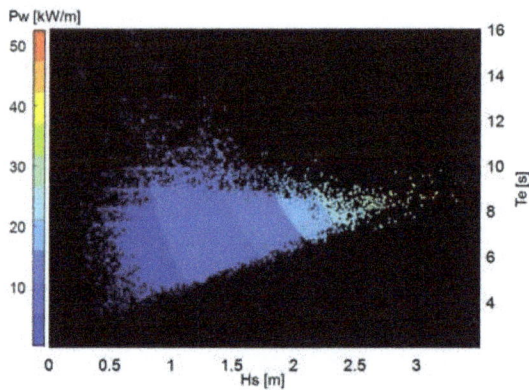

Figure 5. Combined scatter plot showing the occurrence of H_s, T_e and P_w in Kilifi.

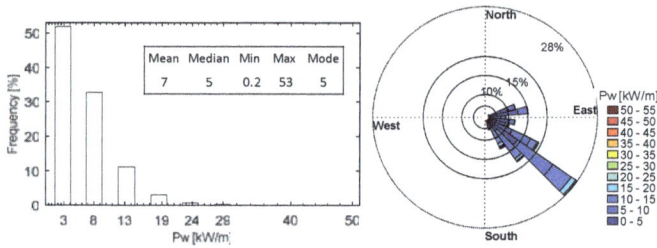

Figure 6. Total probability distribution of wave power P_w versus mean wave direction MD_{ir}. The frequencies of occurrence of sea states are given in percentage in Kilifi.

4.2. Diurnal Variability

The results show diurnal variations on the wave power resource in Kilifi. In Figure 7, there are the mean, mode and median graphs referent to observations at midnight, six, twelve and eighteen hours. From this date it was estimated that wave power values in evening and night are in average higher than in the day. However, the highest value of mode is observed in the morning. The difference between midday and mid night mean values is about 0.3 kW/m.

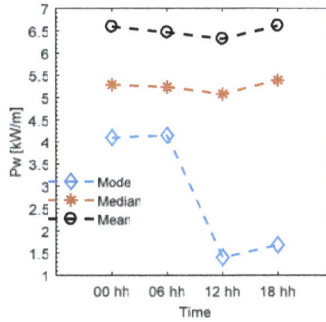

Figure 7. Diurnal variability of wave power in Kilifi based on 6–hourly wave data, showing lower values during day comparing to night.

4.3. Seasonal Variability

The seasonal cycle was obtained by extracting the total average for each month, over the entire time series (Figure 8: red line). There is a unimodal distribution of P_w over a period of 12 months, from January to December. Peak values are observed in August, when the mean is 10.5 kW/m followed by September with 10.2 kW/m. The box plot results shows that August and September had similar median values of approximately 10 kW/m. Although August has the highest mean, July has the highest occurrence of extreme values.

January–May had the lowest H_s mean values of approximately 1 m (Figure 9a). While July, August and September had higher mean values up to 1.5 m with a peak in August when H_s is 1.7 m. Considering T_e, there are three peaks occurring in April, September and December, being the highest mean of 7.5 s. The lowest T_e value, 6.8 s, was observed in July (Figure 9b). The wave mean direction MD_{ir} also had a seasonal variability signal. From February to April MD_{ir} is from east, then the waves gradually propagate from southeast from May to August, and again the mean direction reverts towards east (Figure 9c).

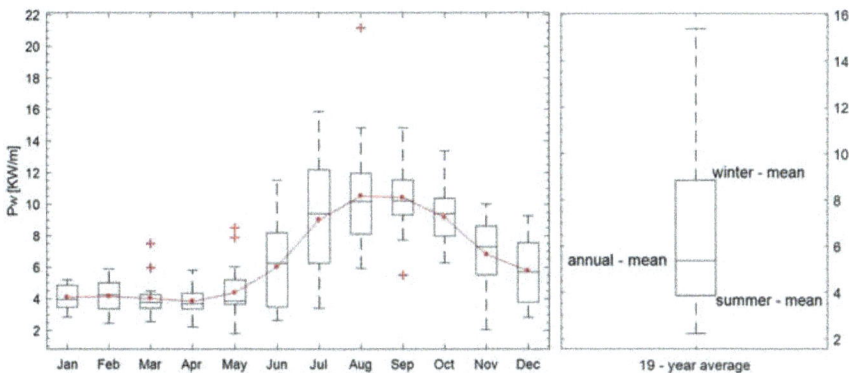

Figure 8. The seasonal variability of P_w in Kilifi. The red line represents monthly mean values of P_w, the box plot contains the median-middle mark, the 25th and 75th percentiles—the bottom and top edges of the box, respectively. The red crosses are outliers of P_w, representing for example storms and tropical cyclones.

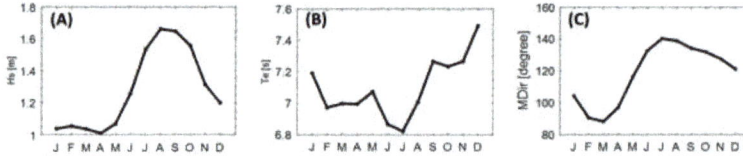

Figure 9. The seasonal variability in Kilifi: (**A**) for H_s; (**B**) for T_e; and (**C**) for MD_{ir}. T_e (**B**) has a rather different pattern with three-modal seasonal variability compared with a unimodal pattern observed in both H_s (**A**) and MD_{ir} (**C**) respectively.

4.4. Interannual and Long-Term Variability

The interannual variations of P_w along the 19-year time-series are evident. The total mean is 6.5 kW/m, total standard deviation is 4.6 kW/m and the mean absolute deviation is 3.4 kW/m. For this period of time, the anomaly of P_w was calculated from the difference between the total mean and the mean of each month. Results show that there is an oscillatory pattern in the anomaly of P_w with periods of 3–5 years (Figure 10). Probably, with a longer data set, it would be possible to better detect the variability.

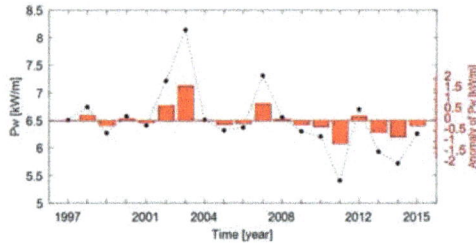

Figure 10. P_w from 1997 to 2016 in Kilifi, plotted as annual mean (black dots) and respective anomalies (red bars).

4.5. Wave Powered Desalination

Results showed that not many WECs are needed to supply water for small communities, assuming a capacity factor of 30% (Figure 11). For example, only 10 UU-WECs with installed capacity of 20 kW per unit would be required for daily production of freshwater for about 5000 inhabitants.

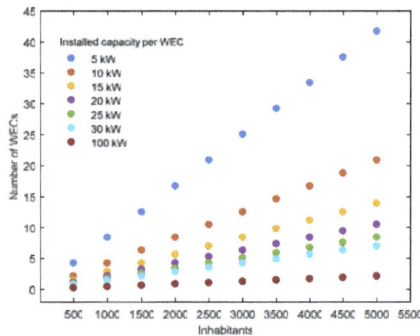

Figure 11. Number of WECs necessary for daily freshwater production as a function of number of inhabitants and installed capacity per unit WEC assuming a capacity factor of 30% and fixed energy consumption of 3 kWh/m^3.

5. Discussion

5.1. Wave Climate

There is a relatively high frequency of occurrence of waves with simultaneously small H_s and longer T_e and vice-versa, see Figure 5. This may explain the reason why P_w occurs densely within 5 kW/m to 25 kW/m. It is also observed that the typical sea sates are determined by values of T_e rather than H_s (Figures 4–6). For instance, the frequency of occurrence of waves of H_s of 1–1.25 m are noticeable the highest and corresponds to values of T_e that widely range from 3–9 s.

Diurnal variability of wave power may be related to the coastal wind dynamics, such as offshore and sea breezes triggered by pressure gradients due to uneven heating and cooling of the land and sea masses. In this study, the wave power's time-variation signal share similarities in shape with the wind field's time-variations obtained by, for example Mahongo et al. [47]. Studies on coastal wind patterns such as [47], revealed the existence of similar diurnal variations on wind intensity using three-hourly wind data of four locations along the coast of Tanzania. The wind is strongest in the afternoon (14:00–15:00), and weakest in early morning (4:00). From the wind pattern described by [47], it is possible to infer that wave power values in the second quarter of the day occur when the wind direction is from north-northeast in the austral summer and from southwest during austral winter. Peak diurnal values of wave power do occur when the wind direction are dominantly form east-northeast (austral winter) and south-southeast (austral winter).

With regard to the seasonal variability, April has the lowest mean P_w of 3.8 kW/m. The minimum monthly mean coincides with calm north-easterly wind conditions in Tanzania and Kenya, which occur in March–April and November [48–50]. Moreover, low values of P_w observed during the same period of the northeast monsoon (November to March). The monthly mean values of P_w differ from what was found by other authors, e.g., Barstow 2008 [51], who estimated 5–10 kW/m for January and 20–30 kW/m for July. The reason for this discrepancy may be due to the fact that they used altimetry-model data from deep water points, while the present study used data from a shallower water point. The P_w and H_s have similar seasonal variability signals (Figure 11). The maximum H_s is in contra-phase with the maximum T_e may justify the reason to why there are considerably low values of P_w in this region. Other contributing factors can be latitude and weather systems including the inter-tropical convergence zone that oscillates within 15° S and 15° N and the monsoon regime [51,52]. The presence of a large land mass-Madagascar may also block larger and longer southerly waves from reaching the Kenya.

Interannual and long term variability of the wave power resources in Kilifi is influenced by local, regional and mesoscale weather systems. For example, previous studies have shown that the coastal wind in Kenya is associated with the monsoon wind system [53]. By looking at the coastal wind intensity, it is found that the shape of its seasonal cycle signal is similar to the one observed in the wave power's signal, with peaks in July and dips in March–April. The peaks are influenced by the austral seasonality of wind and wave fields generated in the southern ocean. With the absence of such influence, the minima of wave power should be registered around December–January. However, for this particular region of the Indian Ocean, the climate dynamics are strongly influenced by the Indian Ocean Dipole, that is an oscillatory mode that couples the variability of both atmosphere and the ocean. The lowest negative mode of the Indian Ocean Dipole occurs in March–April, that coincides with the minima in the magnitude of both wind and wave power.

Calculated anomalies of P_w correlate well with monthly mean sea level pressure at the surface linked to the Southern Ocean Mode (with correlation coefficient $r < -0.6$), Indian Ocean Dipole ($r > 0.4$) and slight influenced by the El Niño ($r \sim 0.4$). The 3 years–4 years variability can be observed with El Niño Southern Oscillation ($r \sim -0.5$). There is a strong correlation between P_w and sea surface temperature of tropical and southern ocean basins (Figure 9). Studies done by Reguero et al., 2015 [54] also found significant correlation of monthly mean wave power with the North Atlantic Oscillation

($r \sim 0.2$), El Niño Southern Oscillation ($r \sim 0.2$), Atlantic Multidecadal Oscillation ($r \sim -0.2$) and Dipole Mode Index ($r \sim -0.3$).

5.2. Wave Powered Desalination

The estimative results in Figure 10 were achieved assuming that wave power has an annual capacity factor near 30%. Several studies suggest that the capacity factor of wave power is ca. 20–40% at the most [55–58]. Being so, the number of WECs (20 kW) required for a daily supply of freshwater to 5000 inhabitants would increase or decrease in function of the capacity factor. Moreover, the capacity factor would vary seasonally, being lower during austral summer months (January to May) and higher during July to November. On the other hand, the number of required WECs would also vary according to the efficiency of the desalination system in terms of energy consumption. For example, similarly to above, assuming a capacity factor of 30%, but with an energy consumption of 2 kWh/m^3, the required number of WECs (20 kW) for 5000 inhabitants would decrease to seven units.

The results indicate a mild wave climate in Kilifi. This is particularly suitable for the UU-WEC technology that is versatile to variety of wave steepness. Due to few extreme wave days and lower forces acting on the system—reducing the risk of technical failure, and lowering its predicted maintenance. The UU-WEC was designed to optimally operate in depths within 20 m and 50 m, in rather flat seabed, conditions which normally occur in Kilifi and on other coastal areas across the globe. Several layout designs of wave power farms can be used to maximize energy abortion, as described in [59–61]. In practice, these wave power farm layouts should also consider the distance between devices, substation and onshore cable connections which are key elements for cost reduction and feasibility, as described in [62,63]. In the case of Kilifi where a rather small number of WECs is required, a rather simple and cost effective wave power farm would be required. For example, it would include an array of WECs deployed along the same isobaths in order to maximize energy absorption.

Compared to the other wave climates, such as of the North Atlantic [54], the waves in Kenya are more predictable, which may enhance the reliability and predictability of the power or freshwater production from a wave powered desalination plant. Therefore, the use of UU-WEC for desalination purposes would be suitable to similar coastal areas where the water need is severe, requiring a reliable source of drinking water, irrigation among other purposes. Apart from the resource assessments aspects, the costs of electricity, water production and the entire life cycle of a project should be holistically analyzed.

6. Conclusions

The wave climate in Kilifi is mild, with mean values of 1.3 m, 7.1 s, and 6.5 kW/m for H_s, T_e, and P_w, respectively. The difference in monthly mean values between the lower and higher seasons was in the order of ± 0.7 m for H_s and of about ± 6.5 kW/m for P_w. This is due to the entire region is located within the tropical and equatorial zone where the weather patterns experience only gradual changes and the variations are rather small. The dynamics of the sea breezes and coastal winds influence the local wave power resource which is higher in the evening and night comparing to daytime values, with an estimated difference of 0.3 kW/m. The seasonal signal indicated that P_w and H_s share the same pattern in which both have peaks in August, while the dip occurred between January and May. In contrast, T_e have maxima in February, May and September, whereas the minima are in February, July and October. This seasonal variability of wave power is also similar in pattern with the coastal winds, and both are conditioned by the seasonality of the monsoon system and the inter-tropical convergence zone. The interannual variability has anomalies within 2 kW/m with periodicity of 3 years to 5 years, which can be influenced by mesoscale weather and climate systems such as Indian Ocean Dipole, Southern Ocean Mode, El Niño Southern Oscillation.

The wave power resource available at the Kenyan coast is estimated to be enough for wave power conversion throughout an entire annual cycle. The peak of wave power availability and conversion would be in the austral winter that match the dry season when freshwater is needed the most. This fact highlights the importance of a desalination plant powered by WECs in arid and remote coastal areas. Assuming a capacity factor of 30% and energy consumption of 3 kWh/m^3, a rather small coastal community of about five thousand inhabitants can be provided of freshwater by only ten WECs with installed capacity of 20 kW a unit, and certainly the number of WECs would decrease with the decrease of energy consumption and/or with an increase of installed capacity. There are still grey areas in what regards techno-economic and life cycle analysis of such projects involving renewable energy technologies. Uncertainties are even higher with non-stablished technologies such as wave power. Notwithstanding, near future projections appoints to reduction of energy consumption to below 3 kWh/m^3 [23], and a possible decrease of water production costs [64]. On the other hand, there will be an increasing freshwater consumption which reflects on growing investments on desalination plants around the world. It is only left to the renewable energy sector to promote itself to competitive standards within the energy-water nexus. This brings us to a final conclusion that the electricity needed to run a desalination plant or any other type of electric water treatment system, can be technically supplied by renewable energy sources such as wave power.

7. Future Work

The present work is part of a larger study being undertaken by Uppsala University on use of wave power for freshwater production through desalination, with focus on the UU-WEC technology. Therefore, after investigating the wave climate in Kilifi and introducing the idea of utilizing WECs as potential power supplier, the following steps would proceed: Full description of the power system with several UU-WECs including voltage amplitude and frequency variations; Then, coupling of the power system with a RO desalination plant in Kenya that will include an energy storage system and freshwater storage seen as buffer for shifting power rates; Techno-economic and life cycle analysis would be investigated; assessment of local environmental parameters such as seabed conditions, residue management among other aspects required by law would precede a possible deployment of WECs that would culminate with an implementation of wave powered desalination system. The aforementioned steps may not follow the presented sequential order. Even so, research is needed before producing freshwater using the UU-WEC or similar technologies. The success of projects such as this, would revolutionize rural electrification in remote and coastal areas across the globe.

Supplementary Materials: The following are available online at http://www.mdpi.com/1996-1073/11/4/1004/s1, the wave data used was provided by Fugro (document No. 162030-1-R0) and can be accessed upon a request to the authors.

Acknowledgments: This project has received funding from the European Union's Seventh Framework Programme for research, technological and demonstration under grant agreement No. 607656. This work was conducted within the STandUP for Energy strategic research framework and was also supported by the Swedish Research Council (VR) grant No. 2015-03126.

Author Contributions: Francisco Francisco wrote the paper, contributed with conception of the work, data analysis and interpretation; Jennifer Leijon contributed with conception of the work, data interpretation, drafting and revision of the manuscript; Cecilia Boström contributed with data acquisition conception of the work and revision of the manuscript; Jens Engström contributed with data acquisition, conception of the work and revision of the manuscript; Jan Sundberg contributed with conception of the work, and revision of the manuscript.

Conflicts of Interest: The authors declare no conflict of interest.

References

1. UN-Water. *Water for a Sustainable World*; The United Nations World Water Development Report 2015; UN-Water: Geneva, Switzerland, 2015.
2. Kaniaru, W. From scarcity to security: Water as a potential factor for conflict and cooperation in Southern Africa. *S. Afr. J. Int. Aff.* **2015**, *22*, 381–396. [CrossRef]
3. Almer, C.; Laurent-lucchetti, J.; Oechslin, M. Water scarcity and rioting: Disaggregated evidence from Sub-Saharan Africa. *J. Environ. Econ. Manag.* **2017**, *86*, 193–209. [CrossRef]

4. United Nations Educational, Scientific and Cultural Organization. *Water and Jobs*; The United Nations World Water Development Report 2016; United Nations Educational, Scientific and Cultural Organization: Paris, France, 2016.

5. World Health Organization (WHO) and the United Nations Children's Fund (UNICEF). Progress on Drinking Water, Sanitation and Hygiene. 2017. Available online: http://www.who.int/mediacentre/news/releases/2017/launch-version-report-jmp-water-sanitation-hygiene.pdf (accessed on 20 August 2017).

6. WHO/Unicef (JMP). *WASH in the 2030 Agenda*; World Health Organization: Geneva, Switzerland, 2016.

7. Gulati, M.; Jacobs, I.; Jooste, A.; Naidoo, D.; Fakir, S. The water-energy-food security nexus: Challenges and opportunities for food security in South Africa. *Aquat. Procedia* **2013**, *1*, 150–164. [CrossRef]

8. Shikuku, K.M.; Winowiecki, L.; Twyman, J.; Eitzinger, A.; Perez, J.G.; Mwongera, C.; Läderach, P. Smallholder farmers' attitudes and determinants of adaptation to climate risks in East Africa. *Clim. Risk Manag.* **2017**, *16*, 234–245. [CrossRef]

9. Ochieng, J.; Kirimi, L.; Mathenge, M. Effects of climate variability and change on agricultural production: The case of small scale farmers in Kenya. *NJAS-Wagen. J. Life Sci.* **2016**, *77*, 71–78. [CrossRef]

10. Ouma, C. *Tewa Training Centre Report: Field Report for Tewa Training Centre as a Case Study under "Water Desalination Using Wave Power" Uppsala University Research Coordinated through Strathmore Energy Research Centre*; Internal report at Uppsala University: Uppsala, Sweden, 2013.

11. González, D.; Amigo, J.; Suárez, F. Membrane distillation: Perspectives for sustainable and improved desalination. *Renew. Sustain. Energy Rev.* **2017**, *80*, 238–259. [CrossRef]

12. El-Dessouky, H.T.; Ettouney, H.M. *Fundamentals of Salt Water Desalination*; Elsevier: New York, NY, USA, 2002. [CrossRef]

13. Shenvi, S.S.; Isloor, A.M.; Ismail, A.F. A review on RO membrane technology: Developments and challenges. *Desalination* **2015**, *368*, 10–26. [CrossRef]

14. Scarazzato, T.; Panossian, Z.; Tenório, J.A.S.; Pérez-Herranz, V.; Espinosa, D.C.R. A review of cleaner production in electroplating industries using electrodialysis. *J. Clean. Prod.* **2017**, *168*, 1590–1602. [CrossRef]

15. Nayar, K.G.; Sundararaman, P.; Schacherl, J.D.; O'Connor, C.L.; Heath, M.L.; Orozco Gabriel, M.; Wright, N.C.; Winter, A.G. Feasibility study of an electrodialysis system for in-home water desalination in urban India. *Dev. Eng.* **2016**, *2*, 38–46. [CrossRef]

16. Khoshrou, I.; Nasr, M.R.J.; Bakhtari, K. New opportunities in mass and energy consumption of the Multi-Stage Flash Distillation type of brackish water desalination process. *Sol. Energy* **2017**, *153*, 115–125. [CrossRef]

17. Van der Bruggen, B.; Vandecasteele, C. Distillation vs. membrane filtration: Overview of process evolutions in seawater desalination. *Desalination* **2002**, *143*, 207–218. [CrossRef]

18. Khayet, M. Solar desalination by membrane distillation: Dispersion in energy consumption analysis and water production costs (a review). *Desalination* **2013**, *308*, 89–101. [CrossRef]

19. Will, M.; Klinko, K. Optimization of seawater RO systems design. **2001**, *138*, 299–306.

20. Banat, F.; Jwaied, N.; Rommel, M.; Koschikowski, J. Performance evaluation of the 'large SMADES' autonomous desalination solar-driven membrane distillation plant in Aqaba, Jordan. *Desalination* **2007**, *217*, 17–28. [CrossRef]

21. Miller, J.E. *Review of Water Resources and Desalination Technologies*; U.S. Department of Commerce: Washington, DC, USA, 2003.

22. Glueckstern, P. The impact of R & D on new technologies, novel design concepts and advanced operating procedures on the cost of water desalination. *Desalination* **2001**, *139*, 217–228.

23. Energy, D.; Elimelech, M.; Phillip, W.A. The Future of Seawater and the Environment. *Science* **2011**, *333*, 712–718.

24. Tzen, E.; Morris, R. Renewable energy sources for desalination. *Sol. Energy* **2003**, *75*, 375–379. [CrossRef]

25. Charlier, R.H.; Justus, J.R. *Ocean. Energies: Environmental, Economic and Technological Aspects of Alternative Power Sources*; Elsevier Science: New York, NY, USA, 1993.

26. Doukas, H.; Karakosta, C.; Psarras, J. {RES} technology transfer within the new climate regime: A 'helicopter' view under the {CDM}. *Renew. Sustain. Energy Rev.* **2009**, *13*, 1138–1143. [CrossRef]

27. Hammar, L. Towards Technology Assessment of Ocean. Energy in a Developing Country Context. Licentiate Thesis, Chalmers University of Technology, Gothenburg, Sweden, 2011.

28. de O. Falcão, A.F. Wave energy utilization: A review of the technologies. *Renew. Sustain. Energy Rev.* **2010**, *14*, 899–918.
29. Davies, P.A. Wave-powered desalination: Resource assessment and review of technology. *Desalination* **2005**, *186*, 97–109. [CrossRef]
30. Khan, N.; Kalair, A.; Abas, N.; Haider, A. Review of ocean tidal, wave and thermal energy technologies. *Renew. Sustain. Energy Rev.* **2017**, *72*, 590–604. [CrossRef]
31. Cornett, A.M. A Global Wave Energy Resource Assessment. In Proceedings of the International Offshore and Polar Engineering Conference, Busan, Korea, 15–20 June 2014.
32. Gunn, K.; Stock-Williams, C. Quantifying the global wave power resource. *Renew. Energy* **2012**, *44*, 296–304. [CrossRef]
33. Lejerskog, E.; Boström, C.; Hai, L.; Waters, R.; Leijon, M. Experimental results on power absorption from a wave energy converter at the Lysekil wave energy research site. *Renew. Energy* **2015**, *77*, 9–14. [CrossRef]
34. Ekström, R.; Baudoine, A.; Rahm, M.; Leijon, M. Marine substation design for grid-connection of a research wave power plant on the Swedish West coast. In Proceedings of the 10th European Wave and Tidal Conference, Aalborg, Denmark, 2–5 September 2013.
35. Leijon, M.; Boström, C.; Danielsson, O.; Gustafsson, S.; Haikonen, K.; Langhamer, O.; Strömstedt, E.; Stålberg, M.; Sundberg, J.; Svensson, O.; et al. Wave energy from the North Sea: Experiences from the lysekil research site. *Surv. Geophys.* **2008**, *29*, 221–240. [CrossRef]
36. Bostrom, C.; Ekergard, B.; Waters, R.; Eriksson, M.; Leijon, M. Linear generator connected to a resonance-rectifier circuit. *IEEE J. Ocean. Eng.* **2013**, *38*, 255–262. [CrossRef]
37. Langhamer, O.; Haikonen, K.; Sundberg, J. Wave power-Sustainable energy or environmentally costly? A review with special emphasis on linear wave energy converters. *Renew. Sustain. Energy Rev.* **2010**, *14*, 1329–1335. [CrossRef]
38. Haikonen, K. Underwater Radiated Noise from Point Absorbing Wave Energy Converters. Ph.D. Thesis, Uppsala University, Uppsala, Sweden, 2014.
39. Wang, L.; Isberg, J. Nonlinear passive control of a wave energy converter subject to constraints in irregular waves. *Energies* **2015**, *8*, 6528–6542. [CrossRef]
40. Ekström, R.; Ekergård, B.; Leijon, M. Electrical damping of linear generators for wave energy converters—A review. *Renew. Sustain. Energy Rev.* **2015**, *42*, 116–128. [CrossRef]
41. Castellucci, V.; Abrahamsson, J.; Kamf, T.; Waters, R. Nearshore tests of the tidal compensation system for point-absorbing wave energy converters. *Energies* **2015**, *8*, 3272–3291. [CrossRef]
42. Tychsen, J. (Ed.) *KenSea. Environmental Sensitivity Atlas for Coastal Area of Kenya*; Geological Survey of Denmark and Greenland (GEUS): Copenhagen, Denmark, 2006; 76p, ISBN 87-7871-191-6.
43. Obura, D.O. Kenya. *Mar. Pollut. Bull.* **2001**, *42*, 1264–1278. [CrossRef]
44. Liu, P.L.F.; Losada, I.J. Wave propagation modeling in coastal engineering. *J. Hydraul. Res.* **2002**, *40*, 229–240. [CrossRef]
45. Gorrell, L.; Raubenheimer, B.; Elgar, S.; Guza, R.T. SWAN predictions of waves observed in shallow water onshore of complex bathymetry. *Coast. Eng.* **2011**, *58*, 510–516. [CrossRef]
46. World Meteorological Organization. *Guide to Wave Analysis Guide to Wave Analysis*; World Meteorological Organization: Geneva, Switzerland, 1998.
47. Mahongo, S.B.; Francis, J.; Osima, S.E. Wind Patterns of Coastal Tanzania: Their Variability and Trends. *West. Indian Ocean J. Mar. Sci.* **2012**, *10*, 107–120.
48. Mcclanahan, T.R. Seasonality in East Africa's coastal waters. *Mar. Ecol. Prog. Ser.* **1988**, *44*, 191–199. [CrossRef]
49. Hammar, L.; Ehnberg, J.; Mavume, A.; Cuamba, B.C.; Molander, S. Renewable ocean energy in the Western Indian Ocean. *Renew. Sustain. Energy Rev.* **2012**, *16*, 4938–4950. [CrossRef]
50. Camberlin, P. *Oxford Research Encyclopedia of Climate Science Climate of Eastern Africa Geographical Features Influencing the Region's Climate*; Oxford University Press: Oxford, UK, 2018; No. April.
51. Barstow, S.; Mørk, G.; Lønseth, L.; Mathisen, J.P. WorldWaves wave energy resource assessments from the deep ocean to the coast. In Proceedings of the 8th European Wave Tidal Energy Conference, Uppsala, Sweden, 7–10 September 2009; pp. 149–159.
52. Semedo, A. Seasonal Variability of Wind Sea and Swell Waves Climate along the Canary Current: The Local Wind Effect. *J. Mar. Sci. Eng.* **2018**, *6*. [CrossRef]

53. Francis, J. Wind patterns of coastal Tanzania: Their variability and trends Wind Patterns of Coastal Tanzania: Their Variability. *West. Indian Ocean J. Mar. Sci.* **2012**, *10*, 107–120.

54. Reguero, B.G.; Losada, I.J.; Méndez, F.J. A global wave power resource and its seasonal, interannual and long-term variability. *Appl. Energy* **2015**, *148*, 366–380. [CrossRef]

55. Ibarra-Berastegi, G.; Jon, S.; Garcia-soto, C. Electricity production, capacity factor, and plant efficiency index at the Mutriku wave farm (2014–2016). *Ocean Eng.* **2018**, *147*, 20–29. [CrossRef]

56. World Energy Council. *World Energy Resources Marine Energy | 2016*; World Energy Council: London, UK, 2016.

57. David Kavanagh, B.E. Capacity Value of Wave Energy in Ireland. Master's Thesis, University College Dublin, Dublin, Ireland, 2012.

58. Stoutenburg, E.D.; Jenkins, N.; Jacobson, M.Z. Power output variations of co-located offshore wind turbines and wave energy converters in California. *Renew. Energy* **2010**, *35*, 2781–2791. [CrossRef]

59. Göteman, M.; Engström, J.; Eriksson, M.; Isberg, J. Fast Modeling of Large Wave Energy Farms Using Interaction Distance Cut-Off. *Energies* **2015**, *8*, 13741–13757. [CrossRef]

60. Giassi, M.; Malin, G. Parameter Optimization in Wave Energy Design by a Genetic Algorithm. In Proceedings of the 2015 International Congress on Technology, Communication and Knowledge (ICTCK), Mashhad, Iran, 11–12 November 2017; pp. 23–26.

61. Giassi, M.; Malin, G.; Thomas, S.; Engstr, J.; Eriksson, M.; Isberg, J. Multi-Parameter Optimization of Hybrid Arrays of Point Absorber Wave Energy Converters. In Proceedings of the 12th European Wave and Tidal Energy Conference (EWTEC), Cork, Ireland, 27–31 August 2017; pp. 1–6.

62. Chatzigiannakou, M.A.; Dolguntseva, I.; Leijon, M. Offshore Deployments of Wave Energy Converters by Seabased Industry AB. *J. Mar. Sci. Eng.* **2017**, *5*, 15. [CrossRef]

63. Dolguntseva, I. Offshore deployment of marine substation in the Lysekil research site. In Proceedings of the The 25th International Ocean and Polar Engineering Conference, Kona, HA, USA, 21–26 June 2015.

64. Voutchkov, N. Desalination—Past, Present and Future. International Water Association, 2016. Available online: http://www.iwa-network.org/desalination-past-present-future/ (accessed on 16 March 2018).

energies

MDPI

Review

Electrical Power Supply of Remote Maritime Areas: A Review of Hybrid Systems Based on Marine Renewable Energies

Anthony Roy [1,2,*], François Auger [1] , Florian Dupriez-Robin [2], Salvy Bourguet [1] and Quoc Tuan Tran [3]

[1]   Laboratoire IREENA, Université de Nantes, 37 Boulevard de l'Université, 44600 Saint-Nazaire, France; francois.auger@univ-nantes.fr (F.A.); salvy.bourguet@univ-nantes.fr (S.B.)
[2]   CEA-Tech Pays de la Loire, Technocampus Océan, 5 Rue de l'Halbrane, 44340 Bouguenais, France; florian.dupriez-robin@cea.fr
[3]   Institut National de l'Énergie Solaire (INES), CEA/DRT/LITEN/DTS/LSEI, 50 Av. du Lac Léman, 73370 Le Bourget-du-Lac, France; quoctuan.tran@cea.fr
*   Correspondence: anthony.roy1@univ-nantes.fr; Tel.: +33-240-172-687

Received: 14 May 2018; Accepted: 16 July 2018; Published: 20 July 2018

Abstract: Ocean energy holds out great potential for supplying remote maritime areas with their energy requirements, where the grid size is often small and unconnected to a continental grid. Thanks to their high maturity and competitive price, solar and wind energies are currently the most used to provide electrical energy. However, their intermittency and variability limit the power supply reliability. To solve this drawback, storage systems and Diesel generators are often used. Otherwise, among all marine renewable energies, tidal and wave energies are reaching an interesting technical level of maturity. The better predictability of these sources makes them more reliable than other alternatives. Thus, combining different renewable energy sources would reduce the intermittency and variability of the total production and so diminish the storage and genset requirements. To foster marine energy integration and new multisource system development, an up-to-date review of projects already carried out in this field is proposed. This article first presents the main characteristics of the different sources which can provide electrical energy in remote maritime areas: solar, wind, tidal, and wave energies. Then, a review of multi-source systems based on marine energies is presented, concerning not only industrial projects but also concepts and research work. Finally, the main advantages and limits are discussed.

Keywords: hybrid systems; multi-source systems; marine renewable energy; combined platform

1. Introduction

The electricity supply of remote marine areas is mostly generated from solar and wind energy, thanks to their maturity and attractive prices compared to other renewable energies [1]. However, these renewable energy sources are based on the exploitation of intermittent resources. To resolve this drawback, storage systems such as batteries and Diesel generators can be used, but investment costs and induced pollution are often not favorable. Moreover, it is costly and logistically difficult to implement a diesel supply in remote marine areas. However, over the last few years, marine renewable energies have encountered some interest and a genuine development by the industry, because of their potential available energy [2]. Tidal and wave energies are the most developed among the different marine energies [3,4]. Hence, hybrid systems combining solar, wind, and marine energies can now be developed to provide sustainable and reliable electrical energy. Wind and wave hybrid energy systems have already been developed, according to reviews written by different authors [5–9].

Some multipurpose platforms have been studied in terms of feasibility [10,11]. The present paper aims to put forward a review of hybrid systems combining marine energies on the same platform or structure, such as wave, tidal, wind, and solar energies. Firstly, a brief review of these renewable energy sources is detailed, to show the basics and existing technologies. A comparison between the different temporal characteristics of each renewable source is given, to highlight the different temporal scales and forecast abilities. Then, a review of industrial and academic multisource systems is presented, from projects tested under real sea conditions to those still at the concept stage. Finally, some advantages and limitations of multisource systems based on marine energies are listed.

2. A Short Review of Renewable Energy Sources Concerned by This Study

Island areas in maritime environments present the advantage of having several primary resources in their neighborhoods. Concerning the development and maturity of renewable energies over recent years, the main sources that can be used seem to be solar and wind energies, which present a high technical level of maturity and the most interesting cost [1,2]. Furthermore, among the marine energies available from the ocean, tidal and wave energies are currently two of the most advanced and promising technologies, with a better maturity level than other marine energies such as thermal and salinity gradient conversion energies [4]. Marine energies have the advantage of a good predictability and a high available energy level [1,2]. This part of the review aims to briefly present these four renewable energy sources in terms of the operating principle, main technologies existing today, and temporal resource characteristics. An overview of the main technologies currently existing is shown in Figure 1.

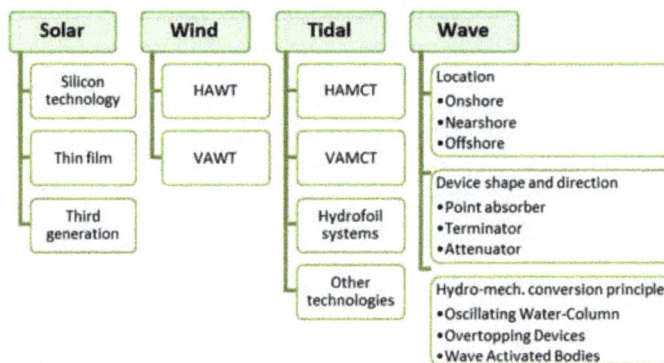

Figure 1. Solar, wind, tidal current, and wave energy converter technologies classification.

2.1. Solar Energy

At present, solar energy is one of the most widely used renewable energies in the world. Photovoltaic panels used to convert solar energy into electrical energy have now reached a high maturity level, with many technologies available on the market for different kinds of application [12].

2.1.1. Fundamentals of Operating Processes

A solar cell uses a semi-conductor material, often silicon, to absorb photons of incident solar radiation received by the cell [12]. A semi-conductor is based on two energy bands. One of them is called the valence band. Electron presence in this band is allowed. In the second energy band, called the conduction band, electrons are absents. The band between the valence band and the conduction band is called the band gap. An electron can move from the valence band to the conduction band if the amount of energy provided by incident solar radiation to the electron is larger than the band gap value. This electron can move into an external circuit due to the p-n junction. This process results in a hole-electron pair formation. The electron moves to the n-region, whereas the hole moves to the *p*-region, resulting in a potential difference. This effect was explained by A. Einstein in 1905 and is called the photovoltaic effect [13]. It consists of transforming solar energy into electrical energy. To produce more power, photovoltaic cells are combined in serial and parallel configurations to reach the desired current and voltage levels, forming a photovoltaic module. The electrical output power depends on the global irradiance and the ambient temperature [14,15].

2.1.2. Main Technologies Currently Used

Different kinds of photovoltaic cells exist on the market at different technology readiness levels. Many articles have discussed the advantages and drawbacks of the different technologies [12,16–18]. Three main categories can be seen at the moment, based on the material used and the maturity level: first, second, and third generation.

The first generation is based on a crystalline structure with a silicon wafer (c-Si). These are the most frequently employed on the market. Two sub-categories can be found: the mono-crystalline (sc-Si) and the poly-crystalline (p-Si). The second generation of photovoltaic cells is characterized by the use of thin film layers. Different sub-categories exist: amorphous silicon (a-Si), micro-amorphous silicon (a-Si/μc-Si), cadmium telluride (CdTe), gallium arsenide (GaAs), copper indium selenide (CIS), and copper indium gallium diselenide (CIGS). Finally, the third generation of photovoltaic cells is based on new technologies [12], for example organic and polymer solar cells, dye-sensitized solar cells, etc. Moreover, concentrated systems are classified in the third generation of photovoltaic cells [17]. Among these technologies, perovskite solar cells, which belong to the dye-sensitized technologies, are one of the most promising technologies [12]. Several advantages are quoted in a previous paper [12], such as flexibility, transparency, and efficiency.

According to several past papers [12,17], the first and second generations are the most employed worldwide, due to their maturity levels, whereas the third category is still at the research state. These authors pointed out that multijunction cells present the highest efficiency level (around 40 to 45%), i.e., around twice as high as the silicon and thin film technologies (between 10% and 25%). A fourth generation exists, based on hybrid organic and inorganic technologies [12]. However, this photovoltaic cells generation is still at the research status.

2.2. Wind Energy

Among all existing energies, along with solar energy wind is one of the most developed renewable energies across the world. Wind energy systems have been greatly developed since the original windmill principle was discovered by the Persians around 200 B.C. High power modern designs were achieved in the 20th and 21st centuries, as described by the timeline of M.R. Islam et al. [19].

2.2.1. Fundamentals of Operating Processes

Wind turbines use wind speed to produce electricity. As explained by several authors, wind turbines convert wind kinetic energy into rotational kinetic energy [20–22]. The wind speed induces the rotation of blades around an axis, which can be either horizontal or vertical. Wind exerts two forces on the blades: a drag force, which is parallel to the airflow, and a lift force, which is perpendicular to it [20,23]. The mechanical energy of this rotation is transformed into electrical energy by an electrical generator, sometimes after a rotational speed increase due to a gearbox. The electrical output power depends on the wind speed cubed [20].

2.2.2. Main Technologies Currently Used

Wind turbine technologies can be separated into two categories, according to the rotating axis position, as has been explained in several references [19,20,23,24]. The most common category with the best technical and economic maturity levels is the horizontal axis wind turbine (HAWT); the second is the vertical axis wind turbine (VAWT). The two technologies are compared by M.R. Islam et al. [19]. The main characteristics of each of them are discussed in the following two paragraphs.

HAWT are characterized by a turbine placed on a nacelle at the top of a hub, with a rotation axis parallel to the ground. As explained in a previous paper [24], different technologies exist. They are classified according to different criteria: the number of blades (two, three or more), the rotor orientation (upwind or downwind), the hub design, the rotor control (stall or pitch), and the yaw orientation system (active or free). The low cut-in speed and the high power coefficient are often cited as advantages of horizontal turbines. However, a nacelle orientation system should be used to follow the wind direction changes and the installation presents more constraints [19,20]. This kind of wind turbine is mostly used in large scale systems. Wind turbines with three blades and upwind rotor orientation are the most widely used technology today [19].

For VAWT, the blades rotate around a vertical axis. The main advantage of this is that there is no need for an orientation system, as this kind of turbine can absorb wind from any direction, and it operates better than HAWT in the case of turbulent winds [19,20]. Among the different sub-technologies of VAWT, two categories are often found: the Darrieus turbines, which are based on lift forces, and the Savonius turbines driven by drag forces [20,24]. VAWT are mostly used for small applications and small power systems, for example residential networks [19,20,23].

The offshore wind turbines can be also distinguished according to the substructure and the foundations. Three categories can be found, according to the water depth and the distance to the shore [5,23,25]:

- In case of shallow water installation (water depth lower than 30 m), several bottom-fixed substructures can be used: the gravity-based substructures and the monopile substructures which are currently the most frequently used [5,25], and the suction bucket still at a development stage [5].
- For transitional water (water depth between 30 m and 80 m), others kinds of bottom-fixed structures are used. The jacket frames structures, the tri-piles structures, and the tripod structures are the most used [5].
- Finally, in case of water depth larger than 80 m, floating structures are used [5]. The mast is mounted on a floating structure moored to the seabed. Three kinds of floating structures exist: the ballast stabilized structures (or spar floaters), the tensioned-leg platforms (also called mooring line stabilized), and the semi-submersible platforms [5,25]. Floating wind turbines are mainly considered for offshore wind farms far from the shore, as the wind resource available is larger than along the coast.

More details related to offshore wind turbine technologies can be found in previous papers [5,23,25].

2.3. Tidal Current Energy

Among all existing marine renewable energy converters, tidal energy converters are one of the most developed technologies [2,4,26]. Belonging to the hydrokinetic type of energy [27], two kinds of tidal energy can be distinguished [28,29]. The first is tidal kinetic energy, which is induced by water movement according to tide cycles, for which turbines are used to produce electricity. The second is the tidal potential energy, where tidal barrages are used to extract the energy resulting from the water elevation cycle (also called tidal range devices). In this paper, only the tidal kinetic energy is studied, as the extracting technology (a turbine), the power density, the size requirements, and the power range are more suitable for coastal areas [29]. Moreover, the geographical areas suitable for tidal range devices are quite rare in the world [29]. Indeed, their installation is possible only in areas with a water level elevation about several meters. Thus, the use of tidal range systems to provide electricity for maritime remote areas, such as islands, is not discussed in this article.

2.3.1. Fundamentals of Operating Processes

Tidal current turbines produce electrical power from the kinetic energy of the rise and fall movements of tides in coastal areas. Indeed, due to the gravitational and rotational forces induced by the Earth, Sun, and Moon positions, ocean water moves horizontally according to cycles that are easily predictable, and which are related to the time and location on the Earth [2,30,31]. Water flow allows the submerged turbine rotation on which blades are mounted, similarly to the process used in wind turbines [2,28]. The turbine drives a generator, for which many technologies can be used [28,32,33]. The output power depends on the tidal current speed [33]. This kind of turbine can also be used to extract kinetic energy from river currents [26]. Tidal turbines are often compared to wind turbines with respect to the turbine operating principles. However, the tidal turbine performs better due to the water density, which is greater than air density, increasing the power density [26,29].

2.3.2. Main Technologies Currently Used

Extracting tidal kinetic energy can be carried out by several techniques, as explained and classified by many past papers [2,26–31,33–35]. The following categories can be distinguished primarily, according to the type of device used.

In Horizontal Axis Marine Current Turbines (HAMCT), the turbine is composed of two or more blades rotating around an axis parallel to the water flow direction [26,28,29,31]. A list of projects is given by Zhang et al. [33]. This is currently the most common tidal turbine on the market [4], due to its good technical and economical maturity level. Thus, HAMCT now reach the megawatt scale [36]. Moreover, floating systems have encountered some interest in the last few years and they are now the subject of active research [36].

Vertical Axis Marine Current Turbines (VAMCT) can harness tidal flow from any direction with two, three, or more blades rotating around an axis that is perpendicular to the current direction [26,29,31,33]. Two main kinds of vertical axis turbines exist: the Darrieus turbine and the helical turbine (also called the Gorlov turbine) [26,31]. However, some disadvantages limit their development. The low self-starting capacity and efficiency, along with torque variations are often cited [36].

Another category exists, that of oscillating hydrofoil systems. Tidal currents make a pressure difference on a foil Section, which creates drag and lift forces on the oscillating foil attached to a lever arm [29]. A linear generator is often used to generate the electrical output power [33]. However, this technology is still at the development level [27].

Other kinds of devices can be found. Among them, ducted turbines, the tidal kite, and helical screw systems (also called the Archimedes screw) can be cited [26,29,31,35]. A further classification of tidal kinetic energy converter technologies exists, based on the water flow harnessing techniques. Axial-flow and cross-flow turbines can be distinguished, and horizontal and vertical designs exist for both of these [29].

2.4. Wave Energy

Concerning other renewable energy sources, wave energy received attention from academics and industrialists, mostly since the 1980s, as the available worldwide resource is considerable [2,37]. Many wave energy converter technologies have been developed up to today, invoking especially a large number of patent publications [4]. Wave energy is sometimes classified among the hydrokinetic energy category [26]. Moreover, wave energy is now considered to be suitable for the electricity supply of small islands and coastal areas [1,3,38]. The Atlantic Ocean is often considered for wave energy projects due to the high wave power density available [39], but some recent articles have analyzed several islands case studies in Mediterranean Sea [40–42], involving some changes in technologies in order to fit the wave characteristics of the considered location [40].

2.4.1. Fundamentals of Operating Processes

A wave energy converter transforms wave energy into electrical energy. Wave energy comes from the effect of wind on the sea surface, creating waves. These follow the wind direction across several thousands of kilometers, creating significant swell, until they reach the narrow waters near the shore where the wave speed decreases. The power output of a wave energy converter depends on the wave height and its peak period [26,37,43]. Harnessing wave energy and converting it into electrical energy is a complex process compared to other renewable energy sources. Indeed, several conversion stages are necessary: primary, secondary, and tertiary conversion stages, according to A.E. Price [44] and several other past papers [39,45,46]. In the case of these papers, short descriptions are given below.

The primary conversion stage aims to convert wave motion into body motion, air-flow or water-flow, using mechanical, pneumatic, or hydraulic systems, called prime movers. This stage converts a low frequency motion (the wave) into a faster motion. The second conversion stage transforms the fluid energy of the first stage into electrical energy. Depending on the fluid used in the primary stage, the converter used in this step can be an air turbine, a hydraulic one, or a hydraulic motor connected to an electrical generator. They are called Power Take-Off systems (PTO). This step converts the low frequency fluid or mechanical motion into high rotational speed with the electrical generator. The tertiary stage conversion aims to adapt the electrical output characteristics of the wave energy converter to the grid requirements with power electronic interfaces. Some wave energy converters show merged primary and secondary conversion stages, where wave energy is directly transformed into electrical energy with a linear generator [39,46].

2.4.2. Main Technologies Currently Used

Many wave energy converter designs currently exist, but this renewable energy source is still at the research and development stage, and the technical maturity level is currently lower than that of other renewable energy sources [4]. Moreover, different classifications can be found in the literature, according to different reviews of recently published wave energy converter technologies [3,4,38,39,43–53]. Often, wave energy converters are classified according to the criteria listed below [3,4,37,39,43–46,50,53].

Location: onshore or shoreline, nearshore or offshore. Onshore systems are placed on a cliff, a dam, or land without a mooring system. Nearshore systems often lie between 0.5 and 2 km from shore, in shallow waters (between 10 and 25 m deep). The first generation of wave energy converters was based on onshore and nearshore systems [54]. Offshore wave energy converters are at several kilometers from the shore in deep water (>40 m), with the ability to harness high wave energy levels [2,37,39,43,45,46,53].

Device shape and direction concerning the swell propagation direction [4,37–39,43,45,46,53]. Three cases are possible:

- Point absorber: they are small with respect to the wavelength and they can absorb energy from any wave direction.
- Terminator: the device axis is perpendicular to the wave propagation direction.
- Attenuator: the device axis is parallel to the wave propagation direction.

Hydro-mechanical conversion principle (primary conversion stage) [4,26,37–39,43,45,46,48,50,53]. Three categories can be found:

- Oscillating Water Columns (OWC) are based on the compression and decompression forces in the air chamber created by water level variations which drive a turbine. OWC devices can be either deployed in shallow water as a stationary structure, or in deep water, for which floating systems can be used [55]. Recently, a new OWC device, called U-OWC has been developed [56]. Based on a vertical U-duct, this new structure avoids the wave to propagate into the inner body as in a traditional OWC device.

- Overtopping Devices (OTD) use the water level difference between the sea and the partially submerged reservoir to produce electricity (potential energy) when the wave overtops the structure and falls into the reservoir. The turbine rotates by releasing the water back into the sea. Some overtopping devices are integrated to a breakwater [57,58]. Moreover, structural design of some overtopping devices can be suitable for other maritime needs [59].
- Wave Activated Bodies (WAB) or Oscillating Devices are based on the use of one or more moving bodies [26,37,48]. Three categories of wave activated bodies can be distinguished: heaving buoy, surface attenuator, and oscillating wave surge converter [43]. The performances of these devices depend on the mooring system, for which different configurations exist [60].

Some references classify wave energy converters according to other criteria. A review of electrical generators used, control methods employed, mechanical and/or electrical controllers applied, wave conditions considered, and power electronic converters used for different projects is proposed by E. Ozkop and I.H. Altas [52]. Classification by the power take-off technology (second conversion stage) is also given in [46], resulting in three main sub-categories: the hydraulic PTO, the turbine PTO, and the all-electric PTO, as discussed in the previous Section. Mooring configurations are also discussed [43,48,51]. A new classification based on the operating principle has recently been carried out [38]. A. Babarit proposed a comparison of the existing technologies based on the so-called capture width ratio [47].

Some development trends concerning the different criteria listed earlier are highlighted in previous papers [4,39]. Offshore application, floating installation, and point absorber technology are the most common aspects considered for the projects reviewed.

2.5. Intermittency and Variability Comparison

The four renewable sources introduced in the previous sections present different temporal characteristics. Indeed, as they are based on the use of different primary resources, their intermittency and variability are different, with more or less predictability, so they cannot be dispatched as conventional sources could be [61]. A limited number of studies have discussed these aspects considering all the sources. In one of these [61], Widén et al. presents the main intermittencies and variabilities of the four sources, with a review of existing forecast methods. The standard deviation of each source according to the different time scales (frequency bands) is studied by J. Olauson et al. at the level of a country [62]. The highest standard deviation rate for the different sources is related to short-term timescales for solar (<2 days), mid/short-term for wind (2 days to 2 weeks), long-term for wave (>4 months), and mid-term for tidal (2 weeks to 4 months) energies [62]. Natural cycle timescales of solar, wind, tidal, and wave resources are also discussed in the International Energy Agency report [63]. Variability of solar, wind, and tidal resources for the UK is studied by P. Coker et al. considering the persistence, statistical distribution, frequency, and correlation with demand [64]. G. Reikard et al. studied the variabilities of solar, wind, and wave energy for integration to the grid, with a forecasting system proposed for the three sources based on a regression method. Wave energy is shown to be more predictable than solar and wind, due to the strong weather impact for the latter two sources [65]. Recently, a review of solar and wind space-time variabilities has been conducted by K. Engeland et al. but this does not include tidal and wave resources [66]. Table 1 presents the main characteristics in terms of variability and intermittency for each source, with the origin and existing methods to evaluate temporal variations according to different publications [61–66].

Table 1. Main variabilities characteristics of solar, wind, tidal, and wave energies.

Source	Kind/Scale of Variability	Origin	Variability Assessment Methods and Models	Determinist or Stochastic Behavior
Solar	Seasonal	Position of Sun and Earth Geographical position on the Earth	Mathematical model	Deterministic
	Daily	Diurnal cycle due to Earth rotation	Mathematical model	Deterministic
	Short-term: from second to hour	Weather conditions	Predicted by ground measurements, satellite data or weather models	Stochastic
Wind	Decadal	Climatic and atmospheric condition changes	Historical climatic observations data analysis	Stochastic
	Yearly and seasonal	Weather conditions depending on the location and the seasonal cycles	- Statistical: autoregressive, Monte-Carlo method with Markov chains, neural network, wavelets … ;	Stochastic
	Weak scale (synoptic peak around 4 days)	Weather conditions	- Physical: historical weather data or weather models;	Stochastic
	Daily and infra-day	Diurnal peak and weather conditions	- Hybrid: statistical and physical methods	Stochastic
	Short-term: from Sub-seconds to few minutes (turbulence peak around 1 min)	Random, caused by turbulences	Hardly predictable	Stochastic
Tidal current	Bi-monthly: depending on the tide cycle (1 cycle = 14.76 days), with spring and neap tides	Tide cycles: depending on the position of the Earth, the Moon and the Sun (tidal currents are the fastest when they are aligned)	Harmonic analysis	Deterministic
	Infra-day	- Tide type: diurnal, semi-diurnal, semi-diurnal with diurnal inequality, or mixed; - Depending on the location on the Earth, and the attraction between the Moon and the Earth and between the Sun and the Earth	Harmonic and geographical analysis	Deterministic
	Short-term: seconds, minutes or hours, due to turbulences	Sea bed, geography of the location, Weather effect: storms, waves …	Geography study and weather forecasts	Stochastic
Wave	Seasonal and monthly	Climate and weather conditions depending on the location	- Scatter diagram get by statistical or empirical methods; - Power variation coefficient; - Seasonal Variability Index (3 months average level)	Stochastic
	Infra-day	Weather conditions depending on the location	Weather forecast	Stochastic

3. Review of Multisource Projects Including Renewable Marine Sources

As explained in the previous section, the most developed marine energies at the current time are tidal kinetics and wave energies. Thus, they can be used to provide electrical power in maritime areas, as for example in floating systems or islands communities. During recent decades, renewable energies used in these applications were often solar and wind energies, but intermittencies of these sources involved the use of Diesel generators or storage capacities. According to the time characteristics of solar, wind, tidal, and wave energies, the development of multisource systems combining several of these sources could bring a sustainable and reliable power level to ensure the load supply in the future.

This Section aims to present a review of projects combining the use of some or all of the four sources presented previously, on the same platform. The details depend on the kind of project and the development status. Firstly, hybrid system projects developed by companies will be reviewed; from hybrid devices tested in offshore conditions to projects that are still at concept status. Also, a review of several energy island concepts will be given. Then, an up-to-date review of studies concerning sizing optimization and energy management systems will be carried out, considering the published papers in these fields. Several projects presented in the following sections have already been more thoroughly reviewed [5–11,67], especially for hybrid wind-wave systems. However, farms and colocated systems such as the independent and combined arrays described by C. Pérez-Collazo et al. [5] are not part of the focus of this article, since they are not considered as combined systems.

3.1. Review of Industrial Hybrid System Projects Including Marine Energies

Multisource systems that include marine energy are still scarce. As wind turbines now reach a high level of maturity, most of these projects consider offshore wind turbine use. Two categories of projects can be identified, according to the maturity level and the development status. Several projects have been tested under real sea conditions (meaning potentially severe environmental conditions) either at a reduced-scale or at full-scale (Section 3.1.1), whereas others have still not progressed beyond the concept step (Section 3.1.2). A review is given below and is summarized in Tables 2 and 3. Technologies are characterized according to the classification given in the previous Section when the technical information is available. An overview of industrial hybrid system projects according to their power scale and to their furthest known development status is given in Figure 2. Finally, some island energy concepts will be presented (Section 3.1.3).

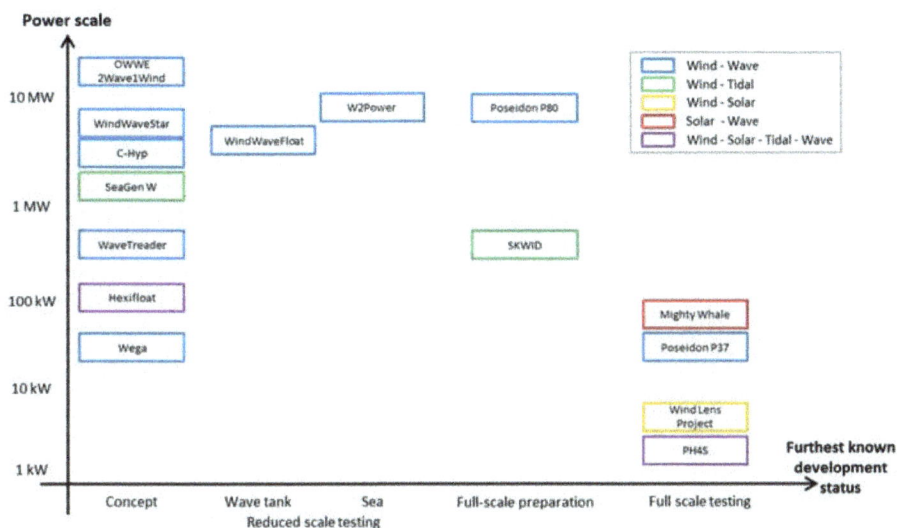

Figure 2. Overview of hybrid systems including marine energies.

3.1.1. Projects Tested under Real Sea Conditions

Despite their current scarcity, multisource systems tested under real sea conditions can be classified according to the sources used. These systems relied on either wind or wave energies combined with one or more renewable sources, whereas other projects only consider wind and wave energies. A single industrial project uses solar, wind, tidal and wave energies on the same platform [68]. Details of these systems are given below, according to the sources used.

- Wind and wave: several projects have considered these sources. The Poseidon P37 product, designed by Floating Power Plant, is currently the most advanced technology in the multisource floating platforms field, as it was the first hybrid system connected to the grid. Twenty months of grid-connected tests were effected successfully on the Danish coasts, with three 11 kW wind turbines and 30 kW of wave energy converters. A Megawatt scale will be reached with the

P80 device, which is expected for 2020 [69,70]. The W2Power device designed by Pelagic Power uses the same energies, with 10 MW installed on the platform [71–73]. However, this project is still at reduced scale test status, as the platform currently tested in the Canary Islands concerns the WIP10+ device, which is a 1:6 scale prototype with only wind turbines [74]. Previously, wave tank tests allowed the mooring system to be validated and the behavior in both operational and survival modes to be assessed [72].

- Wind and solar: although photovoltaic panels and wind turbines now reach a high maturity level, projects combining both energies on a floating platform are still scarce. The Wind Lens hybrid project, developed by the Kyushu University (Fukuoka, Japan), has considered wind turbines (Wind Lens turbine) and solar panels on a floating platform [75–78], connected to batteries to ensure the electrical power supply of measurement and control devices. The total power installed reached 8 kW. Authors have observed that offshore wind turbine production is better than the similar land-based turbine due to higher wind speed values. In winter, the energy produced by the offshore wind turbine is two to three times the energy produced by the land-based wind turbine [76]. A more powerful platform is expected in the future according to [76].

- Wind and tidal: The Skwid system designed by the MODEC company seems currently to be the only project combining wind and tidal turbines at an industrial scale. However, little information is available concerning this project, since the system sank during installation in 2014 [79,80]. The turbines used could harness wind and tidal current flowing from any direction thanks to their vertical axis, avoiding complex orientation systems needed by horizontal axis turbines.

- Wave and solar: The Mighty Whale project is one of the oldest multisource systems which considers the use of ocean energy [81]. During tests at sea between 1998 and 2002, observations showed that combining the use of wave and solar energies allowed the power production to be smoothed and reduced the auxiliary generator use by storing the energy in batteries. However, the results presented in a previous paper [81] are strongly dependent on climatic conditions (Sea of Japan).

- Wind, solar, tidal, and wave: the PH4S device developed by the French company Geps Techno is currently the only platform combining the four renewable sources [68]. A prototype is currently being tested on the French Atlantic coast and the first observations from this company show a reduction of global power intermittency.

The review shown in Table 2 demonstrates that devices tested under real sea conditions are still scarce and often used a few dozen kilowatt systems. All the projects have considered wind and/or wave energies on a floating or fixed platform. These structures often come from a previous wave energy converter platform (e.g., Poseidon P37, Mighty Whale) or an offshore wind turbine system (e.g., Skwid), to which another renewable source has been added. All of the offshore projects tested report that energies used present a positive complemental aspect, bringing a smoother electrical power output. When they are not connected to the grid, power sources are used to supply the platform measurement and control devices. However, projects tested under real sea conditions are still scarce. Most of them were tested at a reduced scale, initially in water tanks before sea installation.

3.1.2. Projects Still at the Concept Status

The industrial project review can be supplemented by projects that have remained at the concept status without sea installation. As detailed in Table 3, most of these concepts concern wind and wave systems, even if a wind-tidal concept [11] and a floating platform concept combining all of the sources considered in this study exist [82,83]. According to all these projects, the following points can be highlighted:

- Wind and wave: many wind-wave system concepts exist. Some of these have been partially tested, either in water tanks or at sea for one of both renewable sources. For example, Principle Power Ltd. (Emeryville, CA, USA) has designed a hybrid device called WindWaveFloat. To date, a 2 MW wind turbine was successfully tested at sea in 2011 with grid connections. However the different wave energy converter technologies initially planned were not included in the tests [84,85]. WindWaveStar and Wega devices, developed respectively by Wavestar and Sea for Life companies, have never been tested with both energy sources. For the first device, tests only concerned the WaveStar wave energy converter in offshore conditions for a reduced scale prototype, whereas the Wega wave energy converter has been studied in wave tank tests. Other wind-wave hybrid system concepts have never surpassed the concept status (WaveTreader, OWWE 2Wave1Wind and C-Hyp).

- Wind and tidal: MCT has considered a wind turbine mounted on the tidal turbine structure in the SeaGen W device. However, this project seems only to be a concept according to the large scale tidal projects without the wind turbine recently developed by the company [11,86].

- Wind, solar, tidal, and wave: In 2012 Hann Ocean (Singapore)patented the layout and design of the Hexifloat device, a platform concept allowing four energies to be harnessed [83], but this is still at the concept status today according to the company's website [82].

Other concepts have been developed in the MARINA Platform framework (Marine Renewable Integrated Application Platform), a European project undertaken between 2010 and 2014 to study different aspects of combined offshore platforms, such as feasibility, economical profitability, engineering etc. Thus, several partners have worked on tools, methods, and protocols to ease multipurpose platform design. Among the different platforms proposed [87], three wind-wave hybrid system concepts have been considered: the Spar Torus Combination (STC) [88], the Semi-submersible Flap Combination (SFC), and the large floater with multiple Oscillating Water Columns and one wind turbine (OWC Array) [11].

The concepts reviewed here often considered wind and wave energies. This trend could be explained by the fact that some companies have already developed a wind or wave energy converter and would like to share their structure with another kind of renewable energy converter. Then costs could be reduced (design, equipment, installation, operation, maintenance, etc.) and power production could be increased with a smoother output level, as explained by Pérez-Collazo et al. [5], M. Karimirad [6], and Casale et al. [10]. Positive aspects of combined wind-wave devices are presented in these references. However, the review carried out in this section shows that many of these concepts have not gone beyond the idea step. High development costs can explain this trend. Also, as offshore tidal and wave energy converters alone are still scarce in the world, their maturity level is not as high as land-based renewable energies and offshore wind turbines. Casale et al. have suggested [10] building hybrid systems around proven and mature offshore systems, for example wind turbines after these technologies have been individually validated and tested. Thus, this consideration could help concepts to overcome this step, which is seen in a few cases where wave or wind energy converters have been tested on wave tanks or at sea [84,85,89,90]. For several projects listed in our review, little information is available to explain their current status and perspectives. It is supposed that some companies have cancelled their hybrid device concept, focusing on separated technologies.

Table 2. Summary of industrial hybrid systems concepts including marine energies tested under real sea conditions.

Reference	Project or Product Name Company/Lab/Institution	Sources Considered and Specifications	Used Storage	Grid Connection Load Considered in Off-Grid Case	Main Outcomes	Current Status
[69,70]	Poseidon P37 and P80 Floating Power Plant (Denmark)	- P37: wind energy: 2 blades HAWT (3 × 11 kW) + wave energy: attenuator, OWC (10 × 3 kW); - P80: wind energy: 3 blades HAWT (from 2.3 to 5 MW) + wave energy: attenuator, OWC (2.6 MW)	No storage	Grid connected	One of the highest efficiency rates among the wave energy converters existing on the market [10]	P37 tested between 2009 and 2013 P80 version expected for 2020
[71–73]	W2Power Pelagic Power (Norway)	- Wind: HAWT (2 × 3.6 MW); - Wave: point absorber, WAB (2 to 3 MW)	No storage	Grid connected and off-grid configurations are possible	Tested on tank at reduced scale (1:40) to study wind and wave interaction, mooring system, and physical limits	2017: Sea conditions tests at 1:6 scale for the WIP10+ device [74]
[75–78]	Wind Lens Project Kyushu University (Japan)	- Solar (2 kW); - Wind: Wind Lens HAWT (2 × 3 kW)	Battery	Off-grid Measurement and air-conditioning devices	Wind speed in offshore conditions is higher than in land case	1st version ended in 2012 after one year of offshore conditions tests, but a 2nd is expected according to [76]
[79,80]	SKWID MODEC (Japan)	- Wind: VAWT-Darrieus; - Tidal: VAMCT-Savonius Total power of 500 kW	No storage	No available information	- Wind and tidal turbines can harness energy from wind and tidal current of any direction; - Prototype sank during offshore installation in 2014	Cancelled in 2014
[81]	Mighty Whale JAMSTEC: Japan Marine Science and Technology Center (Japan)	- Wave: nearshore, terminator, OWC (2 × 30 kW); - Solar: mono-crystalline (10 kW); - Auxiliary generator (20 kW)	Battery (500 Ah)	Off-grid Measurement and control devices	Complementarity of wave and photovoltaic energies	Ended in 2002
[68]	PH4S Geps Techno (France)	- Solar (1 kW); - Wind: VAWT (1.5 kW); - Tidal: VAMCT (500 W); - Wave (500 W)	Battery and supercapacitors	Off-grid	Complementarity of the four sources	2017: offshore tests

Table 3. Summary of industrial hybrid system concepts including marine energies.

Reference	Project or Product Name Company/Lab/Institution	Sources Considered and Specifications	Application and Load Considered	Main Outcomes	Current Status
[85,91]	WindWaveFloat *Principle Power Ltd. (USA)*	- Wind: 3 blades HAWT (5 MW); - Wave: depending on the considered wave converter [85]	No information	- Study of several wave converter technologies; - Numerical tool development [91]; - Test at 1:78.5 scale on wave tank	Only a 2 MW floating wind turbine was tested in 2011 in offshore conditions, now removed after 5 years of grid-connected tests [84]
[89]	WindWaveStar *Wavestar (Denmark)*	- Wind: 3 blades HAWT; - Wave: near shore, point absorber, WAB (6 MW at full scale)	Grid connected	In 2010, tests were conducted at sea for a 1:2 scale prototype only composed of wave converters (total of 600 kW, connected to the grid)	The hybrid wind-wave system is only a concept today
[90]	Wega *Sea for Life (Portugal)*	- Wind: HAWT; - Wave: point absorber, WAB	No information	The power take-off system of WEC device is placed above water, reducing the corrosion risk and improving accessibility. Tests were done in 2010 with only one wave energy converter in a wave tank.	Hybrid wind-wave system is only a sharing infrastructure possibility of the WEC device, still at concept status
[92]	WaveTreader *Green Ocean Energy (Scotland)*	- Wind: 3 blades HAWT; - Wave: offshore, point absorber, WAB surface attenuator Total power of 500 kW	No information	WEC device is mounted on the monopile offshore WT	No information available since 2011
[9,93]	OWWE 2Wave1Wind *Ocean Wave and Wind Energy (Norway)*	- Wind: 3 blades HAWT; - Wave: offshore, point absorber, OTD (depending on the concept)	No information	This concept is known to be one of the largest wave energy platforms (600 m), that allows to harness a large amount of energy (1 TWh per year for 10 units)	Only a concept
[94]	C-Hyp *LHEEA, EOSEA, Technip (France)*	- Wind: 3 blades HAWT (5 MW); - Wave: oscillating wave surge converter (1.89 MW)	Grid connected	- Feasibility analysis with numerical tool development; - The angular spatial share of WECs can smooth the electrical power produced. - WEC power seems to be higher than WT power for low wind speeds. - Costs are high and the size of this concept (100 m diameter) is challenging for the building phase.	Only at concept status in the MARINA Platform project framework [11], but no further development
[11]	SeaGen W *MCT-Atlantis (UK)*	- Wind: 3 blades HAWT (3 MW); - Tidal: HAMCT (1.2 MW)	Grid connected	Concept of SeaGen W consists of a wind turbine added on the top of the existing Seagen tidal device	Only a concept
[82,83]	Hexifloat Renewable Energy Platform *Hann Ocean (Singapore)*	Solar (48 kWp); - Wind: 3 blades HAWT (18 kWp); - Tidal: HAMCT (23 kWp) Wave (45 kWp)	No information	Platform design	Patented in 2012 [83] concerning the design aspects, but not deployed today

3.1.3. Energy Island Concepts

Energy islands [10] or island systems [5] are considered to be large multipurpose platforms including several renewable energy sources and, in contrast to projects reviewed in the previous sections, infrastructures for other activities and functionalities [5,10]. C. Pérez-Collazo et al. divided this kind of project into two categories: artificial islands built on a reef or dyke, and floating islands, considered as very large floating platforms [5]. However, all projects in this field show that they are only at the steps of concepts and ideas. Among them, the following projects can be quoted:

- Kema Energy Island (by KEMA-DNV GL and Lievense): placed in an ocean, this artificial island concept consists of a large scale water tank used for pumped storage, surrounded by dykes on which wind turbines are placed to produce electrical power. According to the figures shown in [95], the KEMA Energy Island project encompasses a large scale storage capacity, with a power of 1.5 GW and an energy capacity of 20 GWh, to store surplus wind electrical power production. Other functionalities are proposed, such as the chemical industry, harbors, tourism, etc. This project has not seen further development than the preliminary design and evaluation steps, but it is still shown in a previous paper [96].
- Offshore Ocean Energy System (by Float Inc.): this concept can be classified in either the floating island or floating platform categories, according to its medium size. Wind, tidal, and wave energies have been considered as the heart of the structure. Moreover, other services have been proposed, such as aquaculture, fishing, and desalination facilities [97].
- OTEC Energy Island (by Energy Island Ltd.: London, UK): the four renewable energy sources discussed in this paper (solar, wind, tidal, and wave energies) have been considered in this floating island concept, along with ocean thermal energy conversion and geothermal energy. Moreover, several infrastructures and services such as a harbor and a water desalination system have been proposed at the design phase. The power considered is about 250 MW [10]. A patent was filed in 2003 [98], but as of today, no further development is known.
- TROPOS project concepts: three research programs have been integrated in "The Ocean of Tomorrow" European call: the TROPOS Project (2012–2015), the H2OCEAN Project (2012–2014), and the MERMAID project (2012–2015). Several research programs designed for the TROPOS project have seen a focus concerning innovative multi-use floating islands: the Leisure Island, the Green & Blue, and the Sustainable Service Hub [99]. The last of these seems to have the highest potential for near-term development. Economic, environmental, logistical requirements, social, and design aspects have been considered. In addition to the renewable energy converters used in these concepts (solar energy, wind energy, and OTEC), other infrastructures and services have been proposed, such as leisure (Leisure Island) or aquaculture (Green & Blue) [100].

All of these island concepts have apparently not gone beyond the idea stage. Also, the powers considered are higher than the floating platform power scales reviewed in previous sections. Thus, island projects seem to be far from reaching industrial and commercial status, concerning high costs, technical challenges, and the facilities required to build such projects [100]. Financial support should be found to overcome the concept status. However, sharing infrastructures with other concerns could help project development by involving different industrial and economic sectors [10,100].

3.2. Review of Academic Research Concerning Hybrid Systems with Marine Energies

Multisource systems with marine energies are still at early stages of developmental processes. Thus, several academic analyses studied combined renewable energies exploited in the sea. These analyses are often at an earlier stage than industrial development processes and they study above all theoretical hybrid systems. Among the different papers describing such systems from the electrical engineering point of view, two categories can be found. The first discusses energy management system and control aspects, whereas the second concerns sizing optimization aspects with method and tool design. Several papers are reviewed in the following sections according to this classification.

3.2.1. Energy Management System and Control Studies

Hybrid systems using marine energies have been modeled and simulated by several authors to design appropriate energy management systems and control strategies. As in the industrial project review presented in the previous section, these academic works can be subclassified with respect to the considered sources, as the requirements and specifications can be different according to the renewable source. Technical information and main outcomes are summarized in Table 4.

- Wave and wind: an off-grid wind-wave system with battery storage and variable AC load has been studied by S.Y. Lu et al. [101]. The converter control schemes developed allow ensuring current and voltage stabilities in transient load phases, concerning the 500 W to 1 kW situation validated by simulation and laboratory tests.

- Wave and solar: as solar energy has been widely used for island electrical power supply, several articles have considered wave energy to compensate the solar energy fluctuations. For example, the Perthian Island (Terengganu, Malaysia), studied by N.H. Samrat et al. [102], did not present sufficient solar resource for the load power required. To ensure the system reliability and power quality, appropriate converter controls have been developed. Thus, the DC voltage link is kept constant, even in the cases of resource or load fluctuations. Similar systems and studies were considered by S. Ahmad et al. [103]. A grid-connected solar-wave hybrid system was studied by L. Wang et al. [104], considering the generated power injected on the DC-link smoothed by a supercapacitor. The converter control schemes developed allow the maximum wave and solar powers to be harnessed. Grid injected power fluctuations are smoothed by inverter control, whereas the DC-link voltage is controlled by a DC/DC converter connected to the supercapacitor.

- Wind and tidal: many articles deal with hybrid wind-tidal systems. For example, Y. Da and A. Khaligh [105] have presented appropriate control schemes for tidal and wind turbines to optimize harnessed powers, considering mega-watt scale generators. Tidal current and wind speed fluctuations have been taken into consideration to validate the proposed strategies. Another wind-tidal hybrid concept called HOTT (Hybrid Off-shore and Tidal Turbine) has been studied in several papers concerning wind power fluctuation compensation [106–109]. Thus, M.L. Rahman et al. proposed [106] the use of a tidal generator as a flywheel storage system, with a one-way clutch ensuring mechanical separation. The tidal generator produces or stores electrical power depending on the inverter control. In a previous paper [107], wind power fluctuations are compensated by tidal generator control for the lowest frequencies and by battery control for the highest ones. The authors stated that tidal compensation reduced the battery capacity, whereas the highest long-term fluctuation compensations required a tidal turbine power increase. The battery storage system was studied in a previous paper [108]. Tidal generator control for wind power fluctuation is also considered in a previous paper [109]. Concerning the grid connection, two solutions for large-scale turbines have been studied by S. Pierre [110]. The DC-link connection between the two generators before the grid-tied inverter brings an easier fluctuation smoothing ability. The separated solution consisting of two back-to-back converters for the AC grid connection allows the extracted power to be maximized. Finally, Y. Fan et al. presented [111] a novel hybrid wind-tidal architecture, where a hydraulic accumulator is used as a storage and balance system, placed between both hydraulic pumps and the electrical generator. Hydraulic pumps transform the output turbine mechanical energy into hydraulic energy. Fluctuations of output turbine mechanical powers are limited by hydraulic pumps control, while the hydraulic accumulator is controlled according to the load demand.

- Wind, tidal, and wave: C. Qin et al. [112] simulated the compensation of short-term output power fluctuations induced by intermittent wind and wave energies (from seconds to minutes). Thus, the tidal generator was used to smooth the output power, according to the tidal current speed. When the tidal turbine cut-in speed is surpassed, a tidal generator produces electrical power.

Thus, its pitch angle and rotational speed are controlled simultaneously to reduce output power fluctuations. If the tidal current speed is lower than the cut-in speed, the tidal generator is used as a flywheel storage system to compensate for variations, after tidal turbine mechanical separation.

According to the articles reviewed in this Section, wind and tidal energies seem to be widely considered. Wind energy fluctuations are often cited as a weakness and a challenge to improve the renewable development in island areas. Different solutions have been investigated to limit these fluctuations. Among them, tidal energy attracted attention, concerning tidal generator control [105–107,109,111,112] and the possibility to use it as a flywheel storage system [106,112]. Another point of interest observed in academic research is the transient state system stability, not only for resource fluctuations but also for load change [101,102,104,107]. Tidal energy has also been considered to smooth wave energy fluctuations [112]. Storage solutions such as batteries [101,102,107] or supercapacitors [104] are sometimes used to smooth generated power fluctuations.

3.2.2. Sizing Optimization Studies

To ensure a high reliability level, the hybrid system should be designed carefully. A storage solution allows the load requirements to be met in terms of power and energy. To avoid an over-sized or under-sized system and to ensure reasonable costs, a sizing optimization must be carried out. Wind/solar systems with a battery and/or Diesel generator have been widely studied in terms of sizing optimization, as described in recent reviews [113–116]. As ocean energies have only been considered recently, such studies for marine energy hybrid systems are still rare. Several authors have proposed sizing optimization studies for both the renewable source sizes and the storage solutions considered to supply island systems.

Hybrid photovoltaic, wind, and tidal system sizing optimization was proposed in a few articles. In previous papers [117,118], O.H. Mohammed et al. considered the case of the remote Ouessant French Island, where the energy load is estimated at around 16 GWh per year, for a maximum power demand of 2 MW. To find the best sources and storage combinations according to the equivalent loss factor reliability index [117,118] and economic constraints [118], several sizing optimization algorithms have been developed: cascade calculation, genetic algorithms and particle swarm optimization. The combination of the three renewable sources is found to be more reliable than cases where only one source is considered [117]. In a previous paper [118], the levelized cost of energy is divided by seven between a configuration based only on solar energy (763.7 $/MWh) and a solution based on PV, wind and tidal energies (127.2 $/MWh). Also, the levelized cost of energy is lower when artificial intelligence approaches are considered for the sizing optimization, as the obtained values reach 94 $/MWh with a genetic algorithm and a particle swarm optimization, whereas a cascade algorithm results in a 149 $/MWh cost [118]. A metaheuristic solution called the crow-search algorithm has been proposed by A. Askarzadeh [119] to optimize a hybrid wind/solar/battery system into which tidal energy is included. Concerning the results, the author concludes that a hybrid solar/wind/tidal system is more cost-effective than a partial combination of these three sources. In the simulation conducted for a one year period, tidal turbines generate almost 25% of the total generated energy and the resulting tidal turbines net present cost for the optimized system represents 20% of the total cost. Moreover, batteries can reduce the cost and improve the reliability index. The net present cost related to a battery reaches 13% of the total net present cost, to ensure a maximum unmet load ratio of 10%. For the study carried out, the proposed crow-search algorithm is reputed to be more efficient than the particle swarm optimization and the genetic algorithm, giving the fastest convergence rate.

Table 4. Summary of academics hybrid systems including marine energies, concerning energy management and control aspects.

Reference	Sources Considered and Specifications	Storage Used	Application and Load Considered	Kind of Study	Main Outcomes
[101]	- Wind: HAWT (2 kW); - Wave: point absorber, WAB-AWS (1.5 kW)	Battery	Tests done in off-grid configuration with DC bus and adjustable AC resistive load	Modeling, simulation, and lab. scale platform tests	The considered DC micro-grid remains stable during load transient phases, observed in both simulated and measured results.
[102]	- Solar: p-Si (400 W); - Wave: OWC (3 kW)	Battery (14 Ah)	Connected to an island grid (Perhentian Island in Malaysia)	Modeling and simulation	The simulated DC link voltage controller (bi-directional buck-boost) ensures the voltage stability in case of generated power fluctuations and load variations, by charging and discharging the battery. Load side voltage presents a low voltage and current THD rates thanks to the inverter control and the passive L-C filter.
[104]	- Solar: sc-Si (50 kW); - Wave: point absorber, WAB-AWS (160 kW)	Supercapacitor (95 kW and 0.5 kWh)	Grid connected	Modeling, control strategies development, and simulation	The developed control scheme allows the power fluctuations to be smoothed with the supercapacitor, ensuring stability and extracting the maximum available power from wave and PV sources.
[105]	- Wind: HAWT (1.5 MW); - Tidal: HAMCT (1 MW)	No storage	Grid connected	Modeling, control strategies development, and simulation	The developed control schemes for both generators allow extraction of the maximum available power. Tidal energy is said to be more predictable and more available than wind energy.
[106]	- Wind: HAWT (1.5 kW); - Tidal: HAMCT (750 W)	Tidal generator as a flywheel storage system	Grid connected	Lab. scale platform tests	The induction generator of tidal energy chain conversion is used as a flywheel storage system, with appropriate rotation speed control and mechanical separation by a one-way clutch. Thus, wind turbine generated power fluctuations can be smoothed.
[107]	- Wind: HAWT (300 kW); - Tidal: HAMCT (100 kW)	Battery	Grid connected	Modeling, control strategies development, and simulation	The proposed tidal turbine control and battery control are able to reduce wind turbine power fluctuations and keep frequency stability. The lowest wind power fluctuation frequencies are compensated by the tidal generator control, whereas the highest frequencies are compensated by battery control.
[110]	- Wind: HAWT (5 MW); - Tidal: HAMCT (5 MW)	No storage	Grid connected	Modeling, control strategies development, and simulation	Two coupling modes have been considered for the tidal and wind system AC grid connection. The first one considers a DC-link coupling before a grid connected inverter. The second one considers two separated AC-DC-AC converters between source generators and AC grid.
[111]	- Wind: HAWT (5 MW); - Tidal: HAMCT (1 MW)	Hydraulic accumulator	Grid connected	Modeling, control strategies development, and simulation	The output power is balanced with the hydraulic accumulator storage, according to the requested and generated powers. Fluctuations are damped by the hydraulic pumps and accumulator control.
[112]	- Wind: HAWT (1.5 MW); - Tidal: HAMCT (1.5 MW); - Wave: point absorber, WAB: Archimedes Wave Swing (200 kW)	Tidal generator as a flywheel storage system	Grid connected	Simulation of the control strategies	The tidal generator is used as a flywheel storage system when the tidal current speed is lower than the cut-in speed. Tidal generator control can smooth the short-term wind and wave power fluctuations.

The sizing optimization for a wind/tidal hybrid system with battery storage has also been described in several articles. Among them, S.G. Mousavi proposed [120] the use of a genetic algorithm to determine the optimal size of a wind/tidal/micro-turbine/battery system, according to an economic analysis, i.e., evaluating for a year the capital cost, the battery replacement cost, the fuel cost and the operation, and maintenance costs. The objective function aims to find the optimal size with the lowest total annual system cost (sum of all the costs), considering the maximum load demand of the standalone system. The optimal configuration is based on a power capacity of 315 kW for wind turbine, 175 kW for tidal turbine, 290 kW for microturbine, and a capacity of 3.27 kAh for lead acid battery, leading to $312,080 total cost. M.B. Anwar et al. presented [121] a methodology to size grid-connected large scale marine current and wind turbines (mounted on the same monopile), with a battery storage station to meet the grid code requirements. Sizing optimization aims to maximize the available power, at the same time respecting the injected power fluctuation requirements given by the grid.

Sizing optimization studies for systems dealing with marine energies are less numerous than studies carried out for solar/wind/battery systems. The first trends of these studies show that optimizing the size of a hybrid system which includes marine energies is necessary, as it allows the cost to be reduced and the reliability index to be improved [117–119]. The amount of power generated is expected to be higher and the intermittency to be reduced, but battery storage is still required to ensure the load energy requirements.

4. Overview of Multisource Systems Based on Marine Renewable Energies

The review of industrial projects and academic research dealing with hybrid systems based on marine energies has shown that such systems have not yet reached commercial status. The interest of industries and researchers in this kind of multisource system is now clear. However, no significant results and operating experience exist to date and projects have often remained at a concept status [10]. Most projects considered a combination of two renewable energy sources. Although the advantages are numerous, some obstacles limit their development. This section aims to summarize the aspects found overall across the different projects and studies reviewed in previous sections, according to synergies and positive aspects, weaknesses and obstacles, and finally feasibility aspects.

4.1. Positive Aspects, Synergies, and Applications

Combining several renewable energy sources in maritime areas presents many advantages and highlights some possible synergies. Thus, further developments in forthcoming years are expected, as potential applications are numerous.

According to several authors [5,6,9,10] and to the projects reviewed in previous sections, positive aspects and benefits brought about by marine energy hybrid systems concern many fields, as they can:

- Increase the energy production rate of an area (area share);
- Reduce the non-production hours, by managing the power flows harnessed from energies presenting different intermittency and variability characteristics (output power smooth). A storage solution can improve the reliability index and ensure the load requirements. Thus, the use of Diesel generators can be reduced;
- Provide sustainable electrical energy for maritime activities, such as fishing, aquaculture, water desalination, oil and gas industries, etc.;
- Share the infrastructure and equipment, allowing the global weight to be reduced;
- Attenuate the platform movement and improve its stability;
- Reduce some costs, with initial savings (infrastructure, mooring and anchoring systems, transmission, connection equipment, etc.) and lifetime savings linked to the operation and maintenance costs, compared to a separate device solution;
- Reduce the visual impact by placing the platform far from the coast (offshore systems).

Moreover, the design of multisource systems based on marine energies presents some positive aspects of synergies which could improve and accelerate their development. According to the synergies explained in several references, four categories can be found [5,6,10]:

- Areas sharing synergies: between renewable energy systems and other facilities (aquaculture, desalination, fishing etc.). Sharing areas allows the sea use densification to be improved, sharing the power produced for the surrounding activities and limiting the studies to a single place.
- Infrastructures, installation, and equipment sharing synergies: this kind of synergy concerns the installation equipment, the logistics (port and vessels), the grid connection, the supervisory control system, the storage and the operation, and maintenance. For each of these items, costs could be reduced by combining different kinds of sources.
- Process engineering synergies: hybrid systems based on marine energies can be combined with several marine activities, such as desalination, hydrogen production, aquaculture, breakwaters, algae production, oil and gas sector, etc.
- Legislative synergies: a common regulation is necessary to develop such hybrid systems. Thus, a legal regulatory framework, maritime spatial planning, a simplified licensing procedure, and a grid and auxiliary infrastructures planning are needed, as explained in a previous paper [5].

Hybrid systems including marine energies can be used in numerous applications in remote and maritime areas, allowing the use of Diesel generators to be reduced by replacing them with sustainable energy sources and/or storage. Among all of them, the following overall categories can be defined:

- Floating buoys: such as mooring or drifting buoys, usually used to measure meteorological or oceanographic parameters. Most of these buoys are currently based on solar energy and battery;
- Floating platforms: larger than floating buoys, they are used to produce electrical power, either for an island or for local use (aquaculture, oil and gas, fishing, etc.). Most of the projects presented in Tables 2 and 3 are based on floating platforms [68,69,71,75,79,81–83];
- Islands or coastal areas: several energy resources could be harnessed by onshore sources, such as PV panels and wind turbines, and by offshore systems, for example by the use of offshore wind turbines, tidal turbines, and wave energy converters [1,40–42,117];
- Artificial islands: as presented in several concepts [10,95,97,100], these are built on a reef or dyke. However, no further developments beyond the concept stage exist;
- Transport: maritime transport could use marine energy for their energetic needs [100].

4.2. Obstacles, Weaknesses and Issues

The reviews presented in the previous sections have shown a mismatch between the number of projects that led up to sea test conditions and those that remained at a status concept. Indeed, hybrid system development requires careful consideration of several aspects to avoid premature project shutdown, by events such as financial, installation, logistical, equipment, environmental, legislation, etc. Several possible obstacles and weaknesses are cited in previous papers [5,6,122]:

- Unbalanced renewable energy converter maturity levels, such as photovoltaic panels which present a higher maturity level than wave energy converters, for which a lot of technologies exist;
- Lack of experience and data: as hybrid systems including marine energies are recent, they are still at an early developmental stage. Information which could help to avoid development or operation issues is still limited;
- Development time: as requirements of such systems are numerous, a lot of development time is needed before commercial status is reached;
- High costs: although several savings can be found concerning previously presented synergies, other categories still present high costs, such as insurance, development time, technologies, etc.;

- Marine environmental constraints: floating systems should undergo severe conditions when they are placed offshore, such as weather (storm, hurricane), strong waves, salinity, biofouling, corrosion, etc.;
- Mooring and anchoring system reliability, which should be able to resist local environmental conditions.

Several projects have encountered issues either at an early stage of development or during the operational test phase, sometimes involving the premature project end. However, little information is available concerning the reasons of the end of a project. The following points can be highlighted, according to several publications:

- Damage or failure during the installation or operation phases, as happened for the SKWID wind/tidal hybrid system [80]. For example, failure can concern the structure, the power take-off technology, or the mooring and anchoring systems [122];
- Project ended prematurely due to high costs and lack of funding. This aspect has been seen at different steps, and it is thought that some companies cease to exist since there is a lack of information concerning recent activities. Also, some concepts appeared to be ambitious and thus costly. This could explain the lack of further development.

4.3. Feasibility and Design Methodology

To overcome some of the obstacles previously listed and make a system sustainable, feasibility aspects should be carefully studied. Thus, design methodologies and recommendations have been proposed by J.S. Martinez et al. [11] and B. Zanuttigh et al. [122] for the integration of energy converters in multipurpose platforms. The following methodology has been proposed previously [11] during the MERMAID project:

- Resource assessment according to the selected site;
- Power take-off technology selection allowing the power production to be maximized;
- Offshore structure technology selection (fixed or floating);
- Technology integration, by either platform sharing or area sharing (offshore energy farms);
- Environmental impact assessment, concerning pollution, recycling, etc.;
- Feasibility of combining with other activities.

Thus, the feasibility of such hybrid systems should start by a local evaluation, as the available resources can differ significantly [10]. Moreover, it has been advised [10] to use mature technologies, to avoid technology failure during the operational phase. Social acceptance must be considered by involving all the actors concerned in the project, including industries, political groups, investors, local communities, etc. Some authors advised developing individual renewable energy systems in the same area (this was for offshore wind farms) [123–125], then developing hybrid platforms that share the same structure [10,11].

5. Conclusions

Ocean energy can provide sufficient energy for the electricity supply of remote maritime areas, since the worldwide resources are major. Thus, combined systems including photovoltaic, wind, tidal current, and wave energies, which harness several kinds of energy, are a possible solution to replace the traditionally used genset-based systems to supply islands or floating systems. These four resources currently demonstrate the best maturity levels among all existing renewable energy sources, even if tidal kinetic energy and wave energy are still earlier in their development process than photovoltaic and wind energy converters.

After an overview of these four energy resources, this paper reviewed the industrial and academic hybrid systems based on marine energies. It appears that the development of such systems is still at an early stage in the development process, as shown by the number of projects that have remained at concept status. Several projects are currently close to full-scale mega-watt operational tests, such as the Poseidon P80 device [58,59] and the W2Power device [74]. Other projects have reached sea tests with small-scale prototypes. This review has also shown a lot of concepts that are more or less realistic given the limited amount of available information concerning further development. On the one hand, concerning possible obstacles for the development of hybrid systems based on marine energies, the required long development times and high costs, especially of insurance, can explain this situation. Moreover, the severe marine environment constraints make the design of hybrid systems more complex, especially for the mooring system which requires a high reliability level. On the other hand, the review of research dealing with energy management aspects and sizing optimization shows the promising aspects of such systems. Indeed, combining different renewable energy resources reduces the output power variations as their temporal characteristics differ, so less storage capacity is needed and Diesel generator use can be reduced. Other positive points have been listed in this article, such as sharing area, equipment, infrastructure, etc. The process engineering synergies should help the development of hybrid systems based on marine energies, with respect to all possible combinations with other sectors and activities: desalination, aquaculture, transport, oil and gas, etc. As a result, a development of hybrid systems based on marine energies is expected in forthcoming years, following the improvement of both tidal kinetic current and wave energy converter maturity levels.

Author Contributions: A.R. made the review and wrote the paper, F.A., F.D.-R., S.B., and Q.T.T. gave helpful comments and revised the paper. All authors read and approved the final manuscript.

Funding: This work was supported by the project "Monitoring and management of marine renewable energies" granted by the Pays de Loire region.

Acknowledgments: The authors would like to thank Katherine Kean who helped us writing this article with more correct English.

Conflicts of Interest: The authors declare no conflicts of interest.

References

1. Kuang, Y.; Zhang, Y.; Zhou, B.; Li, C.; Cao, Y.; Li, L.; Zeng, L. A review of renewable energy utilization in islands. *Renew. Sustain. Energy Rev.* **2016**, *59*, 504–513. [CrossRef]
2. Hussain, A.; Arif, S.M.; Aslam, M. Emerging renewable and sustainable energy technologies: State of the art. *Renew. Sustain. Energy Rev.* **2017**, *71*, 12–28. [CrossRef]
3. Fadaeenejad, M.; Shamsipour, R.; Rokni, S.D.; Gomes, C. New approaches in harnessing wave energy: With special attention to small islands. *Renew. Sustain. Energy Rev.* **2014**, *29*, 345–354. [CrossRef]
4. Mofor, L.; Goldsmith, J.; Jones, F. *Ocean Energy–Technology Readiness, Patents, Deployment Status and Outlook*; International Renewable Energy Agency (IRENA): Abu Dhabi, UAE, 2014.
5. Pérez-Collazo, C.; Greaves, D.; Iglesias, G. A review of combined wave and offshore wind energy. *Renew. Sustain. Energy Rev.* **2015**, *42*, 141–153. [CrossRef]
6. Karimirad, M. *Offshore Energy Structures: For Wind Power, Wave Energy and Hybrid Marine Platforms*; Springer: Cham, Switzerland, 2014; ISBN 978-3-319-12175-8.
7. Van Riet, T. Feasibility of Ocean Energy and Offshore Wind Hybrid Solutions. Master's Thesis, Delft University of Technology, Delft, The Netherlands, 2017.
8. Ding, S.; Yan, S.; Han, D.; Ma, Q. Overview on Hybrid Wind-Wave Energy Systems. In Proceedings of the 2015 International conference on Applied Science and Engineering, Jinan, China, 30–31 August 2015.
9. O'Sullivan, K.P. Feasibility of Combined Wind-Wave Energy Platforms. Ph.D. Thesis, University College Cork, Cork, Ireland, 2014.
10. Casale, C.; Serri, L.; Stolk, S.; Yildiz, I.; Cantù, M. *Synergies, Innovative Designs and Concepts for Multipurpose Use of Conversion Platforms*; Off-shore Renewable Energy Conversion platforms—Coordination Action (ORECCA): Munich, Germany, 2012.
11. Martinez, J.S.; Guanche, R.; Belloti, G.; Cecioni, C.; Cantu, M.; Franceschi, G.; Suffredini, R. *Integration of Energy Converters in Multi-Use Offshore Platforms*; MERMAID: Lyngby, Denmark, 2015.

12. Sampaio, P.G.V.; González, M.O.A. Photovoltaic solar energy: Conceptual framework. *Renew. Sustain. Energy Rev.* **2017**, *74*, 590–601. [CrossRef]

13. Kleissl, J. *Solar Energy Forecasting and Resource Assessment*; Academic Press: Cambridge, MA, USA, 2013; ISBN 9780123977724.

14. Duffie, J.A.; Beckman, W.A. *Solar Engineering of Thermal Processes*; John Wiley & Sons, Inc.: Hoboken, NJ, USA, 2013; ISBN 9781118671603.

15. Singh, G.K. Solar power generation by PV (photovoltaic) technology: A review. *Energy* **2013**, *53*, 1–13. [CrossRef]

16. El Chaar, L.; Lamont, L.A.; El Zein, N. Review of photovoltaic technologies. *Renew. Sustain. Energy Rev.* **2011**, *15*, 2165–2175. [CrossRef]

17. Khan, J.; Arsalan, M.H. Solar power technologies for sustainable electricity generation—A review. *Renew. Sustain. Energy Rev.* **2016**, *55*, 414–425. [CrossRef]

18. Parida, B.; Iniyan, S.; Goic, R. A review of solar photovoltaic technologies. *Renew. Sustain. Energy Rev.* **2011**, *15*, 1625–1636. [CrossRef]

19. Islam, M.R.; Mekhilef, S.; Saidur, R. Progress and recent trends of wind energy technology. *Renew. Sustain. Energy Rev.* **2013**, *21*, 456–468. [CrossRef]

20. Aubrée, R. Stratégies de Commande sans Capteur et de Gestion de l'énergie Pour les aérogénérateurs de Petite Puissance. Ph.D. Thesis, Université de Nantes, Nantes, France, 2014. (In French)

21. Burton, T.; Jenkins, N.; Sharpe, D.; Bossanyi, E. *Wind Energy Handbook*; John Wiley & Sons: Hoboken, NJ, USA, 2011; ISBN 9781119993926.

22. Nichita, C.; Dakyo, B. Conversion Systems for Offshore Wind Turbines. In *Marine Renewable Energy Handbook*; Multon, B., Ed.; John Wiley & Sons, Inc.: Hoboken, NJ, USA, 2012; pp. 123–172. ISBN 9781118603185.

23. Kumar, Y.; Ringenberg, J.; Depuru, S.S.; Devabhaktuni, V.K.; Lee, J.W.; Nikolaidis, E.; Andersen, B.; Afjeh, A. Wind energy: Trends and enabling technologies. *Renew. Sustain. Energy Rev.* **2016**, *53*, 209–224. [CrossRef]

24. Manwell, J.F.; McGowan, J.G.; Rogers, A.L. *Wind Energy Explained: Theory, Design and Application*; John Wiley & Sons: Hoboken, NJ, USA, 2010; ISBN 9780470686287.

25. Oh, K.-Y.; Nam, W.; Ryu, M.S.; Kim, J.-Y.; Epureanu, B.I. A review of foundations of offshore wind energy convertors: Current status and future perspectives. *Renew. Sustain. Energy Rev.* **2018**, *88*, 16–36. [CrossRef]

26. Yuce, M.I.; Muratoglu, A. Hydrokinetic energy conversion systems: A technology status review. *Renew. Sustain. Energy Rev.* **2015**, *43*, 72–82. [CrossRef]

27. Laws, N.D.; Epps, B.P. Hydrokinetic energy conversion: Technology, research, and outlook. *Renew. Sustain. Energy Rev.* **2016**, *57*, 1245–1259. [CrossRef]

28. Chen, H.; Aït-Ahmed, N.; Zaïm, E.H.; Machmoum, M. Marine tidal current systems: State of the art. In Proceedings of the 2012 IEEE International Symposium on Industrial Electronics, Hangzhou, China, 28–31 May 2012; pp. 1431–1437.

29. Roberts, A.; Thomas, B.; Sewell, P.; Khan, Z.; Balmain, S.; Gillman, J. Current tidal power technologies and their suitability for applications in coastal and marine areas. *J. Ocean Eng. Mar. Energy* **2016**, *2*, 227–245. [CrossRef]

30. Khan, N.; Kalair, A.; Abas, N.; Haider, A. Review of ocean tidal, wave and thermal energy technologies. *Renew. Sustain. Energy Rev.* **2017**, *72*, 590–604. [CrossRef]

31. Segura, E.; Morales, R.; Somolinos, J.A.; López, A. Techno-economic challenges of tidal energy conversion systems: Current status and trends. *Renew. Sustain. Energy Rev.* **2017**, *77*, 536–550. [CrossRef]

32. Benelghali, S.; Benbouzid, M.E.H.; Charpentier, J.F. Generator Systems for Marine Current Turbine Applications: A Comparative Study. *IEEE J. Ocean. Eng.* **2012**, *37*, 554–563. [CrossRef]

33. Zhang, J.; Moreau, L.; Machmoum, M.; Guillerm, P.E. State of the art in tidal current energy extracting technologies. In Proceedings of the 2014 First International Conference on Green Energy ICGE 2014, Sfax, Tunisia, 25–27 March 2014; pp. 1–7.

34. Benbouzid, M.; Astolfi, J.A.; Bacha, S.; Charpentier, J.F.; Machmoum, M.; Maitre, T.; Roye, D. Concepts, Modeling and Control of Tidal Turbines. In *Marine Renewable Energy Handbook*; Multon, B., Ed.; John Wiley & Sons, Inc.: Hoboken, NJ, USA, 2012; pp. 219–278. ISBN 9781118603185.

35. EMEC Ltd. Tidal Devices. Available online: http://www.emec.org.uk/marine-energy/tidal-devices/ (accessed on 22 August 2017).

36. Zhou, Z.; Benbouzid, M.; Charpentier, J.-F.; Sciuller, F.; Tang, T. Developments in large marine current turbine technologies—A review. *Renew. Sustain. Energy Rev.* **2017**, *71*, 852–858. [CrossRef]

37. Aubry, J.; Ben Ahmed, H.; Multon, B.; Babarit, A.; Clément, A.H. Houlogénérateurs. In *Energie thermique, Houlogénération et Technologies de Conversion et de Transport des Energies Marines Renouvelables*; Multon, B., Ed.; Lavoisier: Cachan, France, 2012.

38. Manasseh, R.; Sannasiraj, S.; McInnes, K.L.; Sundar, V.; Jalihal, P. Integration of wave energy and other marine renewable energy sources with the needs of coastal societies. *Int. J. Ocean Clim. Syst.* **2017**, *8*, 19–36. [CrossRef]

39. López, I.; Andreu, J.; Ceballos, S.; Martínez de Alegría, I.; Kortabarria, I. Review of wave energy technologies and the necessary power-equipment. *Renew. Sustain. Energy Rev.* **2013**, *27*, 413–434. [CrossRef]

40. Franzitta, V.; Curto, D. Sustainability of the Renewable Energy Extraction Close to the Mediterranean Islands. *Energies* **2017**, *10*, 283. [CrossRef]

41. Franzitta, V.; Catrini, P.; Curto, D. Wave Energy Assessment along Sicilian Coastline, Based on DEIM Point Absorber. *Energies* **2017**, *10*, 376. [CrossRef]

42. Franzitta, V.; Curto, D.; Milone, D.; Rao, D. Assessment of Renewable Sources for the Energy Consumption in Malta in the Mediterranean Sea. *Energies* **2016**, *9*, 1034. [CrossRef]

43. Kovaltchouk, T. Contributions à la co-Optimisation Contrôle-Dimensionnement sur Cycle de vie sous contrainte réseau des houlogénérateurs directs. Ph.D. Thesis, École normale supérieure de Cachan—ENS Cachan, Cachan, France, 2015. (In French)

44. Price, A.A.E. New Perspectives on Wave Energy Converter Control. Ph.D. Thesis, University of Edinburgh, Edinburgh, UK, 2009.

45. Olaya, S. Contribution à la Modélisation Multi-Physique et au contrôle Optimal d'un générateur houlomoteur: Application à un système "deux corps". Ph.D. Thesis, Université de Bretagne occidentale—Brest, Brest, France, 2016. (In French)

46. Wang, L.; Isberg, J.; Tedeschi, E. Review of control strategies for wave energy conversion systems and their validation: The wave-to-wire approach. *Renew. Sustain. Energy Rev.* **2018**, *81*, 366–379. [CrossRef]

47. Babarit, A. A database of capture width ratio of wave energy converters. *Renew. Energy* **2015**, *80*, 610–628. [CrossRef]

48. Falcão, A.F.O. Wave energy utilization: A review of the technologies. *Renew. Sustain. Energy Rev.* **2010**, *14*, 899–918. [CrossRef]

49. Hong, Y.; Waters, R.; Boström, C.; Eriksson, M.; Engström, J.; Leijon, M. Review on electrical control strategies for wave energy converting systems. *Renew. Sustain. Energy Rev.* **2014**, *31*, 329–342. [CrossRef]

50. Mustapa, M.A.; Yaakob, O.B.; Ahmed, Y.M.; Rheem, C.-K.; Koh, K.K.; Adnan, F.A. Wave energy device and breakwater integration: A review. *Renew. Sustain. Energy Rev.* **2017**, *77*, 43–58. [CrossRef]

51. Titah-Benbouzid, H.; Benbouzid, M.; Titah-Benbouzid, H.; Benbouzid, M. An Up-to-Date Technologies Review and Evaluation of Wave Energy Converters. *Int. Rev. Electr. Eng.* **2015**, *10*, 52–61. [CrossRef]

52. Ozkop, E.; Altas, I.H. Control, power and electrical components in wave energy conversion systems: A review of the technologies. *Renew. Sustain. Energy Rev.* **2017**, *67*, 106–115. [CrossRef]

53. Aubry, J. Optimisation du Dimensionnement d'une chaîne de Conversion électrique Directe Incluant un système de lissage de production par Supercondensateurs: Application au houlogénérateur SEAREV. Ph.D. Thesis, École normale supérieure de Cachan—ENS Cachan, Cachan, France, 2011. (In French)

54. Falcão, A.F.O. First-Generation Wave Power Plants: Current Status and R&D Requirements. In Proceedings of the ASME 2003 22nd International Conference on Offshore Mechanics and Arctic Engineering, Cancun, Mexico, 8–13 June 2003; pp. 723–731. [CrossRef]

55. Elhanafi, A.; Macfarlane, G.; Fleming, A.; Leong, Z. Experimental and numerical measurements of wave forces on a 3D offshore stationary OWC wave energy converter. *Ocean Eng.* **2017**, *144*, 98–117. [CrossRef]

56. Malara, G.; Arena, F. Analytical modelling of an U-Oscillating Water Column and performance in random waves. *Renew. Energy* **2013**, *60*, 116–126. [CrossRef]

57. Contestabile, P.; Iuppa, C.; Di Lauro, E.; Cavallaro, L.; Andersen, T.L.; Vicinanza, D. Wave loadings acting on innovative rubble mound breakwater for overtopping wave energy conversion. *Coast. Eng.* **2017**, *122*, 60–74. [CrossRef]

58. Han, Z.; Liu, Z.; Shi, H. Numerical study on overtopping performance of a multi-level breakwater for wave energy conversion. *Ocean Eng.* **2018**, *150*, 94–101. [CrossRef]

59. Buccino, M.; Vicinanza, D.; Salerno, D.; Banfi, D.; Calabrese, M. Nature and magnitude of wave loadings at Seawave Slot-cone Generators. *Ocean Eng.* **2015**, *95*, 34–58. [CrossRef]

60. Sergiienko, N.Y.; Rafiee, A.; Cazzolato, B.S.; Ding, B.; Arjomandi, M. Feasibility study of the three-tether axisymmetric wave energy converter. *Ocean Eng.* **2018**, *150*, 221–233. [CrossRef]

61. Widén, J.; Carpman, N.; Castellucci, V.; Lingfors, D.; Olauson, J.; Remouit, F.; Bergkvist, M.; Grabbe, M.; Waters, R. Variability assessment and forecasting of renewables: A review for solar, wind, wave and tidal resources. *Renew. Sustain. Energy Rev.* **2015**, *44*, 356–375. [CrossRef]

62. Olauson, J.; Ayob, M.N.; Bergkvist, M.; Carpman, N.; Castellucci, V.; Goude, A.; Lingfors, D.; Waters, R.; Widén, J. Net load variability in Nordic countries with a highly or fully renewable power system. *Nat. Energy* **2016**, *1*, 1–8. [CrossRef]

63. IEA. *Variability of Wind Power and Other Renewables: Management Options and Strategies*; International Energy Agency: Paris, France, 2005.

64. Coker, P.; Barlow, J.; Cockerill, T.; Shipworth, D. Measuring significant variability characteristics: An assessment of three UK renewables. *Renew. Energy* **2013**, *53*, 111–120. [CrossRef]

65. Reikard, G.; Robertson, B.; Bidlot, J.-R. Combining wave energy with wind and solar: Short-term forecasting. *Renew. Energy* **2015**, *81*, 442–456. [CrossRef]

66. Engeland, K.; Borga, M.; Creutin, J.-D.; François, B.; Ramos, M.-H.; Vidal, J.-P. Space-time variability of climate variables and intermittent renewable electricity production—A review. *Renew. Sustain. Energy Rev.* **2017**, *79*, 600–617. [CrossRef]

67. Astariz, S.; Vazquez, A.; Iglesias, G. Evaluation and comparison of the levelized cost of tidal, wave, and offshore wind energy. *J. Renew. Sustain. Energy* **2015**, *7*, 053112. [CrossRef]

68. GEPS Techno Le premier hybride marin combine 4 sources d'énergies—PH4S. Available online: https://www.geps-techno.com/ph4s/ (accessed on 7 September 2017).

69. Floating Power Plant Products & Services of Floating Power Plant. Available online: http://www.floatingpowerplant.com/products/ (accessed on 7 September 2017).

70. Yde, A.; Pedersen, M.M.; Bellew, S.B.; Køhler, A.; Clausen, R.S.; Wedel Nielsen, A. *Experimental and Theoretical Analysis of a Combined Floating Wave and Wind Energy Conversion Platform*; DTU Wind Energy: Lyngby, Denmark, 2014.

71. Pelagic Power as W2Power. Available online: http://www.pelagicpower.no/ (accessed on 7 September 2017).

72. Mayorga, P. *W2Power: Wind Integrated Platform for 10+ MW Power per Foundation*; EnerOcean SL: Málaga, Spain, 2017.

73. Hanssen, J.E.; Margheritini, L.; O'Sullivan, K.; Mayorga, P.; Martinez, I.; Arriaga, A.; Agos, I.; Steynor, J.; Ingram, D.; Hezari, R.; et al. Design and performance validation of a hybrid offshore renewable energy platform. In Proceedings of the 2015 Tenth International Conference on Ecological Vehicles and Renewable Energies (EVER), Monte Carlo, Monaco, 31 March–2 April 2015; pp. 1–8.

74. WIP10+ Wind Integrated Platform for 10+ MW Power per Foundation. Available online: http://wip10plus.eu/ (accessed on 25 September 2017).

75. Sato, Y.; Ohya, Y.; Kyozuka, Y.; Tsutsumi, T. *The Floating Offshore Wind Turbine with PC Floating Structure—Hakata Bay Floating Offshore Wind Turbine*; Kyushu University: Fukuoka, Japan, 2014.

76. Ohya, Y.; Karasudani, T.; Nagai, T.; Watanabe, K. Wind lens technology and its application to wind and water turbine and beyond. *Renew. Energy Environ. Sustain.* **2017**, *2*. [CrossRef]

77. 4C Offshore Kyushu University Wind Lens Project—Phase 1. Available online: http://www.4coffshore.com/windfarms/kyushu-university-wind-lens-project---phase-1-japan-jp12.html (accessed on 18 September 2017).

78. 4C Offshore Kyushu University Wind Lens Project—Phase 2. Available online: http://www.4coffshore.com/windfarms/kyushu-university-wind-lens-project---phase-2-japan-jp08.html (accessed on 18 September 2017).

79. MODEC Offshore Wind Power—Floating Production Solutions | MODEC. Available online: http://www.modec.com/fps/offshorewind/index.html (accessed on 18 September 2017).

80. 4C Offshore Savonius Keel & Wind Turbine Darrieus (SKWID). Available online: http://www.4coffshore.com/windfarms/savonius-keel-%26-wind-turbine-darrieus-[skwid]-japan-jp24.html (accessed on 18 September 2017).

81. Osawa, H.; Miyazaki, T. Wave-PV hybrid generation system carried in the offshore floating type wave power device "Mighty Whale". In Proceedings of the OCEANS '04. MTTS/IEEE TECHNO-OCEAN '04, Kobe, Japan, 9–12 November 2004; Volume 4, pp. 1860–1866.

82. Han Ocean Product–Han-Ocean. Available online: http://www.hann-ocean.com/en/product.html (accessed on 7 September 2017).

83. Han, H. *A Modular System for Implementation of Solar, Wind, Wave, and/or Current Energy Convertors*; Hann-Ocean Technology Pte Ltd.: Singapore, 1 March 2012.

84. Principle Power Inc. Principle Power Inc.—Globalizing Offshore Wind. Available online: http://www. principlepowerinc.com/en/windfloat (accessed on 25 September 2017).

85. Weinstein, A.; Roddier, D.; Banister, K. *WindWaveFloat (WWF): Final Scientific Report*; Principle Power Inc.: Emeryville, CA, USA, 2012.

86. Atlantis Resources Commercial Tidal Power Generation. Available online: https://www.atlantisresourcesltd. com/ (accessed on 25 September 2017).

87. Sojo, M.; Gunther, A. *Marine Renewable Integrated Application Platform—Deliverable D1.12: Final Summary Report*; Acciona Energia S.A.: Sarriguren, Spain, 2013.

88. Muliawan, M.J.; Karimirad, M.; Moan, T. Dynamic response and power performance of a combined Spar-type floating wind turbine and coaxial floating wave energy converter. *Renew. Energy* **2013**, *50*, 47–57. [CrossRef]

89. Wave Star A/S Concept | Wavestar. Available online: http://wavestarenergy.com/concept (accessed on 7 September 2017).

90. Sea for Life WEGA Future. Available online: http://www.seaforlife.com/EN/FrameWEGAfuture.html (accessed on 7 September 2017).

91. Peiffer, A.; Roddier, D. Design of an Oscillating Wave Surge Converter on the WindFloat Structure. In Proceedings of the 2012 4th International Conference on Ocean Energy (ICOE), Dublin, Ireland, 17–19 October 2012.

92. Green Ocean Energy Green Ocean Energy Wave Treader. Available online: http://www.renewable-technology.com/projects/green-ocean-wave-treader/ (accessed on 7 September 2017).

93. OWWE Ltd Ocean Wave and Wind Energy. Available online: http://www.owwe.net/ (accessed on 7 September 2017).

94. Soulard, T.; Babarit, A.; Borgarino, B.; Wyns, M.; Harismendy, M. C-HyP: A Combined Wind and Wave Energy Platform with Balanced Contributions. In Proceedings of the ASME 2013 32nd International Conference on Ocean, Offshore and Arctic Engineering, Nantes, France, 9–14 June 2013; p. V008T09A049. [CrossRef]

95. KEMA. *Lievense Energy Island: Opportunities and Social Faisability*; KEMA: Arnhem, Netherlands, 2009.

96. DNV GL Large-scale Electricity Storage—Energy. Available online: https://www.dnvgl.com/services/ large-scale-electricity-storage-7272 (accessed on 7 September 2017).

97. Float Incorporated Offshore Ocean Energy System. Available online: http://floatinc.com/OOES.aspx (accessed on 26 September 2017).

98. Energy Island Ltd. Energy Island. Available online: http://www.energyisland.com/technologies/patents/ patents.html (accessed on 7 September 2017).

99. Brito, J.H. *Modular Multi-use Deep Water Offshore Platform Harnessing and Servicing Mediterranean, Subtropical and Tropical Marine and Maritime Resources—Project Final Report*; TROPOS: Telde, Spain, 2015.

100. Quevedo, E.; Cartón, M.; Delory, E.; Castro, A.; Hernández, J.; Llinás, O.; de Lara, J.; Papandroulakis, N.; Anastasiadis, P.; Bard, J.; et al. Multi-use offshore platform configurations in the scope of the FP7 TROPOS Project. In Proceedings of the 2013 MTS/IEEE OCEANS—Bergen, Bergen, Norway, 10–14 June 2013; pp. 1–7.

101. Lu, S.Y.; Wang, L.; Lo, T.M.; Prokhorov, A.V. Integration of Wind Power and Wave Power Generation Systems Using a DC Microgrid. *IEEE Trans. Ind. Appl.* **2015**, *51*, 2753–2761. [CrossRef]

102. Samrat, N.H.; Ahmad, N.B.; Choudhury, I.A.; Taha, Z.B. Modeling, Control, and Simulation of Battery Storage Photovoltaic-Wave Energy Hybrid Renewable Power Generation Systems for Island Electrification in Malaysia. *Sci. World J.* **2014**. [CrossRef] [PubMed]

103. Ahmad, S.; Uddin, M.J.; Nisu, I.H.; Ahsan, M.M.U.; Rahman, I.; Samrat, N.H. Modeling of grid connected battery storage wave energy and PV hybrid renewable power generation. In Proceedings of the 2017 International Conference on Electrical, Computer and Communication Engineering (ECCE), Cox's Bazar, Bangladesh, 16–18 February 2017; pp. 375–380.

104. Wang, L.; Vo, Q.S.; Prokhorov, A.V. Dynamic Stability Analysis of a Hybrid Wave and Photovoltaic Power Generation System Integrated into a Distribution Power Grid. *IEEE Trans. Sustain. Energy* **2017**, *8*, 404–413. [CrossRef]

105. Da, Y.; Khaligh, A. Hybrid offshore wind and tidal turbine energy harvesting system with independently controlled rectifiers. In Proceedings of the 2009 35th Annual Conference of IEEE Industrial Electronics, Porto, Portugal, 3–5 November 2009; pp. 4577–4582.

106. Rahman, M.L.; Oka, S.; Shirai, Y. Hybrid Power Generation System Using Offshore-Wind Turbine and Tidal Turbine for Power Fluctuation Compensation (HOT-PC). *IEEE Trans. Sustain. Energy* **2010**, *1*, 92–98. [CrossRef]

107. Shirai, Y.; Minamoto, S.; Yonemura, K.; Rahman, M.L. Output power control of hybrid off-shore-wind and tidal turbine generation system with battery storage system. In Proceedings of the 2016 19th International Conference on Electrical Machines and Systems (ICEMS), Chiba, Japan, 13–16 November 2016; pp. 1–6.

108. Rahman, M.L.; Nishimura, K.; Motobayashi, K.; Fujioka, S.; Shirai, Y. Characteristic of small-scale BESS for HOTT generation system. In Proceedings of the 2014 IEEE Innovations in Technology Conference, Warwick, RI, USA, 16 May 2014; pp. 1–9.

109. Rahman, M.L.; Nishimura, K.; Motobayashi, K.; Fujioka, S.; Shirai, Y. Tidal Turbine Control System for Hybrid Integration and Automatic Fluctuation Compensation of Offshore-wind Turbine Generation System. *Int. J. Power Renew. Energy Syst.* **2014**, *1*, 1–11.

110. Pierre, S. Contribution au Développement d'un Concept D'hybridation Energétique: Structures de Commande d'un système intégré éolien-Hydrolien. Ph.D. Thesis, Université du Havre, Le Havre, France, 2015. (In French)

111. Fan, Y.; Mu, A.; Ma, T. Modeling and control of a hybrid wind-tidal turbine with hydraulic accumulator. *Energy* **2016**, *112*, 188–199. [CrossRef]

112. Qin, C.; Ju, P.; Wu, F.; Jin, Y.; Chen, Q.; Sun, L. A coordinated control method to smooth short-term power fluctuations of hybrid offshore renewable energy conversion system (HORECS). In Proceedings of the 2015 IEEE Eindhoven PowerTech, Eindhoven, The Netherlands, 29 June–2 July 2015; pp. 1–5.

113. Al-falahi, M.D.A.; Jayasinghe, S.D.G.; Enshaei, H. A review on recent size optimization methodologies for standalone solar and wind hybrid renewable energy system. *Energy Convers. Manag.* **2017**, *143*, 252–274. [CrossRef]

114. Siddaiah, R.; Saini, R.P. A review on planning, configurations, modeling and optimization techniques of hybrid renewable energy systems for off grid applications. *Renew. Sustain. Energy Rev.* **2016**, *58*, 376–396. [CrossRef]

115. Tezer, T.; Yaman, R.; Yaman, G. Evaluation of approaches used for optimization of stand-alone hybrid renewable energy systems. *Renew. Sustain. Energy Rev.* **2017**, *73*, 840–853. [CrossRef]

116. Sinha, S.; Chandel, S.S. Review of recent trends in optimization techniques for solar photovoltaic–wind based hybrid energy systems. *Renew. Sustain. Energy Rev.* **2015**, *50*, 755–769. [CrossRef]

117. Mohammed, O.H.; Amirat, Y.; Benbouzid, M.; Haddad, S.; Feld, G. Optimal sizing and energy management of hybrid wind/tidal/PV power generation system for remote areas: Application to the Ouessant French Island. In Proceedings of the IECON 2016—42nd Annual Conference of the IEEE Industrial Electronics Society, Florence, Italy, 23–26 October 2016; pp. 4205–4210.

118. Mohammed, O.H.; Amirat, Y.; Benbouzid, M.E.H.; Feld, G. Optimal Design and Energy Management of a Hybrid Power Generation System Based on Wind/Tidal/PV Sources: Case Study for the Ouessant French Island. In *Smart Energy Grid Design for Island Countries*; Springer: Cham, Switzerland, 2017; pp. 381–413. ISBN 978-3-319-50196-3.

119. Askarzadeh, A. Electrical power generation by an optimised autonomous PV/wind/tidal/battery system. *IET Renew. Power Gener.* **2017**, *11*, 152–164. [CrossRef]

120. Mousavi, G.S.M. An autonomous hybrid energy system of wind/tidal/microturbine/battery storage. *Int. J. Electr. Power Energy Syst.* **2012**, *43*, 1144–1154. [CrossRef]

121. Anwar, M.B.; Moursi, M.S.E.; Xiao, W. Novel Power Smoothing and Generation Scheduling Strategies for a Hybrid Wind and Marine Current Turbine System. *IEEE Trans. Power Syst.* **2017**, *32*, 1315–1326. [CrossRef]

122. Zanuttigh, B.; Angelelli, E.; Kortenhaus, A.; Koca, K.; Krontira, Y.; Koundouri, P. A methodology for multi-criteria design of multi-use offshore platforms for marine renewable energy harvesting. *Renew. Energy* **2016**, *85*, 1271–1289. [CrossRef]

123. Astariz, S.; Perez-Collazo, C.; Abanades, J.; Iglesias, G. Co-located wind-wave farm synergies (Operation & Maintenance): A case study. *Energy Convers. Manag.* **2015**, *91*, 63–75. [CrossRef]

124. Astariz, S.; Iglesias, G. Enhancing Wave Energy Competitiveness through Co-Located Wind and Wave Energy Farms. A Review on the Shadow Effect. *Energies* **2015**, *8*, 7344–7366. [CrossRef]

125. Veigas, M.; Iglesias, G. Potentials of a hybrid offshore farm for the island of Fuerteventura. *Energy Convers. Manag.* **2014**, *86*, 300–308. [CrossRef]

energies

MDPI

Article

A Novel Method for Estimating Wave Energy Converter Performance in Variable Bathymetry Regions and Applications

Kostas Belibassakis [1], Markos Bonovas [1] and Eugen Rusu [2],*

[1] School of Naval Architecture & Marine Engineering, National Technical University of Athens, 15780 Athens, Greece; kbel@fluid.mech.ntua.gr (K.B.); markosbonovas@hotmail.gr (M.B.)

[2] Department of Mechanical Engineering, University Dunarea de Jos of Galati, 800008 Galati, Romania

* Correspondence: eugen.rusu@ugal.ro; Tel.: +40-740-205-534

Received: 24 July 2018; Accepted: 10 August 2018; Published: 11 August 2018

Abstract: A numerical model is presented for the estimation of Wave Energy Converter (WEC) performance in variable bathymetry regions, taking into account the interaction of the floating units with the bottom topography. The proposed method is based on a coupled-mode model for the propagation of the water waves over the general bottom topography, in combination with a Boundary Element Method for the treatment of the diffraction/radiation problems and the evaluation of the flow details on the local scale of the energy absorbers. An important feature of the present method is that it is free of mild bottom slope assumptions and restrictions and it is able to resolve the 3D wave field all over the water column, in variable bathymetry regions including the interactions of floating bodies of general shape. Numerical results are presented concerning the wave field and the power output of a single device in inhomogeneous environment, focusing on the effect of the shape of the floater. Extensions of the method to treat the WEC arrays in variable bathymetry regions are also presented and discussed.

Keywords: renewable energy; marine environment; wave energy converters; variable bathymetry effects; arrays

---

1. Introduction

Interaction of the free-surface gravity waves with floating bodies, in water of intermediate depth and in variable bathymetry regions, is an interesting problem with important applications. Specific examples concern the design and evaluation of the performances of special-type ships and structures operating in nearshore and coastal waters; see, e.g., [1,2]. Also, pontoon-type floating bodies of relatively small dimensions find applications as coastal protection devices (floating breakwaters) and they are also frequently used as small boat marinas; see, e.g., [3–7]. In all these cases, the estimation of the wave-induced loads and motions of the floating structures can be based on the solution of the classical wave-body-seabed hydrodynamic-interaction problems; see, e.g., [8,9]. In particular, the performance of the Wave Energy Converters (WECs) operating in nearshore and coastal areas, characterized by variable bottom topography, is important for the estimation of the wave power absorption, determination of the operational characteristics of the system and could significantly contribute to the efficient design and layout of the WEC farms. In this case, wave-seabed interactions may have a significant effect; see [10,11].

In the above studies the details of the wave field propagating and scattered over the variable bathymetry region could be important in order to consistently calculate the responses of the floating bodies. For rapidly varying seabed topographies, including steep bottom parts, local or evanescent modes may have a significant impact on the wave phase evolution during propagation. Such a fact was demonstrated through the interference process in one-directional wave propagation as observed for either varying topographies (see e.g., [12,13]) or abrupt bathymetries including coastal structures (see e.g., [14–16]). For such problems, the consistent coupled-mode theory has been developed in [17], for the water waves propagation in variable bathymetry regions. Furthermore, it was subsequently extended for 3D bathymetry in [18], and applied successfully to treat the wave transformation over nearshore/coastal sites with steep 3D bottom features, like underwater canyons; see, e.g., [19,20].

In recent works [21,22] the coupled-mode model is further extended to treat the wave-current-seabed interaction problem, with application to the wave scattering by non-homogeneous current over general bottom topography. The problem of the directional spectrum transformation of an incident wave system over a region of strongly varying three-dimensional bottom topography is further studied in [22]. The accuracy and efficiency of the coupled-mode method is tested, comparing numerical predictions against experimental data by [23] and calculations by the phase-averaged model SWAN [24,25]. Results are shown in various representative test cases demonstrating the importance of the first evanescent modes and the additional sloping-bottom mode when the bottom slope is not negligible.

In this work, a methodology is presented to treat the propagation-diffraction-radiation problem locally around each WEC, supporting the calculation of the interaction effects of the floating units with variable bottom topography at a local scale. The method is based on the coupled-mode model developed by [17], and extended to 3D by [18], for water wave propagation over general bottom topography, in conjunction with the Boundary Element Method (BEM) for the hydrodynamic analysis of floating bodies over general bottom topography [15] and the corresponding 3D Green's function [26]. An important feature of the present method is that it is free of mild-slope assumptions and restrictions and it is able to resolve the 3D wave field all over the water column, in variable bathymetry regions including the interactions of floating bodies of general shape. Numerical results are presented and discussed concerning simple bodies (heaving vertical cylinders) illustrating the applicability of the present method.

2. Formulation

We consider here the hydrodynamic problem concerning the behavior of a number N of identical cylindrical-shaped WECs, $D_B^{(k)}, k = 1, N$, of characteristic radius a and draft d, operating in the nearshore environment, as shown in Figure 1.

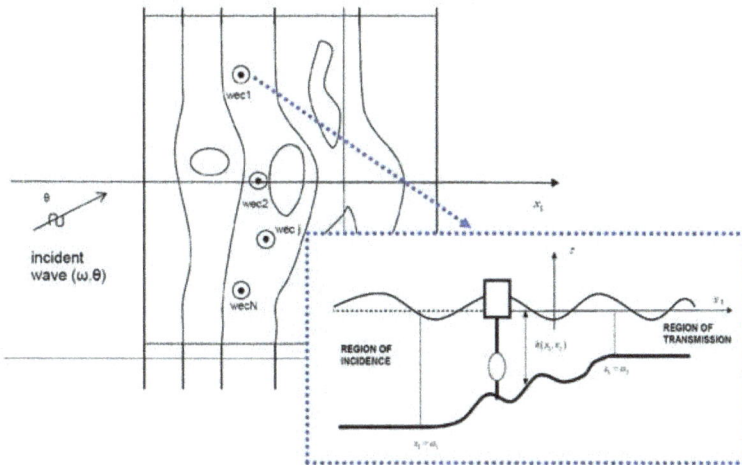

Figure 1. Array of WECs in variable bathymetry region.

The variable bathymetry region is considered between two infinite sub-regions of constant, but possibly different depths $h = h_1$ (region of incidence) and $h = h_3$ (region of transmission). In the middle sub-region, it is assumed that the depth h exhibits an arbitrary variation. The wave field is excited by a harmonic incident wave of angular frequency ω, propagating with direction θ; see Figure 1. Under the assumptions that the free-surface elevation and the wave velocities are small, the wave potential is expressed as follows:

$$\Phi(\mathbf{x}, z; t) = \mathrm{Re}\left\{ -\frac{igH}{2\omega} \varphi(\mathbf{x}, z; \mu) \cdot exp(-i\omega\, t) \right\}, \tag{1}$$

where $\mathbf{x} = (x_1, x_2)$, and satisfies the linearized water wave equations; see [27]. In the above equation H is the incident wave height, g is the acceleration due to gravity, $\mu = \omega^2/g$ is the frequency parameter, and $i = \sqrt{-1}$. The free surface elevation is then obtained in terms of the wave potential as follows:

$$\eta(\mathbf{x}; t) = -\frac{1}{g} \frac{\partial \Phi(\mathbf{x}, z = 0)}{\partial t}. \tag{2}$$

Using standard floating-body hydrodynamic theory [8], the complex potential can be decomposed as follows:

$$\varphi(\mathbf{x}, z) = \varphi_P(\mathbf{x}, z) + \varphi_D(\mathbf{x}, z) + \frac{2\omega^2}{gH}\varphi_R(\mathbf{x}, z), \quad \varphi_R(\mathbf{x}, z) = \sum_{k=1}^{N}\sum_{\ell=1}^{6} \zeta_{k\ell}\varphi_{k\ell}(\mathbf{x}, z), \tag{3}$$

where $\varphi_P(\mathbf{x}, z)$ is the normalized propagation wave potential in the variable bathymetry region in the absence of the WECs, $\varphi_D(\mathbf{x}, z)$ is the diffraction potential due to the presence fixed (motionless) bodies $D_B^{(k)}, k = 1, N$, that satisfies the boundary condition $\partial\varphi_D(\mathbf{x}, z)/\partial n_k = -\partial\varphi_P(\mathbf{x}, z)/\partial n_k$ on k-WEC, where $\mathbf{n}_k = (n_1, n_2, n_3)_k$ the normal vector on the wetted surface of the k-body, directed outwards the fluid domain (inwards the body). Furthermore, $\varphi_{k\ell}(\mathbf{x}, z), k = 1, N$, denotes the radiation potential in the non-uniform domain associated with the ℓ-oscillatory motion of the k-body with complex amplitude $\zeta_{k\ell}$, satisfying $\partial\varphi_{k\ell}(\mathbf{x}, z)/\partial n = n_{k\ell}$, equal to the ℓ-component of generalized normal vector on the wetted surface of the k-WEC ($n_{k\ell} = (\mathbf{r} \times \mathbf{n}_k)_{\ell-3}$ for $\ell = 4, 5, 6$).

In the case of simple heaving WECs, only the vertical oscillation of each body is considered $\zeta_k = \zeta_{k3}$, which is one of the most powerful intensive modes concerning this type of wave energy systems. In the present work we will concentrate on this simpler configuration, leaving the analysis of the more complex case to be examined in future work. For an array of heaving WECs the hydrodynamic response is obtained by:

$$\zeta_{k3} = (A_{km})^{-1}(X_{Pm} + X_{Dm}), \quad k, m = 1, \dots N, \tag{4}$$

where $X_{Pm} + X_{Dm}$ denote the exciting vertical force on each WEC due to propagating and diffraction field, respectively, and the matrix coefficient A_{km} is given by:

$$A_{km} = -\omega^2(M + a_{km}) - i\omega(B_m\delta_{km} + b_{km}) + (C_m + c)\delta_{km}, \tag{5}$$

where δ_{km} denotes Kronecker's delta and M is the body mass (assumed the same for all WECs). The hydrodynamic coefficients (added mass and damping) are calculated by the following integrals:

$$a_{km} - \frac{1}{i\omega} b_{km} = \rho \iint_{\partial D_{Bm}} \varphi_{k3} n_{m3}\ dS, \quad k, m = 1, \dots N, \tag{6}$$

of the heaving-radiation potential of the k-WEC on the wetted surface ∂D_{Bm} of the m-WEC. Moreover, $c = \rho g A_{WL}$ is the hydrostatic coefficient in heaving motion with A_{WL} the waterline surface, and B_m, C_m are characteristic constants of the Power Take Off (PTO) system associated with the m-th degree of freedom of the floater. The components of the excitation (Froude-Krylov and diffraction) forces are calculated by the following integrals of the corresponding potentials:

$$X_{Pm} = \frac{\rho g H}{2} \iint_{\partial D_{Bm}} \varphi_P n_m \, dS, \, X_{Dm} = \frac{\rho g H}{2} \iint_{\partial D_{Bm}} \varphi_D n_m \, dS, \, m = 1, \ldots N, \tag{7}$$

on the wetted surface ∂D_{Bm} of the m-WEC. The total power extracted by the array is obtained as:

$$P(N; \omega, \theta) = \frac{1}{2} \omega^2 \left| \sum_{k=1}^{N} \eta_{eff}^m B_m(\xi_k)^2 \right|, \tag{8}$$

where η_k indicates the efficiency of the PTO associated with the k-th degree of freedom (that could be a function of the frequency ω). Finally, the q-index can be estimated by:

$$q_N(\omega, \theta) = N^{-1} P(N; \omega, \theta, H) / P_1(\omega, \theta, H), \tag{9}$$

where $P_1(\omega, \theta, H)$ indicates the output of a single device operating in the same environment and wave conditions. Obviously, the calculated performance depends on the frequency, direction and height of the incident wave, as well as on the physical environment and the positioning of the WECs in the array (farm layout). Finally, the operational characteristics of the farm, in general multi chromatic wave conditions, characterized by directional wave spectrum, could be obtained by appropriate spectral synthesis; see, e.g., [20,22].

3. Propagating Wave Field

The wave potential $\varphi_P(\mathbf{x}, z)$ associated with the propagation of water waves in the variable bathymetry region, without the presence of the scatterer (floating body), can be conveniently calculated by means of the consistent coupled-mode model developed [17], as extended to three-dimensional environments by [18]. This model is based on the following enhanced local-mode representation:

$$\varphi_P(\mathbf{x}, z) = \varphi_{-1}(\mathbf{x}) \, Z_{-1}(z; \mathbf{x}) + \sum_{n=0}^{\infty} \varphi_n(\mathbf{x}) \, Z_n(z; \mathbf{x}). \tag{10}$$

In the above expansion, the term $\varphi_0(\mathbf{x}) Z_0(z; \mathbf{x})$ denotes the propagating mode of the generalized incident field. The remaining terms $\varphi_n(\mathbf{x}) \, Z_n(z; \mathbf{x})$, $n = 1, 2, \ldots$, are the corresponding evanescent modes, and the additional term $\varphi_{-1}(\mathbf{x}) Z_{-1}(z; \mathbf{x})$ is a correction term, called the sloping-bottom mode, which properly accounts for the satisfaction of the Neumann bottom boundary condition of the non-horizontal parts of the bottom. The function $Z_n(z; \mathbf{x})$ represents the vertical structure of the n-th mode. The function $\varphi_n(\mathbf{x})$ describes the horizontal pattern of the n-th mode and is called the complex amplitude of the n-th mode. The functions $Z_n(z; \mathbf{x})$, $n = 0, 1, 2 \ldots$, are obtained as the eigenfunctions of local vertical Sturm-Liouville problems formulated in the local vertical intervals $-h(\mathbf{x}) \leq z \leq 0$, and are given by:

$$Z_0(z; \mathbf{x}) = \frac{\cosh[k_0(\mathbf{x})(z + h(\mathbf{x}))]}{\cosh(k_0(\mathbf{x})h(\mathbf{x}))}, \, Z_n(z; \mathbf{x}) = \frac{\cos[k_n(\mathbf{x})(z + h(\mathbf{x}))]}{\cos(k_n(\mathbf{x})h(\mathbf{x}))}, \, n = 1, 2, \ldots \tag{11}$$

In the above equations the eigenvalues $\{ik_0(\mathbf{x}), k_n(\mathbf{x})\}$ are obtained as the roots of the local dispersion relations:

$$\mu \, h(\mathbf{x}) = k_0(\mathbf{x}) \, h(\mathbf{x}) \tanh[k_0(\mathbf{x})h(\mathbf{x})], \, \mu \, h(\mathbf{x}) = -k_n(\mathbf{x}) \, h(\mathbf{x}) \, \tan[k_n(\mathbf{x})h(\mathbf{x})]$$
$$n = 1, 2, \ldots \tag{12}$$

The function $Z_{-1}(z; \mathbf{x})$ is defined as the vertical structure of the sloping-bottom mode. This term is introduced in the series in order to consistently satisfy the Neumann boundary condition on the non-horizontal parts of the seabed. It becomes significant in the case of seabottom topographies with non-mildly sloped parts and has the effect of significant acceleration of the convergence of the local mode series Equation (10); see [17]. In fact, truncation of the series (10) keeping only a small number 4–6 totally terms have been proved enough for calculating the propagating wave field in variable bathymetry regions with bottom slopes up to and exceeding 100%. For specific convenient forms of $Z_{-1}(z; \mathbf{x})$ see the discussion ([17]). By following the procedure described in the latter work, the coupled-mode system of horizontal equations for the amplitudes of the incident wave field propagating over the variable bathymetry region is finally obtained:

$$\sum_{n=-1} A_{mn}(\mathbf{x}) \, \nabla^2 \varphi_n(\mathbf{x}) + B_{mn}(\mathbf{x}) \, \nabla \varphi_n(\mathbf{x}) + C_{mn}(\mathbf{x}) \varphi_n(\mathbf{x}) = 0, \, m = -1, 0, 1 \ldots \tag{13}$$

where the coefficients A_{mn}, B_{mn}, C_{mn} of the coupled-mode system (13) are defined in terms of the vertical modes $Z_n(z; \mathbf{x})$. The coefficients are dependent on $\mathbf{x}$ through $h(\mathbf{x})$ and the corresponding expressions can be found in Table 1 of [17]. The system is supplemented by appropriate boundary conditions specifying the incident waves and treating reflection, transmission and radiation of waves. It is worth mentioning here that if only the propagating mode ($n = 0$) is retained in the expansion (11) the above CMS reduces to an one-equation model which is exactly the modified mild-slope Equation derived in [13,28]. So, the present approach could be automatically reduced to mild-slope model in the subregions where such a simplification is permitted, saving a lot of computational cost. On the other hand in subregions where bottom variations are strong the extra (evanescent) modes are turned on and have substantial effects concerning the 3D wave field all over the water column, as illustrated in [17,18].

Table 1. Optimum PML parameters.

$\omega < 2$	$\widetilde{\sigma}_o = 1$	$R/\lambda = 2$	$N/\lambda = 15$	$n = 5$
$2 \le \omega < 7$	$\widetilde{\sigma}_o = 1$	$R/\lambda = 3$	$N/\lambda = 20$	$n = 3$
$7 \le \omega < 8$	$\widetilde{\sigma}_o = 1$	$R/\lambda = 3$	$N/\lambda = 15$	$n = 3$
$8 \le \omega \le 9$	$\widetilde{\sigma}_o = 1$	$R/\lambda = 3$	$N/\lambda = 10$	$n = 3$

4. A Novel BEM for the Diffraction and Radiation Problems in 3D Environments

The corresponding problems on the diffraction and radiation potentials $\varphi_D(\mathbf{x}, z)$ and $\varphi_k(\mathbf{x}, z)$, associated with the operation of the floating WECs, are treated by means of low-order Boundary Element Method, based on simple singularity distributions and 4-node quadrilateral boundary elements ([29]), ensuring continuity of the geometry approximation of the various parts of the boundary. The potential and velocity fields are approximated by:

$$\varphi(\mathbf{r}) = \sum_p F_p \Phi_p(\mathbf{r}), \, \nabla \varphi(\mathbf{r}) = \sum_p F_p \mathbf{U}_p(\mathbf{r}), \tag{14}$$

where the summation ranges over all panels and $F_p(\mathbf{r})$ and $\mathbf{U}_p(\mathbf{r})$ denote induced potential and velocity from the p-th element with unit singularity distribution to the field point r; see, e.g., [30] and the references cited there. We mention here that a minimum number of 10–20 elements per wavelength is used in discretizing the free surface, in order to eliminate errors due to damping and dispersion associated with the above discrete scheme. In order to eliminate the infinite extent of the domain and treat the radiating behaviour of the diffraction and radiation fields at far distances from the bodies, an absorbing layer technique is used, based on a matched layer all around the fore and side borders of the computational domain on the free surface; see, e.g., [31]. The thickness of the absorbing layer is of the order of 1–2 characteristic wavelengths and its coefficient is taken increasing within the layer;

see Figure 2. The efficiency of this technique to damp the outgoing waves with minimal reflection is dependent on the thickness of the layer.

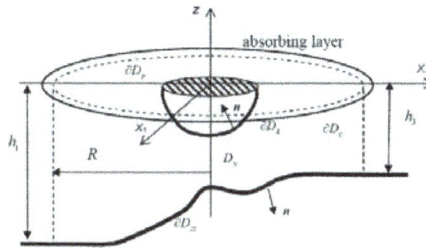

Figure 2. Formulation of diffraction and radiation problems in variable bathymetry regions.

Thus, the diffraction and radiation potentials are represented by integral formulations with support only on the wetted surface of the floating body (ies) ∂D_C, the bottom surface ∂D_Π and free surface ∂D_F; see Figure 2. In accordance with the present absorbing layer model, the free surface boundary condition is modified as follows:

$$\frac{\partial \varphi}{\partial n} - \mu \, \sigma \varphi = 0 \, , \, r \in \partial D_F, \tag{15}$$

where $\mu = \omega^2/g$ and the coefficient $\sigma = 1$ everywhere on ∂D_F, except in the absorbing layer (indicated in Figure 2), where this is given by:

$$\sigma = \left(1 + i \, \sigma_0 \frac{(R - R_a)^n}{\lambda^n}\right), \, R = \sqrt{x_1^2 + x_2^2} > R_a \tag{16}$$

where λ is the local wavelength.

Also, it is assumed for the starting radius of the absorbing layer that $R_a \gg \lambda$. The discrete solution is then obtained using collocation method, by satisfying the boundary conditions at the centroid of each panel on the various parts of the boundaries. Induced potentials and velocities from each panel to any collocation point are calculated by numerical quadratures, treating the self-induced quantities semi-analytically.

4.1. Investigation of the Optimal Parameters of the Absorbing Layer

The radiation condition expresses the weakening behavior of the outgoing waves at the far field, and it formulates the final solution. In complicated problems, where analytical or even semi-analytical solutions are unreachable, this condition cannot be formed a priori and further investigation is needed, in order to obtain its final form. One common way to overcome this obstacle, is the implementation of an absorbing layer from a specific length of activation and with defined characteristics, based on the Perfectly Matched Layer (PML) model [32,33]. In this specific approach, the wave absorbing is induced by an imaginary part of frequency, expressed by Equations (15) and (16), which operates as a damping filter for the waves without significant reflections. The formulation of the optimal PML is a multiparametric problem, mainly based on five parameters. The objective functions of this optimization procedure is the avoidance of any influence of the PML in the region before its appliance, due to reflections, and the progressive nullification of the wave field in the region after its activation. The effectiveness of the PML can be tested by comparing the numerical and the analytical solution in case of a cylindrical WEC body in steady depth regions [34]. The requirement for the PML not to disturb the solution in the computational domain before its activation is quantified with the usage of the Chebyshev Norm. Thus, the PML tunning parameters, discussed above, are:

- Dimensionless frequency $\left(\tilde{\omega} = \omega\sqrt{\frac{h}{g}} \right)$
- Coefficient $\left(\tilde{\sigma}_o = \sigma_o \frac{\lambda^n}{\omega} \right) \hat{\sigma} = \sigma\lambda^n/\omega$
- The activation length R/λ
- The exponent n
- The number of panels per wave length (N/λ)

Aiming to the minimization of the Chebyshev Norm, 64 different PMLs, corresponding to different sets of these parameters, are investigated. The final configuration of the optimum PML is described in Table 1. The efficient operation of the PML, especially in medium frequency bandwidths, where WEC devices operate most of the time and absorb the largest amount of energy, is illustrated in Figure 3, for different values of the non-dimensional frequency $\omega\sqrt{a/g}$ $\tilde{\omega} = \omega\sqrt{\frac{a}{g}}$.

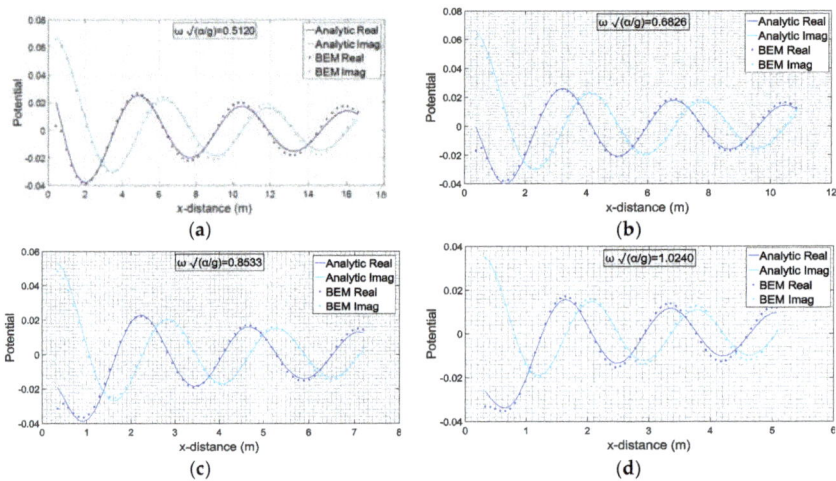

Figure 3. Comparison of present BEM results with the analytic solution in case of cylindrical WEC $a/h = 1/3.5$ and $d/a = 1.5$dr$/a = 1.5/1$, where α: radius, h: local depth and d: draft, for different values of the non-dimensional frequency $\omega\sqrt{a/g}$ $\tilde{\omega} = \omega\sqrt{\frac{a}{g}}$: (**a**) 0.5120; (**b**) 0.6826; (**c**) 0.8533; (**d**) 1.0240.

4.2. Power Output in the Case of Cylindrical WEC

A cylindrical heaving WEC is widely used in offshore installations of the devices for harnessing wave energy [35]. The numerical treatment of the wave-body interaction problem by means of BEM, described in this study, constitutes from three separate regions, namely the free surface, the body of the WEC and the bottom. Appropriate mesh generation in all these surfaces is crucial for obtaining reliable solution. For this purpose, the free surface is discretized in $4 \times (N/\lambda) \times 88$ elements, expanding for 4 wavelengths, where the first number indicates discretization along the radial direction and the other along azimuthal direction, respectively, while the bottom mesh is 26×88 elements, spatially and azimuthally respectively. The WEC mesh is 10×88 elements, in depth and in azimuthal direction, as illustrated in Figure 4. It should be noticed the demand for consistency between the lengths of the elements, those of the WEC and these of the free surface, at the matching position of the body's boundary. Very fine meshes only on the body and not on the free surface, which binds most of the computational capacity and therefore has its limitations, may cause worse approximation of the analytic solution on account of inconsistency.

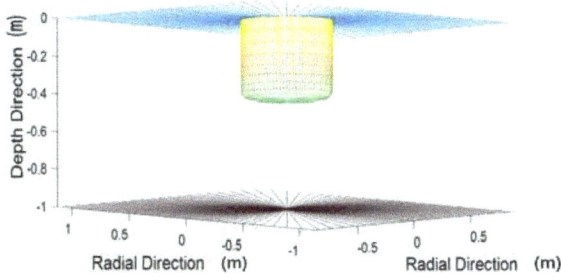

Figure 4. Illustration of the computational mesh in the near field. For clarity only the radial lines of the mesh on the free surface and on the bottom surface under the floater are shown.

Focusing on the power output coming from a single device, the first step for its calculation is the evaluation of the Froude-Krylov and total forces, which are the summation of Froude-Krylov and Diffraction forces, and the related hydrodynamic coefficients of added mass and damping. In Figures 5 and 6 are illustrated the results for these aspects, as they calculated both from the analytical and the numerical treatment of the problem ([34,36]).

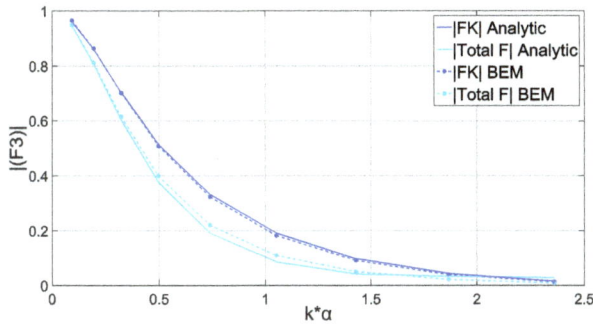

Figure 5. Non-dimensionalized cylinder hydrodynamic Froude-Krylov and total forces for various values of non-dimensionalized wavenumber (kα). Cylindrical WEC with $a/h = 1/3.5$ and $d/a = 1.5$.

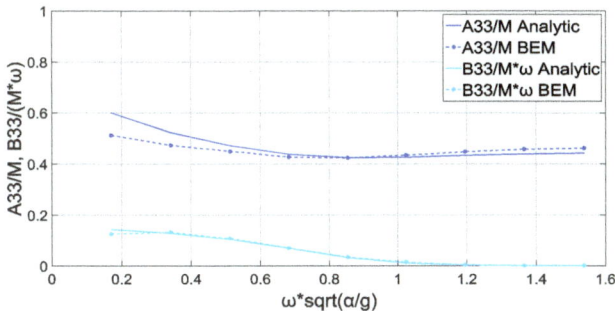

Figure 6. Non-dimensionalized cylinder hydrodynamic coefficients for various values of non-dimensionalized frequency. Cylindrical WEC with $a/h = 1/3.5$ and $d/a = 1.5$.

Furthermore, the WEC responses and the power output are evaluated and plotted in Figure 7, assuming typical PTO damping values, equal to 5, 10 and 20 times a mean value of hydrodynamic damping b_m. This value is estimated as $2\pi b_m/m\omega_R = 0.12$, where the resonance frequency is $\omega_R\sqrt{a/g} = 0.7$, also described in [22].

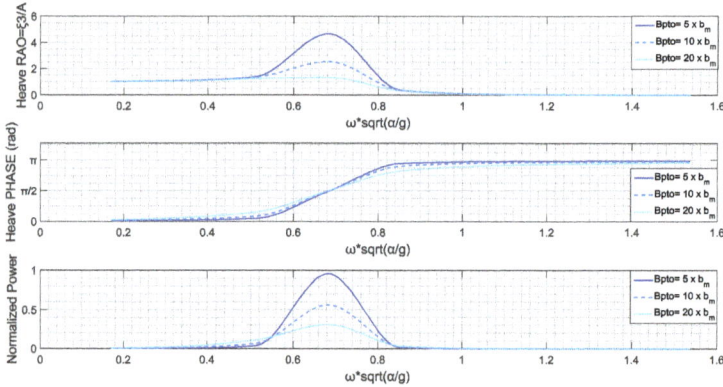

Figure 7. Heave Response amplitude operator (RAO) RAO, Phase RAO and Normalized Power Output for Cylindrical WEC with $a/h = 1/3.5$ and $d/a = 1.5$.

The output power of the WEC by this PTO is normalized with respect to the incident wave powerflux and is defined as $P/(0.5\rho C_g H^2 a)$, for $\eta_{eff} = 1$. It can be observed the fact that maximization of power output occurs at the resonance frequency. In addition, higher values of PTO damping are reasons for the observed decrease of peak values of heave Response Amplitude Operator (defined as $RAO = \xi_{k3}/(H/2)$, where a is the amplitude of the incident wave). However, at the same time they are causing wider frequency spreads of energy productive function of the device.

5. Examination of Other Shapes of Axisymmetric Floaters

Regarding the examination of other WEC shapes, eight different axisymmetric geometries, including the cylinder, are tested. This is made with the conviction of efficiency improvement, in comparison with the reference cylindrical shape. Upon mesh generation, the elements used on the bodies, except cylinder, are 18 × 88, in order to achieve a better approximation of the shape, avoiding gaps and discontinuities of the geometry. For these shapes, there are no analytic solutions, and furthermore, not any prospect for validation by comparing this numerical model with analytical results. The reference cylindrical WEC has a ratio of radius to local depth, equal to 1/3.5 and a ratio of draft to radius equal to 1.5/1. In every other design test, the radius and the draft of each geometry are calculated with the assumption of constant mass. In other words, the area of the submerged vertical cross section of the tested geometry is equal to the area of the submerged vertical cross section of the cylindrical WEC, keeping with this approach the value of the mass unchangeable.

As referred previously, eight different shapes are put under investigation. Heave response and power output are evaluated by the BEM computational code. These geometries, illustrated in Figure 8, are strongly related with the current design trends of the industry and present similarities with many already installed WEC devices [35,37–39].

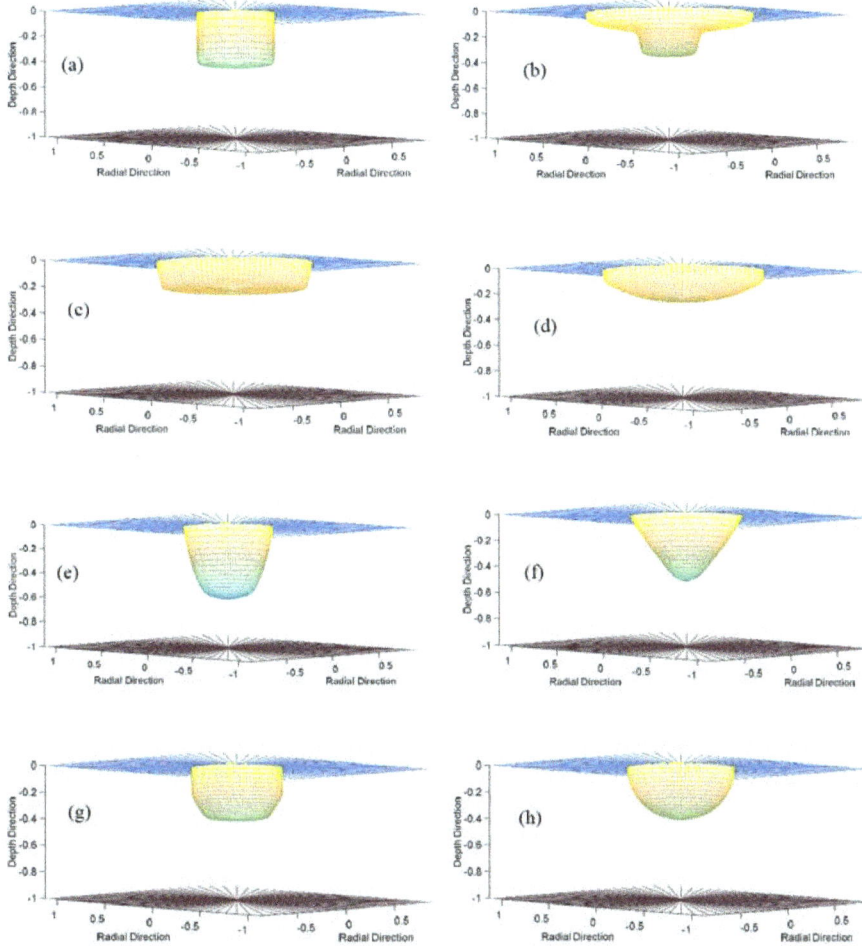

Figure 8. Various WEC shapes and near field computational mesh: (**a**) Cylindrical, (**b**) Nailhead-shaped, (**c**) Disk-shaped, (**d**) Elliptical, (**e**) Egg-shaped, (**f**) Conical, (**g**) Floater-shaped, (**h**) Semi-spherical.

Using as an efficiency index, the area under the curve of the normalized power, which expresses the maximum values so as the functional frequency bandwidth, three of the above geometries are qualified and their heave response and power output are presented in Figures 9–11. The qualified geometries are namely the nailhead-shaped, which also presents further interest due to its unconventional design for studies of multi-dof WECs, the conical and the floater-shaped.

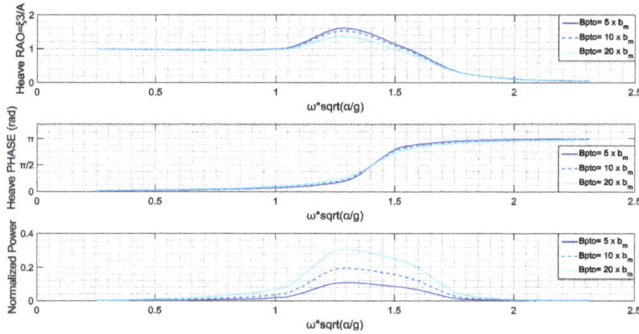

Figure 9. Heave RAO, Phase RAO and Normalized Power Output-Nailhead-shaped WEC.

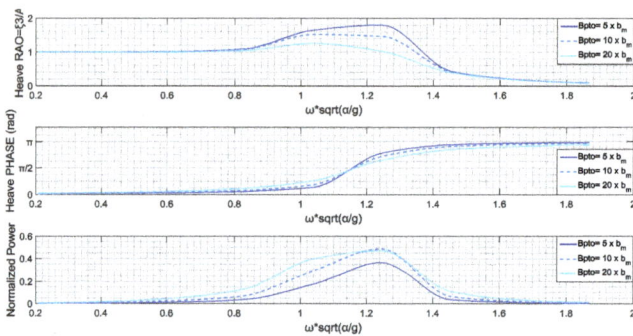

Figure 10. Heave RAO, Phase RAO and Normalized Power Output-Conical WEC.

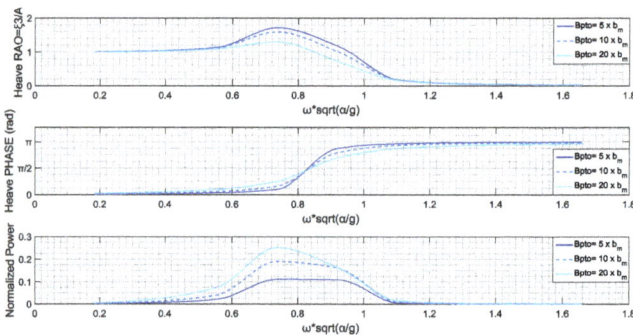

Figure 11. Heave RAO, Phase RAO and Normalized Power Output-Floater-shaped WEC.

On the assessment of these geometries, according to the figures above, despite the fact that the PTO with higher damping is responsible for lower values of heave RAO, the power output appears to be higher and with a higher frequency spreading. This dissimilar behavior of the device, in terms of heave RAO and power output, is very intense in case of the nailhead-shaped WEC, where RAOs are closely oriented, while the power output is far higher in the case of the "harder" PTO. A point of interest in the study of the conical WEC is that a switch of efficiency occurs in $\omega\sqrt{a/g} = 1.25\ \tilde{\omega} = 1.25$, when the medium ranked PTO is more efficient than higher ranked. Furthermore, the floater-shaped WEC is shown to be a little more efficient in lower frequencies, where the maximum of the power output curve is located, while the nailhead-shaped and the conical are more efficient in bandwidths of $11 < \omega\sqrt{a/g} < 1.4$. The conical WEC is far more efficient in frequencies of $0.9 < \omega\sqrt{a/g} < 1.4$, however, the final choice of the device and the PTO is depended on the sea climate and the dominant frequencies in the area of installation.

6. Extensions to Treat the WEC Arrays in Variable Bathymetry Regions

An important part of the present BEM implementation deals with the construction of the mesh on the various parts of the boundary. The details of the mesh generator are illustrated in Figure 12. More specifically, the mesh on the free surface around a single WEC is plotted. The latter consists of two subparts, the one close to the waterline of the floating unit and the far (outer) part. The near mesh is based on the cylindrical distribution of the panels around the waterline of the WEC that gradually deforms in order to end in a rectangular boundary. This permits the continuous junction of the near mesh around one floater with the adjacent one, as illustrated in Figure 12a. After the rectangular boundary, the mesh again deforms to become a cylindrical arrangement on the outer part. Taking into account that in 3D diffraction and radiation fields associated with floating bodies the far field behaves like essentially cylindrical outgoing waves [26], the cylindrical mesh in the outer part of the free surface boundary is considered to be optimum for the numerical solution of the studied problems. The discretization is accomplished by the incorporation corresponding meshes on each floating body and on the bottom variable bathymetry surface, as shown in Figure 12b–d. An important feature is the continuous junction of the various parts of the mesh, which, in conjunction with the quadrilateral elements, ensures global continuity of the geometry approximation of the boundary. It is remarked here that the present BEM is free of any kind of interior meshes or artificial intesection(s) of boundaries. Global continuity of geometry is important concerning the convergence of the numerical results in BEM.

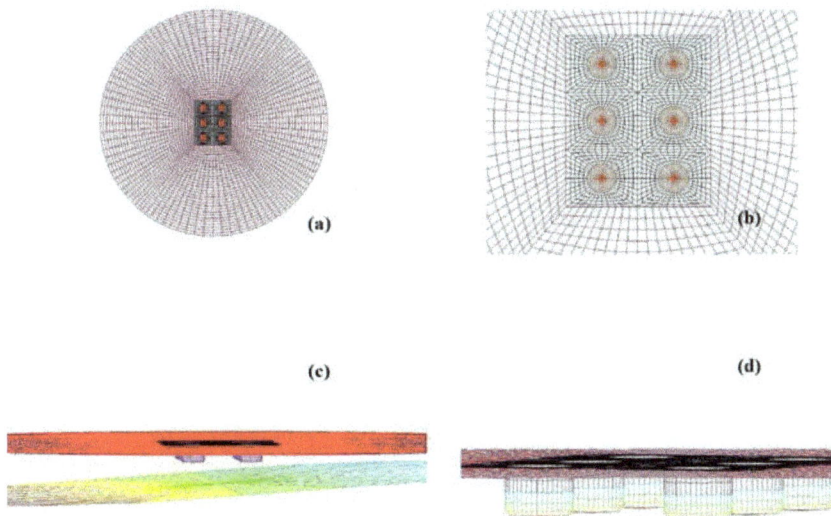

(a)

(b)

(c)

(d)

Figure 12. Computational grid for a WEC array of 3×2 WECs: (**a**) plot of the whole mesh on the free surface, (**b**) zoom in the subregion of floaters, (**c**,**d**) 3D view of the mesh in the vicinity of the WECs.

As an example, we consider the array of 3×2 cylindrical heaving WECs of radius $a = 10$ m and draft $d/a = 1.5$, arranged as illustrated in Figure 12, in the middle of the variable bathymetry region (a smooth upslope with max bottom slope 7%), and operating in waves at the same frequency as before $\omega\sqrt{h_m/g} = 1.5$, $\omega\sqrt{a/g} = 0.8$. The horizontal spacing of the floaters along the x_1 and x_2 axes is $s_1/a = 5$, $s_2/a = 4$. In this case the ratio of the WEC spacing with respect to the characteristic wavelength in the area of the array is small (less than 50%) and thus, the interaction between the floaters is strong. In order to illustrate the applicability of the present BEM, a mesh is used, consisting of 6×40 elements on each WEC and 5×40 elements on the nearest part of the free surface around each WEC and 30×100 elements on the outer part. This includes the absorbing layer, and 14×20 elements on the bottom surface (see Figure 12). The total number of elements is 5920.

The propagating field over the shoaling region, for normally and for 45deg obliquely incident waves is shown in Figure 13. This field represents the available wave energy in the domain for possible extraction. The responses of the above array of cylindrical heaving WECs are then calculated, using, as before, the values of $B_S/b_m = 5, 10, 20$ (where $2\pi b_m/m\omega_R = 0.12$) to model the Power Take Off system for heving floaters. The results calculated by the present BEM approach, at the same frequency as before $\omega\sqrt{h_m/g} = 1.5$, $\omega\sqrt{a/g} = 0.8$, both for normal and 4deg incident waves over the variable bathymetry region are represented in Figure 14. In the specific arrangement and operating conditions the q-factor decreases with increasing PTO damping and ranges from 90–70%, for normal incident waves, and drops down to 75–60%, for 45 degrees obliquely incident waves.

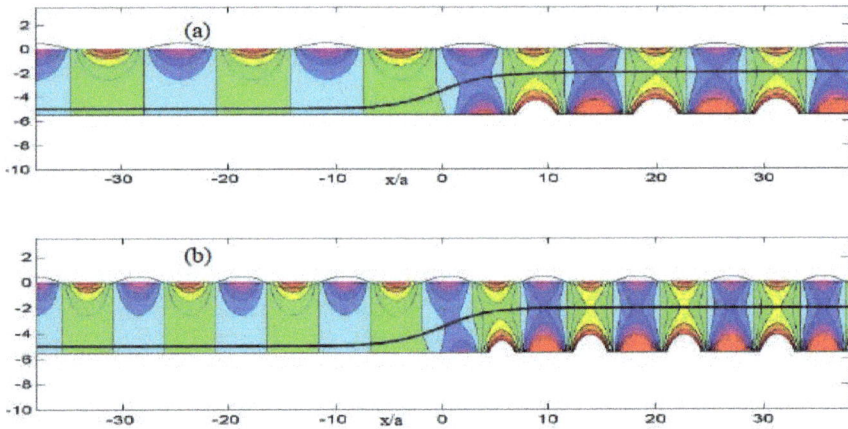

Figure 13. (a) Propagating field over a shoaling region, for normally incident waves of nondimensional frequency $\omega\sqrt{h/g} = 1.5$. (b) Same as before, but for 45 degrees obliquely incident waves.

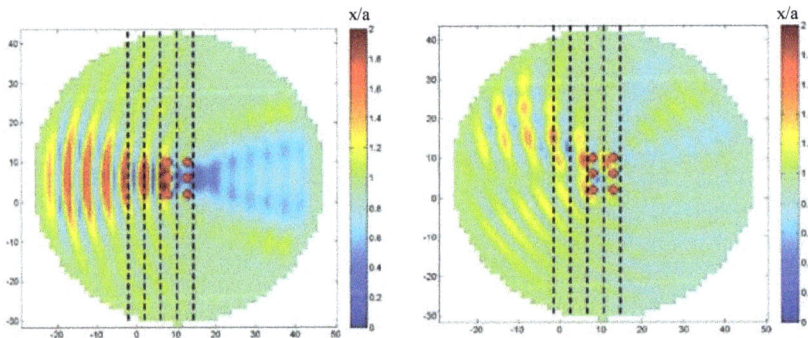

Figure 14. Total field of the WEC array for (a) normal and (b) 45 degrees incidence (right) and $\omega\sqrt{h_m/g} = 1.5$, $\omega\sqrt{a/g} = 0.8$, in the variable bathymetry region (dashed lines represent depth contours). The colorbar indicates relative intensity of the wavefield.

7. Conclusions

In this work a numerical method is presented for the hydrodynamic analysis of the floating bodies over general seabed topography supporting the calculation of the wave power absorption by single WECs and the performance of arrays of devices in nearshore and coastal regions. As a first step, in order to subsequently formulate and solve 3D diffraction and radiation problems for floaters in the inhomogeneous domain, the present approach is based on the coupled-mode model for the calculation of the wave field propagating over the variable bathymetry region. The results are subsequently used for the hydrodynamic analysis of floating bodies over general bottom topography by means of a low-order BEM. Extensions of the present method supporting the estimation of single WEC and WEC array performance in variable bathymetry regions have been discussed. Future work will be focused on the validation of the present method by comparisons with other methods and experimental laboratory data. Moreover, phase-averaged models like SWAN can treat macroscopically WEC-array effects by including energy pumping in the energy balance equations by using sinks with specific intensity; see, e.g., [40]. Future work will examine the possibility of coupling phase-averaged with the present phase-resolving model, by using coupling schemes as the ones presented in [41,42] and present comparisons at the scale of an array in a nearshore region.

Author Contributions: The main ideas of this work as well as the first draft of the text belong to Kostas Belibassakis. Markos Bonavos participated in performing the numerical simulations and writing some parts of the text. Eugen Rusu suggested improvements and performed several corrections and verification of the text and of the Figures. He was also the corresponding author and his work is included in a Romanian national research project as stated in Acknowledgments. All authors agreed with the final form of this article.

Funding: This research received no external funding.

Acknowledgments: The work of the corresponding author was carried out in the framework of the research project REMARC (Renewable Energy extraction in MARine environment and its Coastal impact), supported by the Romanian Executive Agency for Higher Education, Research, Development and Innovation Funding – UEFISCDI, grant number PN-III-P4-IDPCE-2016-0017.

Conflicts of Interest: The authors declare no conflict of interest.

Abbreviations

BEM   Boundary Element Method
CMS   Coupled-Mode System
RAO   Response amplitude operator
PML   Perfectly Matched Layer
PTO   Power Take off
SWAN   Simulating Waves Nearshore
WEC   Wave Energy Converters

References

1. Mei, C.C. *The Applied Dynamics of Ocean Surface Waves*, 2nd ed.; World Scientific: Singapore, 1996.
2. Sawaragi, T. *Coastal Engineering-Waves, Beaches, Wave-Structure Interactions*, 1st ed.; Elsevier: Amsterdam, The Netherlands, 1995.
3. Williams, K.J. The boundary integral equation for the solution of wave–obstacle interaction. *Int. J. Numer. Method. Fluids* **1988**, *8*, 227–242. [CrossRef]
4. Drimer, N.; Agnon, Y.; Stiassnie, M. A simplified analytical model for a floating breakwater in water of finite depth. *Appl. Ocean Res.* **1992**, *14*, 33–41. [CrossRef]
5. Williams, A.N.; Lee, H.S.; Huang, Z. Floating pontoon breakwaters. *Ocean Eng.* **2000**, *27*, 221–240. [CrossRef]
6. Drobyshevski, Y. Hydrodynamic coefficients of a two-dimensional, truncated rectangular floating structure in shallow water. *Ocean Eng.* **2004**, *31*, 305–341. [CrossRef]
7. Drobyshevski, Y. An efficient method for hydrodynamic analysis of a floating vertical-sided structure in shallow water. In Proceedings of the 25th International Conference on Offshore Mechanics and Arctic Engineering (OMAE2006), Hamburg, Germany, 4–9 June 2006.

8. Wehausen, J.V. The motion of floating bodies. *Ann. Rev. Fluid. Mech.* **1971**, *3*, 237–268. [CrossRef]
9. Wehausen, J.V. Methods for Boundary-Value Problems in Free-Surface Flows: The Third David W. Taylor Lecture, 27 August through 19 September 1974. Available online: https://www.researchgate.net/publication/235068177_Methods_for_Boundary-Value_Problems_in_Free-Surface_Flows_The_Third_David_W_Taylor_Lecture_27_August_through_19_September_1974 (accessed on 15 July 2018).
10. Charrayre, F.; Peyrard, C.; Benoit, M.; Babarit, A. A coupled methodology for wave body interactions at the scale of a farm of wave energy converters including irregular bathymetry. In Proceedings of the 33th International Conference on Offshore Mechanics and Arctic Engineering (OMAE2014), San Francisco, CA, USA, 8–13 June 2014.
11. McCallum, P.; Venugopal, V.; Forehand, D.; Sykes, R. On the performance of an array of floating energy converters for different water depths. In Proceedings of the 33th International Conference on Offshore Mechanics and Arctic Engineering (OMAE2014), San Francisco, CA, USA, 8–13 June 2014.
12. Guazzelli, E.; Rey, V.; Belzons, M. Higher-order bragg reflection of gravity surface waves by periodic beds. *J. Fluid Mech.* **1992**, *245*, 301–317. [CrossRef]
13. Massel, S. Extended refraction-diffraction equations for surface waves. *Coast. Eng.* **1993**, *19*, 97–126. [CrossRef]
14. Rey, V. A note on the scattering of obliquely incident surface gravity waves by cylindrical obstacles in waters of finite depth. *Eur. J. Mech. B/Fluid* **1995**, *14*, 207–216.
15. Belibassakis, K.A. A Boundary Element Method for the hydrodynamic analysis of floating bodies in general bathymetry regions. *Eng. Anal. Bound. Elem.* **2008**, *32*, 796–810. [CrossRef]
16. Touboul, J.; Rey, V. Bottom pressure distribution due to wave scattering near a submerged obstacle. *J. Fluid Mech.* **2012**, *702*, 444–459. [CrossRef]
17. Athanassoulis, G.A.; Belibassakis, K.A. A consistent coupled-mode theory for the propagation of small-amplitude water waves over variable bathymetry regions. *J. Fluid Mech.* **1999**, *389*, 275–301. [CrossRef]
18. Belibassakis, K.A.; Athanassoulis, G.A.; Gerostathis, T. A coupled-mode system for the refraction-diffraction of linear waves over steep three dimensional topography. *Appl. Ocean. Res.* **2001**, *23*, 319–336. [CrossRef]
19. Magne, R.; Belibassakis, K.A.; Herbers, T.; Ardhuin, F.; O'Reilly, W.C.; Rey, V. Evolution of surface gravity waves over a submarine canyon. *J. Geogr. Res.* **2007**, *112*, C01002. [CrossRef]
20. Gerostathis, T.; Belibassakis, K.A.; Athanassoulis, G.A. A coupled-mode model for the transformation of wave spectrum over steep 3D topography: Parallel-Architecture Implementation. *J. Offshore Mech. Arct. Eng.* **2008**, *130*. [CrossRef]
21. Belibassakis, K.A.; Athanassoulis, G.A. A coupled-mode system with application to nonlinear water waves propagating in finite water depth and in variable bathymetry regions. *Coast. Eng.* **2011**, *58*, 337–350. [CrossRef]
22. Belibassakis, K.A.; Athanassoulis, G.A.; Gerostathis, T. Directional wave spectrum transformation in the presence of strong depth and current inhomogeneities by means of coupled-mode model. *Ocean Eng.* **2014**, *87*, 84–96. [CrossRef]
23. Vincent, C.L.; Briggs, M.J. Refraction–diffraction of irregular waves over a mound. *J. Waterw. Port Coast. Ocean Eng.* **1989**, *115*, 269–284. [CrossRef]
24. Booij, N.; Ris, R.C.; Holthuijsen, L.H. A third-generation wave model for coastal regions, 1. Model description and validation. *J. Geophys. Res.* **1999**, *104*, 7649–7666. [CrossRef]
25. Ris, R.C.; Holthuijsen, L.H.; Booij, N. A Third-Generation Wave Model for Coastal Regions. Verification. *J. Geophys. Res.* **1999**, *104*, 7667–7681. [CrossRef]
26. Belibassakis, K.A.; Athanassoulis, G.A. Three-dimensional Green's function of harmonic water-waves over a bottom topography with different depths at infinity. *J. Fluid Mech.* **2004**, *510*, 267–302. [CrossRef]
27. Massel, S.R. *Ocean Surface Waves. Their Physics and Prediction*; World Scientific: Singapore, 2013.
28. Chamberlain, P.G.; Porter, D. The modified mild-slope equation. *J. Fluid Mech.* **1995**, *291*, 393–407. [CrossRef]
29. Beer, G.; Smith, I.; Duenser, C. *The Boundary Element Method with Programming for Engineers and Scientists*; Springer: New York, NY, USA, 2008; ISBN 978-3-211-71576-5.
30. Teng, B.; Gou, Y. BEM for wave interaction with structures and low storage accelerated methods for large scale computation. *J. Hydrodyn.* **2017**, *29*, 748–762. [CrossRef]
31. Sclavounos, P.; Borgen, H. Seakeeping analysis of a high- speed monohull with a motion control bow hydrofoil. *J. Ship Res.* **2004**, *48*, 77–117.

32. Berenger, J.P. A perfectly matched layer for the absorption of electromagnetic waves. *J. Comput. Phys.* **1994**, *114*, 185–200. [CrossRef]

33. Turkel, E.; Yefet, A. Absorbing PML boundary layers for wave-like equations. *Appl. Numer. Math.* **1998**, *27*, 533–557. [CrossRef]

34. Sabuncu, T.; Calisal, S. Hydrodynamic coefficients for vertical circular cylinders at finite depth. *Ocean Eng.* **1981**, *8*, 25–63. [CrossRef]

35. Falcão, A.F. Wave energy utilization. A review of the technologies. *Renew. Sustain. Energy Rev.* **2010**, *14*, 899–918. [CrossRef]

36. Filippas, E.S.; Belibassakis, K.A. Hydrodynamic analysis of flapping-foil thrusters operating beneath the free surface and in waves. *Eng. Anal. Bound. Elem.* **2014**, *41*, 47–59. [CrossRef]

37. Drew, B.; Plummer, A.R.; Sahinkaya, M.N. A review of wave energy converter technology. *Proc. Inst. Mech. Eng. Part A J. Power Energy* **2009**, *223*, 887–902. [CrossRef]

38. Faizal, M.; Ahmed, M.R.; Lee, Y. A Design Outline for Floating Point Absorber Wave Energy Converters. *Adv. Mech. Eng.* **2014**, *6*. [CrossRef]

39. Rusu, E.; Onea, F. A review of the technologies for wave energy extraction. *Clean. Energy* **2018**, *2*. [CrossRef]

40. Ruehl, K.; Porter, A.; Chartrand, C.; Smith, H.; Chang, G.; Roberts, J. "Development, Verification and Application of the SNL-SWAN Open Source Wave Farm Code". In Proceedings of the 11th European Wave and Tidal Energy Conference EWTEC, Nantes, France, 6–11 September 2015.

41. Athanassoulis, G.A.; Belibassakis, K.A.; Gerostathis, T. The POSEIDON nearshore wave model and its application to the prediction of the wave conditions in the nearshore/coastal region of the Greek Seas. *Glob. Atmos. Ocean Syst. (GAOS)* **2002**, *8*, 101–117. [CrossRef]

42. Rusu, E.; Guedes Soares, C. Modeling waves in open coastal areas and harbors with phase resolving and phase averaged models. *J. Coast. Res.* **2013**, *29*, 1309–1325. [CrossRef]

energies

MDPI

Article

On the Accuracy of Three-Dimensional Actuator Disc Approach in Modelling a Large-Scale Tidal Turbine in a Simple Channel

Anas Rahman [1], Vengatesan Venugopal [2,*] and Jerome Thiebot [3]

[1] Mechanical Engineering Programme, School of Mechatronic Engineering, University Malaysia Perlis, Pauh Putra Campus, Perlis 02600, Malaysia; anasrahman@unimap.edu.my
[2] Institute for Energy Systems, School of Engineering, The University of Edinburgh, Edinburgh EH9 3DW, UK
[3] Laboratoire Universitaire des Sciences Appliquées de Cherbourg (LUSAC), Normandie Univ, UNICAEN, LUSAC, 14000 Caen, France; jerome.thiebot@unicaen.fr
* Correspondence: V.Venugopal@ed.ac.uk; Tel.: +44-(0)131-650-5652

Received: 17 July 2018; Accepted: 10 August 2018; Published: 17 August 2018

Abstract: To date, only a few studies have examined the execution of the actuator disc approximation for a full-size turbine. Small-scale models have fewer constraints than large-scale models because the range of time-scale and length-scale is narrower. Hence, this article presents the methodology in implementing the actuator disc approach via the Reynolds-Averaged Navier-Stokes (RANS) momentum source term for a 20-m diameter turbine in an idealised channel. A structured grid, which varied from 0.5 m to 4 m across rotor diameter and width was used at the turbine location to allow for better representation of the disc. The model was tuned to match known coefficient of thrust and operational profiles for a set of validation cases based on published experimental data. Predictions of velocity deficit and turbulent intensity became almost independent of the grid density beyond 11 diameters downstream of the disc. However, in several instances the finer meshes showed larger errors than coarser meshes when compared to the measurements data. This observation was attributed to the way nodes were distributed across the disc swept area. The results demonstrate that the accuracy of the actuator disc was highly influenced by the vertical resolutions, as well as the grid density of the disc enclosure.

Keywords: tidal energy; actuator disc; turbulence; wake analysis; Telemac3D

1. Introduction

Flow perturbation due to the deployment of tidal current devices has been extensively studied and discussed in the recent past, as it is expected to have an influence on the power capture and may also alter the physical environment [1–3]. The analytical and computational studies are often validated with small-scale experiments before using them for large-scale implementations. In the current literature, most of the three-dimensional (3D) numerical study of wake characteristics have been executed using computational fluid dynamics (CFDs) models. In these models, a tidal device is represented either as a complete structure with blades, or as an actuator disc. Often, the output from these numerical models were compared with results from experiments conducted in a flume, where porous discs are commonly employed to simulate the effects of a turbine on a fluid flow. Despite the assumption that the actuator disc approach may not accurately produce the vortices from the rotating blades, the concept seems to be able to accurately compute the wake decay as well as the turbulence intensity [4,5]. Although the actuator disc approximation has been widely used in predicting the performance of tidal stream devices, its implementation so far has been restricted to studies involving an extremely small-scale actuator disc (e.g., rotor diameter of 0.1 m). The drawback of a small-scale turbine model includes overestimation of essential parameters such as the mesh density and also the resolution of the vertical layers, making them impractical to be replicated in a large-scale model. As the application of the actuator disc approximation for a full-size turbine is yet to be tested, this work made an attempt to model a full-scale rotor by the actuator disc method within an ocean scale numerical flow model.

Contrary to fully meshed rotating turbines, the simplicity of the actuator disk concept permits to use it for ocean scale modelling [6]. This method does not demand detailed discretization of the turbine structure as required for high fidelity simulations and does not need the use of computer cluster to run. Several recent studies have utilized the actuator disc method in investigating various flow characteristics in the presence of tidal devices. Sun et al. [4] used the commercial CFD software package ANSYS FLUENT to simulate tidal energy extraction for both two and three-dimensional models by applying a retarding force on the flow. This study found that free surface variations may have an influence on the wake and turbine performance. Daly et al. [7] examined the methods of defining the inflow velocity boundary condition using ANSYS CFX, and showed that the 1/8th power law profile was superior than others in replicating the wake region. The same software was also used by Harrison et al. [5] in comparing the wake characteristics of the Reynolds-Averaged Navier-Stokes (RANS) model against the experimental data measured behind a disc with various porosities, where a detailed methodology on the implementation of the momentum sink was presented. In addition, the actuator disc approach was also used by Lartiga and Crawford [8] in correcting the wall interference for their tunnel testing facilities, with the aid of ANSYS CFX. Roc et al. [9] on the other hand proposed an adaption of the actuator disc method by accounting appropriate turbulence correction terms to improve near wake performance for the ROMS (Regional Ocean Modelling System) model. A more recent study by Nguyen et al. [10] also demonstrated the needs to adapt the turbulence models when modelling a tidal turbine with an actuator disc to account for the near wake losses due to unsteady flow downstream of the disc.

With the exception of Reference [9], all of these CFD studies were conducted on a small-scale domain, where a 0.1 m diameter disc was commonly employed. As the implementation of the momentum sink for a full-scale turbine has not been undertaken in the past, the purpose of this study is to demonstrate that the simulation of wake effects from a full-scale actuator disc using a tidal flow model is possible. The experimental data published in Reference [5] is used for validation and comparison purposes. In contrast with other studies where the model dimensions matched the size of the flume experiment, here we compare the results of the scaled experiment to the full-scale model output. Further, the model-experiment comparison is done by using dimensionless variables. This paper presents detailed sensitivity analysis conducted on the disc enclosure and their influence on the wake profiles behind the turbine. The actuator disc is implemented in the open source software-Telemac3D [11] where the effects of a 20-m diameter turbine is modelled and validated with data from literature. It is aspired that the knowledge from this study can be of use in applying the actuator disc for realistic ocean scale simulations.

2. Background

2.1. Description of Telemac3D

Telemac3D is a finite element model that solves the Navier–Stokes equations with a free surface, along with the advection-diffusion equations of salinity, temperature, and other parameters, and has been widely used for regional scale modelling. The numerical scheme also comprises the wind stress, heat exchange with the atmosphere, density, and the Coriolis effects. The 3D flow simulation (with hydrostatic assumption) is calculated by solving the following equations [11]:

$$\frac{\partial U}{\partial x} + \frac{\partial V}{\partial y} + \frac{\partial W}{\partial z} = 0 \tag{1}$$

$$\frac{\partial U}{\partial t} + U\frac{\partial U}{\partial x} + V\frac{\partial U}{\partial y} + W\frac{\partial U}{\partial z} = -g\frac{\partial Z_s}{\partial x} + v\Delta(U) + F_x \tag{2}$$

$$\frac{\partial V}{\partial t} + U\frac{\partial V}{\partial x} + V\frac{\partial V}{\partial y} + W\frac{\partial V}{\partial z} = -g\frac{\partial Z_s}{\partial y} + v\Delta(V) + F_y \tag{3}$$

where U, V, and W are the three-dimensional components of the velocity, v is velocity and tracer diffusion coefficient, F_x and F_y are the source terms of the process being modelled (e.g., sediment, wind, the Coriolis force etc.), Z_s is the bottom depth, and g is the acceleration due to the gravity. As in most regional scale models, Telemac3D offers the choice of using either the hydrostatic or the non-hydrostatic pressure code. The hydrostatic assumption implies that the vertical velocity (W) can be derived using only the mass-conservation of momentum equation, without directly solving the vertical momentum equation. Since this assumption ignores the diffusion and advection term, it is then not possible to simulate any vertical rotational motion. Elaboration on theoretical aspects of Telemac3D can be referred to these articles [11–13].

2.2. Theory of the RANS Actuator Disc

Actuator disc model can be implemented using RANS equations by inserting a momentum sink term in the region where the turbine is to be located. It then works by mimicking the effects a turbine would have on the surrounding regions without the need to implement detailed features of a turbine. This method is adapted from the wind energy industry [14] where it has been widely used to model wind turbines. The implementation of the RANS actuator disc approach on several distinct numerical models have been elaborated and discussed in References [4,5,15], where the approximated forces exerted by the disc to the surrounding flow are applied as source terms in the RANS equations of momentum Equation (4) and mass conservation Equation (5).

$$\frac{\delta(\rho U_i)}{\delta t} + \frac{\delta(\rho U_i U_j)}{\delta x_j} = -\frac{\delta P}{\delta x_i} + \frac{\delta}{\delta x_j}\left[\mu\left(\frac{\delta U_i}{\delta x_j} + \frac{\delta U_j}{\delta x_i}\right)\right] + \frac{\delta}{\delta x_j}\left(-\rho\overline{u_i' u_j'}\right) + \rho g_i + S_i \tag{4}$$

$$\frac{\delta U_i}{\delta x_i} = 0 \tag{5}$$

where U_i ($i = u, v, w$) is the fluid's velocity component averaged over time t, P is the mean pressure, ρ is the fluid density, μ is the dynamic viscosity, u' is an instantaneous velocity fluctuation in time during the time step δt, x_i ($i = x, y, z$) is the spatial geometrical scale, $-\rho\overline{u_i' u_j'}$ is the Reynolds stresses that must be solved using turbulence model, g_i is the component of the gravitational acceleration, and S_i is an added source term for the ith (where $i = x$, y or z) momentum equations. In the present paper, the k-ε turbulence model is utilized to close the RANS equations and solve for the Reynolds stresses.

The momentum source term is imposed to the turbine location by discretizing the RANS equations using a finite volume approach [5,15]. The standard RANS momentum equations will apply to the overall flow domain, while the additional source term, S_i is added using Equation (6) at the specified disc location, as shown in Figure 1:

$$S_i = \frac{1}{2}\rho\frac{K}{\Delta x_t}U_i|U_i| \tag{6}$$

where Δx_t is the thickness of the disc and K is the resistance coefficient. Moreover, the actuator disc concept implies that the turbine is represented by applying a constant resistance to the incoming flow, which causes a thrust to act on the disc. In theory, this thrust should be close to the one acting on the turbine being simulated. The relationship between the resistance coefficient, K, thrust coefficient, C_T, open area ratio, θ, and induction factor, a have been discussed in [16–18] where:

$$C_T = \frac{K}{(1 + 0.25K)^2} = 4a(1 - a) \tag{7}$$

$$\theta^2 = \frac{1}{(1+K)} \tag{8}$$

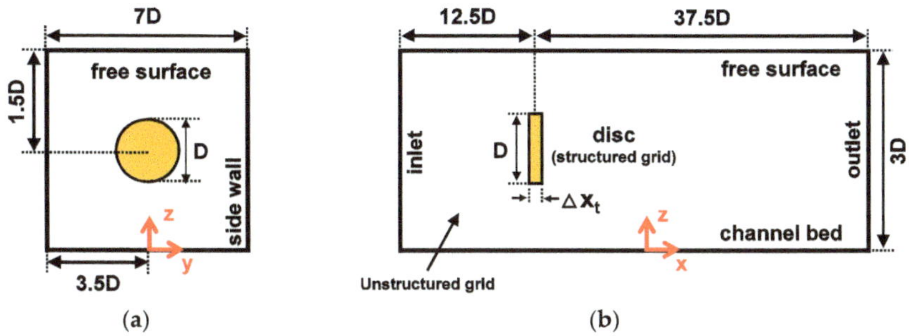

Figure 1. Schematic diagram showing views from the inlet (**a**) and channel side (**b**). The front view displays the swept area of the turbine of diameter, D = 20 m.

2.3. Limitation of the Actuator Disc Approach

Although the RANS actuator disc concept has been successfully employed as a means to imitating a tidal energy converter, the method is not without flaws. Some of the limiting aspects of this approach have been highlighted by References [5,6,19–21], and summarized below:

- The overall turbine structure is not being represented and thus affecting the turbulence in the near wake region, known to be 2–5 rotor diameters downstream of the turbines.
- Kinematics of turbulence, such as vortices trailing from the edges of a blade cannot be replicated, and thus, they must be properly parameterized.
- Energy extraction due to mechanical motion of the turbine rotor cannot be reproduced, instead the energy removed from the disc will be converted into small scale turbulence eddies behind the disc. However, the influence of swirl on the far region is assumed to be minimal.
- This concept cannot be used to investigate the performance of a turbine (e.g., maximum power produced) since it does not include the blades.

Since most swirls and vortices components of a real tidal device would have dissipated beyond the near wake, the actuator disc should exhibit similar flow characteristics in the far wake region as the device, which reflects the principal assumption of the actuator disc approach. Furthermore, since RANS simulations only show the mean flow characteristic, this method is ill suited for application that requires detail of the flow behind the disc as it cannot account for the physical phenomena caused by the rotating blades. However, for studies that focus on a simplified model to explore the interaction between turbines in arrays, the actuator disc model is favoured since it is can reproduce the wake mixing which generally occurs in the far wake region. Previous studies conducted by References [17,18] have verified that turbulence due to the blades has negligible influence on the flow far downstream.

2.4. Benchmarking and Data Validation

In order to validate the models produced in this study, a comparative study has been conducted against data from the physical scale setup published by Harrison et al. [5]. Additional details of the experimental setup are presented in Myers and Bahaj [19]. The experimental work by Reference [19] focused on examining the wake structure and its recovery in the downstream region by using a scale mesh disc rotor. Furthermore, the flume setup was designed to respect both the Froude and Reynolds number, as well as to exhibit fully turbulent flow. Similar experimental setup was also used by Harrison et al. [5] to validate their simulations using ANSYS CFX. Their numerical model utilised the RANS solver to analyse the characteristic of the wake of an actuator disc model, which was used as the principal reference in this study. Highlights of the experimental and numerical scale study from Reference [5] are as follows:

- Flume dimensions are: 21 m long, 1.35 m wide, and tank depth of 0.3 m.
- Perforated disc with diameter of 0.1 m was used, where the porosity ranged from 0.48 m to 0.35 m (corresponding to coefficient of thrust, $C_T = 0.61$ to 0.97).
- The flow speed was approximately set to 0.3 m/s, with mean U velocity component of 0.25 m/s.
- The vertical velocity profile was developed to closely match the 1/7th power law, with uniform velocity near the open surface.
- An acoustic Doppler velocimeter was employed to measure downstream fluid velocities, starting from 3 to 20-disc diameters in a longitudinal direction, as well as up to a 4-disc diameter in the lateral axis.

3. Methodology

3.1. Actuator Disc Representation in Telemac3D

Apart from hydrodynamic simulations of flows in three-dimensional space, Telemac3D software also allow for the user to program specific functions that are beyond the code's standard structure. Every installation comes with a comprehensive library of programming subroutines for executing additional processes, in which the user can modify to suit the objectives of any particular simulations. Table 1 summarises the adopted and modified subroutines used in this study. In Telemac3D, the momentum source term is implemented into the model by modifying and activating the "HYDROLIENNE" keyword in the TRISOU subroutine. Since the actuator disc employs the same geometry as the swept area of the turbine and requires a reduction in momentum of the passing fluid, it is crucial that the calculation of the forces is appropriately appended into the discretised RANS equations. The principal methodology for executing the approach in Telemac3D is summarised as follows:

(a) The turbine arrangement and the overall dimensions of the domain used in this study is presented in Figure 1, where the disc was located 250 m from the channel inlet. Additionally, its z and y axis centreline were fixed at a 30-m mid-depth and 70-m from the side wall.

(b) The use of a structured grid at the turbine position was chosen as it would allow for a better representation of the turbine shape, as well as maintaining the distance between nodes for refinement purposes. Figure 2 provides the graphical information on the dimensions and pertinent parameters concerning the implementation of the structured grid in the domain. The size of the structured grid (i.e., $lx = 26$ m, $ly = 40$ m and $lz = 60$ m) were deliberately set to be larger than the turbine diameter ($D = 20$ m) and its width ($\Delta x_t = 2$ m) to allow for numerical tolerance upon the execution of the momentum sink in the TRISOU subroutine. The grid element spacing within this structured grid are denoted by Δx, Δy and Δz in the x, y, and z directions, respectively. For the simulations, lx, ly, and lz were kept constant, but the dimensions of Δx, Δy, and Δz were varied and their impact on the wake characteristics was investigated and the results are presented in Section 4.

(c) The location of the disc (i.e., the turbine) in the domain was specified by four nodes in the horizontal plane (see Figure 2a), denoted as a, b, c, and d, which will act as the enclosure for the turbine. The coordinate of each node was represented by a pair of x and y. The distance between $y(4)$ and $y(1)$ refers to the turbine diameter, D, while the distance between $x(1)$ and $x(2)$ corresponds to the disc thickness, Δx_t.

(d) Although several mesh transformation options are available in the Telemac3D module, the sigma coordinate system was chosen to represent the depth due to its simplicity, as shown in Figure 3a. In fact, the interval between the vertical planes, Δz as well as the mesh density in the y direction,

Δy must be carefully selected since the intersections between the z and y axis nodes will determine the accuracy of disc frame. Coarser mesh density in both the y and z axis will result in a limited number of nodes available within the disc surface area, as shown in Figure 3b. Whereas, Figure 3c portrays unbalanced concentration of the nodes when one of the axis uses a very fine grid resolution compare to the other. Section 4.3.2 will elaborate further on this subject matter.

(e) Once the optimal resolution for both Δz and Δy was established, the momentum source term (see Equation (6)) was applied into the model through the existing nodes within the 10-m radius from the disc centre, Figure 3b. For this purpose, Equation (9) was employed to locate all the relevant nodes that formulate the disc's 20-m frame.

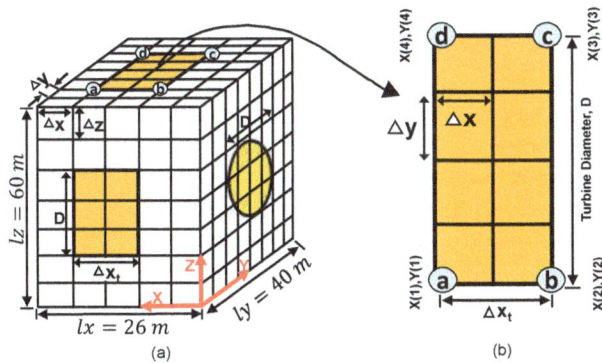

Figure 2. Graphical information of the implementation of the actuator disc approach on the structured grid. Δx, Δy, and Δz are the grids spacing used at the turbine location in x, y and z, directions respectively: (**a**) dimensions of the embedded structured grid in the domain; (**b**) x–y plane displaying the "enclosure" of the disc, where the momentum sink is applied within the specified quadrants (i.e., a, b, c, and d). Note that these figures are not drawn to scale.

Sigma Layer	Depth (meter)
24	0
23	2.61
22	5.22
21	7.83
20	10.44
19	13.05
18	15.66
17	18.27
16	20.88
15	23.49
14	26.1
13	28.71
12	31.32
11	33.93
10	36.54
9	39.15
8	41.76
7	44.37
6	46.98
5	49.59
4	52.2
3	54.81
2	57.42
1	60.00

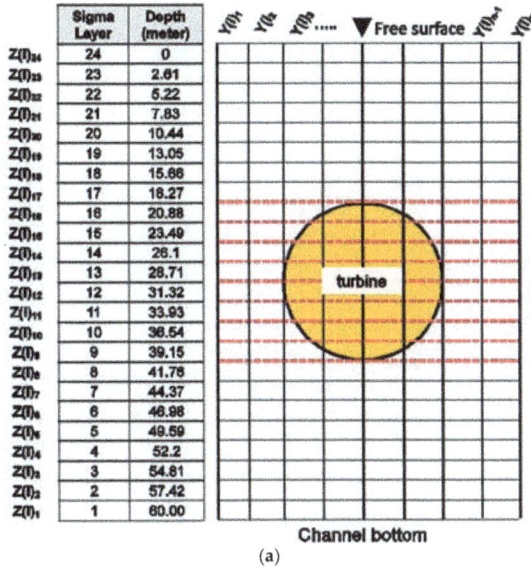

Channel bottom

(a)

Figure 3. *Cont.*

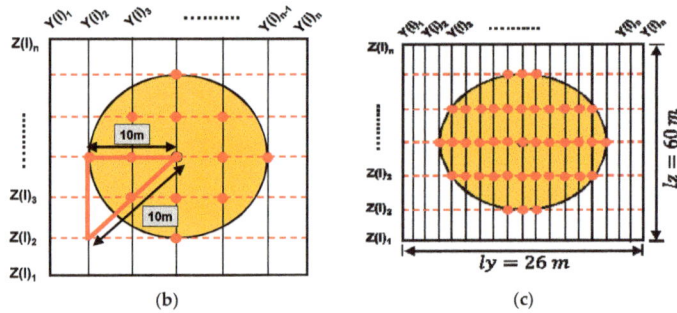

Figure 3. Graphical information exhibiting the influence of the vertical resolutions (24 layers) on the model. Sigma layers and their corresponding depth are presented in drawing (**a**), where the respective planes that cross the disc's surface area are clearly shown. Detailed illustration on the comparison of the structured grid density nodes at the turbine swept area (*y–z* axis) are respectively given in the (**b**) structured grid with minimal resolution and in the (**c**) structured grid with higher density. $Y(I)_i$ refers to the nodes in the *y* orientation of the structured grid, while $Z(I)_i$ corresponds to the *i*th vertical planes imposed on the model. Note that these figures are not drawn to scale.

$$\text{Turbine radius, } r = \sqrt{(Y_i - Y_c)^2 + (Z(I)_i - Z_c)^2} \tag{9}$$

$$Y_c = \frac{y(4) - y(1)}{2} \tag{10}$$

where Y_i refers to the node in *y* orientation within the turbine enclosure, Y_c is the turbine centreline in the *y* direction (70 m), $Z(I)_i$ is the vertical plane in *z* direction where *i* = 24 sigma layers, and Z_c is the depth where the turbine centre is located (30 m).

Table 1. List of Telemac3D subroutines used in this study and their corresponding functions.

Telemac3D Subroutines	Function
CONDIM	To set the initial condition for the model's depth and velocity profile
VEL_PROF_Z	To specify the vertical velocity profile at the channel inlet
KEPCL3 and KEPINI	To be used with k-epsilon turbulence model
TRISOU	To implement the source terms for the momentum equations

3.2. Model Set-Up

For a realistic modelling condition of a full-size turbine, the authors refer to the report by Legrand [22] to get an idea of the generic characteristic of a tidal energy device. A 20-m diameter disc was used in this study since it is a reasonable size for a standard horizontal axis turbine in operation. Moreover, the deployment of devices at any particular location is site and depth dependent, where a depth between 40 m to 80 m are often quoted. For this numerical study, the flow at the free surface and the channel depth was set to 3 m/s and 60 m respectively to resemble the general condition of the Pentland Firth area [23]. The disc *z*-axis centreline is positioned at the mid depth (30 m) to allow for a sufficient bottom clearance to minimize turbulence and shear loading from the bottom boundary layer. Also, the actuator disc is placed 250 m from the channel inlet to ensure that the flow is fully developed upon reaching the turbine area.

The resistance coefficient, *K* is one of the most important variables in simulating the actuator disc since it characterises the thrust exerted by the turbine on the flow. Changing the value of *K* will invariably influence the wake downstream of the disc. The relationship between C_T and *K* has been shown previously using Equation (7). In this paper, *K* is set to a constant value of 2, which correspond to $C_T = 0.89$.

As mentioned previously, structured grid was imposed at the location of the turbine to give an accurate illustration of the disc shape. Conversely, unstructured mesh was used for the surrounding area with a maximum edge length of 10 m, while the edge growth ratio was set to 1.1 for smoother mesh distribution. Figure 4 displays the unstructured mesh used in the computational domain, and it also shows the location of structured grid where the turbine is housed. Note that unstructured mesh was applied for the outer region since it is more flexible in capturing complex coastline features for realistic ocean scale simulations. The number of mesh elements vary significantly, depending on the value of Δx, Δy and Δz chosen for different models, ranging from 70,000 to more than 900,000 cells. The model was run for 1500 s, at which the results indicated that a steady state had been achieved. The time step used for the simulation was ensured to meet the CFL (Courant–Friedrichs–Lewy) criterion. The simulations were computed on an i7 3770 quad core computer with 16 GB memory, and depending on test cases, it took between 3–5 h for simulations to be completed.

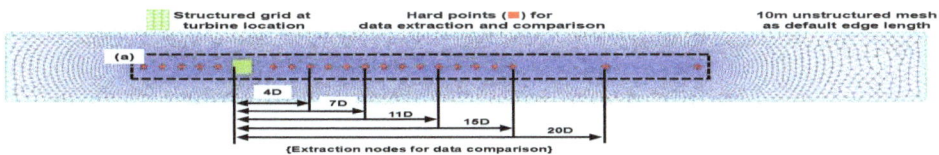

Figure 4. *x–y* horizontal plane illustrating the geometry mesh. The structured grid is used to define the enclosure of the actuator disc, while the rest of the domain is enforced using the unstructured mesh. The turbine is located at 250-m from the channel inlet. Selected downstream nodes that were used in the data extraction and validation purposes are represented by the red points along the turbine centreline in terms of the turbine diameter, D. The dotted rectangle outlines the zone (**a**) where the mesh refinement (min = 1 m, max = 10 m) was administered.

3.3. Boundary Condition

A constant volume flowrate, Q of 21,840 m³/s was imposed at the channel inlet, where Q is equal to the surface area of the inlet (60 m × 140 m) multiply by the mean flow velocity (2.6 m/s). Next, the downstream boundary was set to equal the channel water depth of 60 m to enable flow continuity. Initial condition was set to "PARTICULAR" (an option available in Telemac3D for defining the initial condition), where the initial depth of 60 m was specified in the CONDIM subroutine. A commonly used method of defining inflow velocities is to use a power-law (1/*n*th) profile. Although it is possible to use any value for *n* to approximate the flow conditions, a comparison of several values for *n* was considered to be beyond the scope of this study. Thus, the vertical velocity profile in the domain was imposed using one of the most commonly used power laws, 1/7th, so as to be similar to a full-scale tidal site [24]. In addition, the Chezy formulation with a friction coefficient of 44 was applied to the bottom to reduce the flow velocity as well as to increase the shear near the bed. This value was chosen as it has been used previously in [25] to represent the bottom roughness in the Pentland Firth region. Although not shown in the present paper, a wide range of friction coefficient values have been examined, and their influence on the model's output were shown to be negligible. In this study, both hydrostatic and non-hydrostatic code were tested, and are discussed in Section 4.5.

Boundary condition on the bottom was set to a Neumann condition, which is a slip boundary condition. An attempt to implement a non-slip condition, where the bottom velocities can be set to 0 was not possible in this study since the model would require a very fine mesh near the bottom level to satisfy the wall function. Wall function, y^+ is a non-dimensional distance used to describe the ratio between the turbulent and laminar influences in a grid cell, as well as to indicate the mesh refinement for near wall region in a flow model [26]. The rationale behind the wall function is to reduce computational time and also to increase both numerical stability and convergence speed in resolving the boundary layers. However, because of the strong gradients of the flow and also turbulence variables that exist in the viscous layer, highly-refined grids are needed near the bottom [27]. Small-scale models are known to use a no-slip wall at the bed (refer to References [4,5,7]) since it is still computationally feasible to satisfy the y^+ requirement. Nonetheless, the flow details in the boundary layer were not of specific interest in this paper. Further, the use of a very fine vertical resolution (e.g., 50 planes or more) for modelling a full-size turbine is both impractical and not computationally feasible without the use of computer cluster.

3.4. Turbulence Input

Different values of stream wise turbulence intensities, *TI*, have been reported in the literature for both in situ measurements and flume experiments, which ranges from 3% to 25% [19,28–36]. *TI* is defined as the ratio of the turbulent fluctuations, σ_i of the velocity fluctuation components ($i = u, v, w$) to the mean streamwise velocity of each sample, $\overline{U}$, and is one of the most common metrics utilised to quantify turbulence [37]. This parameter provides a quantification of the magnitude of the turbulent fluctuations and is considered to be a dominant driver of the fatigue loads on tidal turbine blades [38]. In Telemac, the equations for the standard k-ε turbulence model is given as follows:

$$\frac{\partial k}{\partial t} + U\frac{\partial k}{\partial x} + V\frac{\partial k}{\partial y} + W\frac{\partial k}{\partial z} = \frac{\partial}{\partial x}\left(\frac{v_t}{\sigma_k}\frac{\partial k}{\partial x}\right) + \frac{\partial}{\partial y}\left(\frac{v_t}{\sigma_k}\frac{\partial k}{\partial y}\right) + \frac{\partial}{\partial z}\left(\frac{v_t}{\sigma_k}\frac{\partial k}{\partial z}\right) + P - G - \varepsilon \tag{11}$$

$$\begin{aligned}\frac{\partial \varepsilon}{\partial t} + U\frac{\partial \varepsilon}{\partial x} + V\frac{\partial \varepsilon}{\partial y} + W\frac{\partial \varepsilon}{\partial z} &= \frac{\partial}{\partial x}\left(\frac{v_t}{\sigma_\varepsilon}\frac{\partial \varepsilon}{\partial x}\right) + \frac{\partial}{\partial y}\left(\frac{v_t}{\sigma_\varepsilon}\frac{\partial \varepsilon}{\partial y}\right) + \frac{\partial}{\partial z}\left(\frac{v_t}{\sigma_\varepsilon}\frac{\partial \varepsilon}{\partial z}\right) \\ &\quad + C_{1\varepsilon}\frac{\varepsilon}{k}[P + (1 - C_{3\varepsilon})G] - C_{2\varepsilon}\frac{\varepsilon^2}{k}\end{aligned} \tag{12}$$

where

$k = \frac{1}{2}\overline{u_i' u_i'}$ denotes the turbulent kinetic energy of the fluid,

$u_i' = U_i - \overline{u_i}$ denotes the *i*th component of the fluctuation of the velocity (u, v, w),

$\varepsilon = v\overline{\frac{\partial u_i'}{\partial x_j}\frac{\partial u_i'}{\partial x_j}}$ is the dissipation of turbulent kinetic energy,

$P = v_t\left(\frac{\partial \overline{U_i}}{\partial x_j} + \frac{\partial \overline{U_j}}{\partial x_i}\right)\frac{\partial \overline{U_i}}{\partial x_j}$ is a turbulent energy production term, in which $v_t = C_\mu\frac{k^2}{\varepsilon}$

and $G = -\frac{v_t}{Pr_t}\frac{g}{\rho}\frac{\partial \rho}{\partial z}$ is a source term due to the gravitational forces.

To properly validate the numerical simulations, the imposed turbulence intensity on the models should be as close as to the one used by Harrison et al. [5] in their experimental work. Using an online tool [39] to extract the published data from Reference [5], the measured turbulence intensities rate at the domain inlet can be approximated to vary from 5 to 15%. With the value of *TI* now known, the k-ε turbulence model can be adopted to define the turbulence at the channel inlet, where the turbulent kinetic energy, k and energy dissipation, ε are calculated using Equations (13) and (14):

$$k = \frac{3}{2}TI^2U^2 \tag{13}$$

$$\varepsilon = C_\mu^{3/4}\frac{k^{3/2}}{l} \tag{14}$$

TI is the turbulence intensity rate of 5%, *U* refers to the velocity across the water column that follows the 1/7th power law, C_μ is a dimensionless constant, equal to 0.09, and *l* is the turbulence length scale. In the study, the value of *l* was set to 20 m, which corresponded to one-third of the channel depth. Table 2 summarises the input/value of the parameters of the actuator disc model adopted in the numerical simulations.

Table 2. Default values of the numerical parameters employed in the simulation of the actuator disc.

Numerical Parameters	Input/Values
Law of the bottom friction and the corresponding friction coefficient	Chezy (44)
Turbulence model	k-ε turbulence models
Hydrostatic assumption	True
Initial condition	"PARTICULAR" where the initial elevation is set to 60 m
Vertical resolutions	24 sigma layers
Boundary condition on the bottom	Slip condition
Boundary forcing	Inlet: prescribed flowrate, Q = 21,840 m³/s Outlet: prescribed elevation, H = 60 m
Resistance coefficient, K	2 (corresponding to C_T = 0.89)

4. Models' Sensitivity and Validation

4.1. Validation Metric

Appropriate metrics need to be chosen for the model–data comparisons. As this study employs a full-size turbine for the model, a dimensionless metric was needed for the validation against physical scale data. All parameters were made dimensionless by dividing it by a characteristic homogeneous quantity. A commonly used method employed to characterize the wake recovery is the rotor velocity deficit, $U_{deficit}$ (see Equation (15)). U_{wake} refers to the velocity at any points downstream of the disc, while $U_{free\ stream}$ is the unperturbed flow velocity. In this study, $U_{free\ stream}$ corresponds to 3 m/s as defined by the maximum velocity of the incoming stream into the channel. Next, turbulence intensity (*TI*) has been chosen as the benchmark quantity for turbulence behaviours (see Equation (16)).

$$U_{deficit} = 1 - \frac{U_{wake}}{U_{free\ stream}} \tag{15}$$

$$TI = \frac{\sqrt{\frac{2}{3}k}}{U} \tag{16}$$

To examine the accuracy of the 3D models, focus will be placed on the modelled velocity reduction as well as the turbulence characteristics along the actuator disc centreline. The centreline is defined as the horizontal line that passes through the turbine centre along the x orientation of the flow direction. For this study, the z and y disc centreline axis were located at 30-m mid-depth, and at 70-m mid-channel width, respectively. Furthermore, to facilitate data extraction and for comparison purposes, hard points were applied into the geometry mesh during pre-processing stage to establish the positions of the fixed nodes within the domain. Figure 4 demonstrates the implementation of the hard points (specified by the red nodes) along the x centreline orientation at a desired longitudinal distance (in terms of turbine diameter, D).

4.2. Mesh Dependence Test

Because of the approximations and averaging used in the RANS equations, the size as well as the number of cells in any given CFD domain can directly affect the results of a numerical model. A larger number of cells may result in a more accurate solution to a problem for a given set of boundary conditions and solver settings. However, increasing mesh density also requires greater processing time. In order to verify the robustness of the unstructured mesh used in the model, four mesh (i.e., the dimension of the edge length of the unstructured mesh) with varying resolutions were tested; Case 1 = 1 m, Case 2 = 2 m, Case 3 = 5 m, and Case 4 = 10 m. The refinement zone starts at 5D upstream of the disc, continues up to 25D downstream as this region is selected for data extraction for comparisons with published literature. Figure 4 outlines the mesh refinement zone as well as the fixed nodes that are used in validating the model. It was decided to extract the output parameters at five locations of the computational domains, which are located at 4D, 7D, 11D, 15D, and 20D from the centre of the turbine, where D is the rotor diameter (refer to Figure 4). Elsewhere, a default edge length of 10 m was applied. For this mesh dependency test, the structured grid at the turbine area was constructed using elements of size $\Delta x = 2$ m and $\Delta y = 2$ m for all models, while the disc thickness, Δx_t was set to 2 m. Also, the model was run using only the hydrostatic assumption.

Table 3 highlights the models used in the grid dependency study, where four refinements values were tested. The most refined model (Case 1) consists the highest number of elements at 925,536, while the model which has the minimal resolution (Case 4) comprises the least number of elements at about 70,000. The velocity data were then extracted at five distinct nodes (refer to Figure 4) and are listed in Table 3. The results, in general, indicate that increasing mesh density decreases the velocity; however, the difference was not high. This observation is supported by Figure 5 which compares the models' velocity reduction (top plots) and *TI* (bottom graphs) against the laboratory measurement data from [5]. Overall, the results show that all models manage to produce velocities more or less the same as the experimental values behind a disc. Interestingly, although a higher resolution domain was expected to give a better correlation against the experimental data, contradictory results were obtained. That is, the 1 m grid refinement seems to overestimate the velocity reduction by almost 10% in the near wake, before declining slowly in the far wake regions. Likewise, these scenarios are also reflected in the observation of *TI*, where the 1 m model appears to amplify the turbulence below the disc centreline in the 4D and 7D regions. Several possible reasons may contribute to this behaviour; a higher resolution model contains a large number of points in the domain, which signify that the cells are more sensitive to the flow characteristics in the surrounding region.

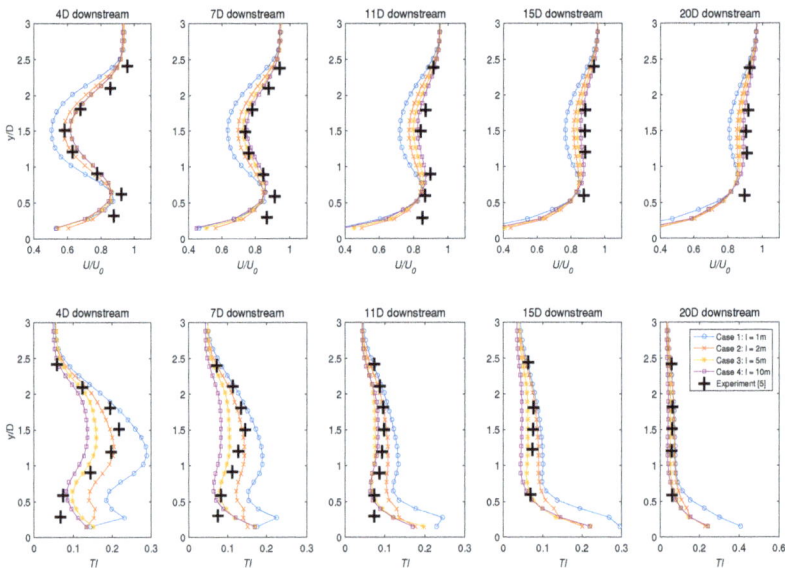

Figure 5. Mesh dependency study for the administered refinement zone at increasing distances downstream of the turbine. The mesh edge length within the refinement zone is varied according to the cases being explored.

On the other hand, the 2 m mesh refinement (Case 2) agrees well with the measurement data, although a slightly higher velocity was observed in the far wake regions. Meanwhile, models with a resolution of 5 m and 10 m were shown to be less accurate in the *TI* prediction in comparison with experimental data, where both models significantly underestimated the *TI* in the 4D and 7D regions. These results illustrate that the coarser resolution models fail to properly represent the flow-rotor interaction, indicating that the models output (i.e., velocity, *TI*) is susceptible to changes in mesh density. To conclude, the use of a very fine mesh alongside hydrostatic assumption may not necessarily provide superior numerical output as demonstrated by test Case 1 using the 1 m mesh density. Consequently, as the simulation results for the 2 m mesh (Case 2) closely matched with the measurement, this has been implemented as the optimal mesh discretisation for the following simulations.

Table 3. Grid dependency study.

Refinement Details		Centreline Velocity at Various Longitudinal Positions from Actuator Disc (m/s)				
Mesh Size	Number of Elements	5D Upstream	4D Downstream	7D Downstream	11D Downstream	20D Downstream
Case 1 (1 m)	925,536	2.683	1.515	1.903	2.153	2.418
Case 2 (2 m)	316,584	2.680	1.735	2.092	2.315	2.554
Case 3 (5 m)	101,496	2.690	1.853	2.186	2.391	2.600
Case 4 (10 m)	72,936	2.694	1.846	2.247	2.464	2.659

4.3. Sensitivity of the Structured Grid at Turbine Location

A structured grid was employed to define the area of the actuator disc for two key reasons: one to facilitate in identifying the nodes for momentum sink implementation and the other to improve the accuracy of the approximated turbine forces by correctly asserting the physical shape of the disc. In this section, the influence of turbine grids resolution (i.e., Δx and Δy) on the wake characteristics of a full-scale actuator disc is explored.

4.3.1. Influence of the Disc Thickness, Δx_t

For the test, a 2 m by 2 m (Δx & Δy) structured grid was embedded into the unstructured mesh domain where the turbine was located. The models were then run using three different turbine thickness values, Δx_t =2 m, 4 m, and 8 m. For clarification, when Δx_t is set to 4 m, the flow will have to pass through two Δx grid cells (each cell is 2 m as displayed in Figure 2) in the x direction where the momentum source terms will determine the forces exerted by the turbine. Figure 6 illustrates the comparison of the measurement data for the above three values of Δx_t.

Interestingly, the results show that the thickness of the actuator disc had negligible influences on both the downstream wake, as well as the turbulence intensities. Although not shown here, similar observations were also apparent when the Δx and Δy values were changed to other than 2 m resolutions. Therefore, it can be deduced that the downstream wakes and turbulences behind the turbine are independent of the actuator disc thickness.

4.3.2. Resolution of the Structured Grid (Δx & Δy)

The interaction between the resistance loss coefficient ($K/\Delta x_t$) and the resolution of the structured grids (Δx and Δy) at the actuator disc location is explored here. As previously mentioned, K is the constant resistance coefficient which corresponds to the experimental values of C_T. As the flow progresses through the turbine swept area, it expands and accelerates around the edge or "width" of the disc. In the numerical model, the flow passing through the width or thickness of the turbine literally means that it is travelling across specific nodes in the x direction as defined by the enclosure of the disc. For a very fine mesh with constant Δx_t, the source term will be applied to a larger set of nodes when compared to a coarser grid. To examine the relationship between the nodes and their impact on the wake characteristic, four structured grids with various densities were created and examined. The lowest resolution of Δx and Δy tested was 4 m, while the most refined was set to 0.5 m. Figure 7 highlights enticing differences between the four models using a constant Δx_t of 2 m. Even though Δx_t was found to have no impact on the results as noticed in Section 4.3.1, we selected 2 m thickness here as this would be close to projected thickness of a turbine.

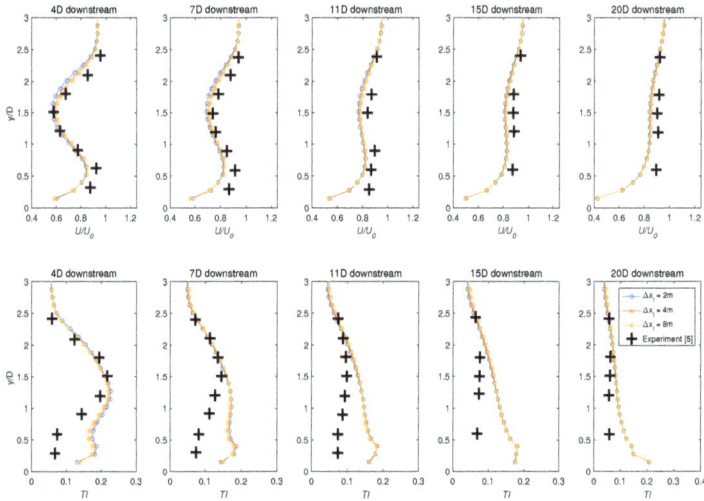

Figure 6. The influence of the disc thickness, Δx_t on the flow at increasing distances downstream of the turbine. The value of Δx_t was varied, while the vertical resolution Δz and the size of the structured grid (Δx & Δy) were maintained at 24 sigma layers and 2 m, respectively for all cases.

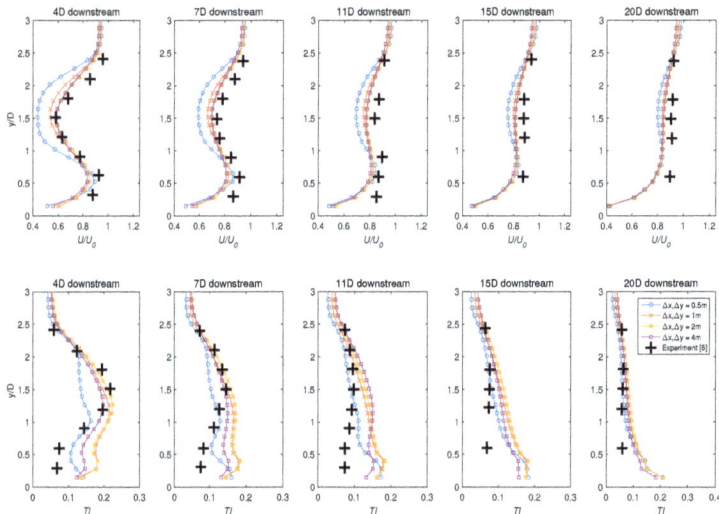

Figure 7. The influence of the structured grid density, (Δx & Δy) on the flow at increasing distances downstream of the turbine. The vertical resolution, Δz, and the disc thickness, Δx_t, were maintained at 24 sigma layers and 2 m, respectively for all cases.

The model with the highest grid density (0.5 m) appeared to overestimate the velocity reduction behind the disc by almost 20% in the near wake region and continued to do so further downstream with a larger velocity deficit compared to other models. The same trend was also observed for the model with a 1 m grid, although the extent of the velocity deficit was less pronounced as the flow started to reach homogeneity. In contrast, models utilizing the 2 m and 4 m grids seemed to be able to simulate the flow-rotor interaction accordingly and compared well with the experimental data. With turbulence intensities, since the 1 m grid produced a greater velocity deficit in the near wake region, this was reflected accordingly in the turbulence plots, where the intensities were less profound behind the disc due to wake mixing. A variation of *TI* up to 5% is seen for the case of 4D regions, then slowly recovers before matching other models at 15D downstream. However, a different trend is noticed for the 4 m grid model, as it shows some discrepancies for the *TI* in the near region, although it is not as dominant as the 0.5 m model.

The results presented in Figure 7 offers interesting insights for discussions. A less accurate output is expected from a coarser grid due to a limited (albeit uniformly scattered) number of nodes within the turbine enclosure as exemplified by Figure 3b. Conversely, one would have thought that the simulation output could probably be improved by using a very fine mesh, although it appears that this is not always the case based on the results attained. As the density of Δx and Δy increases, the nodes that render the turbine swept area, as well as the turbine width will also increase proportionally, as illustrated in Figure 3c. This figure clearly shows an uneven distribution of the nodes at the face of the actuator disc for a higher density model, where they are concentrated prominently at the mid-quarter of the disc. Since increasing the mesh density only applies in the x and y directions, the top and bottom edges were devoid of nodes. As a result, the implementation of the momentum term would only be directed at the centre of the turbine, which consequently will cause inaccuracy in the approximation of the turbine force, as evidenced from the plots in Figure 7.

To solve this, it is recommended that the vertical resolution should also be increased when a fine mesh is employed so that uniform node distribution can be achieved. Nonetheless, increasing the vertical density while using a very fine grid will undoubtedly increase the computational resources, especially when running a large-scale simulation. Thus, finding an optimal ratio for the mesh density and vertical resolution is crucial so that the implementation of the actuator disc can correctly approximate the thrust exerted by a turbine. Based upon the results presented in this case study, $\Delta x = 2$ m and $\Delta y = 2$ m were chosen as the optimal structured grid density since they provide numerical accuracy and computational balance needed for implementing the actuator disc.

4.3.3. Grid Resolution for Δy

It is anticipated that models with coarser Δy will perform poorly when compared with a more refined density, since the force approximation only happens at a considerable node interval. To inspect this hypothesis, four Δy grids (2 m, 2.5 m, 5 m, and 10 m) were tested, while Δx and Δx_t were maintained at 2 m and 2 m, respectively. This hypothesis is substantiated and illustrated by the velocity plots in Figure 8, where the 10 m grid somewhat underestimated the velocity deficit in the near wake region. On the other hand, other Δy models showed good comparison against the laboratory measurements. Furthermore, since the coarser model ($\Delta y = 10$ m) contained a significantly smaller number of nodes, the model was unable to correctly approximate the thrust on the flow, resulting in a distinctly lower turbulence intensity between 4D and 7D downstream regions. The other models, however, showed a relatively similar characteristic for both $U_{deficit}$ and *TI*. Since the size of the disc adopted in this study is 20-m, an Δy value of 10 m or more might not be suitable to accurately model the actuator disc since the number of nodes available are not sufficient for approximating both the forces as well as the size of the swept area. To conclude, based on the results observed, $\Delta y = 2$ m is accepted as the optimal spacing in the y direction, and confirm the remarks made previously in Section 4.3.2.

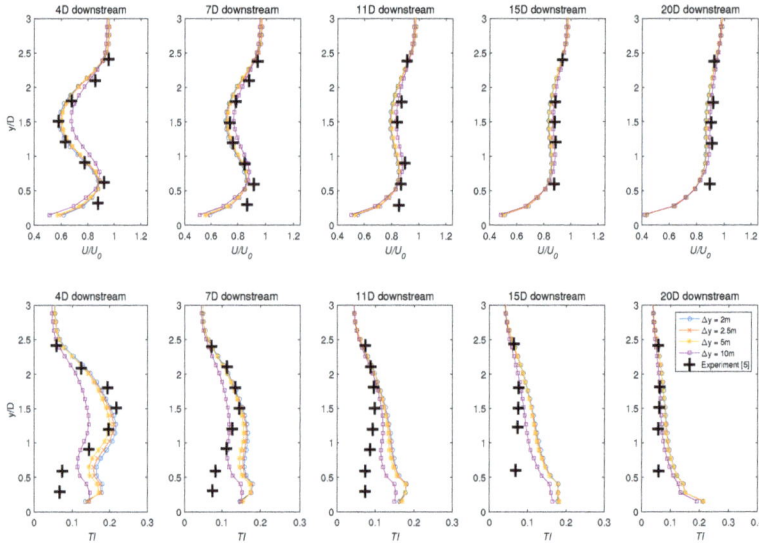

Figure 8. The influence of Δy resolution on the flow at increasing distances downstream of the turbine. The value of Δy was varied, while Δx and Δz were maintained at 2 m and 24 sigma layers, respectively for all cases.

4.4. Sensitivity of the Vertical Resolutions, Δz

To investigate the influence of vertical resolutions on the model, four sigma (σ) layers were put to test; 24 (default value), 18, 15, and 10. Table 4 provides the information of the models used in this study, where the largest (6.66 m) and smallest vertical interval (2.61 m) correspond to 10σ and 24σ layers, respectively.

Table 4. Comparison of the vertical resolutions.

Sigma Layers	Distance between Planes (in Meter)	Number of Mesh Elements
24	2.61	317,928
18	3.53	238,446
15	4.29	198,705
10	6.66	132,470

Further, based on the above observations in Sections 4.2 and 4.3, Δx and Δy was set to 2 m for all models and $\Delta x_t = 2$ m was used. The simulated results are presented in Figure 9, where the model employing 10σ layers underestimated the *TI* as anticipated, since it had the least number of nodes to properly characterise the turbine swept area. However, it is quite interesting to see the same model was able to reproduce the velocity wake that matched the measurement results. One possible reason for this observation could be due to the hydrostatic assumption used in the model. Nonetheless, the poor turbulence correlation observed for the 10σ layers should be approached with caution. It shows that the region between 4D and 11D downstream of the disc was highly turbulent, and the flow fluctuations could not be accurately reproduced using planes with large vertical intervals. To get around this issue, instead of sigma layers, it was possible to utilise other types of mesh transformation for the distribution of vertical planes, such as by fixing planes at desired depths. With this option, the height of the vertical planes can be appropriately adapted to capture the influence of the disc on the flow.

In brief, finding the optimal Δz values are crucial in achieving the balance between computational efficiency and numerical accuracy, more so for simulations involving a very large domain.

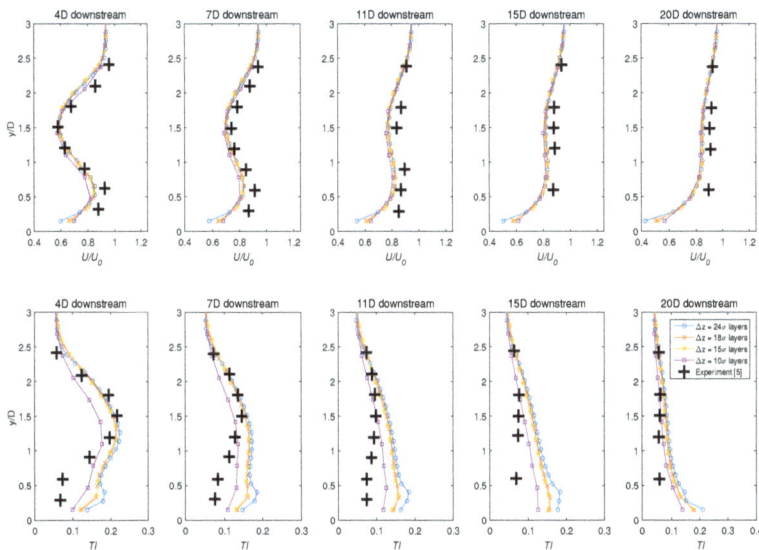

Figure 9. The influence of the vertical resolution Δz on the flow at increasing distances downstream of the turbine. The value of Δz was varied, while the size of the structured grid (Δx & Δy) was set to 2 m.

4.5. Hydrostatic vs. Non-Hydrostatic Pressure Models

Telemac3D offers the choice of using either the hydrostatic or the non-hydrostatic pressure code. The hydrostatic pressure simplifies the vertical velocity (W) assumption, ignoring the diffusion, advection and other terms. Thus, the pressure at a point is the sum of weight of the water column and the atmospheric pressure at the surface. Conversely, the non-hydrostatic option solves the vertical velocity equation and is more computationally intensive. Similar with the 3D flow equations shown previously in Section 2.1 (Equations (1)–(3)), the non-hydrostatic assumption adopted in this section was computed by the following additional equation:

$$\frac{\partial W}{\partial t} + U \frac{\partial W}{\partial x} + V \frac{\partial W}{\partial y} + W \frac{\partial W}{\partial z} = -g \frac{\partial Z_s}{\partial z} + v\Delta(W) + F_Z \tag{17}$$

where W and F_z are the three-dimensional component of the velocity and sink term in the vertical direction, respectively. To examine the influence of both assumptions, models with $\Delta x = 2$ m, $\Delta y = 2$ m, $\Delta x_t = 2$ m, and $\Delta z = 24\sigma$ layers were simulated in both hydrostatic and non-hydrostatic solvers. Interestingly, the impact of the non-hydrostatic code on the velocity component was less pronounced as illustrated in Figure 10, where both models produced a nearly identical velocity reduction at $1.5 < y/D < 3$. However, for the bottom half of the channel, the hydrostatic model somewhat showed a closer agreement with the measurement points than the model using the non-hydrostatic solver, especially in the near bed region. This was apparent in the 4D and 7D regions, where the wake velocity from the non-hydrostatic model showed a slight divergent from the measurement data approximately before $y/D = 0.7$. This trend continued to persist until the far wake regions, where the non-hydrostatic model slightly overestimated the wake velocity up to $y/D = 1.5$.

With regard to the turbulence intensity, the non-hydrostatic model demonstrated excellent agreement with the published data, where the solver seemed to be able to properly resolve the turbulence mixing from the near till far wake regions. In contrast, the computed *TI* from the hydrostatic model illustrated striking disparities from 4D to 20D wake regions, especially in the bottom half of the channel depth (see Figure 10). At 4D downstream of the disc, the hydrostatic model managed to produce good agreement with the measurement at $1.5 < y/D < 3$ before gradually straying from the data points. Similar observations occur further downstream, where the differences in *TIs* between the two models were becoming more apparent. At 15D, the computed *TI* using the hydrostatic solver only matched the experimental data at $2.5 < y/D < 3$, as the model displays an intensity variation of up to 10% against both the measured data and non-hydrostatic model. Ultimately, far downstream of the disc at 20D where the flow eventually reaches homogeneity, the two models presented an almost uniform turbulence characteristic.

To further validate these observations, simulations utilizing different set of structured densities ($\Delta x, \Delta y = 0.5$ m, 1 m, 2 m, 4 m) were run in the non-hydrostatic formulation to examine the influence of the solvers on the wake characteristics. Figure 11 displays the results of these simulations. An excellent agreement against the experimental data was observed, regardless of the resolutions of the structured grid being tested. Moreover, the use of the non-hydrostatic code also eliminated the turbulence variations in the bottom half of the actuator disc, as previously observed in Figure 10 where the hydrostatic assumptions were employed. These findings demonstrate that the non-hydrostatic code has a significant influence in resolving the vertical turbulence mixing downstream of the disc, where the previously seen variations for both $U_{deficit}$ and *TI* from the hydrostatic models were remarkably corrected.

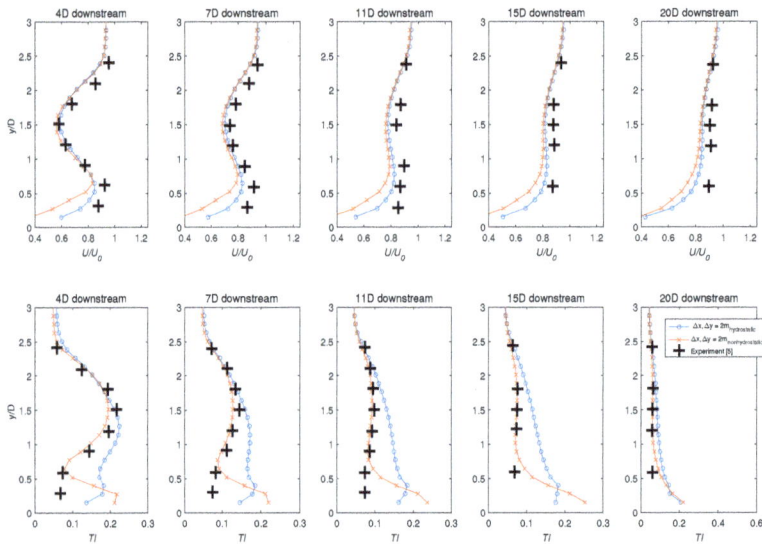

Figure 10. Comparison between hydrodynamic and non-hydrodynamic assumptions on the models. The vertical resolution Δz and the size of the structured grid (Δx & Δy) were maintained at 24 sigma layers and 2 m, respectively for all cases.

The comparisons between the two codes present compelling points for discussions. The output by the non-hydrostatic model exhibited close agreement with the measurement data as expected since it accounts for the gravitational acceleration as well as the w-velocity component, and thus is more accurate. Whereas the flow characteristics observed from the hydrostatic pressure code only matched the scale data for the velocity components, the turbulence comparisons differ greatly. Two possible reasons may be attributed to this observation. First, there existed no stringent guidelines as to what extent could the hydrostatic assumption be safely used for an actuator disc approximation in a simple channel case. A condition which allows for hydrostatic assumption in a system implies that the horizontal length scale is much greater than the water depth. Indeed, it might be possible that the horizontal length of the domain utilised in this study did not satisfy the optimal ratio for hydrodynamic pressure assumption.

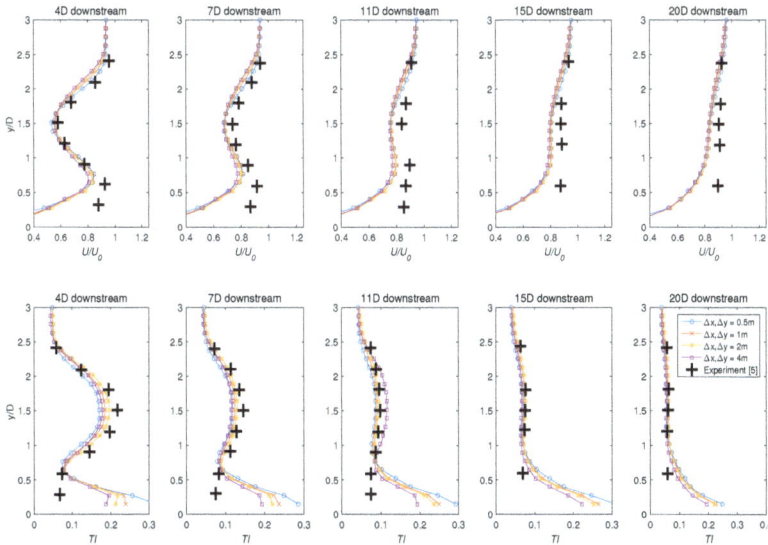

Figure 11. The influence of the structured grid density (Δx & Δy) on the flow using the non-hydrostatic approximations. The vertical resolution Δz and the disc thickness Δx_t were maintained at 24 sigma layers and 2 m, respectively for all cases.

Secondly, the turbulence in the flow may not have fully resolved as the non-slip criteria could not be implemented on the channel bed. The non-slip criteria require the use of a highly-refined mesh to resolve the size of the smallest eddy in the flow as it begins to transition from laminar to turbulence regime. And since the hydrostatic code ignores the advection and diffusion term, the eddies cannot be properly dissipated, and thus influencing the turbulence characteristic in the near wake. Indeed, as the highest turbulence intensity zone was observed in the immediate vicinity behind the disc and then slowly dissipated further downstream, the hydrostatic model can only reproduce the expected intensities in the far wake regions, as shown by the 20D downstream plot. For future references, it may be wise to increase the number of layers near the vicinity of the bed wall to account for the turbulence in the bottom half of the water column.

5. Conclusions

A study to examine the validity of the RANS actuator disc approach for a full-scale tidal turbine has been presented by comparing the model velocities and turbulence intensities against a laboratory measurement data sourced from literature. The numerical model Telemac3D was used for this purpose, and a detailed methodology in the implementation of the momentum source term was introduced and elaborated. The influence of: (i) unstructured mesh sizes used in the computational domain; (ii) change in the turbine thickness (Δx_t); (iii) variation in the structured grid resolutions (Δx and Δy) used to represent the turbine's enclosure; (iv) variation in the vertical layer thickness (Δz); and (v) the effect of hydrostatic and non-hydrostatic formulations on the model output parameters were explored. Key findings from this study are:

- The results demonstrated that the numerical model was highly sensitive to the mesh refinement upstream and downstream of the turbine, where coarser models tended to underestimate both the velocity retardation and *TI*.
- Altering the thickness of the turbine (Δx_t) had negligible impact on the downstream wakes and turbulence mixings.
- Numerical accuracy of the model was found to be highly susceptible to changes in the grid density of the turbine enclosure. The optimal structured grid density of $\Delta x = 2$ m, and $\Delta y = 2$ m satisfactorily modelled both the velocity deficit and the *TI*.
- The importance of the grid spacing in y direction (Δy) in characterising the thrust and also the shape of the disc was also highlighted.
- The influence of vertical resolutions (Δz) in representing the depth of the channel on the model was investigated to find a balance between computational efficiency and numerical accuracy. The findings indicated that appropriate adjustment on both the horizontal and vertical planes must be attained to accomplish the optimal ratio between the nodes resolution in both z and y orientation. Note that the optimal structured grid density found in this study was limited to the computational resources available to the authors. The methodology presented in this article, however, are still valid for a more refined grid implementation.
- The impact of the hydrostatic and non-hydrostatic pressure assumptions on the predicted output were examined, where both models exhibit nearly indistinguishable flow retardation characteristics behind the simulated disc. However, for the *TI*, only the non-hydrostatic model was able to match the experimental result, while the hydrostatic solver failed to properly resolve the turbulence mixing in the wake regions between 4D and 15D.

In summary, this paper provided a preliminary demonstration of the applicability of RANS actuator disc approach for a full-size tidal device. Future work will involve modelling several full-size turbines in an array using the methodology presented in this paper. Additionally, several other turbulence models, such as the k-ω and Reynolds Stress Models (RSM), may also be studied to understand their impact on the models' output. Ultimately, the procedure followed in this study may hence be used as a guideline for the implementation of tidal turbines by the actuator disc method in regional scale simulations.

Author Contributions: A.R. undertook the modelling work and wrote the paper. V.V. conceived the idea of the project, obtained funding, and thoroughly reviewed the paper. J.T. contributed to model configuration and commented on the draft manuscript.

Funding: The first two authors are grateful for the financial support of the UK Engineering and Physical Sciences Research Council (EPSRC) through the FloWTurb: Response of Tidal Energy Converters to Combined Tidal Flow, Waves, and Turbulence (EPSRC Grant ref. number EP/N021487/1) project. The first author is also grateful to the Ministry of Higher Education, Malaysia and also Universiti Malaysia Perlis (UniMAP) for funding the research.

Acknowledgments: The authors gratefully acknowledge the assistance of the staff of the IT department within the School of Engineering, The University of Edinburgh during the course of the project.

Conflicts of Interest: The authors declare no conflict of interest.

References

1. Neill, S.P.; Litt, E.J.; Couch, S.J.; Davies, A.G. The impact of tidal stream turbines on large-scale sediment dynamics. *Renew. Energy* **2009**, *34*, 2803–2812. [CrossRef]
2. Robins, P.E.; Neill, S.P.; Lewis, M.J. Impact of tidal-stream arrays in relation to the natural variability of sedimentary processes. *Renew. Energy* **2014**, *72*, 311–321. [CrossRef]
3. Chatzirodou, A.; Karunarathna, H.; Mainland, S.; Firth, P.; Park, S. Impacts of tidal energy extraction on sea bed morphology. *Coast. Eng.* **2014**, *1*, 33. [CrossRef]
4. Sun, X.; Chick, J.P.; Bryden, I.G. Laboratory-scale simulation of energy extraction from tidal currents. *Renew. Energy* **2008**, *33*, 1267–1274. [CrossRef]

5. Harrison, M.E.; Batten, W.M.J.; Myers, L.E.; Bahaj, A.S. Comparison between CFD simulations and experiments for predicting the far wake of horizontal axis tidal turbines. *IET Renew. Power Gener.* **2010**, *4*, 613–627. [CrossRef]
6. Adcock, T.A.; Draper, S.; Nishino, T. Tidal power generation–A review of hydrodynamic modelling. *Proc. Inst. Mech. Eng. Part A J. Power Energy* **2015**, *299*, 755–771. [CrossRef]
7. Daly, T.; Myers, L.E.; Bahaj, A.S. Modelling of the flow field surrounding tidal turbine arrays for varying positions in a channel. *Philosophical Trans. R. Soc. Lond. A Math. Phys. Eng. Sci.* **2018**, *371*, 20120246. [CrossRef] [PubMed]
8. Lartiga, C.; Crawford, C. Actuator Disk Modeling in Support of Tidal Turbine Rotor Testing. In Proceedings of the 3rd International Conference on Ocean Energy, Bilbao, Spain, 6–8 October 2010; pp. 1–6.
9. Roc, T.; Conley, D.C.; Greaves, D. Methodology for tidal turbine representation in ocean circulation model. *Renew. Energy* **2013**, *51*, 448–464. [CrossRef]
10. Nguyen, V.T.; Guillou, S.S.; Thiebot, J.; Cruz, A.S. Modelling turbulence with an Actuator Disk representing a tidal turbine. *Renew. Energy* **2016**, *97*, 625–635. [CrossRef]
11. Hervouet, J.M. *Hydrodynamics of Free Surface Flows: Modelling with the Finite Element Method*; John Wiley & Sons: West Sussex, UK, 2007.
12. Desombre, J. *TELEMAC-3D OPERATING MANUAL (Release 6.2)*; EDF R&D: Paris, France, 2013.
13. Hervouet, J.M.; Jankowski, J. Comparing numerical simulations of free surface flows using non-hydrostatic Navier-Stokes and Boussinesq equations. Electricité de France; Bauwesen der Universität Hannover. Unpublished work, 2000.
14. Burton, T.; Jenkins, N.; Sharpe, D.; Bossanyi, E. *Wind Energy Handbook*; John Wiley & Sons: Weat Sussex, UK, 2011.
15. Batten, W.M.J.; Harrison, M.E.; Bahaj, A.S. The accuracy of the actuator disc-RANS approach for predicting the performance and far wake of a horizontal axis tidal stream turbine. *Philos. Trans. R. Soc. Lond. A Math. Phys. Eng. Sci.* **2013**, *371*, 20120293. [CrossRef] [PubMed]
16. Taylor, G. *The Scientific Papers of Sir Geoffrey Ingram Taylor*; Cambridge University Press: Cambridge, UK, 1963.
17. Whelan, J.; Thomson, M.; Graham, J.M.R.; Peiro, J. Modelling of free surface proximity and wave induced velocities around a horizontal axis tidal stream turbine. In Proceedings of the 7th European Wave and Tidal Energy Conference, Porto, Portugal, 11–13 September 2007.
18. Belloni, C.; Willden, R.H.J. A computational study of a bi-directional ducted tidal turbine. In Proceedings of the 3rd International Conference on Ocean Energy, Bilbao, Spain, 6–8 October 2010.
19. Myers, L.E.; Bahaj, A.S. Experimental analysis of the flow field around horizontal axis tidal turbines by use of scale mesh disk rotor simulators. *Ocean Eng.* **2010**, *37*, 218–227. [CrossRef]
20. Troldborg, N.; Sørensen, J.N.; Mikkelsen, R.F. Actuator Line Simulation of Wake of Wind Turbine Operating in Turbulent Inflow. *J. Phys. Conf. Ser.* **2007**, *75*, 012063. [CrossRef]
21. Sforza, S.P.M.; Smorto, M. Three-dimensional wakes of simulated wind turbines. *AIAA J.* **1981**, *19*, 1101–1107. [CrossRef]
22. Legrand, C. *Assessment of Tidal Energy Resource: Marine Renewable Energy Guides*; European Marine Energy Centre: London, UK, 2009.
23. Scott, N.C.; Smeed, M.R.; McLaren, A.C. *Pentland Firth Tidal Energy Project Grid Options Study Prepared for: Highlands and Islands Enterprise*; Xero Energy: Glasgow, UK, 2009.
24. Carbon Trust. *PhaseII: UK Tidal Stream Energy Resource Assessment*; Black & Veatch Ltd.: London, UK, 2005.
25. Rahman, A.; Venugopal, V. Inter-Comparison of 3D Tidal Flow Models Applied To Orkney Islands and Pentland Firth. In Proceedings of the 11th European Wave and Tidal Energy Conference (EWTEC 2015), Nantes, France, 6–11 September 2015.
26. Salim, S.M.; Cheah, S.C. Wall y^{+} Strategy for Dealing with Wall-bounded Turbulent Flows. In Proceedings of the International MultiConference of Engineers and Computer Scientists (IMECS), Hong Kong, China, 18–20 March 2009.
27. Kalitzin, G.; Medic, G.; Iaccarino, G.; Durbin, P. Near-wall behavior of RANS turbulence models and implications for wall functions. *J. Comput. Phys.* **2005**, *204*, 265–291. [CrossRef]
28. Osalusi, E.; Side, J.; Harris, R. Reynolds stress and turbulence estimates in bottom boundary layer of fall of warness. *Int. Commun. Heat Mass Transf.* **2009**, *36*, 412–421. [CrossRef]

29. Osalusi, E.; Side, J.; Harris, R. Structure of turbulent flow in emec's tidal energy test site. *Int. Commun. Heat Mass Transf.* **2009**, *36*, 422–431. [CrossRef]

30. Macenri, J.; Reed, M.; Thiringer, T. Influence of tidal parameters on SeaGen flicker performance. *Philos. Trans. A Math. Phys. Eng. Sci.* **2013**, *371*, 20120247. [CrossRef] [PubMed]

31. Myers, L.E.; Bahaj, A.S. An experimental investigation simulating flow effects in first generation marine current energy converter arrays. *Renew. Energy* **2012**, *37*, 28–36. [CrossRef]

32. Giles, J.; Myers, L.; Bahaj, A.; Shelmerdine, B. The downstream wake response of marine current energy converters operating in shallow tidal flows. In Proceedings of the World Renewable Energy Congress, Linkoping, Sweden, 8–13 May 2011.

33. Milne, I.A.; Sharma, R.N.; Flay, R.G.J.; Bickerton, S. Characteristics of the onset flow turbulence at a tidal-stream power site. *Philos. Trans. R. Soc. A* **2013**, *371*, 20120196. [CrossRef] [PubMed]

34. Thomson, J.; Polagye, B.; Durgesh, V.; Richmond, M.C. Measurements of turbulence at two tidal energy sites in puget sound, WA. *IEEE J. Ocean. Eng.* **2012**, *37*, 363–374. [CrossRef]

35. Mycek, P.; Gaurier, B.; Germain, G.; Pinon, G.; Rivoalen, E. Experimental study of the turbulence intensity effects on marine current turbines behaviour. Part I: One single turbine. *Renew. Energy* **2014**, *66*, 729–746. [CrossRef]

36. Barthelmie, R.J.J.; Folkerts, L.; Larsen, G.C.C.; Rados, K.; Pryor, S.C.C.; Frandsen, S.T.T.; Lange, B.; Schepers, G. Comparison of wake model simulations with offshore wind turbine wake profiles measured by sodar. *J. Atmos. Ocean. Technol.* **2006**, *23*, 888–901. [CrossRef]

37. IEC. *Wind Turbines–Part 21: Measurement and Assessment of Power Quality Characteristics of Grid Connected Wind Turbines IEC 61400-21*; IEC: Geneva, Switzerland, 2008.

38. Milne, I.A.; Sharma, R.N.; Flay, R.G.J.; Bickerton, S. The Role of Onset Turbulence on Tidal Turbine Blade Loads. In Proceedings of the 17th Australasian Fluid Mechanics Conference 2010, Auckland, New Zealand, 5–9 December 2010; pp. 1–8.

39. Rohatgi, A. WebPlotDigitizer. 2013. Available online: http://arohatgi.info/WebPlotDigitizer/ (accessed on 20 April 2016).

energies

MDPI

Article

Evaluation of Some State-Of-The-Art Wind Technologies in the Nearshore of the Black Sea

Florin Onea and **Liliana Rusu** *

Department of Mechanical Engineering, Faculty of Engineering, Dunarea de Jos University of Galati,
47 Domneasca Street, 800008 Galati, Romania; florin.onea@ugal.ro
* Correspondence: lrusu@ugal.ro; Tel.: +40-745-399-426

Received: 25 August 2018; Accepted: 12 September 2018; Published: 15 September 2018

Abstract: The main objective of this work was to evaluate the nearshore wind resources in the Black Sea area by using a high resolution wind database (ERA-Interim). A subsequent objective was to estimate what type of wind turbines and wind farm configurations would be more suitable for this coastal environment. A more comprehensive picture of these resources was provided by including some satellite measurements, which were also used to assess the wind conditions in the vicinity of some already operating European wind projects. Based on the results of the present work, it seems that the Crimea Peninsula has the best wind resources. However, considering the current geopolitical situation, it seems that the sites on the western part of this basin (Romania and Bulgaria) would represent more viable locations for developing offshore wind projects. Since there are currently no operational wind projects in this marine area, some possible configurations for the future wind farms are proposed.

Keywords: Black Sea; wind power; nearshore; reanalysis data; satellite measurements

1. Introduction

The energy market and carbon emissions seem to have a strong connection, and since the energy demand is expected to increase in the near future, the negative impacts on the environments will be more noticeable [1,2]. A possible way to tackle this issue is to use natural resources (such as solar or geothermal) to secure a sustainable future and limit the effects of fossil fuels products [3–5]. One of the most successful sectors is wind energy, which has already demonstrated its technical-economic viability in various parts of the world, being possible to develop projects on land or in the marine environment [6–8]. Coastal areas seem to present much higher wind resources than onshore, while the wind turbulences reported in these regions seem to have a lower impact on turbine performance. Furthermore, the diurnal/nocturnal variations of the air masses in the nearshore areas may increase the performance of a wind generator [9].

By looking at the global offshore wind market (at the end of 2017), we can notice that European countries dominate this sector, as in the case of UK (6.8 GW), Germany (5.3 GW), Denmark (1.2 GW), Netherlands (1.1 GW) or Belgium (0.8 GW). PR China (with 2.8 GW) can be also considered in the front line of this industry, while in the second line we may include countries such as Vietnam, Japan, South Korea, USA or Taiwan, which nevertheless report a cumulated value below 0.5 GW [10]. In 2017, Europe upgraded its offshore parks with almost 560 new turbines, including the first floating project and, on average, the water depth for these projects was around 28 m with a distance to the coastline of 41 km. It is expected that by 2020 the installed capacity will be around 25 GW, which can be achieved if we consider that the average turbine capacity is around 6 MW (+23% reported to 2016) and the average size of a farm is estimated close to 493 MW (+34% compared to 2016) [11].

Offshore wind farms operating in the areas located between 30° and 60° latitude (both hemispheres) are expected to have the best performances, considering the action of the prevailing western winds [12,13]. However, the enclosed basins located in those regions have particular wind conditions, as in the case of the Black Sea environment. Thus, the characteristics of the wind conditions in the Black Sea seem to present interest for the scientific communities, as we can see from the previous works focused on meteorological studies, accuracy assessment of various wave models or renewable studies. Rusu et al. [14,15] assessed the performance of a wind-wave modeling system, where among others the wind resources were evaluated by considering in-situ measurements (reported at 10 m height). According to these works, we may expect average wind speeds around 7 m/s and 6 m/s, in the vicinity of Romanian offshore and nearshore area, respectively. Close to the Crimean Peninsula, an average wind speed of about 5.5 m/s was considered to be representative. Valchev et al. [16] evaluated the storm events from the western part of this region, highlighting that, during such an event, extreme values of the wind speed may be encountered, in the range of 21.8–27.8 m/s. Onea and Rusu [17] evaluated the regional wind potential for an interval of 14 years considering various sources of data, such as reanalysis wind models, satellite measurements and data coming from 11 in-situ stations, which are located close to the Romanian and Ukrainian coastal areas. According to these results, the western part of the sea seems to have more consistent wind resources suitable for a wind project. Akpinar and Ponce de Leon [18] considered several reanalysis wind datasets to model the storm occurrences in the Black Sea. They assessed the accuracy of the numerical simulations by comparing them against real measurements. The potential of the offshore wind resources from the Mediterranean and Black Sea were assessed by Koletsis et al. [19] considering the climatological changes. This list can continue, mentioning at the same time the DAMWAVE [20] and ACCWA [21] projects which involve wind data reported on long term.

Regarding the renewable studies involving offshore wind turbines, we can mention that there are fewer studies focused on this region, and probably related to the fact that the Black Sea area is considered less attractive for such projects. Davy et al. [22] briefly discussed the performance of Enercon E-126 wind turbine (7.6 MW) for the Black and Azov Seas, while Onea and Rusu [23] discussed the expected efficiency of a Siemens 2.3 generator which may operate in the northwestern part of this basin. Ilkilic and Aydin [24], in a review study, provided a complete description of wind projects that operate in the Turkish coastal regions (onshore). The expected energy performances of some commercial offshore wind turbines were assessed by Onea and Rusu [25] for several sites distributed along the Black Sea coastline at a water depth of about 50 m. According to these results, during the nocturnal interval, the wind turbines considered may have better performance than in the diurnal period. Argin and Yerci [26] proposed several offshore sites that seem to be suitable for the development of a wind project. Raileanu et al. [27] focused on the performance assessment of two offshore wind turbines (Siemens SWT-3.6-120 and Senvion 6.2M 126) by using satellite measurements recorded between January 2010 and December 2014.

In this context, the objectives of the present work were: (1) to assess the Black Sea wind characteristics by using a reanalysis product defined by a relatively high resolution ($0.125° \times 0.125°$); (2) to estimate the performances of several state-of-the-art wind turbines, including a generator rated at 9.5 MW; and (3) to identify the configurations for wind projects that are suitable for the Black Sea coastal areas.

2. Materials and Methods

2.1. The Target Area

The Black Sea is a semi-enclosed basin defined by an area of 423,000 km^2 and a maximum depth of 2258 m. The geographical coordinates of its extreme points are 41°N/46°N and 27°E/42°E. In the western part, an extended continental shelf defined by a lower water level can be noticed, while for the rest of the sea a steep continental slope ends with a flat sea bed, where the depths easily exceed 2000 m. On a large scale, the wind conditions are under the influence of the Siberian and Azores high-pressure areas and by the Asian low-pressure area. It is expected that during the wintertime, the prevailing wind speeds (predominant values) are around 8 m/s (from northeast), while during the summer time we may expect wind conditions of 2–5 m/s coming from the northwest [28]. In addition, it is important to mention the local winds such as the breezes, which can account for 190 days/year in the southern part of Crimea, while this event is less noticeable in the western part of the sea. Since this area is surrounded by mountains, it is possible to notice some local events, such as the Bora wind which is more visible in the vicinity of the Novorossiysk region, being caused by the strong northeast Arctic wind collapsing through Kolkhida Lowland [17,29].

Figure 1 illustrates the locations of the twenty reference sites (denoted clockwise from P1 to P20) considered for assessment, which are distributed between four sectors (A–D). Each site is associated to a major Black Sea city or harbor, as shown in Table 1. Furthermore, they were defined in water depths of 26–31 m, being the average depth at which we may find most European wind projects [11]. However, the sites Odessa (14 m) and Kerch (11 m) have smaller depths, while the last one is located in the Azov Sea. A maximum distance from the shore of 42 km corresponds to Site P3 (Bilhorod-Dnistrovskyi, Ukraine), while a minimum of 0.9 km is indicated for Site P19 (Primorsko, Bulgaria).

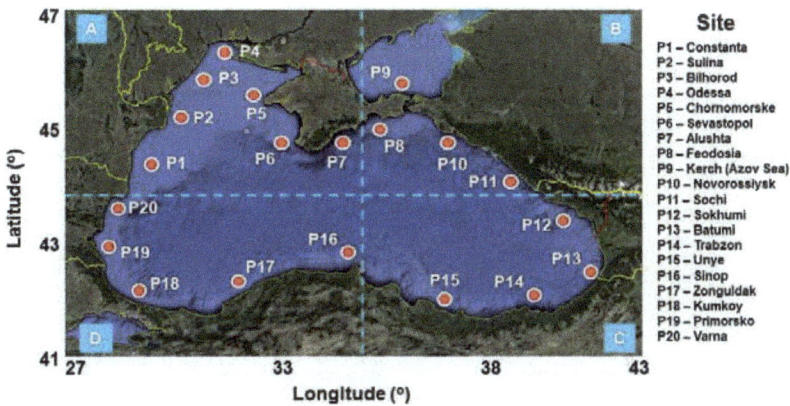

Figure 1. The Black Sea area and the location of the reference sites (map from Google Earth, 2018).

Table 1. The main characteristics of the sites considered.

No.	Site	Country	Sector	Long (°)	Lat (°)	Water Depth (m)	Distance from Coastline (km)
P1	Constanta	Romania	A	28.77	44.15	31	9.32
P2	Sulina		A	29.90	45.09	28	7.17
P3	Bilhorod-Dnistrovskyi	Ukraine	A	30.82	45.78	26	42.37
P4	Odessa		A	31.18	46.45	14	32
P5	Chornomorske	Russia	A	32.66	45.57	28	6.93
P6	Sevastopol		A	33.36	44.58	32	1.46
P7	Alushta		A	34.45	44.66	30	3.16
P8	Feodosia		B	35.52	44.95	30	9.66
P9	Kerch (*Azov Sea*)		B	36.48	45.61	11	18.11
P10	Novorossiysk		B	37.77	44.62	31	3.23
P11	Sochi		B	39.68	43.57	30	1.95
P12	Sokhumi		C	41.02	42.97	29	3.19
P13	Batumi	Georgia	C	41.56	41.64	31	3.59
P14	Trabzon	Turkey	C	39.71	41.02	32	1.4
P15	Unye		C	37.31	41.15	33	3.44
P16	Sinop		D	35.14	42.04	30	2.09
P17	Zonguldak		D	31.76	41.47	31	2.54
P18	Kumkoy		D	29.04	41.27	28	2.14
P19	Primorsko	Bulgaria	D	27.77	42.26	30	0.87
P20	Varna		D	28.21	43.13	30	22.34

2.2. Dataset

The reanalysis wind data considered in the present work are provided by the European Centre for Medium-Range Weather Forecasts (ECMWF) [30,31]. Thus, the considered ERA-Interim product is defined by a high spatial resolution of 0.125° × 0.125°, a temporal resolution of 6 h (associated to 00:00–06:00–12:00–18:00 UTC) and a 20-year time interval from January 1998 to December 2017. The wind fields are reported to a 10 m height above the sea level, and in this case the wind speed is denoted with $U10$ (m/s). For large water areas, the ERA-Interim data are frequently used to assess the wind potential on various regions, such as Global [32,33], Europe [34–36], South China Sea [37], South Korea [38] or Chile [39]. Another dataset used in this work comes from the AVISO (Archiving, Validation and Interpretation of Satellite Oceanographic Data) project, and includes gridded near-real time wind speeds. This is a multi-mission product, which means that data from at least two missions, such as TOPEX/Poseidon, OSTM/Jason-2 or Saral/AltiKa, need to be available [40,41]. The NetCDF files provided by AVISO include measurements of the $U10$ parameter, which are available for the time interval from September 2009 to September 2017 (one measurement per day).

2.3. Wind Turbines

In Table 2, the main characteristics of the offshore wind turbines considered in the present work are presented, which are currently considered for implementation in European offshore wind projects [42]. The selected turbines cover a full spectrum of rated capacities, starting from 3 MW for the V90-3.0MW system and ending with a 9.5 MW for the V164-9.5MW generator, which is expected to be implemented in near-future projects. The power curves of each device can be identified throughout the cut-in, rated speed and cut-out thresholds, while the hub height was considered the lowest value indicated by the manufacturer, for which the performance of each system was assessed.

Table 2. Technical specifications for the technologies considered.

Turbine	Rated Power (MW)	Cut-in Speed (m/s)	Rated Speed (m/s)	Cut-out Speed (m/s)	Hub Height (m)	Reference
V90-3.0 MW	3	4	15	25	80–105	[43]
Areva M5000-116	5	4	12.5	25	90	[44]
Senvion 6.2M126	6.15	3.5	13.5	30	85–95	[45]
V164-8.8 MW	8.8	4	13	25	105–140	[46]
V164-9.5 MW	9.5	3.5	14	25	105–140	[47]

Usually, standard wind datasets are provided for a 10 m height, but, to adjust these values to the hub height of a particular wind turbine, it is possible to use a logarithmic law [48]. In the present work, the wind resources were assessed at 80 m height, which represents the hub height of the V90-3.0 MW system. This logarithmic law is expressed as:

$$U80 = U10 \frac{\ln(z_{80}) - \ln(z_{10})}{\ln(z_{10}) - \ln(z_0)} \tag{1}$$

where $U80$ represents the wind speed at 80 m, $U10$ is the initial wind speed (at 10 m), z_0 represents the roughness of the sea surface (0.01 m), and z_{10} and z_{80} are the reference heights.

To estimate the Annual Electricity Production (AEP) of a wind turbine, several approaches are available [49]. For the present work, a similar method to that used by Hrafnkelsson et al. [50] and Salvação and Guedes Soares [51] was used. It can be defined as:

$$AEP = T \times \int_{cut-in}^{cut-out} f(u)P(u)du \tag{2}$$

where AEP is expressed in MWh, T represents the average hours per year (8760 h/year), $f(u)$ is the Weibull probability density function, $P(u)$ is the power curve of a turbine, and cut-in and cut-out represent the turbine characteristics presented in Table 2. Figure 2 illustrates the $U80$ histogram of four sites and the turbine power curves. From the combination of the wind histograms and power curves, we can easily notice that better performances may be expected for Sites P1 and P9, which present a more consistent presence of the wind conditions higher than 10 m/s.

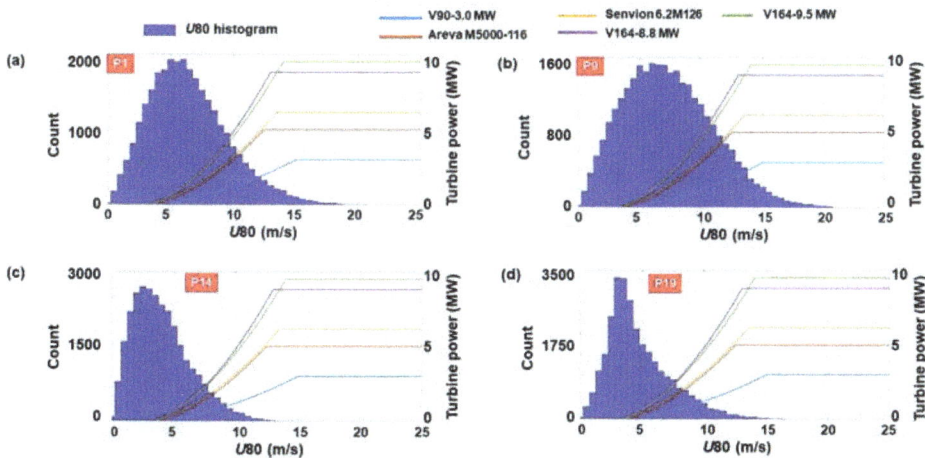

Figure 2. Representation of the wind turbine power curves and wind speed histograms ($U80$). The wind distribution is related to the ERA-Interim project (January 1998–December 2017) and includes the sites: (**a**) P1; (**b**) P9; (**c**) P14; and (**d**) P19.

3. Results

3.1. Analysis of the Wind Data

A first analysis is presented in Figure 3, where a direct comparison between the AVISO measurements and the ERA-Interim data was carried out. Figure 3a presents the percentage of the missing values corresponding to the satellite data for each site, from which we can notice three categories. The first one is related to 100% missing values (no data) which include sites such as P1, P13 and P19. This indicates that for these sites another type of data should be considered. Other sites exceed the 10% limit, which is usually considered acceptable for the accuracy of a dataset [52]. This means that the results are biased for these sites. In the group located below 10%, we may find P5–P11, P16 and P18, while Sites P6–P8 do not exceed 0.35% compared to P9 where a value of 10.5% is reported. Figure 3b illustrates the differences between the two datasets in terms of the 50th and 95th percentiles, where the negative values indicate that the AVISO measurements exceed the ERA-Interim values. In general, it seems that the reanalysis data overestimate the wind resources, which for Sites P5–P10 indicate a maximum difference of 2.9 m/s (95th percentile), while, for Sites P11 (east) and P16 (south), it is possible to notice negative values, which reach a value of 2.1 m/s.

Figure 4 illustrates the distribution of the average $U80$ parameter, reported for various intervals, such as total time (full time distribution), winter season (December–February), spring season (March–May), summer season (June–August) and autumn season (September–November). Usually, the attractiveness of a site for a wind project is indicated throughout wind classes, denoted C1–C7, with higher classes being considered more promising for renewable projects [53]. As expected, the most energetic season is winter, while summer has much lower values. The energy pattern is visible in the case of Sites P1–P10 which indicates values in the range 6–9 m/s, more promising results being expected for Sites P6 (Sevastopol) and P9 (Kerch-Azov Sea), which during the winter present values located close to C6 wind class. In addition, Sites P18 and P20 seem to be defined by relevant wind resources, which nevertheless do not exceed the C4 class. Much lower wind conditions are noticed near Sites P11–P17 (south and southeast), indicating average values located in the C1 class, regardless of the time interval considered for assessment.

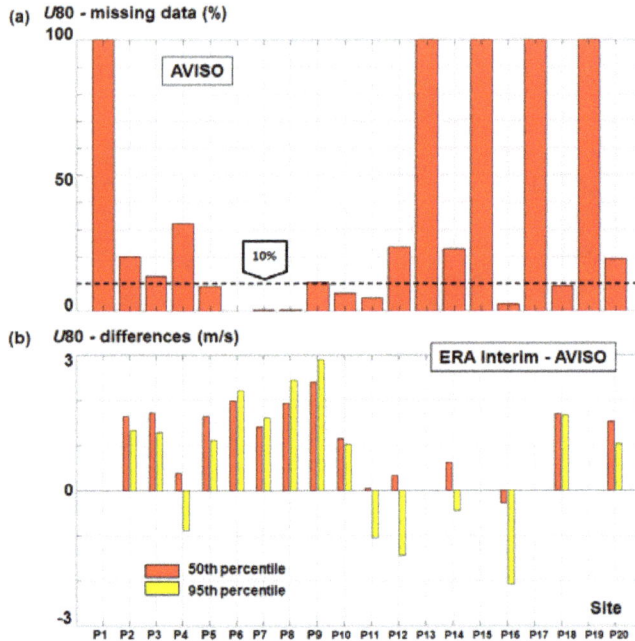

Figure 3. Comparison between the ERA-Interim wind data and the satellite measurements (from AVISO) corresponding to the time interval September 2009–September 2017. Results indicated in terms of: (**a**) missing data reported by AVISO; and (**b**) differences reported between ERA-Interim and AVISO data, considering the 50th an 95th percentiles, respectively.

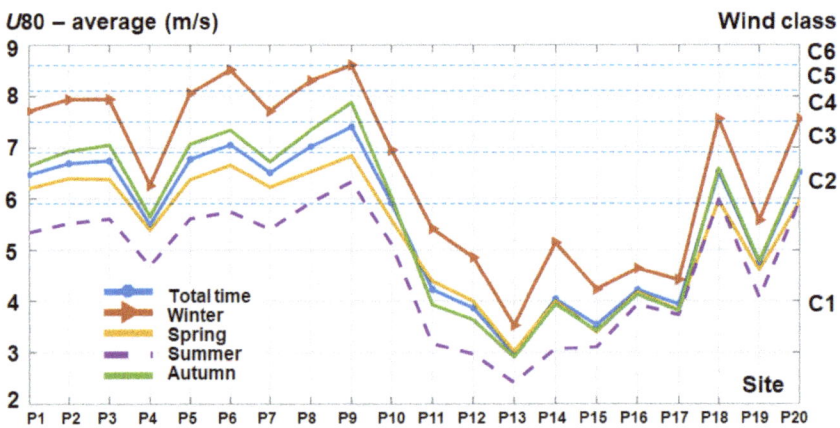

Figure 4. Distribution of the $U80$ average values corresponding to the full time distribution and the representative seasons, considering the ERA-Interim data (January 1998–December 2017). The wind class levels (C1–C6) are also indicated.

In Figure 5, the Weibull distributions are illustrated, considering all the reference sites. These results are based on the ERA-Interim data (total time) and provide some insights regarding the distribution of the wind resources by intervals and the energy potential of a particular site. Much lower performances are expected for Sites P13, P15 and P17 which present peaks around 2–3 m/s.

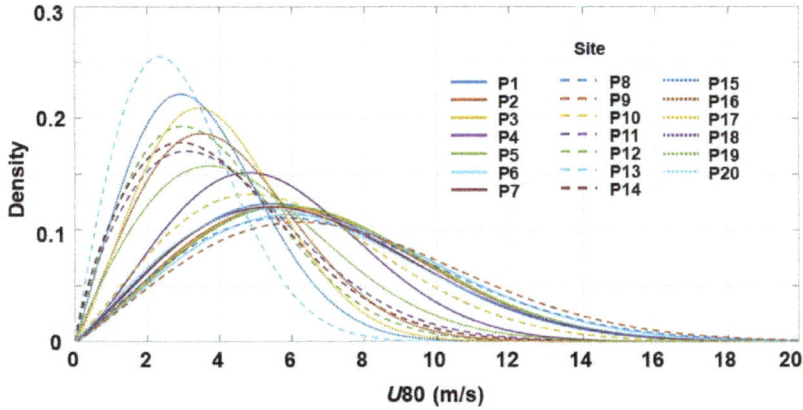

Figure 5. The Weibull distributions corresponding to the ERA-Interim values (January 1998–December 2017).

Another important parameter is the wind direction, which is represented in Figure 6 for Sites P3, P10, P14 and P18 considering only the total time data, distributed by wind classes, from which we can mention that only Site P18 indicates wind conditions coming from the offshore area. Each site has a different pattern indicating for Site P3 a significant distribution from the northern sector, compared to P10 where maximum 8% may be expected from the northeast. Site P18 presents a similar pattern as P10, while in the case of Trabzon site (P14) most of the wind resources are coming from the onshore area, compared to similar ones coming from the sea, which seem to be less energetic.

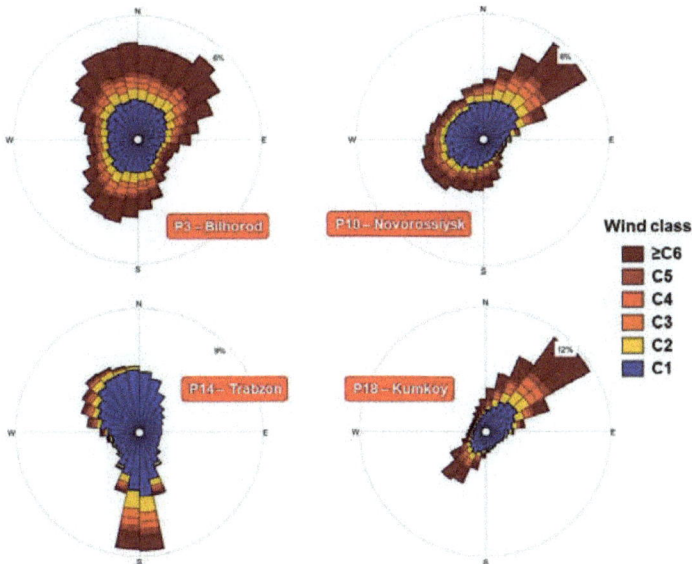

Figure 6. The directional wind distribution, expressed in terms of the wind classes (C1–C6). The full time distribution corresponds to the ERA-Interim data (January 1998–December 2017).

Table 3 summarizes some statistical results presented for each season. Significant variations may occur in the case of the wind direction, as we can see for Sites P13 and P18, which during winter and summer intervals may indicate differences of 56.55° and 62.56°, respectively. In the case of P2, this variation is around 21.39°, which is quite similar to the ones reported between winter and autumn (23.18°). As for the distribution of C3–C6 classes, we may expect that during the winter Sites P2, P9 and P18 reach values in the interval 47–59%, compared to a minimum of 0.9 % reported by P13. These values gradually decrease as we shift to the summer season where a maximum of 34.26% was found at P9, while P13 indicates a value close to zero.

Table 3. *U*80 seasonal statistics indicated for some of the reference sites. The results are related to the ERA-Interim data and cover the interval from January 1998 to December 2017.

	Winter				Spring		
	*U*80 (m/s)	*Dir* (°)	C3–C6 (%)		*U*80 (m/s)	*Dir* (°)	C3–C6 (%)
P2	7.93	189.02	52.64	P2	6.37	172.74	33.99
P9	8.61	169.65	59.00	P9	6.83	168.56	39.55
P13	3.52	183.43	0.88	P13	3.02	206.74	0.42
P18	7.55	160.41	46.99	P18	5.95	143.63	27.39
	Summer				Autumn		
P2	5.52	167.66	21.52	P2	6.93	165.84	39.88
P9	6.32	156.63	34.26	P9	7.88	154.94	50.86
P13	2.41	239.98	0.04	P13	2.91	195.09	0.78
P18	5.98	97.85	29.47	P18	6.58	130.94	36.78

3.2. Evaluation of the Wind Turbine Performance

For the wind turbines, a first indicator considered for assessment is the Capacity Factor (*Cf* in %), which is frequently used to estimate the efficiency of a turbine. This can be defined as [25]:

$$Cf = \frac{P_{turbine}}{P_{rated}} \cdot 100 \tag{3}$$

where $P_{turbine}$ is the theoretical energy output of a wind turbine (in MW) and P_{rated} is the rated power of a wind turbine (in MW) mentioned in Table 2.

Figure 7 illustrates the distribution of this index, where the values are sorted in descending order. The first places include Sites P9, P8, P6 and P5 (Crimea Peninsula) while on the opposite side we may find P12, P17, P15 and P13. In the case of the Areva M5000-116 system, we may notice better performances, which reach a maximum of 35% at P9, 25.7% at P1 and 2% at P13. The turbines Senvion 6.2M126 and V164-8.8 MW have similar values, which are almost identical as the values decrease through P10 (19.1%). For the sites located on the lower end of this chart, it seems that they have the same efficiency regardless of the rated capacity considered, while turbine V90-3.0 MW presents much lower values. By comparing the values reported by Areva M5000-116 with the ones from V90-3.0 MW, we observe the following differences: P9, 11%; P2 and P1, 8%; P11, 3%; and P13, 0.8%.

Another important index is the rated capacity (in %) which expresses the percentage of the time during which a particular turbine will operate at full capacity, being defined as the wind distribution between the rated speed and the cut-out speed. In Figure 8, we can see the evolution of this parameter, which does not exceed 10% and has a similar distribution as in the case of the *Cf* index, where the system Areva M500-116 presents the best performances compared to V90-3.0 MW, which is on the last place. In the first places, we find P9, P8 and P6, which report values in the range 7.42–9.8%. The values are gradually decreasing for the reference points P18 (5.4%) and P5 (5.3%), while as we approach to Site P19 there is a decrease of this index below the 1.2% limit. A maximum value of 2.9% is reported by the system V90-3.0 MW close to Site P9, while turbine V164-9.5 MW, which has the highest rated capacity, presents values in the range 3.85–5.25% for the group Sites P6–P9 and a sharp decrease from 2.5% (Site P5) to 0.5% (Site P19).

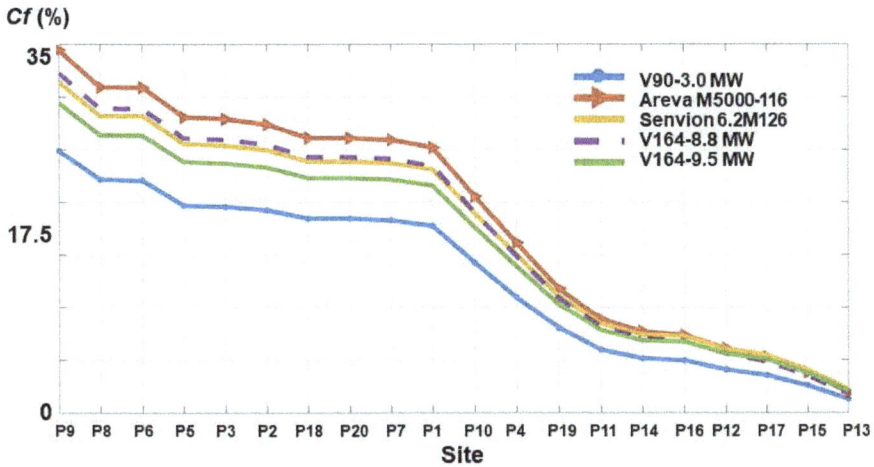

Figure 7. Capacity factor (in %) of the wind turbines considering the ERA-Interim data for the time interval January 1998–December 2017.

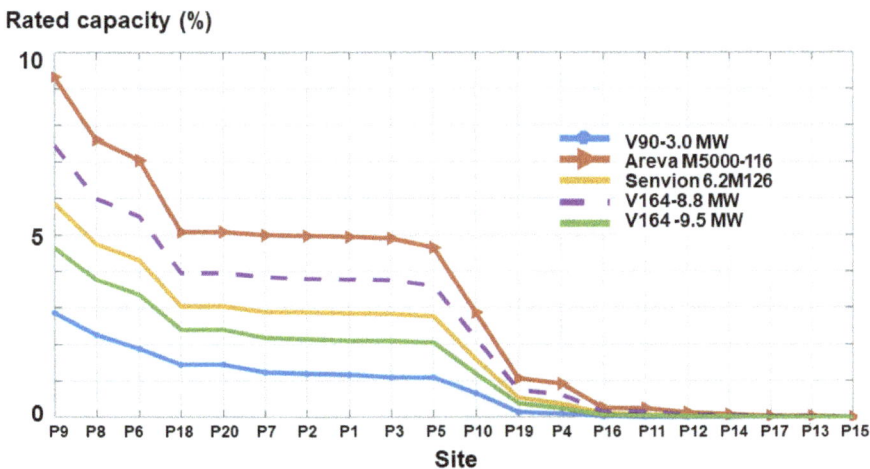

Figure 8. The rated capacity (in %) of the wind turbines which may operate in the Black Sea area. Results based on the ERA-Interim data covering the interval January 1998–December 2017.

Figure 9 presents the evolution of the AEP index, from which we can notice that this is influenced by the local wind resources and by the rated power of the turbine. In this case, better performances are reported by the systems V164-8.8 MW and V164-9.5 MW, which seem to have similar values indicating a maximum of 24,738 MWh near P9 and a minimum of 1747 MWh at P13. From the Areva M5000-116 and Senvion 6.2M126, the second turbine seems to generate more electricity, the differences between them being around: P1, 1319 MWh; P6, 1532 MWh; P11, 620 MWh; P16, 659 MWh; and P20, 1326 MWh. The system V90-3.0MW seems to be the less attractive turbine for the Black Sea environment, presenting much lower AEP values. For the interval P1–P9, this indicates values in the range 4657–6539 MWh, except the peak of 2876 MWh for Site P4. Sites P11–P17 account for the lowest values, indicating a minimum of 337 MWh (P13) and a maximum of 1573 MWh (P11).

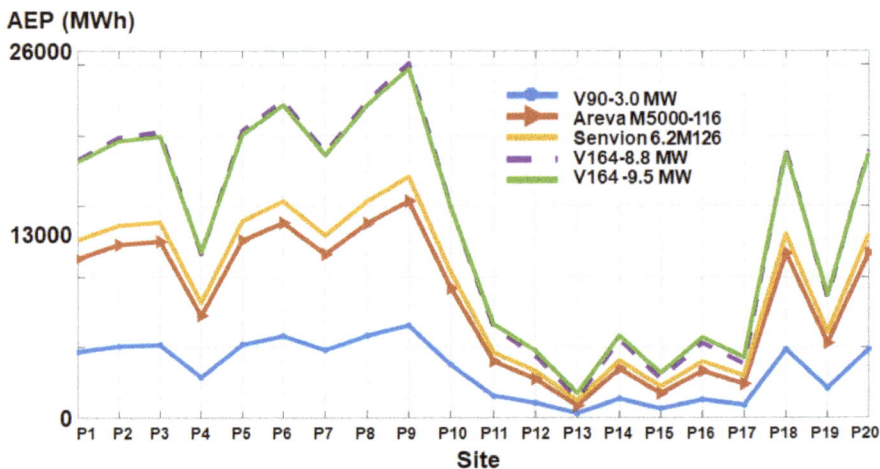

Figure 9. Annual Energy Production (AEP in MWh) of the offshore wind turbines.

Table 4 presents a more detailed assessment of the AEP index, by considering the seasonal fluctuations. As expected, the winter dominates with more impressive values, and at a first look it seems that the differences reported between the systems Areva M5000-116 and Senvion 6.2M126 are more significant than similar ones indicated for the Vestas turbines rated at 8.8 MW and 9.5 MW. In addition, it is important to mention that, in most cases, the system rated at 8.8 MW exceeds the AEP values reported by the 9.5 MW turbine, which indicate that probably for this coastal environment the first turbine will be more suitable. During the summer, we may expect a minimum of 94 MWh in P13, which may easily increase to 726 MWh if a V164-9.5 MW turbine were used.

Table 4. The AEP production (MWh) corresponding to the offshore wind turbines on a seasonal level. Results available for the interval from January 1998 to December 2017 (ERA-Interim wind data).

	Winter						Spring				
Turbine	T1	T2	T3	T4	T5		T1	T2	T3	T4	T5
P2	7325	17,303	19,155	29,187	28,654	P2	4458	11,016	12,188	18,497	18,219
P9	8748	20,042	22,320	33,925	33,431	P9	5561	13,207	14,699	22,274	21,989
P13	593	1537	2053	2622	3129	P13	338	880	1236	1510	1893
P18	6827	15,806	17,637	26,750	26,457	P18	3870	9468	10,608	15,934	15,856
	Summer						Autumn				
P2	3033	7575	8544	12,721	12,785	P2	5453	13,229	14,623	22,254	21,866
P9	4468	11,038	12,210	18,507	18,217	P9	7428	17,225	19,180	29,165	28,716
P13	94	254	458	451	726	P13	320	836	1153	1423	1760
P18	3704	9381	10,369	15,672	15457	P18	5025	12,121	13,472	20,414	20,145

T1, V90-3.0 MW; T2, Areva M5000-116; T3, Senvion 6.2M126; T4, V164-8.8 MW; T5, V164-9.5 MW.

4. Discussion

Since there are currently no operational offshore wind projects in the Black Sea area, the purpose of this section is to identify some suitable projects that may be implemented in this region. One way is to compare the wind resources from the Black Sea sites with the ones reported near some offshore wind farms that are already operating in Europe.

Figure 10 presents the distribution of the *U*80 parameter (average values based on *U*10 satellite measurements) for Sites P1–P20. As noticed, some sites have missing values (*NaN*, Not A Number values) which may indicate that the selected sites are located too close to the coastline and the satellite missions are not able to accurately measure the local wind conditions. Much lower values are reported compared to ERA-Interim data (Figure 4), while in this case the reference location P5 seems to have the best wind resources compared to P9 as indicated by the reanalysis dataset. The sites located in Sector A report relatively small differences, while the sites located in Sector C seem to register moderate wind resources, a minimum of 3.78 m/s being observed close to P14.

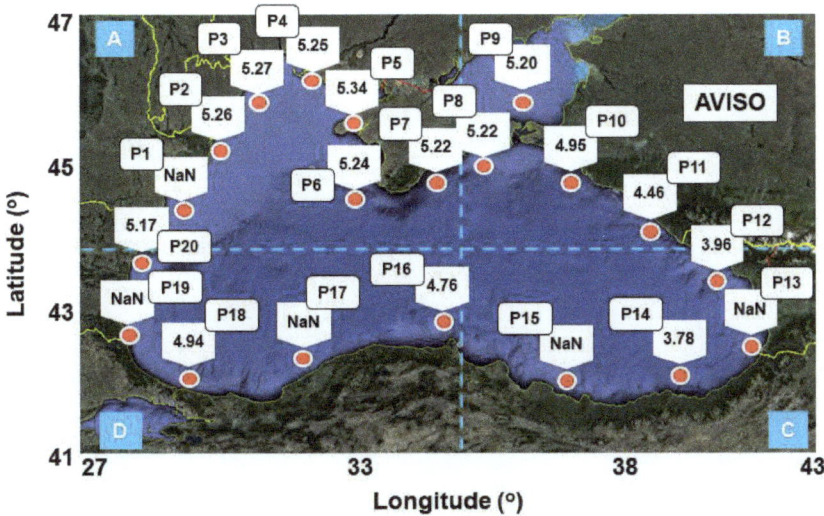

Figure 10. *U*80 average values reported by the AVISO measurements (September 2009–September 2017).

Although the best sites from the point of view of the wind resources seem to be located close to the Crimea Peninsula, considering the current geopolitical issues, it is difficult to believe that a renewable project can be developed in the near future. Therefore, attention probably needs to be shifted to an EU country, such as Romania, which also seems to present better wind resources according to AVISO and ERA-Interim data, thus Site P2 is considered for comparison with some European projects.

The wind conditions in the vicinity of 171 offshore wind projects [42] were considered for assessment, including projects in Belgium, Denmark, France, Germany, Netherlands, Sweden and United Kingdom. By comparing the wind resources at Site P2 (Black Sea, Romania) with the ones in Europe, a top 10 best agreement was identified, as presented in Figure 11. The projects in the early stages of development are denoted with an asterisk symbol (*) and more details about them can be found in Table 5. According to the AVISO values (Figure 11a), all of the European sites seem to have better wind resources than Site P2, on average values are between 5.7 m/s and 5.9 m/s, being noticed a constant distribution between Borkum and Wikinger.

In Figure 11b, we can identify the water depth and the distance from the coastline of the selected sites. From these combinations, it results that Site P2 is located a similar distance from the shore as the Bockstingen and Nenuphar sites, which are used as test sites. As for the water depth, much lower values are reported by two operational projects from Germany (EnBW Baltic1, 48.3 MW; and Borkum Riffgrund 1, 582 MW) and by two early development stage projects from Sweden and Denmark (Kriegers Flak, 590 MW; and Kriegers Flak II, 640 MW).

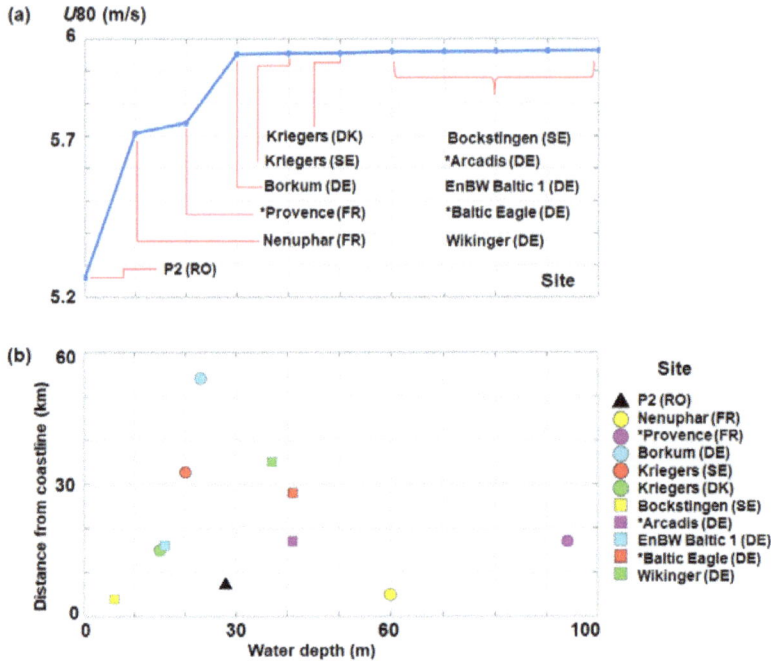

Figure 11. Direct comparison between the wind conditions reported close to reference point P2 and some European wind projects, where the dataset correspond to the AVISO measurements (September 2009–September 2017). The results correspond to: (**a**) *U*80 comparison; and (**b**) main characteristics of the selected sites.

Table 5. The European offshore wind projects considered for comparison [42].

No.	Project	Country	Status	Project Capacity (MW)	Turbine Model	Water Depth (m)	Distance from Coastline (km)
1	Nénuphar (test site)	France	Consent authorized	10	Not decided	60–70	5
2	Provence (floating)	France	Early planning	24	SWT-8.0-154	94–104	17
3	Borkum Riffgrund 1	Germany	Operational	582	SWT-6.0-154	28–34	45
4	Kriegers Flak II	Sweden	Consent authorized	640	Not decided	20–40	32.7
5	Kriegers Flak	Denmark	Pre-construction	590	SG 8.0-167 DD	15–30	15
6	Bockstingen	Sweden	Operational	2.75	Wind World 550 kW	6	4
7	Arcadis Ost 1	Germany	Consent authorized	247	Haliade 150-6 MW	41–46	17
8	EnBW Baltic 1	Germany	Operational	48.3	SWT-2.3-93	16–19	16
9	Baltic Eagle	Germany	Consent authorized	476	Not decided	41–44	28
10	Wikinger	Germany	Operational	350	AD 5-135	37–43	35

According to the CIA World Factbook [54], during 2015–2017, the estimated electricity productions of the countries located around the Black Sea were around: Romania, 62.16 billion kWh; Ukraine, 152.2 billion kWh; Russia, 1.008 trillion kWh; Georgia, 11.57 billion kWh; Turkey, 245.8 billion kWh; and Bulgaria, 45.04 billion kWh. To reach at least 1% of this capacity for Romania with offshore wind turbines, according to the AEP values presented in Figure 9, it would be necessary for at least 123 V90-3.0 MW turbines (denoted with T1) or 32 V164-9.5 MW turbines (denoted with T5) (Site P2) to be installed. For the rest of the countries, we may estimate: Ukraine (Site P3), 296 T1 systems or 76 T5 systems; Georgia (P13), 344 T1 systems or 66 T5 systems; Turkey (Site P18), 507 T1 systems or 131 T5 systems; and Bulgaria (Site P20), 93 T1 systems or 24 T5 systems. The most energetic sites were considered for assessment, while the sites from Russia were not included in this analysis since the necessary number of turbines cannot be supported by a single site.

Referring to similar studies focused on the Black Sea wind energy assessment, such as the one presented ibyn Onea and Rusu [25], we can notice that the western part of this region seems to reveal more important wind resources, in particular close to the Romanian nearshore. In that work, the sites are defined for a water depth located close to 50 m, while the evaluation uses a different wind dataset coming from US National Centers for Environmental Prediction, which is processed for 1979–2010. Regarding the performances of a wind turbine, the same work highlights that the Senvion 6.2M126 generator may reveal close to the Crimean Peninsula (northwest) a maximum of 7.3%, which is much higher than the one reported in this work. Nevertheless, compared to some other previous works, which are focused on the assessment of various wind turbines performances [48,55], the present approach seems to be more exact. This is because it involved the combination of a Weibull probability density function and the power curve of the turbine, while some previous works use a method that considers various values of the Betz coefficient, sometimes chosen close to the ideal threshold (59%), which cannot be considered realistic for a wind project.

5. Conclusions

In the present work, the nearshore wind energy resources from the Black Sea region were assessed by considering some environmental parameters (such as $U80$) and the power curves of some state-of-the-art wind turbines, which operate or are currently implemented in European offshore wind projects. Based on the ERA-Interim reanalysis dataset (corresponding to 1998–2017) and AVISO satellite measurements (corresponding to September 2009–September 2017), it was possible to highlight the dominant energy patterns from this area, revealing some hot-spots, which can be used for developing future wind projects. Both wind datasets indicate the Crimea Peninsula region as being one of the most energetic regions in the Black Sea. While ERA-Interim indicates the sites from the Azov Sea as being more important, AVISO highlights the sites from the western coast of this peninsula. Since there are currently no nearshore or offshore wind projects in this region, another direction of the present work was related to the assessment of the performances provided by some standard wind turbines and to estimate what kind of projects could operate in this region. According to these results, it seems that the Black Sea may be suitable for the development of wind projects, especially in the western part of the basin which is dominated by a lower water depth, in particular for the development of some pilot projects or test sites.

The main findings of this work are:

1. From the literature review, it was highlighted the fact that wind resources in the Black Sea are not very often evaluated from the point of view of a renewable energy project, most of the research being focused on climatological studies or calibrations of the numerical models.
2. The Crimea Peninsula seems to present the most attractive wind resources, but considering the current geopolitical situation, the likelihood of a wind project in the near future in those coastal waters is not high.
3. Considering the expected performances of the V90-3.0 MW, we can say that this is not a viable candidate for the Black Sea area. A lower rated wind speed system, such as Areva M5000-116, seems to be more efficient for this coastal environment.
4. A wind project, including high capacity turbines, seems to be more productive in the western and northern parts of the Black Sea (Sectors A and B). There is a significant difference between the results reported by Senvion 6.2M126 and V164-8.8 MW or V164-9.5 MW, while for latter two turbines the results are almost identical in the case of the AEP index. For most sites, better results are reported by the system rated at 8.8 MW.
5. Based on the electricity production statistics and the computed AEP values, it was possible to estimate the required number of turbines to reach at least 1% of the national share. The results indicate that, for the western part of the Black Sea, this target can be easily achieved throughout one or two wind farms projects.

Author Contributions: F.O. performed the literature review, processed the wind data, carried out the statistical analysis and interpreted the results. L.R. guided this research, wrote the final form of the manuscript and drew the conclusions. The final manuscript was approved by all authors.

Funding: This work was supported by a grant of Ministery of Research and Innovation, CNCS—UEFISCDI, project number PN-III-P1-1.1-PD-2016-0235, within PNCDI III.

Acknowledgments: ECMWF ERA-Interim data used in this studywere obtained from the ECMWF data server. The altimeter products were generated and distributed by Aviso (http://www.aviso.altimetry.fr/) as part of the SSALTO ground processing segment.

Conflicts of Interest: The authors declare no conflict of interest.

Nomenclature

ACCWA	Assessment of the Climate Change effects on the WAve conditions in the Black Sea
AEP	Annual Electricity Production
AVISO	Archiving, Validation and Interpretation of Satellite Oceanographic Data
DAMWAVE	Data Assimilation Methods for improving the WAVE predictions in the Romanian nearshore of the Black Sea
ECMWF	European Centre for Medium-Range Weather Forecasts
NaN	Not A Number
Cf	capacity factor
T	average hours per year (8760 h/year)
*U*10	wind speed reported at 10 m above sea level
*U*80	wind speed reported at 80 m above sea level
z_0	roughness of the sea surface (0.01 m)
P(u)	power curve of a turbine
f(u)	Weibull probability density function
z_{10}; z_{80}	reference heights

References

1. Ji, Q.; Zhang, D.; Geng, J. Information linkage, dynamic spillovers in prices and volatility between the carbon and energy markets. *J. Clean. Prod.* **2018**, *198*, 972–978. [CrossRef]
2. Wang, Y.; Guo, Z. The dynamic spillover between carbon and energy markets: New evidence. *Energy* **2018**, *149*, 24–33. [CrossRef]
3. Child, M.; Koskinen, O.; Linnanen, L.; Breyer, C. Sustainability guardrails for energy scenarios of the global energy transition. *Renew. Sustain. Energy Rev.* **2018**, *91*, 321–334. [CrossRef]
4. Armeanu, D.Ş.; Vintilă, G.; Gherghina, Ş.C. Does Renewable Energy Drive Sustainable Economic Growth? Multivariate Panel Data Evidence for EU-28 Countries. *Energies* **2017**, *10*, 381. [CrossRef]
5. Karakosta, C. A holistic approach for addressing the issue of effective technology transfer in the frame of climate change. *Energies* **2016**, *9*, 503. [CrossRef]
6. Castro-Santos, L.; Martins, E.; Guedes Soares, C. Methodology to calculate the costs of a floating offshore renewable energy farm. *Energies* **2016**, *9*, 324. [CrossRef]
7. Willis, D.J.; Niezrecki, C.; Kuchma, D.; Hines, E.; Arwade, S.R.; Barthelmie, R.J.; DiPaola, M.; Drane, P.J.; Hansen, C.J.; Inalpolat, M. Wind energy research: State-of-the-art and future research directions. *Renew. Energy* **2018**, *125*, 133–154. [CrossRef]
8. Poulsen, T.; Hasager, C.B. The Revolution of China: Offshore Wind Diffusion. *Energies* **2017**, *10*, 2153. [CrossRef]
9. Lapworth, A. The diurnal variation of the marine surface wind in an offshore flow. *Q. J. R. Meteorol. Soc.* **2005**, *131*, 2367–2387. [CrossRef]
10. Global Statistics. Available online: http://gwec.net/global-figures/graphs/ (accessed on 15 August 2018).
11. The European Offshore Wind Industry–Key Trends and Statistics 2017. Available online: https://windeurope.org/about-wind/statistics/offshore/european-offshore-wind-industry-key-trends-statistics-2017/ (accessed on 15 August 2018).

12. Rusu, E.; Onea, F. Joint Evaluation of the Wave and Offshore Wind Energy Resources in the Developing Countries. *Energies* **2017**, *10*, 1866. [CrossRef]
13. Rusu, L.; Onea, F. The performance of some state-of-the-art wave energy converters in locations with the worldwide highest wave power. *Renew. Sustain Energy Rev.* **2017**, *75*, 1348–1362. [CrossRef]
14. Rusu, L.; Bernardino, M.; Guedes Soares, C. Wind and wave modelling in the Black Sea. *J. Oper. Oceanogr.* **2014**, *7*, 5–20. [CrossRef]
15. Rusu, L.; Ganea, D.; Mereuta, E. A joint evaluation of wave and wind energy resources in the Black Sea based on 20-year hindcast information. *Energy Explor. Exploit.* **2018**, *36*, 335–351. [CrossRef]
16. Valchev, N.N.; Trifonova, E.V.; Andreeva, N.K. Past and recent trends in the western black sea storminess. *Nat. Hazards Earth Syst. Sci.* **2012**, *12*, 961–977. [CrossRef]
17. Onea, F.; Rusu, E. Wind energy assessments along the Black Sea basin. *Meteorol. Appl.* **2014**, *21*, 316–329. [CrossRef]
18. Akpinar, A.; Ponce de León, S. An assessment of the wind re-analyses in the modelling of an extreme sea state in the black sea. *Dyn. Atmos. Ocean.* **2016**, *73*, 61–75. [CrossRef]
19. Koletsis, I.; Kotroni, V.; Lagouvardos, K.; Soukissian, T. Assessment of offshore wind speed and power potential over the mediterranean and the black seas under future climate changes. *Renew. Sustain. Energy Rev.* **2016**, *60*, 234–245. [CrossRef]
20. Damwave. Available online: http://www.im.ugal.ro/DAMWAVE/index_engleza.htm (accessed on 25 March 2018).
21. ACCWA. Available online: http://193.231.148.42/accwa/index_en.php# (accessed on 25 March 2018).
22. Davy, R.; Gnatiuk, N.; Pettersson, L.; Bobylev, L. Climate change impacts on wind energy potential in the European domain with a focus on the Black Sea. *Renew. Sustain. Energy Rev.* **2018**, *81*, 1652–1659. [CrossRef]
23. Onea, F.; Rusu, E. An Evaluation of the Wind Energy in the North-West of the Black Sea. *Int. J. Green Energy* **2014**, *11*, 465–487. [CrossRef]
24. Ilkiliç, C.; Aydin, H. Wind power potential and usage in the coastal regions of turkey. *Renew. Sustain. Energy Rev.* **2015**, *44*, 78–86. [CrossRef]
25. Onea, F.; Rusu, E. Efficiency assessments for some state of the art wind turbines in the coastal environments of the black and the caspian seas. *Energy Explor. Exploit.* **2016**, *34*, 217–234. [CrossRef]
26. Argın, M.; Yerci, V. The assessment of offshore wind power potential of Turkey. In Proceedings of the 2015 9th International Conference on Electrical and Electronics Engineering (ELECO), Bursa, Turkey, 26–28 November 2015; pp. 966–970.
27. Raileanu, A.B.; Onea, F.; Rusu, E. Evaluation of the Offshore Wind Resources in the European Seas Based on Satellite Measurements. In Proceedings of the Conference: International Multidisciplinary Scientific GeoConferences SGEM, Albena, Bulgaria, 16–25 June 2015; pp. 227–234.
28. Arkhipkin, V.S.; Gippius, F.N.; Koltermann, K.P.; Surkova, G.V. Wind waves in the Black Sea: Results of a hindcast study. *Nat. Hazards Earth Syst. Sci.* **2014**, *14*, 2883–2897. [CrossRef]
29. Kosarev, A.N. *The Black Sea Environment*; Springer-Verlag: Berlin, Germany, 2008; pp. 11–30, ISBN 978-3-540-74291-3.
30. ERA-Interim. Available online: https://www.ecmwf.int/en/forecasts/datasets/archive-datasets/reanalysis-datasets/era-interim (accessed on 18 August 2018).
31. Dee, D.P.; Uppala, S.M.; Simmons, A.J.; Berrisford, P.; Poli, P.; Kobayashi, S.; Andrae, U.; Balmaseda, M.A.; Balsamo, G.; Bauer, P. The ERA-Interim reanalysis: Configuration and performance of the data assimilation system. *Q. J. R. Meteorol. Soc.* **2011**, *137*, 553–597. [CrossRef]
32. Zheng, C.; Xiao, Z.; Peng, Y.; Li, C.; Du, Z. Rezoning global offshore wind energy resources. *Renew. Energy* **2018**, *129*, 1–11. [CrossRef]
33. Dupont, E.; Koppelaar, R.; Jeanmart, H. Global available wind energy with physical and energy return on investment constraints. *Appl. Energy* **2018**, *209*, 322–338. [CrossRef]
34. Onea, F.; Rusu, E. Sustainability of the Reanalysis Databases in Predicting the Wind and Wave Power along the European Coasts. *Sustainability* **2018**, *10*, 193. [CrossRef]
35. Carvalho, D.; Rocha, A.; Gómez-Gesteira, M.; Silva Santos, C. Offshore wind energy resource simulation forced by different reanalyses: Comparison with observed data in the iberian peninsula. *Appl. Energy* **2014**, *134*, 57–64. [CrossRef]

36. Gallagher, S.; Tiron, R.; Whelan, E.; Gleeson, E.; Dias, F.; McGrath, R. The nearshore wind and wave energy potential of Ireland: A high resolution assessment of availability and accessibility. *Renew. Energy* **2016**, *88*, 494–516. [CrossRef]

37. Wan, Y.; Fan, C.; Dai, Y.; Li, L.; Sun, W.; Zhou, P.; Qu, X. Assessment of the Joint Development Potential of Wave and Wind Energy in the South China Sea. *Energies* **2018**, *11*, 398. [CrossRef]

38. Kim, H.-G.; Kim, J.-Y.; Kang, Y.-H. Comparative Evaluation of the Third-Generation Reanalysis Data for Wind Resource Assessment of the Southwestern Offshore in South Korea. *Atmosphere* **2018**, *9*, 73. [CrossRef]

39. Mattar, C.; Borvarán, D. Offshore wind power simulation by using WRF in the central coast of Chile. *Renew. Energy* **2016**, *94*, 22–31. [CrossRef]

40. MSWH/MWind: Aviso+. Available online: https://www.aviso.altimetry.fr/en/data/products/windwave-products/mswhmwind.html (accessed on 18 August 2018).

41. Pujol, M.-I.; Faugère, Y.; Taburet, G.; Dupuy, S.; Pelloquin, C.; Ablain, M.; Picot, N. DUACS DT2014: The new multi-mission altimeter data set reprocessed over. *Ocean Sci.* **2016**, *12*, 1067–1090. [CrossRef]

42. Global Offshore Wind Farms Database–4C Offshore. Available online: https://www.4coffshore.com/windfarms/ (accessed on 19 May 2018).

43. Vestas V90-3.0-300 MW-Wind Turbine. Available online: https://en.wind-turbine-models.com/turbines/603-vestas-v90-3.0 (accessed on 29 July 2018).

44. AREVA M5000-116-500 MW-Wind Turbine. Available online: https://en.wind-turbine-models.com/turbines/23-areva-m5000-116 (accessed on 29 July 2018).

45. 6.2M126-Wind Turbine 6.2 MW. Available online: https://www.senvion.com/global/en/products-services/wind-turbines/6xm/62m126/ (accessed on 19 August 2018).

46. MHI Vestas Offshore V164-8.8 MW-880 MW-Wind Turbine. Available online: https://en.wind-turbine-models.com/turbines/1819-mhi-vestas-offshore-v164-8.8-mw (accessed on 19 August 2018).

47. MHI Vestas Offshore V164/9500–Manufacturers and turbines–Online Access–The Wind Power. Available online: https://www.thewindpower.net/turbine_en_1476_mhi-vestas-offshore_v164-9500.php (accessed on 19 August 2018).

48. Onea, F.; Deleanu, L.; Rusu, L.; Georgescu, C. Evaluation of the wind energy potential along the Mediterranean Sea coasts. *Energy Explor. Exploit.* **2016**, *34*, 766–792. [CrossRef]

49. Manwell, J.F.; McGowan, J.G.; Rogers, A.L. Wind Energy Explained: Theory, design and application. *Wind Eng.* **2006**, *30*, 169–170. [CrossRef]

50. Carballo, R.; Iglesias, G. A methodology to determine the power performance of wave energy converters at a particular coastal location. *Energy Convers. Manag.* **2012**, *61*, 8–18. [CrossRef]

51. Salvacao, N.; Guedes Soares, C. Wind resource assessment offshore the Atlantic Iberian coast with the WRF model. *Energy* **2018**, *145*, 276–287. [CrossRef]

52. Dong, Y.; Peng, C.-Y.J. Principled missing data methods for researchers. *SpringerPlus* **2013**, *2*, 222. [CrossRef] [PubMed]

53. Archer, C.L.; Jacobson, M.Z. Spatial and temporal distributions of US winds and wind power at 80 m derived from measurements. *J. Geophys. Res.* **2003**, *108*, 4289. [CrossRef]

54. The World Factbook. Available online: https://www.cia.gov/library/publications/the-world-factbook/ (accessed on 15 Jun 2018).

55. Onea, F.; Rusu, E. The expected efficiency and coastal impact of a hybrid energy farm operating in the Portuguese nearshore. *Energy* **2016**, *97*, 411–423. [CrossRef]

energies

MDPI

Article

A Shape Optimization Method of a Specified Point Absorber Wave Energy Converter for the South China Sea

Yadong Wen, Weijun Wang *, Hua Liu, Longbo Mao, Hongju Mi, Wenqiang Wang and Guoping Zhang

Department of Electrical Engineering, Army Logistics University of PLA, Chongqing 401331, China; elecrivalry@163.com (Y.W.); liuhua6753@163.com (H.L.); mlb84@163.com (L.M.); mimihj_123@163.com (H.M.); 15223330879@163.com (W.W.); zgp064@126.com (G.Z.)
* Correspondence: wjwang636@126.com; Tel.: +86-023-8673-6189

Received: 9 September 2018; Accepted: 30 September 2018; Published: 3 October 2018

Abstract: In this paper, a shape optimization method of a truncated conical point absorber wave energy converter is proposed. This method converts the wave energy absorption efficiency into the matching problem between the wave spectrum of the South China Sea and the buoy's absorption power spectrum. An objective function which combines these two spectra is established to reflect the energy absorbing efficiency. By applying Taguchi design, the frequency domain hydrodynamic analysis and the response surface method (RSM), the radius, cone angle and draft of the buoy are optimized. Since the significant influence of power take-off system (PTO) on energy absorption, the optimal PTO damping under random wave conditions is also studied. The optimal shape is acquired by maximizing the energy absorbing efficiency. Four types of performance and the influence of each geometrical parameter are also obtained. In addition, the cause of the trend of performance as well as the effects of adjusting the input parameters are analyzed. This study can provide guidance for the shape optimization of multi-parameter buoys.

Keywords: wave energy converter; point absorber; shape optimization; Taguchi design; RSM; South China Sea; absorption power spectrum

---

1. Introduction

In recent decades, in order to obtain wave energy with a high energy density, the development of wave energy converters (WECs) has received increased attention. The first model patent for a WEC was acquired by Girard and his son in 1799 [1]. Since then, thousands of WEC prototypes have been developed, which can be classified into three types according to Falcão [2]: oscillating water column (OWC), overtopping device and oscillating body device. Most WECs are currently in the pre-commercial stage; thereforethe major concern at this stage is efficiency and several works have been carried out with respect to design optimization. For example, since physical modelling is widely applied in the optimization of OWC devices, Viviano [3,4] studied the difference between the nonlinear results of small scale model and largescale model, which contributes to correct the scale effect of the results obtained from physical model experiments. Falcão [5] and Gomes [6] concentrate on improving the performance of a floating OWC spar buoy by optimizing the tube geometry and turbine characteristic. As for overtopping devices, Martins [7] performed a numerical study for evaluation of the geometry influence over the dimensionless available power of nearshore overtopping WECs. Han [8] proposed a multi-level breakwater for overtopping wave energy conversion, which optimize the opening width of the lower reservoir, the sloping angle and the height ratio by means of numerical and experimental tests.

As a type of oscillating body device, a point absorber (PA) generally operates in a heave or surge motion to harness wave energy via a buoy, whose diameter is small in comparison to the wavelength.

For the purpose of improving the efficiency of PAs, a few approaches are presented in different energy conversion stages. Usually, PAs can be divided into four stages [9]: the absorption stage, the transmission stage, the generation stage and the conditioning stage. At the energy absorption stage, geometric optimization is usually used to improve the wave energy capture efficiency [10–18]. At the transmission stage, many studies focus on optimizing the PTO system configuration [19–21] and proposing new PTO types for different transmission systems, such as nonlinear power capture mechanism [22,23], fluidic flexible matrix composite PTO pump [24] and gyroscopic PTO [25]. At the generation and conditioning stage, system performance can be improved by optimizing the power conversion quality or applying electrical controllers [26,27].

In addition, due to the randomness and volatility of the wave, control strategies play an important role in the WEC system, which can be applied at multiple stages. Corresponding to the aforementioned three types of approaches, the strategies can be divided into three types as Wang [28] recommended: hydrodynamic control, PTO control and Grid/Load side control. Nowadays many researches focus on latching control [29], declutching control [30], model predictive control [31,32] and so forth.

For the ultimate practical operation of WECs, a precise nonlinear model is imperative. Nonlinear factors, e.g., viscosity, friction, nonlinear wave force and nonlinear PTO system, have received wide attention to improve the reliability of the evaluation results [33,34].

Most of the above approaches are applied to components of WEC, while others focus on the study of wave-to-wire (W2W) models, which take into account all the components, from ocean waves to the electrical network. Through W2W model, designers can simultaneous investigate the coupling relationship between components, improve the overall efficiency and apply advanced control strategies [9,28,35].

The wave energy absorption performance of the buoy has a significant impact on the total efficiency of the PA. Thus, the optimization of the buoy in order to maximize the absorbed wave energy is the first necessary step, which is the focus of this paper. The geometric optimization of traditional marine structures aims to minimize dynamic response to maintain stability [36,37]. Conversely, the geometric optimization of wave energy aims to maximize the dynamic response, thereby improving the energy absorption efficiency [38–41]. In addition, it is well known that, as an input condition for WEC design, sea characteristics directly determine the efficiency and feasibility of WEC. Therefore, a high-performing PA is required to maximize the wave energy absorption based on the sea characteristics of the intended deployment location.

For the purpose of improving the energy absorption efficiency, some studies have been carried out on the optimization of the design of different types of PA in recent years. Based on the sea conditions near the Shetland Islands, McCabe [10,11] used a genetic algorithm to optimize the design of the surging WEC buoy. The WEC shape, parametrically described by bi-cubic B-spline surfaces, was assessed using three cost functions within four different constraint regimes defined by two displacements and two power delivery limits. Under the condition of regular waves, Koh [12] used a multi-objective optimization algorithm to optimize the height and radius of a heave–pitch buoy under the constraint of a cost function. Kurniawan [13] applied a multi-objective algorithm to optimize the size and draft of three pitch buoys under the constraints of two cost functions. Danial Khojasteh [14] and Pastor [15] optimized the radius and draft of hemispherical and conical buoys using Advanced Quantitive Wave Analysis (AQWA) and exhaustive search algorithms based on sea state data from the Iranian coast and the North Sea, respectively. Goggins [16] explored the methodology involved in the shape optimization of several types of heave PAs at the Atlantic marine energy test site. The optimal geometrical configuration was established by defining the significant velocity as the objective function which needed to be maximized. Shami [17] and Shadman [18] used the Design of Experiment (DOE) method to optimize the radius and draft of a heave cylindrical buoy. The objective functions used are the maximum absorbed power, resonance frequency and absorption bandwidth of the buoy's absorption power spectrum.

In this paper, a geometry optimization methodology for a one-body PA buoy is proposed. Based on the sea characteristics of the South China Sea, the radius, cone angle and draft of the truncated conical buoy are optimized. The optimization process is developed by combining Taguchi design, frequency domain analysis based on linear potential flow theory and the response surface method. The optimization goal is to maximize the energy absorption efficiency. An objective function that relates to the wave spectrum and the absorption power spectrum is established, which also reflects the energy absorption efficiency. Since the solution of wave energy absorption efficiency is deemed as the matching problem between wave spectrum and the absorption power spectrum of buoy, the key step is to acquire these two spectra. According to the joint probability distribution of wave height and period, the wave spectrum is calculated. The absorption power spectrum of each buoy, which relies on hydrodynamic and PTO parameters, is obtained. The PTO system is considered as a pure damping system, and the calculation method of the optimal damping coefficient is determined. The hydrodynamic analysis is conducted by the boundary element software AQWA (19.0, ANSYS, Canonsburg, PA, USA), while the Taguchi design and response surface design are carried out by the statistical analysis software Minitab (17.0, Minitab Inc., Commonwealth, PA, USA); other related calculations are realized by MATLAB (R2010b, The MathWorks, Inc, Natick, MA, USA). The influence of geometric parameters on the results and the reason for the trend of each output parameter are analyzed. The methodology proposed in this paper can be extended to multiple geometric parameters or a multiple-degree buoy optimization process.

This paper is organized as follows, Section 1 gives the introduction. Section 2 describes mathematical model of PA and the establishment of objective function. The geometry optimization methodology is elaborated in Section 3. The calculation of wave spectrum and absorption power spectrum are presented in Sections 4 and 5, respectively. Section 6 analyzes parameter optimization based on RSM. Results and discussions are presented in Section 7. In the end, Section 8 summarizes the conclusion.

2. Theoretical Analysis

The form of WEC considered here is a PA, whose equivalent schematic diagram is shown in Figure 1. F_{wave} denotes the wave exciting force, and m is the mass of buoy. K_{PTO} and R_{PTO} represent the damping and elastic properties of the PTO system, respectively. K is on behalf of the spring characteristics of the hydrostatic force. In this paper, K_{PTO} is ignored since the PTO is deemed as a pure damping system. The PA is considered to be a two-body WEC when the reference is a submerged body but a one-body WEC when the reference is the sea bed, and this paper concentrates on the latter case.

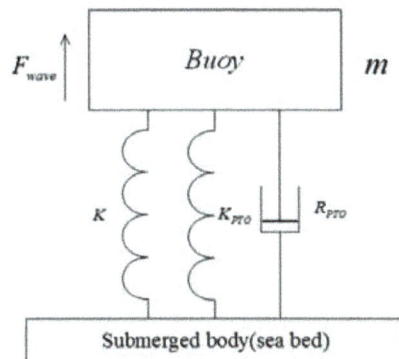

Figure 1. Equivalent schematic diagram of point absorber.

2.1. Mathematical Model

Based on the potential flow theory, buoys are mainly subjected to wave force, hydrostatic restoring force and the force of the PTO system; other external forces such as mooring force are ignored. The external forces cause motion in six degrees of freedom, whereas only heave motion is investigated on account of the fact that wave energy absorption predominately occurs in heave. Ignoring the viscous effects [33,34,40], the force equation of the buoy can be written as

$$m\ddot{z} = f_e + f_r + f_h + f_{ext} = f_e - \left(A\ddot{z} + B\dot{z}\right) - \rho g S z + f_{PTO}, \tag{1}$$

where m is the mass of the buoy and z is the dynamic heave motion response. The wave force is composed of the wave excitation force f_e and radiation force f_r. f_r can be expressed as $-A\ddot{z} - B\dot{z}$, where A and B denote the added mass coefficient and radiation damping coefficient, respectively. f_h is the hydrostatic restoring force represented by $-\rho g S z$, S is the buoy's water plane area. For conical buoy, S slightly shifts from the value of stationary state in heave motion. However, in this paper the water plane area is limited to be constant due to insignificant effects. f_{ext} represents the external force, which in the current work is deemed to be equal to the PTO force f_{PTO}.

2.2. Objective Function

In order to optimize the buoy's geometry, an objective function must first be established as the evaluation criteria. In some studies, the energy absorption performance under random wave conditions was evaluated by analyzing the maximum absorbed power, the resonance frequency and the absorption bandwidth (half-power bandwidth). In fact, these three performance parameters are the spectral parameters of the absorption power spectrum of a buoy. The essence of the absorption power spectrum is the dynamic response characteristics of the buoy excited by the incident wave with unit wave amplitude ($a(\omega) = 1$). The energy absorption of the buoy under random spectral conditions can be written as

$$P_R = R_{PTO} \sum_{j=1}^{M} \omega_j^2 |H(i\omega_j)|^2 S(\omega_j) \Delta\omega, \tag{2}$$

where $H(\omega)$ is the response amplitude operator (RAO), R_{PTO} is the PTO damping coefficient and $S(\omega)$ is the wave spectrum density function. Meanwhile, $a(\omega)$ and $S(\omega)$ have the following relationship:

$$a(\omega) = \sqrt{2S(\omega)\Delta\omega}. \tag{3}$$

Substituting the above equation into Equation (2), the energy absorption at the frequency ω_j may be written as

$$P_R(\omega_j) = \frac{1}{2} R_{PTO} \omega_j^2 |H(i\omega_j) a(\omega_j)|^2 = \frac{1}{2} R_{PTO} \omega_j^2 |H(i\omega_j)|^2. \tag{4}$$

The above formula is the energy absorption spectrum function, which formulates the absorption power spectrum of the buoy. The spectral parameters (resonance frequency, absorption bandwidth and maximum absorbed power) can be obtained by analyzing the spectral characteristics. It is worth noting that the absorption power spectrum is the self-property of the buoy and does not contain sea characteristics. Currently, the main method designs the resonance frequency close to the peak frequency of the target sea state and then maximizes the absorption power and absorption bandwidth. However, this is often faced with the problem of a qualitative trade-off between parameters, which is difficult to quantitatively evaluate. It is the wave spectrum that can fully reflect the energy distribution characteristics of the random spectrum; thus, only analyzing the absorption power spectrum will reduce the reliability under real sea conditions. Hence, it is necessary to establish an objective function that can simultaneously contain information about the energy absorption performance of the buoy and the energy distribution characteristics of the sea state. As Price [42] stated, the absorption power spectrum is similar to a window that captures energy on a wave spectrum. Goggins [16] and Clauss [37] suggested using the significant velocity associated with both the dynamic motion response of WEC and the wave spectrum as the objective function. Referring to the definition of the significant wave height, the significant velocity is defined as

$$V_S^2 = 16 \int_0^\infty S_s(\omega) d\omega, \tag{5}$$

where $S_s(\omega)$ is the dynamic velocity response spectrum and is defined as

$$S_s(\omega) = (|H(i\omega)|\omega)^2 S(\omega). \tag{6}$$

This gives

$$V_S^2 = 16 \int_0^\infty (|H(i\omega)|\omega)^2 S(\omega) d\omega. \tag{7}$$

In discrete form, this becomes

$$V_S^2 = 16 \sum_{j=1}^N (|H(i\omega)|\omega)^2 S(\omega) \Delta\omega. \tag{8}$$

The above equation contains the dynamic response performance of the buoy $|H(i\omega)|\omega|$ and also the information of the wave spectrum ($S(\omega)\Delta\omega$) and so it is more complete than only analyzing the absorption power spectrum. However, it is worth noting that there is no PTO damping parameter in the above equation but it contains the dynamic response $H(i\omega_j)$, which is affected by the PTO damping coefficient. The result is that different results will be obtained under different PTO damping coefficient and it is well known that PTO damping coefficient have a direct impact on the energy absorption, which indicates that using significant velocity as the objective function is still not perfect. Based on this, the objective function proposed in this paper is

$$f = \frac{R_{PTO} V_S^2}{16D}. \tag{9}$$

where D is the waterline diameter. The above equation contains the PTO damping coefficient and the significant velocity, which can compensate for the lack of PTO damping information. Meanwhile, this is also divided by the waterline diameter to help compare the performance of buoys of different sizes.

To fully understand the meaning of the proposed objective function, the following analysis may be conducted. On the one hand, considering Equations (4) and (8), the above equation can be rewritten as

$$f = \frac{R_{PTO} V_S^2}{16D} = \frac{2}{D} \sum_{j=1}^N P_R(\omega_j) S(\omega_j) \Delta\omega_j. \tag{10}$$

It can be seen that the objective function is the sum of the products of the absorption power spectrum and the wave spectrum at each frequency, which means that the degree of matching between the two spectra is reflected.

On the other hand, the relationship between the objective function and the wave energy absorption efficiency may be derived. The absorption efficiency is generally determined by the capture width, which is defined as the ratio of the energy extracted by the buoy to the wave input energy and is given by

$$C_w(\omega) = \frac{R_{PTO} \sum_{j=1}^M \omega_j^2 |H(i\omega_j)|^2 S(\omega_j) \Delta\omega}{\rho g \sum_{j=1}^M v_g(\omega_j, h) S(\omega_j) \Delta\omega_j}, \tag{11}$$

where h is the water depth and v_g is the wave group velocity. Considering Equations (4), (8) and (10), it can be found that

$$C_w(\omega) = \frac{f}{\rho g \sum_{j=1}^M v_g(\omega_j, h) S(\omega_j) \Delta\omega_j} D. \tag{12}$$

According to Cheng [43], the denominator is the wave input energy of the unidirectional long-crested random sea. When the random spectrum is given, its annual average wave input energy is constant and so the objective function is proportional to the capture width when the size is constant, which indicates that the objective function can simultaneously reflect the absorption efficiency under random waves.

Till then, conclusion can be drawn that by establishing the objective function, the wave energy absorption efficiency is converted into the matching problem of the wave energy and absorption power spectrum.

3. Geometry Optimization Methodology

From the established objective function, the key to evaluating the performance of the buoy is to obtain the wave spectrum and the absorption power spectrum. The Taguchi design, hydrodynamic analysis and response surface method (RSM) are combined to achieve optimization. The design flowchart is shown in Figure 2. First, according to the joint probability distribution of significant wave height (SWH) and wave average period (T_{av}), the South China Sea's wave spectrum is obtained. Secondly, given the range of the buoy geometry parameters, the candidate buoy library is obtained by the Taguchi method. Then, the hydrodynamic performance of the candidate buoy is analyzed, along with the optimal PTO damping coefficient of each buoy being calculated to obtain the absorption power spectrum. On the basis of obtaining these two spectra, the objective function is applied to evaluate the performance so as to obtain the approximate range of the optimal buoy parameters. Finally, the RSM is used for local optimization to obtain the optimal buoy parameter configuration.

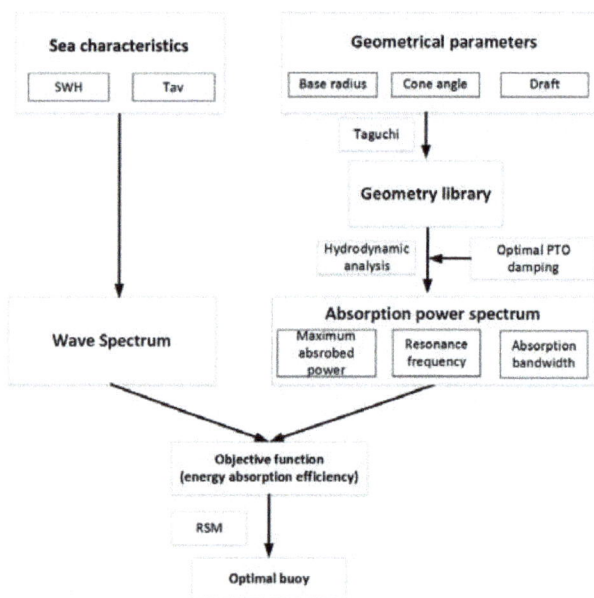

Figure 2. Flowchart of geometrical optimization.

The wave spectrum calculation is presented in Section 4, while the Taguchi design and hydrodynamic analysis are utilized in Section 5 for the sake of calculating absorption power spectrum. The parameter optimization based on RSM is elaborated in Section 6.

4. Wave Spectrum

The purpose of the sea state analysis is to obtain the wave spectrum. The intended deployment site is the South China Sea, which contains a large amount of potential wave energy. Zheng [44,45] has performed a large number of works over the last several years. In one study of his, the wave field in the China Sea from January 1988 to December 2011 was simulated using the WAVEWATCH-III (WW3) model, with a Cross-Calibrated, Multi-Platform (CCMP) wind field as the driving field. The joint probability distributions of average annual significant wave height (SWH) and wave average period (T_{av}) are presented in Table 1. SWH represents the mean wave height of the top one-third of the waves. The background color scale illustrates the occurrence probability level, showing that the most probable SWH values are below 4.5 m and the majority of SWH values fall in the 0.5–2.5 m interval. Most of the wave periods are distributed within the range of 3.5–9.5 s. The wave power level, P, per unit width in a wave can be calculated as follows [46]:

$$P = \frac{\rho g^2 H_s^2 T_{av}}{64\pi}, \tag{13}$$

where ρ is the density of the fluid, g is acceleration due to gravity and H_s and T_{av} denote SWH and average wave peak period, respectively.

Table 1. Joint probability distribution of significant wave height (SWH) and wave average period (T_{av}).

		Wave Average Period (s)												
		1.5	3.5	4.5	5.5	6.5	7.5	8.5	9.5	10.5	11.5	12.5	13.5	14.5
	10.5	0	0	0	0	0	0	0	0	0	0	0	0	0
	9.5	0	0	0	0	0	0	0	0	0	0	0	0	0
	8.5	0	0	0	0	0	0	0	0	0	0	0	0	0
	7.5	0	0	0	0	0	0	0	0	0	0	0	0	0
significant wave	6.5	0	0	0	0	0	0	0	1	9	2	0	0	0
height (SWH, m)	5.5	0	0	0	0	0	0	28	54	24	1	0	0	0
	4.5	0	0	0	0	0	50	373	240	37	7	0	0	0
	3.5	0	0	0	0	163	1317	950	436	86	0	0	0	0
	2.5	0	0	0	874	4743	2891	1459	546	54	0	0	0	0
	1.5	0	19	3720	9300	5011	2925	1131	149	7	1	0	0	0
	0.5	531	4859	11,299	6525	3159	997	189	38	3	2	0	0	0

By combining the occurrence probabilities detailed in Table 1 with Equation (13), the wave power level at each sea state is estimated as shown in Table 2. The background color scale illustrates the wave power level. It can be seen that in this sea state, the largest energy contribution is given by waves with an SWH of 2–3 m and a period of 6–7 s, which account for about 15.2% of the energy. The second-largest energy contribution (10.7%) comes from waves with an SWH of 2–3 m and a period of 7–8 s and the third-largest (9.6%) is from waves with an SWH of 3–4 m and a period of 7–8 s. By normalizing the power level of each sea state in Table 2 (divided by the total number of wave occurrences, 64,210), a single equivalent energy distribution can be obtained as shown in Figure 3, from which the resultant annual average energy density is calculated as approximately 10 kW/m. According to Figure 3, the equivalent wave height for each period (frequency) can be obtained by Equation (13) and then the wave spectrum density function, $S(\omega)$, at the period (frequency) can be calculated by Equation (3). After the $S(\omega)$ at each period (frequency) is acquired, the wave energy spectrum of the South China Sea can be obtained by interpolation method, as shown in Figure 4. It is observed that the peak wave angular frequency is 0.837 rad/s (viz. Tp = 7.5 s) and the highest wave spectrum density is about 0.83 m²/(rad/s).

Table 2. Wave power level and energy contribution distribution.

		1.5	3.5	4.5	5.5	6.5	7.5	8.5	9.5	10.5	11.5	12.5	13.5	14.5
						Wave Average Period (s)								
	10.5	0	0	0	0	0	0	0	0	0	0	0	0	0
	9.5	0	0	0	0	0	0	0	0	0	0	0	0	0
	8.5	0	0	0	0	0	0	0	0	0	0	0	0	0
significant wave height (SWH, m)	7.5	0	0	0	0	0	0	0	0	0	0	0	0	0
	6.5	0	0	0	0	0	0	0	196 0.0%	1959 0.3%	476 0.1%	0	0	0
	5.5	0	0	0	0	0	0	3533 0.6%	7616 1.2%	3741 0.6%	170 0.0%	0	0	0
	4.5	0	0	0	0	0	3727 0.6%	31,510 5.1%	22,660 3.7%	3861 0.6%	800 0.1%	0	0	0
	3.5	0	0	0	0	6370 1%	59,386 9.6%	48,549 7.8%	24,902 4%	5429 0.9%	0	0	0	0
	2.5	0	0	0	14,745 2.4%	94,569 15.2%	66,511 10.7%	38,042 6.1%	15,911 2.6%	1739 0.6%	0	0	0	0
	1.5	0	74 0.0%	18,486 3%	56,485 9.1%	35,969 5.8%	24,226 3.9%	10,616 1.7%	1563 0.3%	81 0.0%	12 0.0%	0	0	0
	0.5	98 0.0%	2087 0.3%	6239 1%	4403 0.7%	2519 0.4%	917 0.1%	197 0.0%	44 0.0%	3 0.0%	2 0.0%	0	0	0

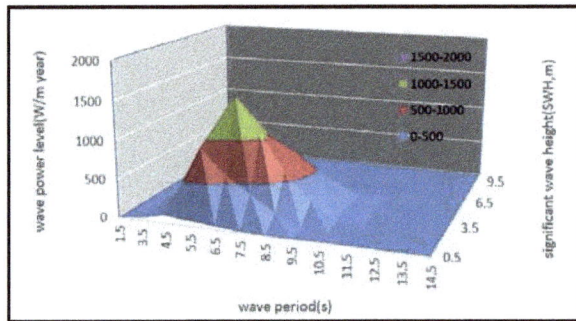

Figure 3. Surface chart of the wave power level for the South China Sea.

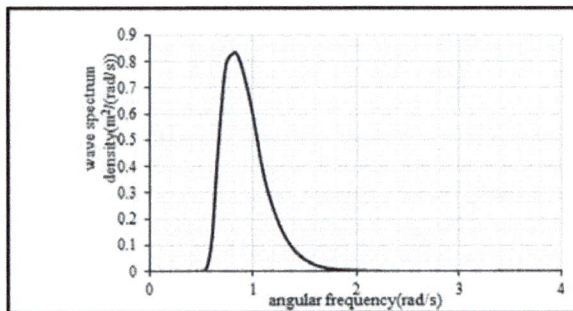

Figure 4. Wave spectrum of the South China Sea.

5. Absorption Power Spectrum

5.1. Geometry Library Generation Based on Taguchi Design

The truncated cone buoy analyzed in this paper is presented in Figure 5, where *r* is the bottom radius, *θ* is the cone angle, *d* stands for the draft and *h* is the height. Axisymmetric buoys [47] can accept the energy of any wave direction and can better dissipate the horizontal component of the wave force than other types of buoys, which is beneficial to improving survivability. As one of the axisymmetric buoys, the truncated cone buoy has the advantage of being able to flexibly adjust its waterline area compared to the cylindrical buoy.

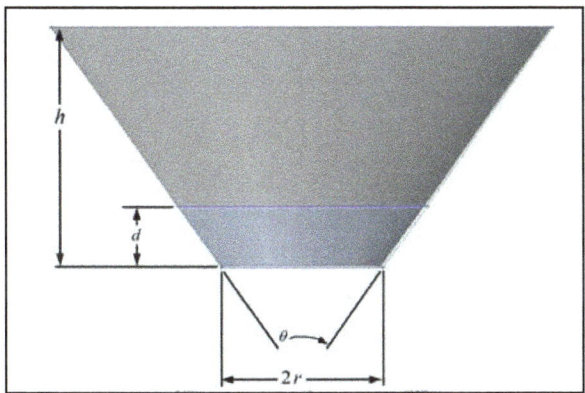

Figure 5. Sketch of the truncated cone buoy.

Taguchi design is an efficient quality management method proposed by Dr. Taguchi in 1950s for optimization of industrial processes. This design approach is based on mathematical statistics and improves product quality/system performance by analyzing the sensitivity of the system response to multiple parameters [48]. The effect of each parameter on the system in Taguchi method is similar to a signal/noise ratio. Taguchi design consists of three sub-designs, namely system design, parametric design and tolerance design. The system design is to determine the technical route or performance parameters used to improve the quality of the product/design; while the parameter design is to determine the optimal level combination of parameters; and the tolerance design is to determine the reasonable tolerance of each parameter, thus achieving the expected quality requirements at the lowest cost.

For the wave energy absorption of PA, corresponding to the system design is to select the truncated cone buoy and determine the three key geometric parameters: base radius, cone angle, draft; corresponding to the parameter design is to determine the optimal level combination of geometric parameters. Since the economic factors are not within the scope of this paper, the tolerance design is not considered at this stage. In addition, Taguchi design can analyze the impact of input parameters on performance to determine the most important parameters and correlation analysis is elaborated in Section 7.2.

The foundation of parameter design is orthogonal arrays, which fulfills the experimental purpose with a minimum number of experiments; it also means that the design is balanced, that is, each parameter is given the same weight. As shown in Figure 5, three parameters which need to be optimized are the base radius (r), cone angle (θ) and draft (d), while the height of the buoy is fixed at three-times the base radius. The range of the geometrical parameters considered in this study is presented in Table 3, in which the draft is considered to be proportional to the base radius.

Table 3. Range of geometrical parameters.

Geometrical Parameter	Minimum	Maximum
Base radius (r)	3 (m)	12 (m)
Cone angle (θ)	40°	120°
Draft (d)	0.5r	2.5r

The above three parameters can be composed into hundreds of buoy shapes within the range of consideration. One feasible method is to take a finite value for each parameter and then combine them into a finite number of buoy representatives. The performance of the buoy representatives is evaluated, while the performance of other shape buoys is obtained by interpolation or fitting. The advantage of this method is that the buoy performance database can be built and the required buoy performance can be conveniently queried during the optimization process but the disadvantage is that the calculation cost is high. Taking five levels for each parameter as an example, up to $5^3 = 125$ tests are required to establish a database. In order to improve the optimization efficiency, Taguchi design is applied in this study. As shown in Table 4, an L_{25} (5^3) orthogonal experiment table is applied and a geometry library which contains 25 buoys is generated.

Table 4. Candidate buoy library.

Buoy ID	Base Radius (m)	Cone Angle (°)	Draft Ratio (d/r)	CoG (m)	CoB (m)
1	3	40	0.5	−0.9	−0.7
2	3	60	1	−1.8	−1.28
3	3	80	1.5	−2.7	−1.7
4	3	100	2	−3.6	−2
5	3	120	2.5	−4.5	−2.26
6	6	40	1	−3.6	−2.7
7	6	60	1.5	−5.4	−3.62
8	6	80	2	−7.2	−4.3
9	6	100	2.5	−9	−4.8
10	6	120	0.5	−1.8	−1.2
11	8	40	1.5	−7.2	−5.2
12	8	60	2	−9.6	−6.1
13	8	80	2.5	−12	−6.9
14	8	100	0.5	−2.4	−1.7
15	8	120	1	−4.8	−2.8
16	10	40	2	−12	−8.3
17	10	60	2.5	−15	−9.2
18	10	80	0.5	−3	−2.2
19	10	100	1	−6	−3.8
20	10	120	1.5	−9	−4.9
21	12	40	2.5	−18	−12
22	12	60	0.5	−3.6	−2.8
23	12	80	1	−7.2	−4.9
24	12	100	1.5	−10.8	−6.4
25	12	120	2	−14.4	−7.5

In addition, in order to maintain the stability of the buoy and avoid the situation of overturning, the center of gravity (CoG) of buoys are set at 0.6 times the draft under the hydrostatic surface, while the center of Buoyancy (CoB) is calculated by hydrostatic analysis in AQWA. The z coordinates of CoG and CoB are also given in Table 4. It can be seen that all the CoG are located below the CoB, which indicates that all buoys in the library are stable.

5.2. Absorption Power Spectrum Calculation

As can be seen from Equation (4), the absorption power spectrum includes the $H(i\omega)$ term and the PTO damping coefficient term. Among them, $H(i\omega)$ is calculated from the hydrodynamic parameters, so the acquisition of the absorption power spectrum relies on the calculation of hydrodynamic parameters and the PTO damping coefficient.

5.2.1. Hydrodynamic Parameters Calculation

The acquisition of hydrodynamic parameters can be solved by the tank experiment or numerical method and in this paper, the numerical method is applied by using the boundary element software

AQWA. The frequency range of interest is divided into 20 equal parts and the wave excitation force, the added mass and the radiation damping coefficient are obtained under unit wave amplitude condition. The dynamic response $H(i\omega)$ is calculated as

$$H(i\omega) = \frac{z(i\omega)}{a(\omega)} = \frac{f_e(i\omega)}{-(m_1 + A_1)\omega^2 + i\omega(B_1 + R_{PTO}) + (\rho g S + K)}. \tag{14}$$

5.2.2. Optimal PTO Damping Determination

The PTO damping coefficient affects the dynamic response performance of the buoy [49,50]. It is assumed that the PTO system is a pure damping system and the mass and spring properties are neglected [43]. Falnes [51] pointed out that, in a regular wave, the maximum absorbed power is obtained when the PTO damping coefficient is equal to the conjugate of the inherent damping of the system, as follows:

$$R_{PTO} = \sqrt{B_1^2 + (\omega m_1 + \omega A_1 - (\rho g S + K)/\omega)^2}. \tag{15}$$

Price [42] indicates that the optimal damping under random wave conditions should be related to the spectral and buoy characteristics and the best case is that the PTO damping is always equal to the conjugate value of the system impedance; however, this is difficult to achieve in practice. The suboptimal condition used in this paper is that the PTO damping is always equal to the inherent damping of the system at the wave peak frequency. Instead of ω, the peak frequency ω_p of the wave spectrum is used and Equation (15) can be rewritten as

$$R_{PTO} = \sqrt{B_1^2 + (\omega_p m_1 + \omega_p A_1 - (\rho g S + K)/\omega_p)^2}. \tag{16}$$

It should be noted that some research determines the PTO damping coefficient as equal to the radiation damping coefficient. The premise of this optimal PTO damping condition is that the buoy is in a resonance state; that is, the buoy resonance frequency is equal to the peak frequency. In fact, when the buoy is in resonance, the second square term under the root of the equation of the above equation is equal to zero; that is, the PTO damping coefficient is equal to the radiation damping, so the above formula is more universal than the radiation damping condition.

When the optimal damping is determined, the RAO of the buoy can be calculated in MATLAB according to Equation (14) and the absorption power spectrum of each buoy can be obtained according to Equation (4).

6. Parameter Optimization Based on RSM

After obtaining the wave spectrum and the absorption power spectrum, the energy absorption efficiency of each buoy can be evaluated according to the objective function. By evaluating the energy absorption efficiencies of all candidate buoys, the optimal level of the three geometric parameters can be determined. Taguchi design can quickly realize a multi-factor optimization process. The only shortcoming is that it can merely analyze isolated design points and the optimization results are not detailed enough. Compared with Taguchi design, the prediction model obtained by the response surface method (RSM) is continuous, which means that it can continuously analyze input parameters. The RSM method was first introduced by Box and Wilson for modelling, problem analysis development, modification and optimization of various processes [52]. Although other experimental design and optimization methods can find the optimal value, it is difficult for the designer to intuitively discriminate the optimization region due to the lack of the intuitive graph. For this reason, the response surface method has emerged. It takes the response of the system (such as the wave energy absorption performance studied in this paper) as a function of multiple parameters. Since the prediction model is continuous, it can use graphical techniques to display this functional relationship for the designer to select the optimal conditions by intuitive observation.

In addition, when the parameter range is wide, RSM yields to a high calculation cost due to the need of a large number of design points. However, when the parameter range is narrow, an ideal result can be achieved which is suitable for analysis after the Taguchi design result and more accurate results could be obtained. To this end, local response surface optimization is performed near the optimal level obtained by Taguchi design and finally the optimal buoy parameters are obtained.

7. Results and Discussion

According to the proposed optimization method, four types of performance, namely the objective function value, absorption bandwidth, resonance frequency and maximum absorbed power of the buoy, are obtained. First, the optimal geometric parameter configuration is analyzed from two perspectives: energy absorption efficiency (objective function) and absorption power spectrum parameters. Secondly, the magnitude of each geometric parameter's effect on performance parameters (objective function and absorption power spectrum parameters) as well as the cause of the trend of each performance parameter are studied. Then, the control effects of adjusting the geometric parameters and PTO damping are discussed. Finally, based on the results of the Taguchi design, the local response surface optimization is conducted and the final buoy design is obtained.

7.1. Optimal Geometry Configuration Analysis

The main effect diagram of the energy absorption efficiency (objective function), the resonance frequency, the absorption bandwidth and the maximum absorbed power of the buoy are shown in Figures 6–9, respectively. It can be seen from Figure 6 that the energy absorption efficiency shows a trend of increasing first and then decreasing with increasing radius and cone angle and decreasing with the increase of draft. The maximum value is obtained when radius is equal to 6, the cone angle is equal to $60°$–$80°$ and the draft is equal to 0.5.

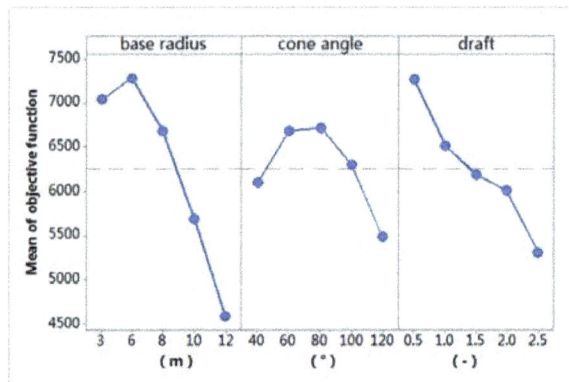

Figure 6. Main effects diagram for objective function.

As seen in Figure 7, the resonance frequency of most of the buoys are lower than the peak frequency of the wave spectrum (0.837 rad/s) and the average resonance frequency of the buoy with a radius of 6, a cone angle of $60°$–$80°$, or a draft ratio of 0.5 is basically near 0.7 rad/s, which is higher than most other candidate buoys. This result stems from the fact that values shown in the main effect diagram is the average of buoys with the same geometrical parameters. For instance, buoy 6 ($r = 6$, $\theta = 40°$, $d/r = 1$) has a resonance frequency of 0.818 rad/s, while buoy 9, with the same radius ($r = 6$, $\theta = 120°$, $d/r = 2.5$), has a resonance frequency of 0.57 rad/s.

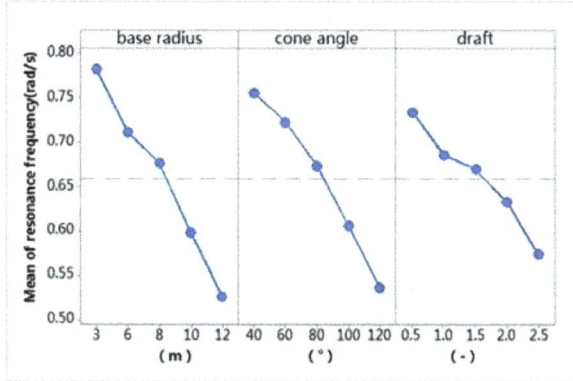

Figure 7. Main effects diagram for resonance frequency.

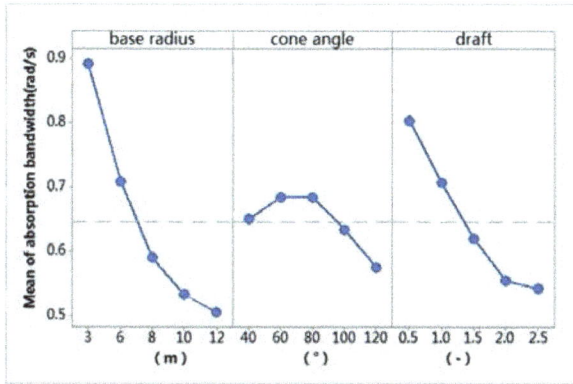

Figure 8. Main effects diagram for absorption bandwidth.

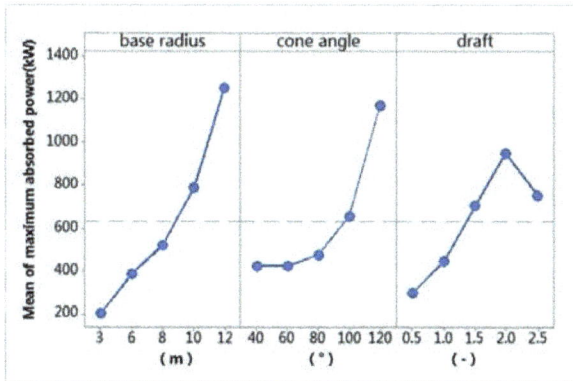

Figure 9. Main effects diagram for maximum absorbed power.

It can be seen from Figure 8 that the absorption bandwidth of the buoy with a cone angle equal to 60°–80° or a draft of 0.5 is optimal. As for the radius, the absorption bandwidth of the buoy with a radius of 6 is next to the buoy with a radius of 3. The bandwidth of the buoy with the above geometric parameters is in the range of 0.67–0.8 rad/s, which is wider than the bandwidth of the wave spectrum (0.42 rad/s).

Figure 9 shows that the maximum absorbed power of the buoy with a radius of 6, an angle of 60°–80°, or a draft of 0.5 is relatively small. However, as discussed previously, the value of the resonance frequency and the absorption bandwidth for these geometrical parameters is greater, viz., the matching with the wave spectrum is more suitable and therefore the energy absorption efficiency is better. This also shows to some extent that the pursuit of a better match with the wave spectrum is more important than the pursuit of maximum absorbed power. In addition, Figure 9 shows that the maximum absorbed power is obtained at the maximum radius (radius = 12 m), the maximum angle (cone angle = 120°) and the larger draft ratio (draft ratio = 2); however, it can be seen from Figure 6 that the energy absorption efficiency is low when each geometric parameter takes a maximum value. That is to say, the geometric parameters at which the larger maximum absorption power is obtained correspond to lower energy absorption efficiency, which results in high cost and low cost-effectiveness. To this point, the conclusion can be drawn that the optimal geometric parameter levels are a radius equal to 6, a cone angle equal to 60°–80° and a draft ratio equal to 0.5.

Based on the above results, the response surface optimization design is performed. The response surface type adopted is the central composite design (CCD). Since the energy absorption efficiency and the draft are always negatively correlated, the draft is fixed at 0.5 and the performance of the buoy with a radius in the range [6, 8] and angle in the range [60, 80] is optimized. The response surface of the buoy's objective function with respect to radius and angle is shown in Figure 10 and the response surface optimizer is used to predict the optimal result, as shown in Figure 11. The optimal geometric configuration with a radius of 6.7 m, an angle of 80° and a draft ratio of 0.5 (the draft depth of 3.35 m) is finally obtained.

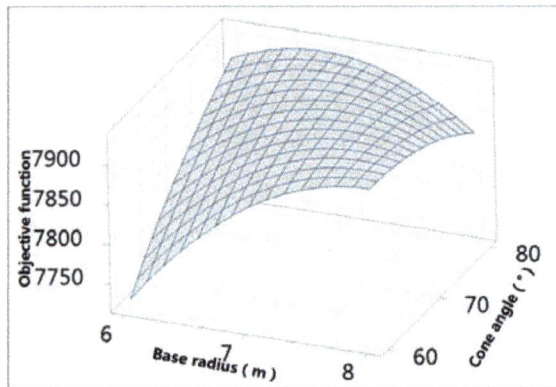

Figure 10. Surface diagram of objective function vs. cone angle and base radius.

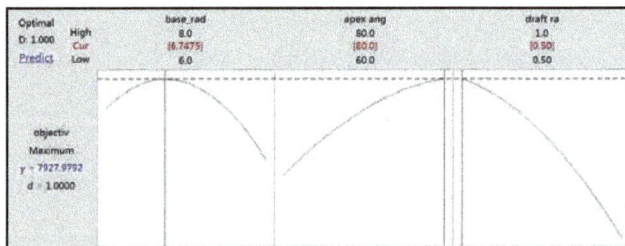

Figure 11. Calculation result of response surface method (RSM) optimizer.

7.2. Performance Characteristic Analysis

7.2.1. Energy Absorption Efficiency

Taguchi design provides range analysis and variance analysis to analyze the influence degree of each geometric parameter on output performance. In this study, the range analysis is adopted due to convenience. It can be seen from Table 5 that the radius has the greatest influence on the energy absorption efficiency, followed by the draft and angle has the least influence.

Table 5. Influence degree of geometric parameters on energy absorption efficiency.

	Base Radius	Cone Angle	Draft
Maximum	7282	6683	7264
Minimum	4591	5484	5310
Range	2691	1237	1954
Percentage	46%	21%	33%
Rank	1	3	2

Furthermore, it was found that the objective function values of all candidate buoys did not exceed 8000, indicating that there is an upper limit to the objective function value. It can be seen from Equation (12) that the objective function is proportional to the capture width when the size is constant. Since the energy capture width of a single-degree-of-freedom axisymmetric buoy has a theoretical maximum of $\lambda/2\pi$ [2], there is also an upper limit to the objective function.

7.2.2. Resonance Frequency

Table 6 illustrates that the radius has the greatest influence on the resonance frequency, followed by the angle and finally the draft.

Table 6. Influence degree of geometric parameters on resonance frequency.

	Base Radius	Cone Angle	Draft
Maximum (rad/s)	0.7820	0.7560	0.7335
Minimum (rad/s)	0.5263	0.5364	0.5735
Range (rad/s)	0.2557	0.2196	0.1600
Percentage	40%	35%	25%
Rank	1	2	3

As can be seen from Figure 7, the increase in geometric parameters results in a decrease in the resonance frequency. To analyze the cause of the trend of resonance frequency, the simplified natural frequency formula may be applied, which is written as

$$\omega \ = \ 2\pi\sqrt{\frac{\rho g S}{m + A1}}. \tag{17}$$

The increase of geometric parameters will lead to an increase of the mass of the buoy. In addition, the added mass will change accordingly. Taking the radius as an example, the trend of the added mass with the radius is shown in Figure 12. It can be seen that the increase of the radius will lead to an increase in the added mass. The hydrodynamic parameters are closely related to the submerged volume and the increase of the geometric parameters will cause the submerged volume to increase, so the resonance frequency decreases as the geometric parameters increase.

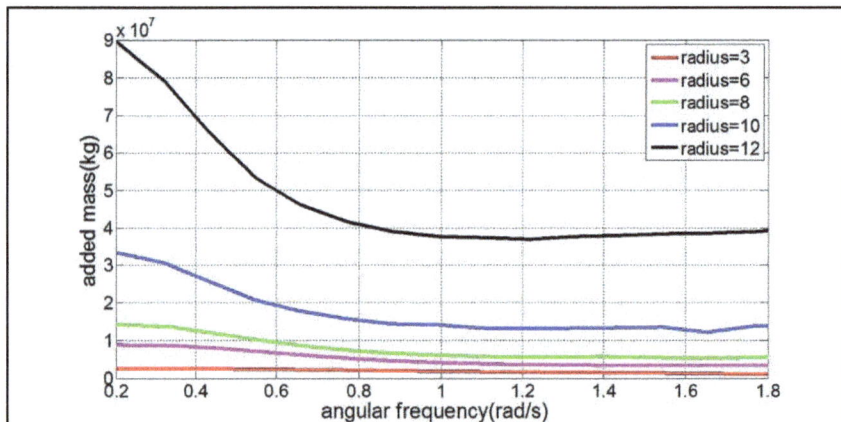

Figure 12. Diagram of added mass vs. base radius.

7.2.3. Absorption Bandwidth

As seen in Figure 8, the absorption bandwidth decreases as the radius and draft increase. As the cone angle changes, the bandwidth tends to become larger and then smaller and the maximum value is obtained at an angle of 60°–80°. The degree of influence of each geometric parameter on the absorption bandwidth is shown in Table 7, which shows that the radius has the greatest influence, followed by the draft and finally the angle.

Table 7. Degree of influence of geometric parameters on absorption bandwidth.

	Base Radius	Cone Angle	Draft
Maximum (rad/s)	0.8919	0.6856	0.8039
Minimum (rad/s)	0.5064	0.5762	0.5439
Range (rad/s)	0.3855	0.1096	0.2600
Percentage	51%	15%	34%
Rank	1	3	2

The absorption bandwidth of a single buoy is related to the damping coefficient of the oscillating system. As a result, for the sake of analyzing the cause of the trend of absorption bandwidth, it is necessary to pay attention to the systematic damping coefficient. According to Falnes [51], the systematic damping coefficient can be calculated as

$$\delta = \frac{R_{pto} + B_1(\omega)}{2(A_1(\omega) + m)}.$$ (18)

Taking the hydrodynamic parameters at the peak frequency as representative, the variation of the systematic damping coefficient with respect to the geometric parameters is plotted in Figure 13. As can be seen, when the radius and the draft change, the systematic damping coefficient and the absorption bandwidth change equally and when the cone angle changes, the systematic damping coefficient is opposite to the absorption bandwidth. The reason for this difference may be that the system damping coefficient calculated here ignores the variation of the hydrodynamic coefficient with frequency. In addition, the damping coefficient of the system does not change much under different angles. It is also known from the previous range analysis that the influence of the angle on the bandwidth is only 15% and the impact is minimal.

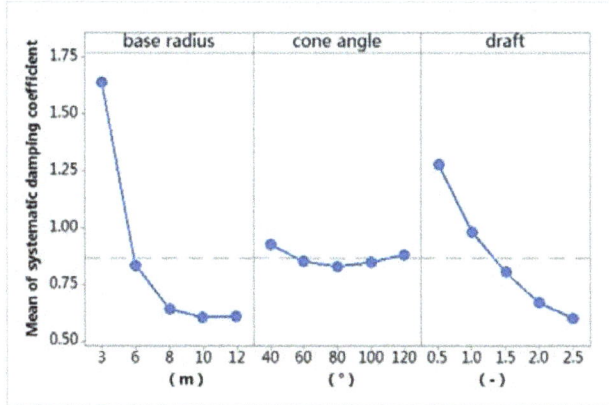

Figure 13. Main effects plot for systematic damping coefficient.

Moreover, Equation (18) indicates that when the geometry of the buoy is specified, the hydrodynamic parameters are also determined and increasing the PTO damping deduces the increase of the absorption bandwidth. Falnes [51] recommended that in order to make the absorption bandwidth larger than the bandwidth of the wave spectrum, R_{PTO} should be greater than the radiation damping. According to Equation (16), the R_{PTO} proposed in this paper is not less than the radiation damping, therefore all candidate buoys have a wider absorption bandwidth than the bandwidth of the wave spectrum (0.42 rad/s).

7.2.4. Maximum Absorbed Power

As can be seen from Figure 9, the maximum absorbed power mostly increases with the increase of the geometrical parameter and decreases only when the draft ratio is 2.5. As shown in Table 8, the radius has the greatest influence, followed by the cone angle and finally the draft ratio.

Table 8. Influence degree of geometric parameters on maximum absorbed power.

	Base Radius	Cone Angle	Draft
Maximum (kW)	1254	1169	951
Minimum (kW)	201	423	297
Range (kW)	1053	746	654
Percentage	43%	30%	27%
Rank	1	2	3

The energy-absorbing ability of the buoy is attributable to the radiation capability. For the purpose of extracting the incident wave energy as much as possible, a radiation wave field is required to dissipate the incident wave field behind the buoy. The radiation capacity is associated with the radiation damping. To this end, the mathematical relationship between absorbed power and radiation damping coefficient needs to be analyzed. The absorbed power formulated by the hydrodynamic parameters is

$$P_R = \frac{R_{PTO}}{2}|\hat{u}|^2 = \frac{(R_{PTO}/2)\left|\hat{f}_e\right|^2}{(B_1 + R_{PTO})^2 + ((m_1 + A_1)\omega - (\rho g S + K)/\omega)^2}. \tag{19}$$

For an axisymmetric buoy with heave motion, the following equation can be obtained by using the Haskind relationship:

$$|f_e|^2 = \frac{4\rho g v_g(\omega, h)}{k} B_1.$$ (20)

Substituting Equation (20) into Equation (19) gives

$$P_R = \frac{R_{PTO}}{2}|\hat{u}|^2 = \frac{R_{PTO}B_1 S(\omega_i)\Delta\omega}{(B_1 + R_{PTO})^2 + ((m_1 + A_1)\omega - (\rho g S + K)/\omega)^2} \cdot \frac{4\rho g v_g(\omega, h)}{k}.$$ (21)

For the above formula, the partial derivative of the radiation damping B_1 is solved as

$$\frac{\partial P_R}{\partial B_1} = \frac{R_{PTO}^2 + ((m_1 + A_1)\omega - (\rho g S + K)/\omega)^2}{\left[(B_1 + R_{PTO})^2 + ((m_1 + A_1)\omega - (\rho g S + K)/\omega)^2\right]^2} \cdot \frac{4R_{PTO}\rho g v_g(\omega, h)S(\omega_i)\Delta\omega}{k} > 0.$$ (22)

The above equation illustrates that, with other parameters being constant, the absorbed power is positively correlated with the radiation damping, which is consistent with the study by Babarit [53]. Thus, as can be seen graphically Figures 9 and 14, the relationship between radiation damping at peak frequency and geometric parameters shows that the change trend of the radiation damping is the same as the change trend of the maximum absorbed power.

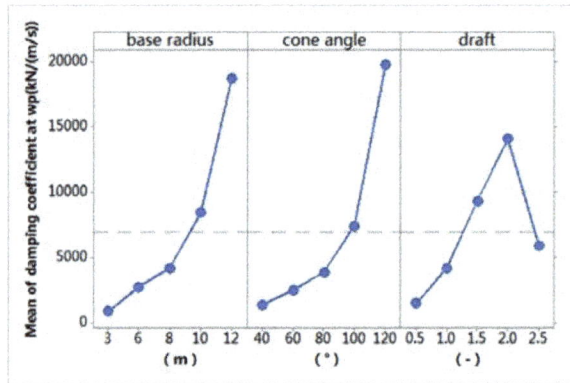

Figure 14. Main effects plot for radiation damping coefficient at peak frequency.

7.2.5. Effects of Adjusting Geometrical Parameters

Studying the relationship between geometric parameters and resonance frequency, the maximum absorbed power and absorption bandwidth are beneficial to guide the optimal process in the future for different target wave conditions. For instance, geometric parameters should be increased to achieve a reduction in the resonance frequency in the case of a lower peak frequency of the wave spectrum. When the wave bandwidth becomes wider, such as from the Joint North Sea Wave Project (JONSWAP) spectrum to the PM spectrum, then the decrease of radius and draft should be considered. The radius and cone angle changes by changing the geometry of the buoy, while changes in draft can be changed by water injection or drainage to change mass. In addition, according to the degree of influence of the geometric parameters analyzed previously on the energy absorption efficiency, the adjustment of radius should be first considered due to its having the greatest impact.

7.2.6. Effects of Adjusting PTO Damping

PTO damping directly affects the absorption performance of the buoy. Taking buoys 6, 7 and 8 as examples, the absorption power spectra under three kinds of PTO damping conditions (100 kNs/m, optimal damping condition and three-times optimal damping) are obtained, as shown in Figure 15.

The wave spectrum is drawn simultaneously and the absorbed power of each buoy is normalized (divided by the maximum absorbed power of buoy 8, viz., 654 kW) for comparison. It can be seen that as the PTO damping increases, the absorption power spectra of buoys shift to the low frequency and the peak values increase. As for spectral parameters, the maximum absorbed power increases with the increase of PTO damping. The resonance frequency decreases as the PTO damping increases, which is inconsistent with the PTO-independent resonance frequency obtained by Equation (17). This inconsistency may be caused by an inconsistent solving method. The former is calculated by the first-order natural frequency formula of solid-state physics, which ignores the change of hydrodynamic parameters with frequency, while the latter is acquired according to the characteristics of the absorption power spectrum, which is related to PTO damping. Besides this, the change of hydrodynamic parameters with frequency is also considered during the calculation process of the absorption power spectrum. As for the absorption bandwidth, it increases with the increase of the PTO damping. To analyze the cause of the trend of the absorption bandwidth, attention need to be paid to Equation (18). It can be seen that the increase of the PTO damping coefficient leads to the increase of the systematic damping coefficient and finally results in the increase of the absorption bandwidth. A larger bandwidth contributes to a reduced need for control strategies.

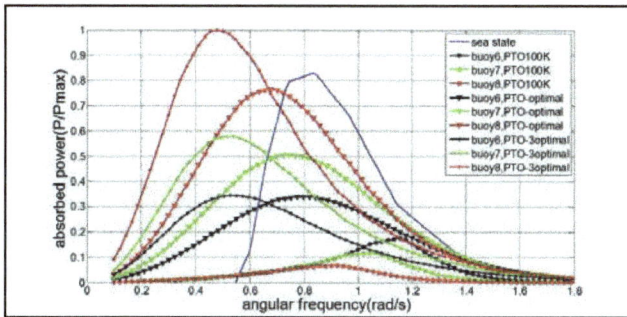

Figure 15. Absorption power spectrum with different power take-off system (PTO) dampings.

8. Conclusions

In this paper, a geometrical optimization method is proposed for a truncated cone buoy of a one-body point absorber wave energy converter. The geometric parameters studied are the radius, cone angle and draft. An objective function, which converts the wave energy absorption efficiency into the matching problem between the wave spectrum and the buoy's absorption power spectrum, is established. The goal of the optimization is to maximize the wave energy absorption efficiency. First, the wave spectrum of the South China Sea was drawn based on sea state data over the past 23 years. Secondly, in order to obtain the buoy absorption power spectrum, the Taguchi design method is applied to generate a buoy library containing 25 candidates. The hydrodynamic parameters are calculated by using the boundary element software AQWA through frequency domain analysis. The PTO system is treated as a pure damping system and the determination of optimal PTO damping is given. According to the hydrodynamic parameters and PTO damping, the absorption power spectrum of each buoy is calculated in MATLAB and the spectral parameters such as resonance frequency, absorption bandwidth and maximum absorbed power are obtained. By analyzing the objective function value and the absorption power spectral parameters, a radius of 6, a cone angle of 60°–80° and a draft ratio of 0.5 are identified as optimal parameter levels. Then, the response surface method was used for detailed optimization and ultimately a buoy with a radius of 6.7 m, a cone angle of 80° and a draft depth of 3.35 m (draft ratio = 0.5) is deemed as optimal. Finally, the influence magnitude of each geometric parameter's effect on the performance of the buoy, as well as the cause of the trend of performance, is studied. Through the range analysis, the radius has the greatest impact on performance and the effects of cone angle and draft vary with performance. In addition, the derivation shows that the resonance frequency is related to the mass, the added mass and the PTO damping; the absorption bandwidth is related to the systematic damping coefficient; the maximum absorbed power is related to the radiation damping. This paper reveals the relationship between and the wave spectrum, the energy absorption spectrum of the buoy and the energy absorption efficiency. The proposed method can quickly achieve shape optimization with limited calculations. Furthermore, the effects of viscosity and control strategies should be considered in the future.

Author Contributions: Y.W. and H.L. designed the main parts of the research, including objective function establishment and optimal methodology designing. W.W. (Weijun Wang) was responsible for guidance, a number of key suggestions and manuscript editing. L.M. and H.M. mainly contributed to the writing of the paper and also responsible for project administration. W.W. (Wenqiang Wang) and G.Z. made contributions to simulations in AQWA and gave the final approval of the version to be published.

Funding: This research received no external funding.

Acknowledgments: The authors would like to acknowledge the financial support from "The National Science and Technology Support Program (supported by the Ministry of Science and Technology of P.R.C. No. 2014BAC01B05)." The authors are grateful that comments and suggestions provided by the anonymous reviewers and editor helped to improve the quality of the paper.

Conflicts of Interest: The authors declare no conflict of interest.

References

1. Clément, A.; Mccullen, P.; Falcão, A. Wave energy in Europe: Current status and perspectives. *Renew. Sustain. Energy Rev.* **2002**, *6*, 405–431. [CrossRef]
2. Falcão, F.D.O. Wave energy utilization: A review of the technologies. *Renew. Sustain. Energy Rev.* **2010**, *14*, 899–918. [CrossRef]
3. Viviano, A.; Naty, S.; Foti, E. Scale effects in physical modelling of a generalized OWC. *Ocean Eng.* **2018**, *162*, 248–258. [CrossRef]
4. Naty, S.; Viviano, A.; Foti, E. Feaseability study of a WEC integrated in the port of Giardini Naxos, Italy. *Coast. Eng. Proc.* **2017**, *1*, 22. [CrossRef]
5. Falcão, A.F.O.; Henriques, J.C.C.; Cândido, J.J. Dynamics and optimization of the OWC spar buoy wave energy converter. *Renew. Energy* **2012**, *48*, 369–381. [CrossRef]
6. Gomes, R.P.F.; Henriques, J.C.C.; Gato, L.M.C. IPS 2Body Wave Energy Converter: Acceleration Tube Optimization. *Int. J. Offshore Polar Eng.* **2010**, *20*, 247–255.
7. Martins, J.C.; Goulart, M.M.; Gomes, M.D.N. Geometric evaluation of the main operational principle of an overtopping wave energy converter by means of Constructal Design. *Renew. Energy* **2018**, *118*, 727–741. [CrossRef]
8. Han, Z.; Liu, Z.; Shi, H. Numerical study on overtopping performance of a multi-level breakwater for wave energy conversion. *Ocean Eng.* **2018**, *150*, 94–101. [CrossRef]
9. Penalba, M.; Ringwood, J.V. A Review of Wave-to-Wire Models for Wave Energy Converters. *Energies* **2016**, *9*, 506. [CrossRef]
10. Mccabe, A.P.; Aggidis, G.A.; Widden, M.B. Optimizing the shape of a surge-and-pitch wave energy collector using a genetic algorithm. *Renew. Energy* **2010**, *35*, 2767–2775. [CrossRef]
11. Mccabe, A.P. Constrained optimization of the shape of a wave energy collector by genetic algorithm. *Renew. Energy* **2013**, *51*, 274–284. [CrossRef]
12. Koh, H.J.; Ruy, W.S.; Cho, I.H. Multi-objective optimum design of a buoy for the resonant-type wave energy converter. *J. Mar. Sci. Technol.* **2015**, *20*, 53–63. [CrossRef]
13. Kurniawan, A.; Moan, T. Optimal Geometries for Wave Absorbers Oscillating About a Fixed Axis. *IEEE J. Ocean. Eng.* **2013**, *38*, 117–130. [CrossRef]
14. Khojasteh, D.; Kamali, R. Evaluation of wave energy absorption by heaving point absorbers at various hot spots in Iran seas. *Energy* **2016**, *109*, 629–640. [CrossRef]
15. Liu, Y.; Pastor, J. Power Absorption Modeling and Optimization of a Point Absorbing Wave Energy Converter Using Numerical Method. *J. Energy Resour. Technol.* **2014**, *136*, 119–129.

16. Goggins, J.; Finnegan, W. Shape optimisation of floating wave energy converters for a specified wave energy spectrum. *Renew. Energy* **2014**, *71*, 208–220. [CrossRef]

17. Shami, E.A.; Wang, X.; Zhang, R. A parameter study and optimization of two body wave energy converters. *Renew. Energy* **2019**, *131*, 1–13. [CrossRef]

18. Shadman, M.; Estefen, S.F.; Rodriguez, C.A. A geometrical optimization method applied to a heaving point absorber wave energy converter. *Renew. Energy* **2018**, *115*, 533–546. [CrossRef]

19. López, M.; Taveira-Pinto, F.; Rosa-Santos, P. Influence of the power take-off characteristics on the performance of CECO wave energy converter. *Energy* **2017**, *120*, 686–697. [CrossRef]

20. Kim, B.H.; Wata, J.; Zullah, M.A. Numerical and experimental studies on the PTO system of a novel floating wave energy converter. *Renew. Energy* **2015**, *79*, 111–121. [CrossRef]

21. Gaspar, J.F.; Calvário, M.; Kamarlouei, M.; Soares, C.G. Design tradeoffs of an oil-hydraulic Power Take-Off for Wave Energy Converters. *Renew. Energy* **2018**, *129*, 245–259. [CrossRef]

22. Zhang, X.; Yang, J. Power capture performance of an oscillating-body WEC with nonlinear snap through PTO systems in irregular waves. *Appl. Ocean Res.* **2015**, *52*, 261–273. [CrossRef]

23. Zhang, X.; Tian, X.; Xiao, L.; Li, X.; Chen, L. Application of an adaptive bistable power capture mechanism to a point absorber wave energy converter. *Appl. Energy* **2018**, *228*, 450–467. [CrossRef]

24. Philena, M.; Squibb, C.; Groo, L.; Hagerman, G. Wave energy conversion using fluidic flexible matrix composite power take-off pumps. *Energy Convers. Manag.* **2018**, *171*, 1773–1786. [CrossRef]

25. Zhang, Z.; Chen, B.; Nielsen, S.R.K.; Olsen, J. Gyroscopic power take-off wave energy point absorber in irregular sea states. *Ocean Eng.* **2017**, *143*, 113–124. [CrossRef]

26. Hong, Y.; Waters, R.; Boström, C. Review on electrical control strategies for wave energy converting systems. *Renew. Sustain. Energy Rev.* **2014**, *31*, 329–342. [CrossRef]

27. Ahmed, T.; Nishida, K.; Nakaoka, M. Grid power integration technologies for offshore ocean wave energy. In Proceedings of the Energy Conversion Congress and Exposition, Atlanta, GA, USA, 12–16 September 2010; pp. 2378–2385.

28. Wang, L.; Isberg, J.; Tedeschi, E. Review of control strategies for wave energy conversion systems and their validation: The wave-to-wire approach. *Renew. Sustain. Energy Rev.* **2018**, *81*, 366–379. [CrossRef]

29. Wu, J.; Yao, Y.; Zhou, L.; Göteman, M. Real-time latching control strategies for the solo Duck wave energy converter in irregular waves. *Appl. Energy* **2018**, *222*, 717–728. [CrossRef]

30. Zhang, X.T.; Yang, J.M.; Xiao, L.F. Declutching control of a point absorber with direct linear electric PTO systems. *Ocean Syst. Eng.* **2014**, *4*, 63–82. [CrossRef]

31. Li, G.; Belmont, M.R. Model predictive control of sea wave energy converters—Part I: A convex approach for the case of a single device. *Renew. Energy* **2014**, *69*, 453–463. [CrossRef]

32. Jama, M.; Wahyudie, A.; Noura, H. Robust predictive control for heaving wave energy converters. *Control Eng. Pract.* **2018**, *77*, 138–149. [CrossRef]

33. Retes, M.P.; Giorgi, G.; Ringwood, J.V. A Review of Non-Linear Approaches for Wave Energy Converter Modelling. In Proceedings of the 11th European Wave and Tidal Energy Conference, Nantes, France, 6–11 September 2015.

34. Bhinder, M.; Babarit, A.; Gentaz, L. Assessment of viscous damping via 3D-CFD modelling of a Floating Wave Energy Device. In Proceedings of the 9th European Wave and Tidal Energy Conference (EWTEC), Southampton, UK, 5–9 September 2011.

35. Bailey, H.; Robertson, B.R.D.; Buckham, B.J. Wave-to-wire simulation of a floating oscillating water column wave energy converter. *Ocean Eng.* **2016**, *125*, 248–260. [CrossRef]

36. Birk, L. Application of Constraint Multi-Objective Optimization to the Design of Offshore Structure Hulls. *J. Offshore Mech. Arct. Eng.* **2009**, *131*, 403–410. [CrossRef]

37. Clauss, G.F.; Birk, L. Hydrodynamic shape optimization of large offshore structures. *Appl. Ocean Res.* **1996**, *18*, 157–171. [CrossRef]

38. Li, H. Ocean Wave Energy Converters: Status and Challenges. *Energies* **2018**, *11*, 1250.

39. Devolder, B.; Stratigaki, V.; Troch, P. CFD Simulations of Floating Point Absorber Wave Energy Converter Arrays Subjected to Regular Waves. *Energies* **2018**, *11*, 641. [CrossRef]

40. Zurkinden, A.S.; Ferri, F.; Beatty, S. Non-linear numerical modeling and experimental testing of a point absorber wave energy converter. *Ocean Eng.* **2014**, *78*, 11–21. [CrossRef]

41. Piscopo, V.; Benassai, G.; Cozzolino, L. A new optimization procedure of heaving point absorber hydrodynamic performances. *Ocean Eng.* **2016**, *116*, 242–259. [CrossRef]
42. Price, A.A.E.; Dent, C.J.; Wallace, A.R. On the capture width of wave energy converters. *Appl. Ocean Res.* **2009**, *31*, 251–259. [CrossRef]
43. Cheng, Z.S.; Yang, J.M.; Hu, Z.Q. Frequency/time domain modeling of a direct drive point absorber wave energy converter. *Sci. China Phys. Mech.* **2014**, *57*, 311–320. [CrossRef]
44. Zheng, C.W.; Zhuang, H.; Li, X. Wind energy and wave energy resources assessment in the East China Sea and South China Sea. *Sci. China Technol. Sci.* **2012**, *55*, 163–173. [CrossRef]
45. Zheng, C.W.; Pan, J.; Li, J.X. Assessing the China Sea wind energy and wave energy resources from 1988 to 2009. *Ocean Eng.* **2013**, *65*, 39–48. [CrossRef]
46. Falnes, J. A review of wave-energy extraction. *Mar. Struct.* **2007**, *20*, 185–201. [CrossRef]
47. Shi, H.; Cao, F.; Liu, Z. Theoretical study on the power take-off estimation of heaving buoy wave energy converter. *Renew. Energy* **2016**, *86*, 441–448. [CrossRef]
48. Taguchi, G.; Cariapa, V. Taguchi on Robust Technology Development: Bringing. *J. Press. Vess.-Technol. Asme* **1993**, *115*, 161–171. [CrossRef]
49. Hansen, A.H.; Asmussen, M.F.; Bech, M. Model Predictive Control of a Wave Energy Converter with Discrete Fluid Power Power Take-Off System. *Energies* **2018**, *11*, 635.
50. Beatty, S.; Ferri, F.; Bocking, B. Power Take-Off Simulation for Scale Model Testing of Wave Energy Converters. *Energies* **2017**, *10*, 973. [CrossRef]
51. Falnes, J. *Ocean Waves and Oscillating Systems: Linear Interactions Including Wave-Energy Extraction*; Cambridge University: Cambridge, UK, 2002.
52. Verma, P.; Sharma, M.P.; Dwivedi, G. Prospects of bio-based alcohols for Karanja biodiesel production: An optimisation study by Response Surface Methodology. *Fuel* **2016**, *183*, 185–194. [CrossRef]
53. Babarit, A. A database of capture width ratio of wave energy converters. *Renew. Energy* **2015**, *80*, 610–628. [CrossRef]

![energies logo]

MDPI

Article

Optimal Design of Rated Wind Speed and Rotor Radius to Minimizing the Cost of Energy for Offshore Wind Turbines

Longfu Luo [1], Xiaofeng Zhang [1], Dongran Song [2,*], Weiyi Tang [2,*], Jian Yang [2], Li Li [2], Xiaoyu Tian [2] and Wu Wen [1]

[1] College of Electrical and Information Engineering, Hunan University, Changsha 410082, China; llf@hnu.edu.cn (L.L.); zxf0303@126.com (X.Z.); 13637408048@139.com (W.W.)
[2] School of Information Science and Engineering, Central South University, Changsha 410083, China; jian.yang@csu.edu.cn (J.Y.); lili112209@163.com (L.L.); txy15388967580@163.com (X.T.)
* Correspondence: humble_szy@163.com (D.S.); 154601016@csu.edu.cn (W.T.); Tel.: +86-181-636-56151 (D.S.); +86-138-6122-2783 (W.T.)

Received: 18 September 2018; Accepted: 1 October 2018; Published: 11 October 2018

Abstract: As onshore wind energy has depleted, the utilization of offshore wind energy has gradually played an important role in globally meeting growing green energy demands. However, the cost of energy (COE) for offshore wind energy is very high compared to the onshore one. To minimize the COE, implementing optimal design of offshore turbines is an effective way, but the relevant studies are lacking. This study proposes a method to minimize the COE of offshore wind turbines, in which two design parameters, including the rated wind speed and rotor radius are optimally designed. Through this study, the relation among the COE and the two design parameters is explored. To this end, based on the power-coefficient power curve model, the annual energy production (AEP) model is designed as a function of the rated wind speed and the Weibull distribution parameters. On the other hand, the detailed cost model of offshore turbines developed by the National Renewable Energy Laboratory is formulated as a function of the rated wind speed and the rotor radius. Then, the COE is formulated as the ratio of the total cost and the AEP. Following that, an iterative method is proposed to search the minimal COE which corresponds to the optimal rated wind speed and rotor radius. Finally, the proposed method has been applied to the wind classes of USA, and some useful findings have been obtained.

Keywords: offshore wind turbines; cost of energy; annual energy production; optimal design

---

1. Introduction

Renewable energy has been very attractive since the end of last century, as there have been ever-growing concerns over limited fossil-fuel resources, serious environmental regulations, and heavy energy demand. Among various types of renewable energy, wind energy is one of the most economical sources. The wind energy development has been rapidly developed in recent years. In 2017, the global cumulative installed wind turbine capacity has reached a new peak value of 539.58 GW [1]. On the other side, there is a new trend for the development of wind energy, that is, the installation of wind turbines has gone from onshore sites to offshore sites [2]. Despite the rapid growth of wind energy utilization, the challenge still exists, especially for the offshore-site turbines. The high cost of energy (COE) for offshore wind power (compared to the onshore wind power and traditional sources) has hindered the utilization of offshore wind energy across the world.

As the turbine COE is relevant to the total annual cost and the annual energy production (AEP) [3], optimizing COE reduces the production cost and increases the production efficiency.

Considering the overall procedure of the wind turbine development, the potential approaches for minimizing COE can be accordingly categorized into three types: design optimization, manufacturing process optimization, and on-site optimization. Since the manufacturing procedure depends on the manufacture technology which is normally scheduled during a certain period, its optimization is hard to employ in practice. By comparison, the on-site optimization is utilized as a common practice in the wind energy industry [4]. The on-site optimization is to optimize the turbine controller and its parameters matching the wind characteristics, so the AEP is enhanced by optimizing the energy capture efficiency below rated wind speeds. Two types of approaches are available for the on-site optimization. One approach is to control the rotor speed to follow the changing wind speed, so that the known principle of the optimal tip speed ratio tracking can be fulfilled [5]. To do this, the advanced wind estimator-based torque controller has been utilized by industrial turbines [6–8], and the Lidar-based previewer controllers have gradually been payed attention to [9]. Meanwhile, some researchers have proposed to optimize the performance by adjusting the controller parameter according to the wind condition [10], as it has been revealed that the controller performance is significantly affected by the wind conditions [11,12]. Furthermore, the energy production efficiency can be improved by considering a hybrid wind-hydro power plant for the isolated power system [13,14]. The other approach of improving the energy capture efficiency is to control the yaw system to track the wind direction, so that the yaw error can be minimized. In some recent studies, the previewed yaw controller and its parameter optimizations have been proposed, which have been proven to be efficient in enhancing the energy capture efficiency [15,16]. Despite these on-site optimizations being cost-effective and convenient to carry out, the achieved profit is quite limited as the energy capture efficiency is only improved in some control regions.

Many efforts have been made towards improving the performance of the wind turbines through design optimization. A comprehensive review of wind turbine optimization technologies is given in [17], in which a few of works have been referred for wind turbines towards minimizing the COE by optimizing the aerodynamics shape of airfoils. As a key wind turbine component, the blade is a determining factor for energy harvesting efficiency and its aerodynamic shape optimization is very momentous. The aerodynamic shape optimization involves many objectives, such as the AEP, the air loads of the blades and rotors, and the blade mass. Improving one objective inevitably deteriorates the others, and thus the aerodynamic shape optimization widely uses multi-objective functions. A numerical optimization method for the design of horizontal axis wind turbines is presented in [18], in which the fatigue and extreme loads and the AEP are considered. A multi-objective optimization method is proposed for the turbine blade using the lifting surface method as the performance prediction model [19]. The first study on the external axis wind turbines is conducted to optimize blade count and operating point to simultaneously maximize power, while minimizing power fluctuating and the peak point reaching time [20]. When these researches optimize the aerodynamic shape, most of them only concerned with the open-loop static aerodynamic performance, that is, the AEP is calculated under the implicit assumption that the turbines can keep operating at the optimal TSR. But in practice, the large-scale wind turbines cannot instantly respond to the wind fluctuation and the performance is influenced by the wind conditions [21]. In this regard, the closed-loop optimal design should be considered, which has been presented in a recent study [22]. Nevertheless, the blade aerodynamic optimization is a small portion of the turbine design, and the optimal design involving the most important parameters of the overall design may achieve a low COE in an effective way.

As the first step of design process of wind turbines, the conception design defines the most important parameters, of which the optimizations have been proven to be highly efficient in minimizing the COE [23–27]. The dominant ingredient of designing a satisfactory turbine with low COE includes the suitable physical and operational parameters, which are determined by the wind conditions on the erected site of the turbines, but there have been very few studies conducted. An optimization method is presented for the concept design of a grid-connected onshore wind turbine, in which the blade number, rotor diameter, tower height, rotor rotational speed, the rated wind speed, and the rated power have been optimized to match the wind condition described by the Weibull parameters [28]. Based on the case study results, it has been shown that some of the existing onshore turbines appear to be well designed, and others do not. An iterative approach is presented to optimize the turbine design based on a simple COE model, which is a function of rotor diameter, tower height, rated power, and the TSR. In their results, it is revealed that the onshore turbines about 1–2 MW can achieve minimum COE for considered cases [29]. Another design study is conducted on onshore turbines, in which the COE model is described relevant to rotor diameter, hub height, capacity factor, rated power, and rotor diameter [30]. Recently, a mathematical approach is proposed to minimize the COE of onshore turbines, in which the COE model is expressed as a function of rated power and rated wind speed [31]. When compared with the referenced turbines, a noticeable profitability has been gained by the optimized turbines. The above references have shown that the site-specific turbine design can achieve a low COE for onshore turbines, but the studies for the offshore wind turbines are lacking.

Currently, the offshore wind turbines are designed towards the large-size trend, but whether the offshore turbines with large capacity and long blades will have a low COE remains unclear. This paper aims at clarifying this issue. For this purpose, the relation among the COE, the rated wind speed, and the rotor radius is established, a method to achieving optimal COE of the offshore turbines is proposed, and the optimal results are obtained and analyzed. Since the cost model is important to determine the COE results, the detail cost model of the bottom-fixed offshore wind turbines developed by NREL is employed in this study [32]. By comparison to the literature, the contribution of this study is twofold: on one hand, a method to minimizing the COE of offshore turbines through optimizing the rated wind speed and rotor radius is proposed, which can be extended to other types of wind turbines; on the other hand, the optimal design parameters achieving the minimal COE of offshore wind turbines are obtained and explored under different wind conditions, which can be used as references for offshore wind turbine designers. The remaining sections are organized as follows: the design process of wind turbines is summarized in Section 2. The COE model of onshore turbines is discussed in Section 3, and Section 4 presents the method of optimizing COE by selecting the optimal rated wind speed and rotor radius, and the optimal results through the case studies. Finally, Section 5 concludes the study.

2. Design Process of Wind Turbines

For a wind turbine, its design process can be divided into six steps [33]:

- Step 1: Conception stage. As the first stage of a turbine design, the conception design involves the definitions of the nominal parameters of the wind turbine, such as the nominal power output, rotor diameter, electrical energy conversion system and so on.
- Step 2: Blade design. The blade design step defines the aerodynamics and structural concepts of the blades, which are determined by the overall conception in the step 1 and the controller development in the step 3, respectively.
- Step 3: Control development and preliminary design models. During this step, the preliminary design models, which are typically based on multibody models, are employed to estimate the loads acting on major components of the turbine during its life cycle. When the loads are beyond limitations, the controller and blade structural designs will be required to be redesigned. Thus, steps 2 and 3 depend on each other, and the final design of blades and controller will be finished after several iterations.
- Step 4: Design engineering and strength calculation. During this step, a strength calculation of the major components is carried out, so that it can be verified that the designed components are able to withstand the loads calculated during the life cycle.
- Step 5: Construction and erection of the prototype. During this step, the prototype of the wind turbine is manufactured and erected on the wind farm site.

- Step 6: Measurements on the prototype. As a final step, it aims at proving the predicted property of the designed turbine within a measurement campaign.

From the above explanation, it is evident that the conception design has a crucial impact on the COE of the designed wind turbine due to its placement at the first stage of the design process. To further illustrate its importance, the time line of the wind turbine's design process in terms of the cost and the possibility of modifications during the development is shown in Figure 1.

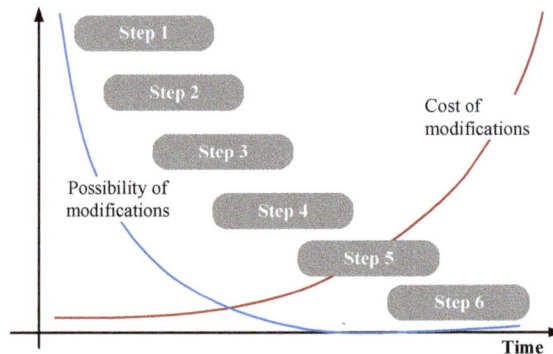

Figure 1. Cost and possibility of modifications during the design process of a wind turbines.

It is obvious that it is the best chance to modify the design at the conception stage rather than other stages to obtain the least cost and the most possibility. This conclusion specially fits the application occasion of the offshore turbines, as the overall COE is very high compared to the onshore turbines. Therefore, minimizing the COE of offshore wind turbines is indispensable and it is the objective of this study.

3. COE Model of Offshore Turbines

The COE of the offshore wind turbine is determined by two parts: the annual energy production (AEP) and the total annual cost, and these two parts are elaborated in the following sections.

3.1. AEP Model of the Offshore Turbines

The AEP model of the offshore turbines is similar to the one of the onshore turbines, which is normally estimated based on the Weibull probability distributions of the wind statistics, a standardized power curve, a physical description of the turbine and physical constants. The difference between the AEP models of offshore and onshore turbines consists in the wind statistics. Specifically, the roughness of the sea surface is different from the one of land-based surface, and thus the wind statistics may be different. The AEP output of turbines can be calculated by using the mean power production P_m during one hour and the total hours of one year:

$$\text{AEP} = 8760(1 - \sigma)P_m \tag{1}$$

where σ is the total power generation loss, which includes the power converter loss, electrical grid loss, availability loss and so on. In this study, σ is assumed to be a constant of 0.17 [34].

When determined by the power curve of the concerned turbine and the wind characteristics on the erected site, which is expressed as a Weibull distribution, the mean power production P_m is calculated by

$$P_m = \int_0^\infty P(v)f(v)dv \tag{2}$$

where v is the wind speed, $P(v)$ and $f(v)$ denote the power curve model and the Weibull distribution as functions of the wind speed, respectively.

3.1.1. Power Curve Modeling of the Offshore Turbines

Wind turbines have different power curves, even the same turbine may produce different power determined by the control curves. At an offshore site, the typical type of wind turbines is the large-scale variable-speed horizontal-axis machine, and its power curve mainly relies on the important characteristics of the wind speed, in which the cut-in wind speed v_c, rated wind speed v_r, cut-out wind speed v_f are involved. At cut-in v_c, the turbines is able to start generating power; at the rated wind speed v_r, the turbine produces the rated power P_r; and at the cut-out wind speed v_f, the turbine stops producing power to avoid its component over-loads. Hence, the power curve model of the turbines can be formulated by

$$P(v) = \begin{cases} 0 & (v < v_c) \\ P_f(v) & (v_c \leq v \leq v_r) \\ P_r & (v_r \leq v \leq v_f) \\ 0 & (v_f < v) \end{cases} \tag{3}$$

where $P_f(v)$ is the power fitted to manufacturer power curve data by using mathematical formulation.

From Equation (3), it is seen that the power curve is mainly determined by P_r and $P_f(v)$, of which the theoretical model is expressed by

$$P_f(v) = \rho \pi R^2 C_{p,\max} v^3 / 2 \tag{4}$$

where ρ and R denote the air density and the rotor radius, respectively; $C_{p,\max}$ is the maximum value of power coefficient of the concerned turbine.

Besides the theoretical model, there are some other models, which are predicted by sampling the turbine output power at various wind speed. In [35], nine power curves models have been presented and compared, including linear model, general model, quadratic model, cubic-I model, cubic-II model, exponential model, power-coefficient model, approximate power-coefficient model, and polynomial model. Through the comparison results, it is observed that the power-coefficient model has the most accurate results. Thus, this study employs this model, which is expressed by

$$P_f(v) = \rho \pi R^2 C_{p,eq} v^3 / 2 \tag{5}$$

where $C_{p,eq}$ is the equivalent power coefficient, which is a constant for the variable-speed wind turbines operating in the range between the cut-in and rated wind speed [35,36]. In this study, it is assumed to be 0.42 [36]. It is noticed that, by setting $v = v_r$, the turbine rated power P_r can be calculated by

$$P_r = \rho \pi R^2 C_{p,eq} v_r^3 / 2 \tag{6}$$

By referring to Equations (1)–(6), the AEP model is determined by the rotor radius, rated wind speed, cut-in wind speed, and cut-out wind speed, and accordingly it is reformulated by

$$\text{AEP} = f_{\text{AEP}}(R, v_r, v_c, v_f) = 1454.16 \rho \pi R^2 \left(\int_{v_c}^{v_r} v^3 f(v) dv + \int_{v_r}^{v_f} v_r^3 f(v) dv \right) \tag{7}$$

3.1.2. Weibull Distribution of the Offshore Wind Statistics

In this study, the Weibull probability density function $f(v)$ is employed to represent the offshore wind statistics, which depends on the parameters k and c that determine the shape and intensity of the wind during one year on a site [28]:

$$f(v) = (k/c)(v/c)^{k-1}e^{-(v/c)^k} \tag{8}$$

where c and k denote the scale parameter and the shape parameter, respectively.

The wind speed becomes stronger along with the increasing altitude, and thus the wind experienced by the turbine reaches a high value at the hub. Accordingly, the parameters c and k are functions of the hub height H [37]. Based on the relationship between the wind speed and the hub height, the scale parameter c at the hub height is calculated by [38]

$$c = c_0(H/H_0)^{\alpha} \tag{9}$$

where c_0 is the value of c at reference altitude H_0, and α is the Hellmann exponent, which depends on serval surface properties of the wind field. The typical value of α is 0.1 over water, and 0.14 over land [39,40].

In this study, the shape parameter k and the annual mean wind speed v_m are considered as known parameters, and c_0 is calculated by [41]

$$c_0 = v_m/\Gamma(1 + 1/k) \tag{10}$$

where $\Gamma(\ )$ is the gamma function.

The shape parameter k follows a law provided by [38]:

$$k = k_0[1 - 0.088\,ln(H_0/10)]/[1 - 0.088\,ln(H/10)] \tag{11}$$

where k_0 is the value of k at reference altitude H_0.

In Equations (9)–(11), c and k depend on c_0, k_0, H_0 and H. According to the European Wind Energy Association, there is the relationship between the hub height and the turbine rotor radius, which is expressed by [42]

$$H = 2.7936 \times (2R)^{0.7633} \tag{12}$$

Therefore, based on Equations (7)–(12), the AEP of the offshore turbines is expressed as a function of the wind statistics factors: c_0, k_0, α, and the turbine characteristics parameters: v_c, v_r, v_f, R, formulated by

$$\text{AEP} = f_{\text{AEP}}(R, v_r, v_c, v_f, c_0, k_0, \alpha) \tag{13}$$

3.2. Cost Model of the Offshore Turbines

When erected at the sea, there have been two types of the support platform for the offshore turbines, namely, bottom-fixed platform and floating platform. Accordingly, the cost model may be constructed based on the platform type. However, there is no cost details of the floating offshore wind turbines, and thus the study considers the bottom-fixed offshore wind turbine, of which the cost model has been developed by NREL, and the cost data was converted to 2002 dollars. The bottom-fixed offshore turbines are widely erected at the shallow-water sea, of which the cost model detailed in the report of NREL is expressed as [32]

$$Cost = FCR \times ICC + AOE \tag{14}$$

where *Cost* is the total turbine cost, *ICC* is the initial capital cost, and *AOE* is the annual operating expense. *FCR* is the fixed charge rate, defined as the annual amount per dollar of initial capital cost needed to cover the capital cost, a return on debt and equity, and various other fixed charges. In this study, it is set to 0.1158 per year.

3.2.1. The Initial Capital Cost

The initial capital cost *ICC* is the sum of the turbine system cost ICC_{turb} and the balance of the platform station cost ICC_{BOP}. The whole turbine system consists of four major subsystems, which are referred as to mechanical system (including blade, gearbox, low-speed shaft, main bearings, and mechanical brake), electrical system (including generator, power converter, and electrical connection), control system (including pitch control, yaw control, torque control, and safety systems), and the auxiliary system (including hydraulic cooling equipment, hub, nose cone, mainframe, nacelle cover, and tower). The detailed mathematical functions of these eighteen components are summarized in Table 1, in which it is obtained that ICC_{turb} is determined by the rated power, rotor radius, and hub height.

Table 1. The turbine system cost from the tower up.

Type	Cost Model (Unit: $)
Mechanical System	
Blade	$(0.4019R^3 + 2.7445R^{2.5025} - 955.24)/0.72$
Gearbox	$16.45P_r^{1.249}$
low speed shaft	$0.1 \times (2R)^{2.887}$
Main bearings	$(0.64768R/75 - 0.0107) \times (2R)^{2.5}$
Mechanical brake	$1.9894P_r - 0.1141$
Electrical System	
Generator	$65P_r$
Power converter	$79P_r$
Electrical connection	$40P_r$
Control System	
Pitch system	$0.48 \times (2R)^{2.6578}$
Yaw system	$0.0678 \times (2R)^{2.964}$
Control safety system	$55,000$
Auxiliary System	
Hydraulic cooling system	$12P_r$
Hub	$2.0 \times (2R)^{2.53} + 24141.275$
Nose cone	$206.69R - 2899.185$
Mainframe	$11.917 \times (2R)^{1.953}$
Nacelle cover	$11.537P_r + 3849.7$
Tower	$0.59595\pi R^2 H - 2121$

The balance of the platform station cost involves the cost of infrastructure (including marinization, support structure, transportation, electrical interface connections, engineering permits, assembly and installation) and offshore engineering (including scour protection, personal access equipment, surety bond, port and staging equipment, and Offshore warranty premium). The detailed mathematical functions of these eleven components are summarized in Table 2, in which it is obtained that ICC_{BOP} is only determined by the rated power.

Table 2. The balance of the platform station cost.

Type	Cost Model (Unit: $)
Marinization	$0.135 \times ICC_{turb}$
Offshore support structure	$300P_r$
Offshore transportation	$P_r(1.581 \times 10^{-5} \times P_r^2 - 0.0375P_r + 54.7)$
Offshore turbine installation	$100P_r$
Offshore electrical interface and connection	$260P_r$
Offshore engineering permits	$37P_r$
Port and staging equipment	$20P_r$
Personnel access equipment	$60,000$
Scour protection	$55P_r$
Surety bond	$0.03 \times (ICC - 0.15ICC_{turb})$
Offshore warranty premium	$0.15 \times ICC_{turb}$

3.2.2. The Annual Operating Expense

The annual operating expense *AOE* consists of the land lease, levelized operation and maintenance (O&M), and the levelized replacement costs. The detailed mathematical functions of these three components are summarized in Table 3, in which it is obtained that *AOE* is only determined by the rated power and the annual energy production.

Table 3. Wind turbine annual operation cost.

Type	Cost Model (Unit: $)
Offshore Levelized replacement cost	$17P_r$
Offshore bottom lease cost	0.00108AEP
Offshore O&M	0.02AEP

3.3. COE Model of the Offshore Turbines

In Tables 1 and 2, *ICC* is determined by P_r, R, and H, while in Table 3, *AOE* is determined by P_r and AEP. Meanwhile, by referring to Equations (6) and (12), P_r is a function of R and H, and H is a function of R. Thus, the cost model of the offshore wind turbines can be formulated as a function relevant to rated wind speed, rotor radius, and the annual energy production:

$$Cost = 0.1158 f_{ICC}(v_r, R) + f_{AOE1}(v_r, R) + 0.02108\text{AEP} \tag{15}$$

Based on Equations (13) and (15), the final COE model of the offshore wind turbines is calculated and formulated by

$$\text{COE} = \frac{Cost}{\text{AEP}} = \frac{f_{cost}(R, v_r)}{f_{AEP}(R, v_r, v_c, v_f, c_0, k_0, \alpha)} + 0.02108 \tag{16}$$

In Equation (16), the COE model of offshore wind turbines has been formulated as a nonlinear function, in which the rated wind speed and the rotor radius are the turbine design parameters, the cut-in and cut-out wind speeds are typically known constant parameters, and the wind statistics of the offshore windfarm are the scale factor, the shape factor, and the Hellmann exponent.

4. COE Optimization for Offshore Turbines

Based on Equation (16), it is obtained that minimizing COE of the offshore wind turbines can be fulfilled by optimizing the two important design parameters, the rated wind speed and the rotor radius. The objective is to minimize the COE when considering the practical constraints. The objective function and the constraint is expressed by

$$\text{COE}^{\min}(R, v_r) = \min(\text{COE}(R, v_r)) \tag{17}$$

$$s.t. \left\{ R^{\min} \leq R \leq R^{\max}, v_r^{\min} \leq v_r \leq v_r^{\max} \right\}$$

where $R^{\min}$ and $R^{\max}$ are the minimum and maximum values of rotor radius, and $v_r^{\min}$ and $v_r^{\max}$ are the minimum and maximum values of the rated wind speed, respectively.

4.1. Optimizaiton Method

After preparing the COE model, the optimization algorithm should be designed. Since the COE model described by Equation (17) is highly nonlinear under constraints, the iterative method is employed in this study to solve the optimization problem. The flow chart of the developed iterative approach is presented in Figure 2, comprising the following steps:

- Step 1: Initialize the wind statistics of the offshore site: v_m, k_0, α;
- Step 2: Define the ranges of the design parameters: v_r, R;

- Step 3: Evaluate the COE using different v_r and R, and save the smallest value of COE and its corresponding v_r and R;
- Step 4: Repeat step 3 until all design parameters sets of v_r and R have been evaluated;
- Step 5: Output minimal COE and the corresponding design parameters.

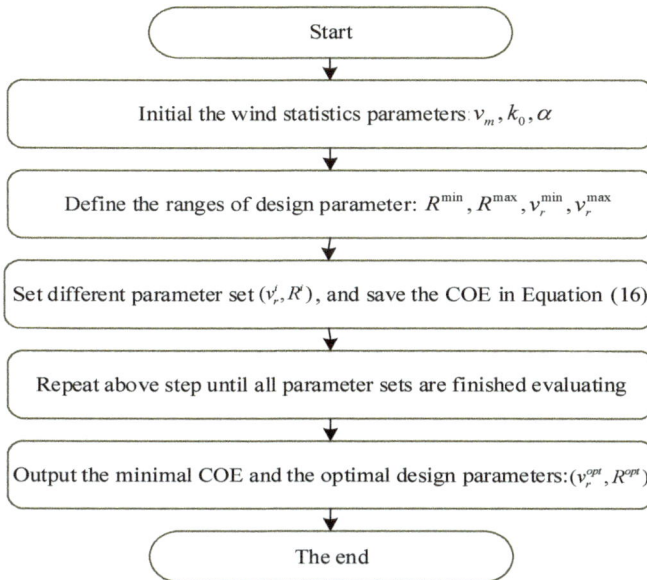

Figure 2. Flow chart of the proposed cost of energy (COE) minimization approach.

4.2. Parameter Settings

As shown in Equations (13) and (16), the wind statistics has an important influence on the obtained AEP and COE. This study concerns the wind statistics in USA, in which seven wind classes are introduced from class I as the lowest wind power class to class VII as the highest one [43]. The annual mean wind speeds and power densities of these seven classes at reference height of 10 m are given in Table 4. Based on the wind statistics of USA, the parameters settings are summarized in Table 5, in which the rated wind speed ranges in 6–16 m/s with a step of 0.2 m/s, and the rotor radius ranges in 10–70 m with a step of 2 m. Accordingly, the minimum and maximum values of the rated power are 17.45 kW and 16.22 MW, which correspond to the designed wind turbines with $v_r = 6$ m/s and $R = 10$ m, and the one with $v_r = 16$ m/s and $R = 70$ m, respectively. Thus, the optimal results can be obtained from by searching this big range of turbine capacity.

Table 4. Wind classification defined in USA [43].

Wind Power Class	Mean Annual Wind Speed (m/s)	Power Density (W/m²)
I	0–4.4	0–100
II	4.4–5.1	100–150
III	5.1–5.6	150–200
IV	5.6–6.0	200–250
V	6.0–6.4	250–300
VI	6.4–7.0	300–400
VII	7.0–9.4	400–1000

Table 5. Parameters used in the proposed COE minimization method.

Parameter	Values or Ranges
v_m	[4, 7] with a step of 1 m/s
k_0	[1.2, 3.6] with a step of 0.4
H_0	10 m
α	0.1
ρ	1.225 kg/m^3
R	[10, 70] with a step of 2 m
v_r	[6, 16] with a step of 0.2 m/s
v_c	3 m/s
v_f	25 m/s
μ	0.17
$C_{p,eq}$	0.42

4.3. Optimizaiton Results and Discussions

The results of the minimum COE are obtained by using the proposed iterative search method and presented for different wind classes of USA according to Table 4. The minimum values of COE versus the annual mean wind speed values and the shape factors, k_0, are presented in Figure 3. As observed from Figure 3, the minimum COE decreases following the increment of the annual mean wind speed and the increase of the shape factor, and the minimum COEs are influenced more obviously by the annual mean wind speed than by the shape factor. To be specific, for the wind classes of the annual wind speeds being 4 m/s, 5 m/s, 6 m/s, and 7 m/s, the minimum COEs range in $0.15 - 0.17$, $0.13 - 0.1$, $0.11 - 0.08$, and $0.1 - 0.07$ \$/kWh, respectively. From the results, it is seen that the minimum COE of installing offshore wind turbines at the site with a low wind power class almost doubles the one at the site of a high wind power class, which fits the known concept that installing the wind turbines at the rich wind site is more profitable than at the poor wind site. When considering 0.1 \$/kWh is an acceptable COE value to exploit the wind energy, the offshore wind turbines should be installed at the sea with the wind class III or above.

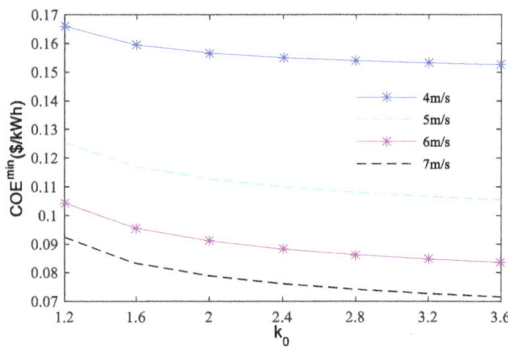

Figure 3. The minimum COEs versus the annual mean wind speeds and the shape factors.

To check the optimal design parameters achieving the minimum COEs, the obtained optimal rotor radius and rated wind speed versus the annual mean wind speed values and the shape factors, k_0, are presented in Figure 4a,b. As seen from Figure 4, the optimal rotor radius is surprisingly kept in a certain range of $38 - 40$ m, while the optimal rated wind speed shows the relevance to the wind statistics. Specifically, for the annual wind speeds being 4 m/s, 5 m/s, 6 m/s, and 7 m/s, the optimal rated wind speed ranges in $6 - 8.5$ m/s, $7.5 - 9$ m/s, $8.5 - 9.5$ m/s, and $9 - 10$ m/s, respectively. Thus, it is obtained that the optimal rated wind speed of the offshore wind turbines should be designed around 1.5 times more than the annual mean wind speed, and the optimal rotor radius may be designed around a certain value regardless of the wind statistics.

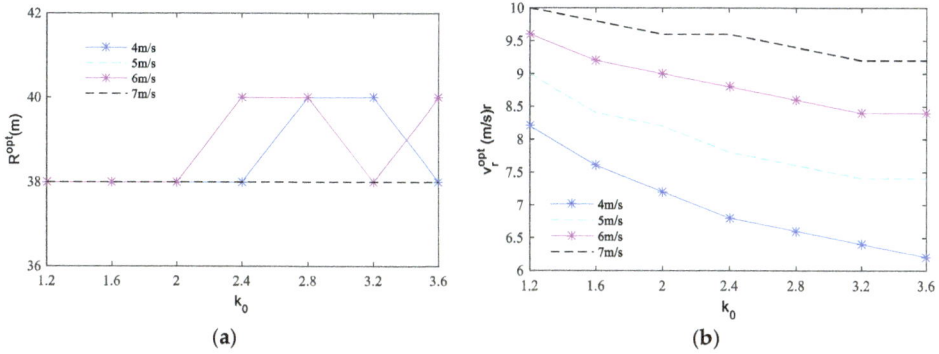

Figure 4. The optimal design parameters of offshore turbines versus the annual mean wind speeds and the shape factors: (**a**) the rated rator radius; (**b**) the rated wind speed.

Since the turbine capacity is determined by the rated wind speed and the rotor radius, the optimal rated power of the offshore turbines having the minimum COE are calculated and presented in Figure 5. It is clearly seen from Figure 5 that, when the annual mean wind speed is increasing or the shape factor is reducing, the optimal capacity of the turbines giving the minimum COEs takes ever-bigger value, from several hundreds of kilo-watt to one megawatt. In detail, for the wind classes of the annual wind speeds being 4 m/s, 5 m/s, 6 m/s, and 7 m/s, the optimal capacity of the turbines range in $0.3 - 0.7$ MW, $0.5 - 0.9$ MW, $0.7 - 1.1$ MW, and $0.9 - 1.2$ MW, respectively. From the results, it is concluded that the offshore wind turbines with a large capacity and long blades may not absolutely bring the COE down, and their capacity should be designed at a proper value depending on the wind conditions at the sea.

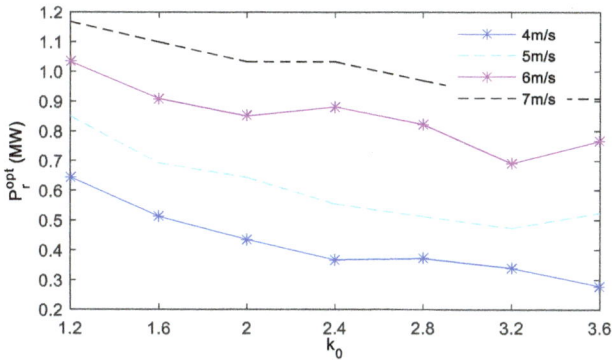

Figure 5. The optimal capacity of offshore turbines versus v_m and k_0.

The numerical results in Figures 3–5 are summarized in Table 6. Based on this data table, the suitable rated wind speed and rotor radius can be chosen to ensure the minimum energy cost of the offshore turbines for a specific wind region. For instance, the rated wind speed and the rotor radius for the shape factor of $k_0 = 1.2$ can be adopted 10 m/s and 38 m, respectively, for the annual mean wind speed around 7 m/s. With these two optimal design parameters, the COE of the offshore turbines is minimized to 0.0905 $/kWh, which is an acceptable value for exploiting the offshore wind energy.

Table 6. Minimal COEs and optimal design parameters at different wind classes and wind statistics.

Wind Class	v_m(m/s)	k_0	COE_{min}($/kWh)	v_r^{opt}(m/s)	R^{opt}(m)	P_r^{opt}(kW)
I, II	4.0	1.2	0.1660	8.2	38	643.45
		1.6	0.1595	7.6	38	512.29
		2.0	0.1566	7.2	38	435.58
		2.4	0.1550	6.8	38	366.94
		2.8	0.1540	6.6	40	371.76
		3.2	0.1532	6.4	40	338.97
		3.6	0.1527	6.2	38	278.13
II, III, IV	5.0	1.2	0.1253	9	38	850.75
		1.6	0.1170	8.4	38	691.69
		2.0	0.1127	8.2	38	643.45
		2.4	0.1099	7.8	38	553.80
		2.8	0.1080	7.6	38	512.29
		3.2	0.1066	7.4	38	472.90
		3.6	0.1055	7.4	40	523.99
IV, V, IV	6.0	1.2	0.1044	9.6	38	1032.49
		1.6	0.0956	9.2	38	908.73
		2.0	0.0911	9	38	850.75
		2.4	0.0883	8.8	40	881.20
		2.8	0.0862	8.6	40	822.47
		3.2	0.0847	8.4	38	691.69
		3.6	0.0835	8.4	40	766.41
VI, VII	7.0	1.2	0.0925	10	38	1167.00
		1.6	0.0833	9.8	38	1098.38
		2.0	0.0789	9.6	38	1032.49
		2.4	0.0761	9.6	38	1032.49
		2.8	0.0741	9.4	38	969.30
		3.2	0.0727	9.2	38	908.73
		3.6	0.0715	9.2	38	908.73

4.4. Result Analysis on a Typical Case

Above results of the minimum COEs and the corresponding optimal design parameters give some information about designing a satisfactory offshore turbine, but they cannot reveal the details. To further examine the relationship between the design parameters and the offshore turbine COE, the results from a specific wind class with $v_m = 7$ m/s and $k_0 = 3.6$ are presented and analyzed.

The two major component costs of offshore turbines, the turbine system cost and the balance of the platform station cost are respectively presented in Figure 6a,b, in forms of cost per kilo-watt, calculated by ICC_{turb}/P_r and ICC_{BOP}/P_r. As seen from Figure 6, the high cost of the turbine system is more than 3000 $/kW and up to 6000 $/kW in the low rated wind speed region, while the normal cost is lower than 1000 $/kW in the other part of the region. By comparison, the high cost of the station balance is less than 4500 $/kW, which appears in the region of the lowest rated wind speed and smallest rotor radius, and in the region of the highest rated wind speed and biggest rotor radius. Thus, these results reveal that the turbines with a very big or with very small capacity are unfavorable for the offshore site, while the rated wind speed should be designed at more than 8 m/s for the wind field with the annual mean wind speed of 7 m/s.

The obtained annual energy production AEP versus the rotor radius and the rated wind speed is presented in Figure 7, from which it is seen that, the AEP is more sensitive to the rotor size than the rated wind speed. Nevertheless, selecting a suitable rated wind speed is important to obtain a high AEP. For instance, for a wind turbine with the rotor radius of 60 m, the AEP is 18,540 MWh with $v_r = 15$ m/s, while it is decreased to 6771 MWh with $v_r = 7$ m/s.

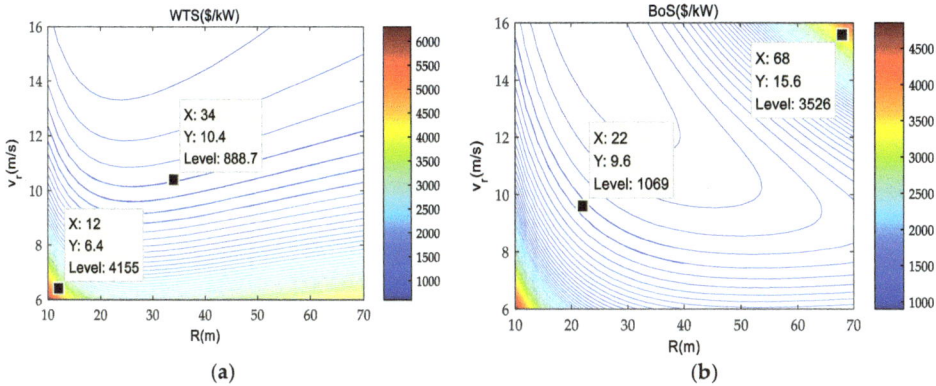

Figure 6. The two major component costs of offshore turbines versus the annual mean wind speeds and the shape factors: (**a**) the turbine system cost; (**b**) the station balance cost.

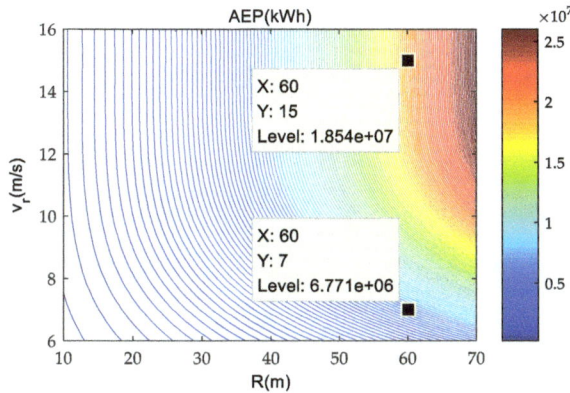

Figure 7. The obtained annual energy production versus the rotor radius and the rated wind speed.

The obtained COE versus the rotor radius and rated wind speed are presented in Figure 8. As seen from Figure 8, the high COE appears in two regions, including one with the high rated wind speed and big rotor radius and the other with the small rotor radius. Thus, it is clear that the turbines with very big capacity or with short blades are unfavorable for the offshore site. Besides, it is worth noticing that the COE is sensitive to the variation of the rated wind speed rather than the rotor radius, which means that the designed turbine can have a big range of rotor radius. For instance, around the rated wind speed of $v_r = 9.2$ m/s, when the values of COE vary in the range of $0.072 - 0.076$ \$/kWh, the designed rotor radius changes from 30 m to 68 m. Therefore, the designed capacity of the favorable offshore turbines at the wind class is with the rated power from 566 kW to 2900 kW. In this regard, a robust optimal design for the offshore wind turbines can be with long blades and large capacity. Nevertheless, the rated wind speed of the designed wind turbines should be carefully selected to achieve a low COE.

Finally, to check the influence of equivalent power coefficient on the obtained COE, the obtained COE versus the rotor radius and rated wind speed using two different equivalent power coefficients, $C_{p,eq} = 0.48$ and $C_{p,eq} = 0.54$ are presented in Figure 9a,b, respectively. By comparing Figure 9 with Figure 8, similar results are obtained, including two aspects: on one side, the turbines with very big capacity or with short blades are unfavorable for the offshore site; on the other side, the favorable turbines having the minimal COE are designed with the rated wind speed ranging in $8 - 10$ m/s and the rotor radius ranging in $25 - 66$ m. These similar results reveal that the optimal design of turbines are less sensitive to the variations of the equivalent power coefficient, which well justify that the outstanding significance of the conception design optimization over the blade design optimization. Nevertheless, it is worth noticing that the COE minimum is slightly decreasing with the increase of the equivalent power coefficient. In this regard, it is also economically beneficial to optimize the blade airfoil, as the power coefficient of wind turbines largely relies on the design of the blade airfoil [17,21].

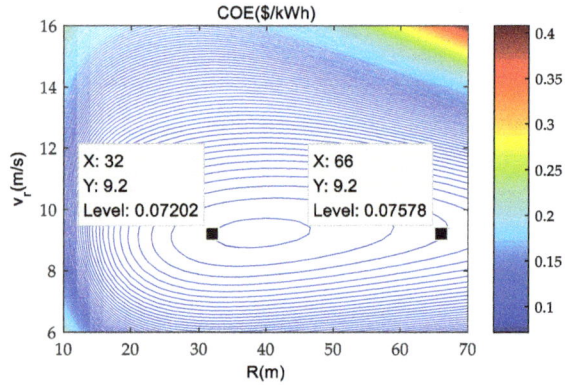

Figure 8. The offshore turbines COE versus the rotor radius and rated wind speed at a specific class with $v_m = 7\,\text{m/s}$, $k_0 = 3.6$ and $C_{p,eq} = 0.42$.

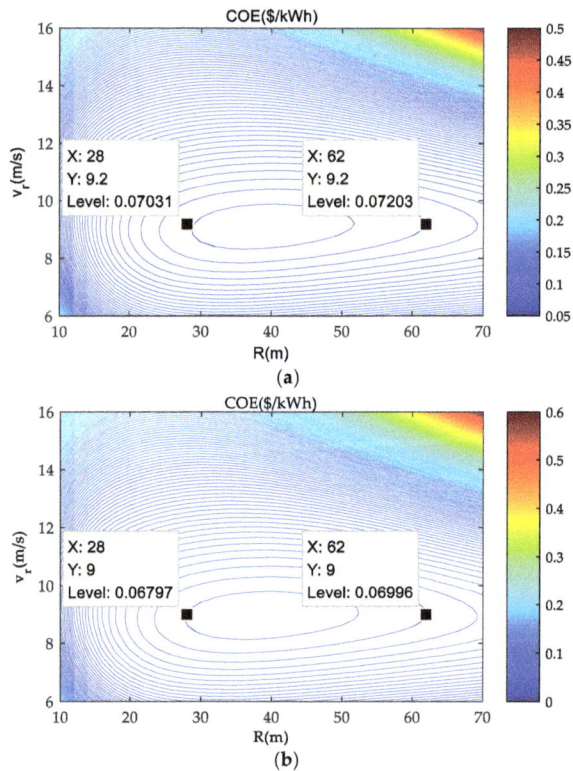

Figure 9. The offshore turbines COE versus the rotor radius and rated wind speed at a specific wind climate with $v_m = 7\,\text{m/s}$, $k_0 = 3.6$ under different power coefficient: (**a**) $C_{p,eq} = 0.48$; (**b**) $C_{p,eq} = 0.54$.

5. Conclusions

This paper has proposed a method to minimize the COE of offshore wind turbines, through optimally designing the rated wind speed and the rotor radius. To do this, both of the annual energy production (AEP) and the offshore turbine cost have been expressed as functions of the wind statistics factors and the two design parameters of offshore turbines, the rated wind speed and the rotor radius. Then, the COE minimization problem has been formulated, and an iterative method has been proposed. After that, the optimal results have been obtained by applying the proposed method to the wind fields of USA, from which the useful findings are concluded as follows:

- Considering 0.1 \$/kWh is an acceptable COE value to exploit the wind energy, the offshore wind turbines should be installed at the sea with the wind class III or above.
- The offshore wind turbines with a large capacity and long blades may not absolutely bring the COE down, as their optimal capacities depend on the offshore wind conditions.
- The COE is more sensitive to the rated wind speed than the rotor size, so it is critical to bring the COE down by selecting the suitable rated wind speed.
- The COE of offshore turbines is less sensitive to the power coefficient than to the rated wind speed and the rotor size, which justify the significance of the conception design optimization.
- The optimal rated wind speed of the offshore wind turbines should be designed around 1.5 times more than the annual mean wind speed, while the optimal rotor can be designed towards big-size, as the offshore wind turbines having a low COE appear with long blades and a large capacity.

Author Contributions: All the authors contributed to this work. X.Z. designed the methodology and wrote the original draft of the paper under the supervision of L.L. (Longfu Luo), D.S. and J.Y. provided the resources and revised the draft. W.T. helped with the writing and reviewing the draft. X.T., W.W. and L.L. (Li Li) contributed to algorithm design, data collection, and visualization.

Funding: This research and The APC were funded by the National Natural Science Foundation of China under Grant 61803393 and 51777217, and by China Postdoctoral Science Foundation under Grant 2017M622605.

Conflicts of Interest: The authors declare no conflict of interest.

References

1. Global Wind Report. 2017. Available online: http://www.gwec.net/global-figures/graphs/ (accessed on 3 September 2018).
2. Dai, J.C.; Yang, X.; Wen, L. Development of wind power industry in China: A comprehensive assessment. *Renew. Sustain. Energy Rev.* **2018**, *97*, 156–164. [CrossRef]
3. Willis, D.J.; Niezrecki, C.; Kuchma, D. Wind energy research: State-of-the-art and future research directions. *Renew. Energy* **2018**, *125*, 133–154. [CrossRef]
4. Burton, T.; Jenkins, N.; Sharpe, D.; Bossanyi, E. *Wind Energy Handbook*; John Wiley & Sons: West Sussex, UK, 2011.
5. Zhu, Y.; Cheng, M.; Hua, W.; Wang, W. A novel maximum power point tracking control for permanent magnet direct drive wind energy conversion systems. *Energies* **2012**, *5*, 1398–1412. [CrossRef]
6. Song, D.; Yang, J.; Cai, Z.L.; Dong, M.; Su, M.; Wang, Y.H. Wind estimation with a non-standard extended Kalman filter and its application on maximum power extraction for variable speed wind turbines. *Appl. Energy* **2017**, *190*, 670–685. [CrossRef]
7. Song, D.; Yang, J.; Su, M.; Liu, A.; Liu, Y.; Joo, Y. A comparison study between two MPPT control methods for a large variable-speed wind turbine under different wind speed characteristics. *Energies* **2017**, *10*, 613. [CrossRef]
8. Song, D.; Yang, J.; Cai, Z.; Dong, M.; Joo, Y.H. Model predictive control with finite control set for variable-speed wind turbines. *Energy* **2017**, *126*, 564–572. [CrossRef]
9. Scholbrock, A.; Fleming, P.; Schlipf, D. Lidar-enhanced wind turbine control: Past, present, and future. In Proceedings of the 2016 American Control Conference (ACC), Boston, MA, USA, 6–8 July 2016; pp. 1399–1406.

10. Zhang, X.L.; Huang, C.; Hao, S.P.; Chen, F.; Zhai, J.J. An Improved Adaptive-Torque-Gain MPPT Control for Direct-Driven PMSG Wind Turbines Considering Wind Farm Turbulences. *Energies* **2016**, *9*, 977. [CrossRef]

11. Huang, C.; Li, F.; Jin, Z. Maximum power point tracking strategy for large-scale wind generation systems considering wind turbine dynamics. *IEEE Trans. Ind. Electron.* **2015**, *62*, 2530–2539. [CrossRef]

12. Yin, M.H.; Li, W.J.; Chung, C.Y.; Chen, Z.Y.; Z, Y. Inertia Compensation Scheme of WTS Considering Time Delay for Emulating Large-Inertia Turbines. *IET Renew. Power Gener.* **2017**, *11*, 529–538. [CrossRef]

13. Martínez-Lucas, G.; Sarasúa, J.I.; Sánchez-Fernández, J.Á. Frequency Regulation of a Hybrid Wind–Hydro Power Plant in an Isolated Power System. *Energies* **2018**, *11*, 239. [CrossRef]

14. Martínez-Lucas, G.; Sarasúa, J.I.; Sánchez-Fernández, J.Á. Eigen analysis of wind–hydro joint frequency regulation in an isolated power system. *Int. J. Electr. Power Energy Syst.* **2018**, *103*, 511–524. [CrossRef]

15. Song, D.; Yang, J.; Fan, X.; Liu, Y.; Liu, A.; Chen, G.; Joo, Y. Maximum power extraction for wind turbines through a novel yaw control solution using predicted wind directions. *Energy Convers. Manag.* **2018**, *157*, 589–599. [CrossRef]

16. Song, D.; Fan, X.; Yang, J.; Liu, A.; Chen, S.; Joo, Y. Power extraction efficiency optimization of horizontal-axis wind turbines through optimizing control parameters of yaw control systems using an intelligent method. *Appl. Energy* **2018**, *224*, 267–279. [CrossRef]

17. Chehouri, A.; Younes, R.; Ilinca, A.; Jean, P. Review of performance optimization techniques applied to wind turbines. *Appl. Energy* **2015**, *142*, 361–388. [CrossRef]

18. Fuglsang, P.; Madsen, H.A. Optimization method for wind turbine rotors. *J. Wind Eng. Ind. Aerodyn.* **1999**, *80*, 191–206. [CrossRef]

19. Shen, X.; Chen, J.G.; Zhu, X.C. Multi-objective optimization of wind turbine blades using lifting surface method. *Energy* **2015**, *90*, 1111–1121. [CrossRef]

20. Ferdoues, M.S.; Ebrahimi, S.; Vijayaraghavan, K. Multi-objective optimization of the design and operating point of a new external axis wind turbine. *Energy* **2017**, *125*, 643–653. [CrossRef]

21. Tang, C.; Soong, W.L.; Freere, P.; Pathmanathan, M.; Ertugrul, N. Dynamic wind turbine output power reduction under varying wind speed conditions due to inertia. *Wind Energy* **2013**, *16*, 561–573. [CrossRef]

22. Yin, M.H.; Yang, Z.Q.; Xu, Y.; Liu, J.K.; Zhou, L.J.; Zou, Y. Aerodynamic optimization for variable-speed wind turbines based on wind energy capture efficiency. *Appl. Energy* **2018**, *221*, 508–521. [CrossRef]

23. Bortolotti, P.; Bottasso, C.L.; Croce, A. Combined preliminary–detailed design of wind turbines. *Wind Energy Sci.* **2016**, *1*, 71–88. [CrossRef]

24. Arias-Rosales, A.; Osorio-Gómez, G. Wind turbine selection method based on the statistical analysis of nominal specifications for estimating the cost of energy. *Appl. Energy* **2018**, *228*, 980–998. [CrossRef]

25. Villena-Ruiz, R.; Ramirez, F.J.; Honrubia-Escribano, A.; Gomez-Lazaro, E. A techno-economic analysis of a real wind farm repowering experience: The Malpica case. *Energy Conver. Manag.* **2018**, *172*, 182–199. [CrossRef]

26. Gualtieri, G. Improving investigation of wind turbine optimal site matching through the self-organizing maps. *Energy Conver. Manag.* **2017**, *143*, 295–311. [CrossRef]

27. Yang, H.; Chen, J.; Pang, X. Wind Turbine Optimization for Minimum Cost of Energy in Low Wind Speed Areas Considering Blade Length and Hub Height. *Appl. Sci.* **2018**, *8*, 1202. [CrossRef]

28. Diveux, T.; Sebastian, P.; Bernard, D. Horizontal axis wind turbine systems: Optimization using genetic algorithms. *Wind Energy Int. J. Prog. Appl. Wind Power Conver. Technol.* **2001**, *4*, 151–171. [CrossRef]

29. Mirghaed, M.R.; Roshandel, R. Site specific optimization of wind turbines energy cost: Iterative approach. *Energy Conver. Manag.* **2013**, *73*, 167–175. [CrossRef]

30. Eminoglu, U.; Ayasun, S. Modeling and design optimization of variable-speed wind turbine systems. *Energies* **2014**, *7*, 402–419. [CrossRef]

31. Chen, J.; Wang, F.; Stelson, K.A. A mathematical approach to minimizing the cost of energy for large utility wind turbines. *Appl. Energy* **2018**, *228*, 1413–1422. [CrossRef]

32. Fingersh, L.J.; Hand, M.M.; Laxson, A. *Wind Turbine Design Cost and Scaling Model*; Technical Report NREL/TP-500-40566; National Renewable Energy Laboratory (NREL): Gold, CO, USA, 2006.

33. Zierath, J.; Jassmann, U.; Abel, D.; Weber, F. Introduction of model predictive control for load reduction on a 3,0 mw wind turbine. In Proceedings of the Brazil Windpower 2015, Rio de Janeiro, Brazil, 1–3 September 2015.

34. Moné, C.; Hand, M.; Bolinger, M. *2015 Cost of Wind Energy Review*; Technical Report NREL/TP-500-40566; National Renewable Energy Laboratory (NREL): Gold, CO, USA, 2017.

35. Teyabeen, A.A.; Akkari, F.R.; Jwaid, A.E. Power Curve Modelling for Wind Turbines. In Proceedings of the 2017 UKSim-AMSS 19th International Conference on Computer Modelling & Simulation (UKSim), Cambridge, UK, 5–7 April 2017.

36. Goudarzi, A.; Davidson, I.E.; Ahmadi, A. Intelligent analysis of wind turbine power curve models. In Proceedings of the 2014 IEEE Symposium on Computational Intelligence Applications in Smart Grid (CIASG), Orlando, FL, USA, 9–12 December 2014; pp. 1–7.

37. Justus, C.G.; Hargraves, W.R.; Mikhail, A. Methods for estimating wind speed frequency distributions. *J. Appl. Meteorol.* **1978**, *17*, 350–353. [CrossRef]

38. Tar, K. Some statistical characteristics of monthly average wind speed at various heights. *Renew. Sustain. Energy Rev.* **2008**, *12*, 1712–1724. [CrossRef]

39. Liu, Y.; Chen, D.; Yi, Q. Wind Profiles and Wave Spectra for Potential Wind Farms in South China Sea. Part I: Wind Speed Profile Model. *Energies* **2017**, *10*, 125. [CrossRef]

40. Hsu, S.A.; Meindl, E.A.; Gilhousen, D.B. Determining the power-law wind-profile exponent under near-neutral stability conditions at sea. *J. Appl. Meteorol.* **1994**, *33*, 757–765. [CrossRef]

41. Seguro, J.V.; Lambert, T.W. Modern estimation of the parameters of the Weibull wind speed distribution for wind energy analysis. *J. Wind Eng. Ind. Aerodyn.* **2000**, *85*, 75–84. [CrossRef]

42. European Wind Energy Association. *Wind Energy-The Facts: A Guide to the Technology, Economics and Future of Wind Power*; Routledge: Landon, UK, 2012.

43. Sedaghat, A.; Hassanzadeh, A.; Jamali, J.; Mostafaeipour, A. Determination of rated wind speed for maximum annual energy production of variable speed wind turbines. *Appl. Energy* **2017**, *205*, 781–789. [CrossRef]

Article

Energy Production Benefits by Wind and Wave Energies for the Autonomous System of Crete

George Lavidas [1,*,†] and **Vengatesan Venugopal** [2,†]

1   Faculty of Maritime, Mechanical & Materials Engineering, Delft University of Technology,
    2628 CD Delft, The Netherlands
2   Institute for Energy Systems, The University of Edinburgh, Edinburgh EH8 9YL, UK; V.Venugopal@ed.ac.uk
*   Correspondence: g.lavidas@tudelft.nl; Tel.: +31-(0)-15-278-3864
†   These authors contributed equally to this work.

Received: 14 September 2018; Accepted: 9 October 2018; Published: 12 October 2018

Abstract: At autonomous electricity grids Renewable Energy (RE) contributes significantly to energy production. Offshore resources benefit from higher energy density, smaller visual impacts, and higher availability levels. Offshore locations at the West of Crete obtain wind availability ≈80%, combining this with the installation potential for large scale modern wind turbines (rated power) then expected annual benefits are immense. Temporal variability of production is a limiting factor for wider adaptation of large offshore farms. To this end multi-generation with wave energy can alleviate issues of non-generation for wind. Spatio-temporal correlation of wind and wave energy production exhibit that wind and wave hybrid stations can contribute significant amounts of clean energy, while at the same time reducing spatial constrains and public acceptance issues. Offshore technologies can be combined as co-located or not, altering contribution profiles of wave energy to non-operating wind turbine production. In this study a co-located option contributes up to 626 h per annum, while a non co-located solution is found to complement over 4000 h of a non-operative wind turbine. Findings indicate the opportunities associated not only in terms of capital expenditure reduction, but also in the ever important issue of renewable variability and grid stability.

Keywords: wave energy; wind energy; renewable energy; co-generation; offshore energy

1. Introduction

Greece is located at the East Mediterranean Sea and among its unique characteristics is the high number of islands that rely on fossil fuels, constituting a wide number of small decentralized energy systems. Crete is the biggest island of Greece, its electricity system can be characterized as a decentralized (autonomous) network which relies for its energy production on a mixture of predominately wind and solar. Renewable energy (RE) such as wind (≈187.6 MW) and solar (≈95.5 MW) have experienced a growth in their installed capacities, although their base load and peak demand is still heavily dependent on conventional fuels [1,2]. The Cretan electrical system is heavily dependent on three oil-fired thermal power plants with ≈700 MW rated capacity.

The stochastic nature of RE often is not able to provide energy and cover demand when needed [3,4]. Higher scenarios of renewable integration have been proposed, with main limitation for higher penetration their perceived effects on electrical grid stability [5]. To assist in the adaptation of renewables alternatives which reduce conventional fuels and increase energy independence such as storage and/or mainland interconnection have been suggested [6–8].

There are significant financial considerations to be taken into account with the future increase of RE expected. They are associated with the infrastructure needed to maintain uninterrupted power supply and the cost of energy [9–12]. Several solutions have been proposed to assist with the energy transition, such as energy storage alternatives [13–16], and development of small autonomous grids, to reduce the curtailments of RE when peak demand is exceeded [17–19]. More specifically [3] explored seasonal variations of wind energy in the Danish system, and assessed the system's response flexibility due to high energy curtailments. An alternative was to considered electricity storage technologies for power system balancing with potential technologies including batteries, flow batteries, electric vehicles, Compressed Air Energy Systems (CAES), and Pumped Hydro Systems (PHS).

For the Greek region Kaldellis et al. [20] explored energy system losses, and the implications of income losses due to restricted access of renewable energy to the local grid. This has prompted numerous studies to explore methods to complement of renewable curtailments and income losses due to system restriction. Alternatives explored have been batteries, Pumped Hydro Storage (PHS), desalination and CAES among other proposed solutions [16,21–24].

This study addresses the temporal combination of two overlapping resources, that can assist in addressing the intermittent production and accelerate the level of RE acceptance in the local grid. Greece has relied heavily on limited number of renewable resource, with others under-investigated [25–27].

The results enhance knowledge in the opportunities of wind and wave multi-generation, and provide tangible evidence for the inclusion of the other untapped resources that exist in the Aegean Sea. Temporal interconnection of RE can reduce the variability issues and adjust generation into a more continuous profile. To ensure such a coupling the power production trends of major renewable components in a system have to be evaluated temporally and coupling alternatives must provide some level of temporal satisfaction.

While a deterministic solution is not feasible at this stage, the methodology, data sourcing/manipulation and analysis for energy production is applicable globally. Resources are firstly quantified independently, and their power performance is analyzed. Subsequently, co-located and non co-located scenarios are assessed in terms of coverage for missing production intervals. Since wind power is more established, proven, and offers higher rated capacity devices it is consider as the "base" energy source. Wave converters act as a substitution mechanism. Temporal overlaps in energy production are assessed and determine the coverage that can be offered based on two installation options, co-located and non co-located.

2. Materials and Methods

2.1. Benefits of Multi-generation

Use of multiple renewable technologies can provide some levels of "storage" through resource dependence, due to resource dependence. Currently, in Crete, the most dominant RE technology is wind followed by photovoltaic (PV) and solar. In terms of operating profiles PV/solar have a specific range of temporal operation. This is associated with hours of sunshine, hence predominately over a period of 8–9 a.m. to 18–19 p.m., on the other hand wind generation is temporally more distributed. While, this is a major benefit in terms of total production hours, the disadvantage of wind is its associated variability i.e. wind tend to change at rapid rates. Bai et al. [28] discussed the necessity for realistic wind forecasting in order to minimize losses due to the variable nature of wind. The study focused on forecasting and presented the complexities that exist in uncertainty reduction.

Another resource available to the region, though under-investigated is the wave energy resource [25,29]. Waves predominately are generated by wind interactions with the upper layer of the sea. This in turn generates and propagates waves, for this reason wave resources are classified as wind-waves (locally generated or enhanced) and swells. Swells represent wave components generated and propagated in a far distance from our interest locations. These components when propagated over large depths are able to amplify their energy content and encompass large fluxes of energy. Indirectly waves, due to their nature and properties act as a "storage" medium for wind energy. Wave velocities can exceed the wind speed, though the majority of times the wave resource is propagated with a time-lag from its originating wind [15,30,31].

2.2. Methodology

Offshore wind data are obtained from the Climate Forecast System Reanalysis (CFSR) wind global datasets [32]. Wave data are extracted by a wave hindcast database produced by the authors with a high–resolution nearshore wave numerical model. Calibration, validation and a resource analysis can be found in Lavidas et al. [33] and Lavidas and Venugopal [25]. Both datasets have a 1-h timestep interval and correspond to the same offshore coordinates (see Figure 1). Distance of locations from nearest coast is also accounted for, to provide realistic and feasible estimate according to international practises, for Point 1 is ≈14 km and for Point 2 ≈12.5 km.

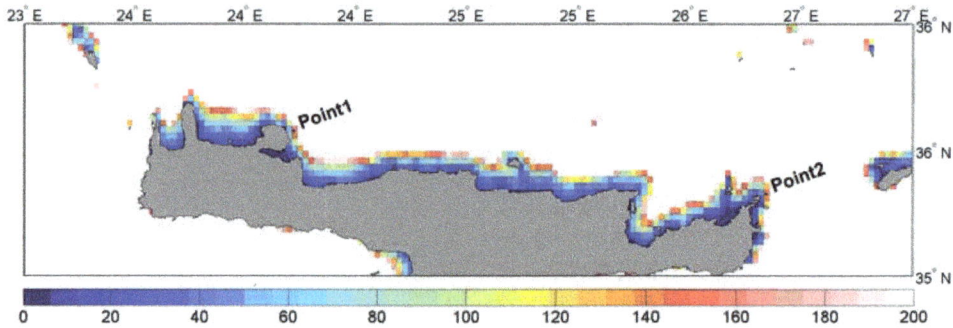

Figure 1. Location and bathymetry in meters.

To estimate energy production capabilities both resources are evaluated against potential devices. The offshore wind turbine is a Vestas V112-3.3 MW [34]. There are numerous wave energy converters (WECs) [35], although their level of maturity varies. In contrast to solar and wind power production, wave energy poses a more complex problem. To estimate power production both wave height (H_{sig}) and appropriate wave period have be combined. Selecting a suitable WEC depends highly on the local environment, and critical decision making. Performance of seven different WEC has been automatically estimated for the location, with the capacity factor indicating the optimal selection. This allowed us to select the optimum operating device based on the characteristics of local resource and corresponding power matrix. A detailed discussion, on how to estimate and perform a coupling for wave energy and resource assessment has been presented in previous studies [25,36,37].

The overlapping production is achieved as there is a dependence of waves by wind resources. The analysis does not considered separation of swell waves, as focus is given on the energy produced by the WEC based on complex mixed sea states. To determine the cross-correlation (*cR*) of wind and wave energy, the following Equation (1) is used:

$$cR(\tau) = \frac{1}{N} \cdot \sum_{t=1}^{N-\tau} \frac{\left(Wind_{el_t} - \mu_{Wind_{el}}\right) \cdot \left(Wave_{el_{t+\tau}} - \mu_{Wave_{el}}\right)}{std_{Wind_{el}} \cdot std_{Wave_{el}}} \tag{1}$$

where τ is the time lag, set at 1 h, *std* the standard deviation, μ the mean electrical power (of wind and/or wave), and N the sample size over time (t). The *cR* provides the correspondence of variables, classifying the instantaneous production. One hour step interval provides important information about the potential use of co-located wind-wave farms, that are expected to reduce variability of production. Fusco et al. [30] suggested that although the time-lag can be subjected to any range, a higher range will deteriorate the correlation for system balancing, and reduces *cR*. Astariz and Iglesias [38] gave similar insights on cross-correlation of resources, both authors suggested a time lag of 1 h. In Table 1 we have classified the levels of cross correlation according to *cR* values.

Table 1. Cross Correlation Classification.

Cross-Correlation Ranges	Value
$cR = 1$	High positive
$cR > 0$ & $cR \leq 0.5$	Moderate Positive
$cR = 0$	No cross-correlation
$cR > 0$ & $cR \leq -0.5$	Moderate Inverse
$cR = -1$	High Inverse

Another metric also considered is the standard deviation of produced electrical power. This examines deviation of energy production, and assesses its "distance" from the nominal installed capacity. Representing a percentage of variability within the sample.

3. Results and Analysis

3.1. Wind Resource and Power Extracted

The wind resource is taken at height of 10 m (h_{10}m), however in order to realistically represent wind turbine production, adjustment of the resource is necessary. The structural characteristics of the turbine provide rotor diameter of 112 m^2 and blade length of 54.65 m, hence hub height considerations require to scale up the resource to 100 m (h_{100}). The energy analysis of wind speeds are subjected to height modification under the power law (see Equation (2)), and all subsequent data correspond to 100 m.

$$\frac{U_{100}}{U_{10}} = \left[\frac{h_{100}}{h_{10}}\right]^a \tag{2}$$

where α representing the power law exponent, that can be considered as quite volatile. However, experimental results and literature review suggest a value of $\alpha = 1/7$ [39]. Figure 2 shows the power curve and characteristics of the wind turbine. Operation starts at U_{wCI} = 3 m/s and stops at U_{wCO} = 25 m/s, nominal power is given at U_{wNO} = 13 m/s. The U_{wCI}, U_{wCO} and U_{wNO} are used to assess availability of production.

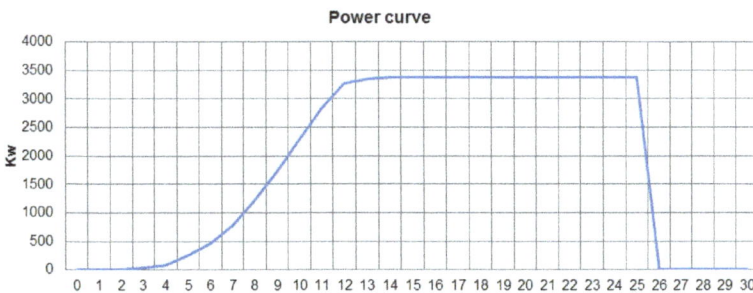

Figure 2. Wind turbine Power curve as adapted from [34].

The power curve has been applied to the scaled wind speeds and the Weibull distribution is fitted to the data (see Figure 3 top right). On the bottom left of Figure 3 the simulated energy production is given. The estimated availability took into account the U_{wCO}. Western location (Point 1) yield availability $\approx$80% and a capacity factor 39.66%. Eastern location (Point 2) has lower availability at $\approx$70% and a reduced capacity factor of 15.86%. In both locations the availability i.e. potential percentage of time for favourable operation is very high ($\geq$70%).

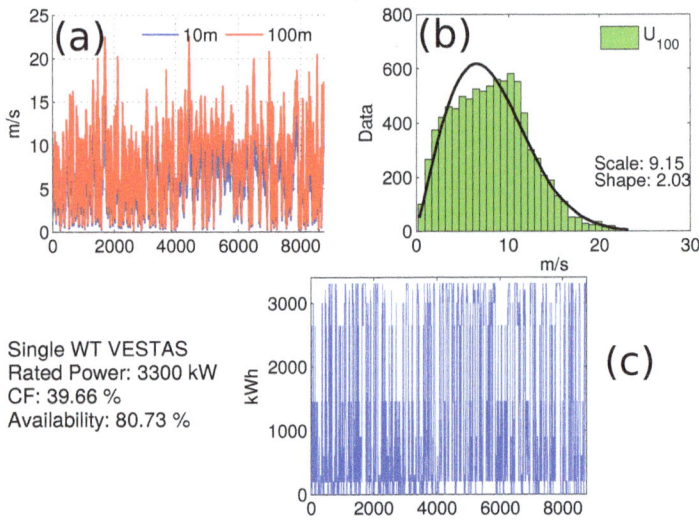

Figure 3. Wind characteristics of Point 1. (**a**) Wind resource at 10m & adjusted at 100m; (**b**) Wind distribution; (**c**) Hourly production, CF: 39.66%.

3.2. Wave Resource and Extracted Power

Selection of a wave energy converter (WEC) depends highly on the location's metocean conditions, depth, WEC characteristics (type of operation, power-take-off (PTO) etc.). Operating principles are vital to the proper selection of WEC, a apply a wide array of WECs that represent different technologies were used to select the device with most suitable characteristics. There are several different technologies based on different principles of operation such as an oscillating water column [40], an over-topping converter [41] and many more. It has to be noted that while all WECs share one parameter H_{sig}, the wave period associated with their operation changes per device. Some devices use peak period T_{peak}, mean zero crossing (T_{m02}), and other the mean absolute wave period (T_{m01}). The wave database includes all periods necessary for numerous converters and applications. Energy quantification for WEC, is done on basis of investigating the joint distribution of wave height-wave period. From there we can estimate the probabilities of occurrence, and apply the power matrix. Each estimated annual-based production, uses the proper wave period, more information on the power matrices used can be found in [35,37].

All power matrices are coupled with bivariate metocean distributions with available WECs as in the process presented also in [25]. From the comparison and based on the location's characteristics, best annual performance was achieved by the WaveStar converter 600 kW [42]. Availability of location and device are assessed according to the power matrix of this converter and H_{sig}. The H_{cut-in} is at 1 m and wave period of 4 s, nominal power is achieved at 3 m and periods from 4–8 s, with stop of operation at wave heights over $H_{cut-off}$ 3 m. Annual analysis indicates that Point 1 has higher availability and capacity factor than Point 2. Specifically, Point 1 availability is ≈31%, and its capacity factor 19.9%. Availability is almost twofold than Point 2 (≈17%) with a lower capacity factor of 11.5%. Instantaneous electrical production (E_{el}) was estimated using a linear interpolation of the power matrix ($P_{PM_{i,j}}$) to provide with specific production corresponding to the hourly (t) components [15] (see Equation (3)).

$$E_{el}(t) = \int_{t=1}^{t=8761} P_{PM_{i,j}}(H_{sig_{i,j}}; T_{m02_{i,j}}) \tag{3}$$

Wave resource and device characteristics are given at Figure 4, in each sub-figure the top right panel shows the interpolated instantaneous power production at hourly timesteps, top left panel displays the bivariate distribution and number of combined occurrences. Lastly the bottom panel shows the expected cumulative production achieved by the device at specified intervals. For Point 1 dominant conditions describing the location are wave heights 1.5–3.5 m and wave periods from 4–9 s. Point 2 has dominant conditions at much lower magnitudes of H_{sig} from 0.5–1.5 m, and higher frequencies 3–7 s.

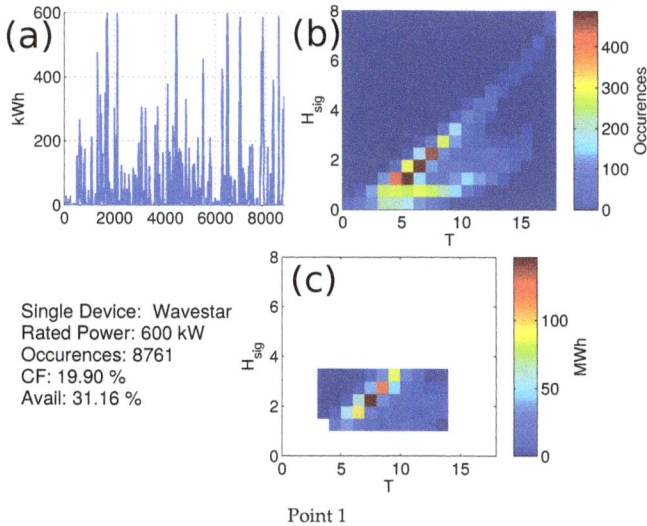

Figure 4. Point 1 wave locations characteristics (**a**) Hourly production (**b**) Joint distribution (**c**) Cumulative production (binned).

3.3. Co-Located Combined Production

As any RE production depends on resource availability there are possibilities for potential overlaps and supplementary production by different technologies. As a "primary" source of contribution wind energy is considered, while wave energy contributions on non-operative hours are assessed in a complementary way. The levels of contribution and cross-correlation are based on the WEC providing additional energy production when the wind turbine is not operating.

At Point 1 the wind-wave co-located device can complement 26.08 days or 626 h of wind non-operation. The average cR between the production is 0.49 moderately positive. The mean annual std_{Wind} = 1239.88 kWh and std_{Wave} = 52.31 kWh. Point 2 wind has a slightly lower operational availability, complemented amounted hours from the wave device are 702 h or ≈29.25% days. The cR between resource production is highly positive at ≈0.6%. Mean annual std_{Wind} = 699.52 kWh and std_{Wave} = 74.63 kWh.

Supplementing many hours is beneficial for grid operators and can improve power quality consideration. Over-lapping production can provide added security by RE generation and assist in the alleviation of non-predictability. While, this is based on an hourly annual timeseries approach it is more interesting to evaluate the short-term benefit, i.e., monthly. Wind resource has a high level of uncertainty, with fast changing wind speeds and directions. On the other hand, the wave resource is less volatile though the energy production is highly dependent on a much more parameters. WEC performance depends on three variable, wave height, period, direction, however the directional matrix is not provided in international literature for the majority of WECs. Therefore, the amount of energy produced is highly dependent on wave height and period joint occurrences (see Figure 4).

The analysis also considers the monthly wave production and associated values of *cR* and *std*. The wave resource is subjected to lower levels during the late spring and summer months, expressed as power per meter unit crest (kW/m). This dictates that during lower energetic months, the magnitude of wave heights is smaller, while the frequency of wave is higher (smaller periods), indicating the selection of the device.

The wind power curve uses the available resource and produces almost at all months its nominal rated power, as supported by the higher levels of availability and capacity factor. On the other hand WEC is able to produce its rated power during some portion of months. For January-February the area is exposed to higher magnitude waves, which do not allow full use but instead push the WEC in cut-off mode. The situation changes during low energy months May-July, where wave heights are not as energetic allows for a much higher operation, from September to December metocean conditions allows the WEC to produce near its nominal values at higher rates.

The *std* of production is expressed as a percentage in regards to nominal rated power, expressing the variability of *std* in regards to potential highest energy production. Cross-correlation considers wind as "base" production, and wave electricity as supplementing energy when no production is given the wind turbine (see Figure 5). Lowest levels of wind standard deviation are achieved in April at 10%. This can be attributed to operational wind speeds that achieve lower levels of nominal production (see also Figure 3). On the other hand, WEC shows significantly lower levels of deviation 5% for April, highest *std* occurs in March, September and November. In this case, higher deviation are attributed to higher magnitude waves which are met during the winter and early spring periods (see also Figure 4).

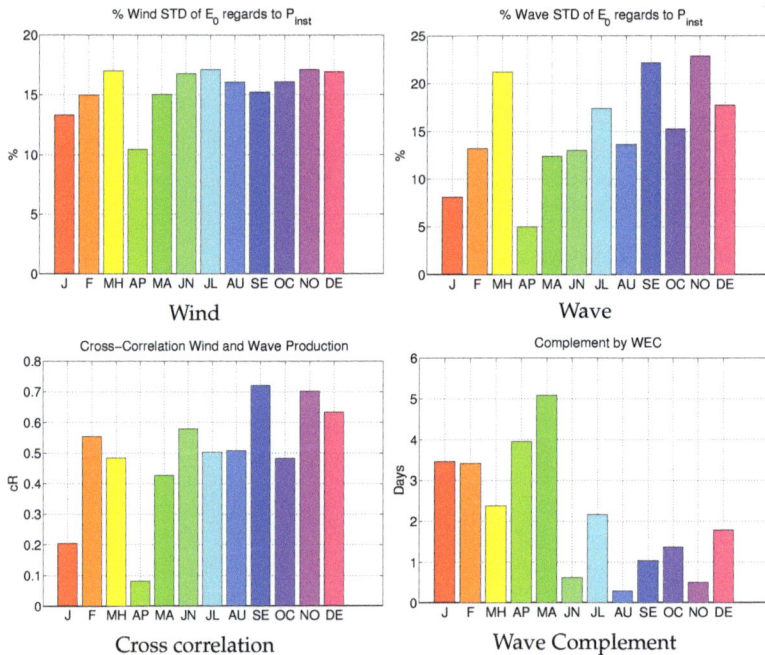

Figure 5. Monthly characteristics at Point 1.

In the case of production cross-correlation (see Figure 5) higher levels are encountered in February, September and November. The summation of hours per month allows to quantify energy complementing benefits from the co-located farm. Highest levels of cross-correlation are achieved during winter months, where wind speeds initiate cut-off states. Contribution of WEC device also varies per month, though significant levels of complemented energy can be achieved (see panel (d) Figure 5). The highest temporal contribution by the WEC is achieved from January to May. In summer months the wind turbine achieves higher operation with August having the highest at 96%. In August WEC availability is also at it highest at 47%. From September-December WEC availability decreases to ≈25–26%, while wind has higher level ranging from 78–85%. In terms of absolute energy contributions, the wind turbine has the highest levels of contribution. While its production fluctuates, its mean value is ≈1 GWh. The WEC production has greater fluctuations, with mean monthly ≈39 MWh.

In terms of wind *std* lower levels are for April and August, while waves have exhibit the lowest at April below 5% (see Figure 6). Cross-correlation has a more diverse profile than Point 1, highest *cR* is similarly achieved at September ≈0.8. Although in April where Point 1 has the lowest *cR*, Point 2 has a stronger correlation 0.47, and maximum complemented days for waves to wind production lower than Point 1. December and March have the highest contributions with ≈3.5 days. Lowest overlap contribution is seen in September with less than a day or 24 h. In terms of energy production wave and wind show the same temporal maximum in March, with 56 and 538 MWh respectively.

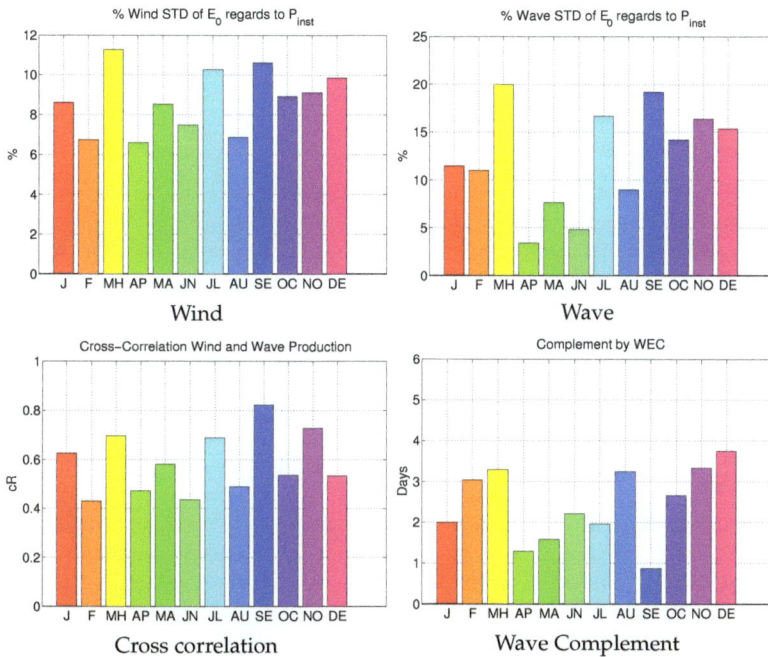

Figure 6. Monthly characteristics at Point 2.

In Table 2, the monthly information on cross correlation and energy production are displayed. In the table we also the availability of the resource for production at least 50% of the nominal power of each converter. For the wind turbine (W/T) operational time is high throughout the months for both locations. Consistently W/T is able to produce energy at above 70% of the time. On the other hand, the WEC has a lower Total Time (TT) production. Point 1 has significantly higher performance almost 20% increased than Point 2, over all months. *cR* is much higher at Point 2 for the majority of the months.

Table 2. Operational Information for Wave Energy Converter (WEC) and Wind turbine (W/T).

	Jan	Feb	Mar	Apr	May	Jun	Jul	Aug	Sep	Oct	Nov	Dec
					Point 1							
cR	0.20	0.55	0.48	0.08	0.43	0.58	0.50	0.51	0.72	0.48	0.70	0.63
Hours Comp	83.00	82.00	57.00	95.00	122.00	15.00	52.00	7.00	25.00	33.00	12.00	43.00
WEC Avail at 50% Rated	0.00%	2.53%	8.20%	0.00%	0.00%	0.83%	4.03%	2.15%	10.83%	3.90%	8.19%	4.56%
W/T Avail at 50% Rated	31.59%	40.18%	48.39%	19.31%	33.60%	46.81%	63.17%	75.81%	45.83%	50.54%	50.97%	44.43%
					Point 2							
cR	0.63	0.43	0.70	0.47	0.58	0.43	0.69	0.49	0.82	0.54	0.73	0.53
Hours Comp	48.00	73.00	79.00	31.00	38.00	53.00	47.00	78.00	21.00	64.00	80.00	90.00
WEC Avail at 50% Rated	1.21%	0.00%	6.85%	0.00%	0.00%	0.00%	4.03%	0.00%	6.94%	2.15%	3.19%	3.36%
W/T Avail at 50% Rated	13.44%	8.18%	24.06%	12.92%	13.98%	8.33%	16.53%	9.68%	15.56%	17.88%	13.61%	11.01%

Additionally, to operational information the table also provides the resource availability that corresponds to at least 50% of nominal rated power production. As expected the wind turbine has higher levels of percentages. Most energetic location in regards to wind is Point 1, which consistently has over 30% opportunities for nominal rated production. In contrast the WEC shows poor results in terms of potential nominal production at the location, with results slightly favoring Point 1. This suggests that the WEC should be adjusted according to local environment and its peak rated power should be re-adjusted to facilitate lower H_{sig}. Thus, while wave % TT is higher, the availability based on nominal suggests that the majority of operational hours the WEC produces less than nominal and seldom achieves rated production.

3.4. Non Co-Located Combined Production

So far co-located temporal configuration showed that production overlap can provide some level of stabilization. In an energy system though, autonomous or not, a consistent flow of energy and reduction of variability is maybe of greater importance than just power contribution, which can be scaled up by increasing installed capacities. For this reason, the study also considers a dispersed spatial option. We consider the installation of wind and wave converters between the two locations, as seen in Figure 1, Points 1 and 2 are positioned in completely different regions. In Sections 3.1 and 3.2, the local characteristics also denoted their differences. For this reason, two different scenarios are taken into account:

- Scenario 1: Point 1 (WEC) with Point 2 (W/T)
- Scenario 2: Point 1 (W/T) with Point 2 (WEC)

The results are assessed in terms of potential overlap in days. The contribution of wave production in regards to non wind operation is potential highly beneficial in reducing variability effects.

In terms of cR the co-located option acquires higher annual mean values. Although, in the non co-located option, cR of production is higher for some months. For example Scenario 1 for January has higher correlation than Point 1, Point 2 has strong positive correlation cR for all months (see Figure 7).

Scenario 2 also shows similar results with specific months of higher cR when compared with Point 1 cR. Interestingly, during the month of April it is the first time that a inverse correlation exist between production of converters. Indicating that during these months WEC benefit from swells.

The scenarios are also assessed for potential contribution by spatially dispersed WEC and W/T, see Figure 8. As presented in Section 3.3, co-located W/T-WEC can provide complemented production which amounts up to ≈5 days (see Figures 5 and 6 and Table 2). The large distances between the scenarios and different hourly resource characteristics, make the contribution of non-operative hours greater. Both scenarios outperform the co-located options examined in the previous section.

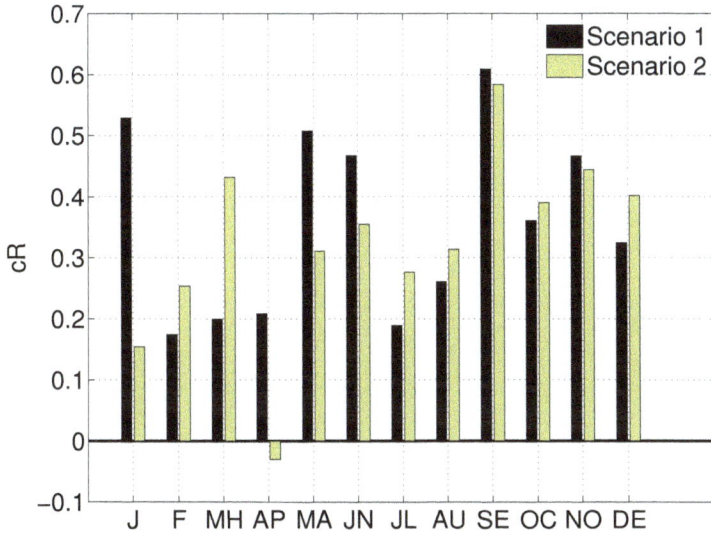

Figure 7. *cR* scenarios for non co-located.

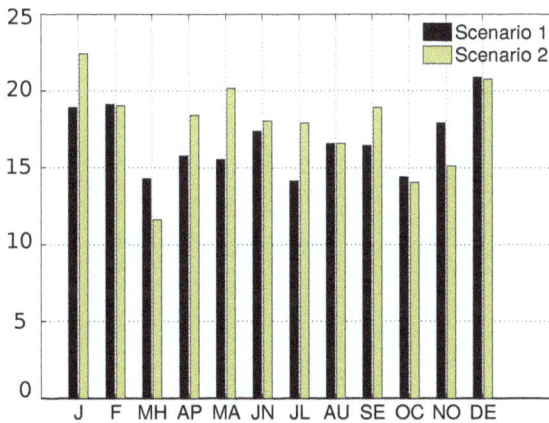

Figure 8. Days complemented by WEC for non co-located.

Interestingly, both scenarios contribute multiple hours of reduced variability. In the co-located examination, maximum hours of WEC complementing are achieved in April for Point 1 (95 h), and December at Point 2 (90 h). In the same months both Scenarios 1–2 contribute ≈>370 h in April and ≈500 h December. Thus, annual potential contribution in terms of days are 201 and 213 for Scenarios 1 and 2, respectively.

4. Discussion

In terms of energy production levels and availability use by the wind turbine all monthly indices are high. Since maturity of wind converters has evolved through offshore installation and have contributed significant energy benefits to the system. On the other hand WECs are an emerging technology with a variety of devices which can be suited to extract maximum benefits.

In regards to resource connectivity, wind and waves can complement each other in temporal terms. Wave energy can provide supplements in production for hours of which a wind turbine (farm) is not producing. That said, one major drawback is that most WECs have been constructed with preferred operational ranges suitable for higher latitudes and more energetic (open ocean) conditions. This proves a significant disadvantage for their applicability in lower latitudes.

The selected device had the best performance from the WECs investigated, this is supported by its operational characteristics. For the dataset of our analysis, the co-located option contributed significant amount of hours in a years, which correspond to ≈26 and ≈29 days for Point 1 and 2 respectively. Greater benefits, in terms of WEC complemented hours and potential reduction of variability are found when there is an non co-located configuration. In these instances the two scenarios tested, provided larger temporal production coverage. Both solution were able to cover non-wind operative hours ≥55% or ≥200 days.

However, some concerns have to be discussed concerning the availability at 50% of rated capacity. The majority of WECs are designed for higher energetic region and oceanic waters. To fully maximize the potential of wave energy extraction rated capacity of a WEC must be adapted for smaller energetic region. Based on its performance characteristics, a scale down approach may be taken to ensure that maximum wave power can be achieved at lower resource wave heights, such as the one found in the Aegean. Such a downscale must be driven through by a hydrodynamic model, which allows the incorporation of spectra by a wave numerical model. A down-scaled converter idea was presented in [43,44], and proved that the use and capacity factor almost doubled in a variety of devices examined. A similar approach can be taken to enhance the production at availability 50% of nominal rated WEC power. It is important though, that such an optimization is based on long-term metocean data, that include the intra-annual, seasonal, and decadal variations of the area.

Such an approach will have multiple benefits for a co-located farm. In terms of energy and variations, the levels of variability are expected to be reduced significantly. In terms of the WEC component, a down-scaled device will accelerate the proof-of-concept for dual platforms. Another benefit, is the added survivability of the WEC, due to the smaller wave heights, structural integrity of the device will not be compromised as much as in oceanic region. Although, to establish the probable extreme values a long-term metocean dataset is vital.

This study did not consider the effects of production by large offshore wind and wave farms, more specifically wind turbine wake, wave directions and wave to wave interactions were not fully accounted for. The reason for the "simplified" single converter were (1) that deployment regions have not been assigned for offshore wind and/or wave farm in Greece, hence deployment strategy needs (2) in regards to WECs directional information are often not published, and the hydrodynamic losses, optimal spacing of WECs is still very device dependent and require different analysis. Finally, cost estimates and amortization periods especially for the emerging wave energy field, are volatile and the pay-back periods will depend on the nature of selling price and/or support scheme. For wave energy development in Greece a detail techno-economic analysis can be found in [25].

5. Conclusions

In this study, the potential temporal benefits from co-located and non co-located wind and converters are examined. Offshore combined farms are expected to reduce capital expenditures for the devices, and allow better spatial planning. This study examined energy production overlap between wind and wave energy converters. The data are extracted by a wind re-analysis and wave database, which offer all major components for energy estimates in hourly intervals. With wind being a volatile resource, the necessity for a RE converter complementing its production attributes is vital, to reduce intermittent nature. With wave energy directly correlated with wind resource, it is evident that potential overlaps will benefit the end user and/or distributor.

For a low energetic wave and a moderate wind resource region, there are significant benefits, specifically through cross-correlation of two resources wave energy can provide production overlap to wind production. The wave device used was selected after comparison of the dataset with several devices, ensuring that the selection will maximize metocean conditions at the location.

Between the two technologies and available resource, wind obtains the highest availability. Similarly, the capacity factor is larger for the selected wind turbine, while the wave converter used follows with ≈50% reduction in regards to the wind capacity factor (≈20%) at Point 1 location. The correlation of energy production patterns, indicated that the WEC even at mild levels is able to contribute production, when the wind turbine was non-operating.

Depending on the selection, co or non co-located temporal effects change, the co-located examined WEC complemented W/T non-production ≈626 h (Point 1) and ≈720 h (Point 2). The highest number of complemented hours by the co-located configuration is predominately in the months of January till May for Point 1. However, Point 2 has also a "high" month contribution in August where the wind resource seems to be reduced and the low operational range of the WEC favors the complemented production. Throughout the winter months the wind resource forces the wind turbine to go into safety (cut-off) mode. In both cases there is a positive moderate to high cross-correlation which assists in the consideration for local co-generation. Highest temporal benefits are established in the non co-located options, where the different scenarios provide WEC overlap ≈≥4000 h within a year. Overlap coverage of production can decrease variability, and allow for smoother energy contribution to autonomous grids. Combination of multiple RE forms of energy, re-affirms the fact that electricity can be supplied in a more predictable manner that can reduce deferral capital expenditures to the grid and ensures sustainable development.

Author Contributions: Conceptualization was done through the collaboration of G.L. and V.V. Formal Analysis was performed by G.L. and reviewed V.V. The writing and review was performed by both authors.

Funding: This research received no external funding.

Acknowledgments: The author would like to thank Ioannis (Yannis) Katsigiannis for his comments on improving the manuscript. The author would like to thank the reviewers for their constructive comments, which helped at improving the manuscript.

Conflicts of Interest: The authors declare no conflict of interest.

Abbreviations

The following abbreviations are used in this manuscript:

RE	Renewable Energy
CAES	Compressed Air Energy Systems
PHS	Pumped Hydro Systems
PV	Photovoltaic
m	metres
Km	kilometres
CFSR	Climate Forecast System Reanalysis
H_{sig}	Significant Wave Height
cR	Cross-Correlation
τ	time lag
std	standard deviation
N	sample size
WEC	Wave Energy Converter
h_m	hub height at metres
PTO	Power-Take-Off
T_{peak}	Peak wave period
T_{m01}	mean absolute wave period
T_{m02}	mean zero crossing wave period
H_{cut-in}	Initiate operation of WEC based on wave height
$H_{cut-off}$	Stop operation of WEC based on wave height

References

1. Regional Energy Agency of Crete. Available online: https://ec.europa.eu/energy/intelligent/projects/en/partners/regional-energy-agency-crete (accessed on 6 July 2016).
2. Karapidakis, E.; Katsigiannis, Y.; Tsikalakis, A.; Maravelakis, E. Current Status and Future Prospects of Crete's Power System. In Proceedings of the International Symposium on Fundamental of Electrical Engineering, Bucharest, Romania, 28–29 November 2014.
3. Hedegaard, K.; Meibom, P. Wind power impacts and electricity storage—A time scale perspective. *Renew. Energy* **2012**, *37*, 318–324. [CrossRef]
4. Michalena, E.; Hills, J.M. Renewable energy issues and implementation of European energy policy: The missing generation? *Energy Policy* **2012**, *45*, 201–216. [CrossRef]
5. Karapidakis, E.S.; Katsigiannis, Y.A.; Georgilakis, P.S.; Thalassinakis, E. Generation expansion planning of Crete power system for high penetration of renewable energy sources. *Mater. Sci. Forum* **2011**, *670*, 407–414. [CrossRef]
6. Jacobson, M.Z.; Delucchi, M.A. Providing all global energy with wind, water, and solar power, Part I: Technologies, energy resources, quantities and areas of infrastructure, and materials. *Energy Policy* **2011**, *39*, 1154–1169. [CrossRef]
7. Zafirakis, D.; Chalvatzis, K.; Kaldellis, J. "Socially just" support mechanisms for the promotion of renewable energy sources in Greece. *Renew. Sustain. Energy Rev.* **2013**, *21*, 478–493. [CrossRef]
8. Vaona, A. The effect of renewable energy generation on import demand. *Renew. Energy* **2016**, *86*, 354–359. [CrossRef]
9. Schaber, K.; Steinke, F.; Hamacher, T. Transmission grid extensions for the integration of variable renewable energies in Europe: Who benefits where? *Energy Policy* **2013**, *43*, 123–135. [CrossRef]
10. Connolly, D.; Lund, H.; Mathiesen, B.; Pican, E.; Leahy, M. The technical and economic implications of integrating fluctuating renewable energy using energy storage. *Renew. Energy* **2012**, *43*, 47–60. [CrossRef]
11. Skea, J.; Anderson, D.; Green, T.; Gross, R.; Heptonstall, P.; Leach, M. Intermittent renewable generation and the cost of maintaining power system reliability. *System* **2008**, *2*, 82–89. [CrossRef]
12. Verbruggen, A.; Fischedick, M.; Moomaw, W.; Weir, T.; Nadaï, A.; Nilsson, L.J.; Nyboer, J.; Sathaye, J. Renewable energy costs, potentials, barriers: Conceptual issues. *Energy Policy* **2010**, *38*, 850–861. [CrossRef]
13. Madlener, R.; Latz, J. Economics of centralized and decentralized compressed air energy storage for enhanced grid integration of wind power. *Appl. Energy* **2013**, *101*, 299–309. [CrossRef]
14. Barbour, E. An Investigation into the Potential of Energy Storage to Tackle Intermittency in Renewable Energy Generation. Ph.D Thesis, University of Edinburgh, Edinburgh, UK, 2013.
15. Friedrich, D.; Lavidas, G. Combining offshore and onshore renewables with energy storage and diesel generators in a stand-alone Hybrid Energy System. In Proceedings of the OSES Offshore Energy Storage Symposium, Edinburgh, UK, 1–3 July 2015.
16. Kaldellis, J.; Zafirakis, D.; Kavadias, K. Techno-economic comparison of energy storage systems for island autonomous electrical networks. *Renew. Sustain. Energy Rev.* **2009**, *13*, 378–392. [CrossRef]
17. Battaglini, A.; Lilliestam, J.; Haas, A.; Patt, A. Development of SuperSmart Grids for a more efficient utilisation of electricity from renewable sources. *J. Clean. Prod.* **2009**, *17*, 911–918. [CrossRef]
18. Giannoulis, E.; Haralambopoulos, D. Distributed Generation in an isolated grid: Methodology of case study for Lesvos-Greece. *Appl. Energy* **2011**, *88*, 2530–2540. [CrossRef]
19. Hammons, T. Integrating renewable energy sources into European grids. *Int. J. Electr. Power Energy Syst.* **2008**, *30*, 462–475. [CrossRef]
20. Kaldellis, J.; Kavadias, K.; Filios, A.; Garofallakis, S. Income loss due to wind energy rejected by the Crete island electrical network—The present situation. *Appl. Energy* **2004**, *79*, 127–144. [CrossRef]
21. Kaldellis, J.; Kavadias, K. Optimal wind-hydro solution for Aegean Sea islands' electricity-demand fulfilment. *Appl. Energy* **2001**, *70*, 333–354. [CrossRef]
22. Kaldellis, J. *Stand-Alone and Hybrid Wind Energy Systems. Technology, Energy Storage and Applications*; Woodhead Publishing Limited: Cambridge, UK, 2011.
23. Zafirakis, D.; Kaldellis, J. Economic evaluation of the dual mode CAES solution for increased wind energy contribution in autonomous island networks. *Energy Policy* **2009**, *37*, 1958–1969. [CrossRef]

24. Zafirakis, D.; Kaldellis, J. Autonomous dual-mode CAES systems for maximum wind energy contribution in remote island networks. *Energy Convers. Manag.* **2010**, *51*, 2150–2161. [CrossRef]

25. Lavidas, G.; Venugopal, V. A 35 year high-resolution wave atlas for nearshore energy production and economics at the Aegean Sea. *Renew. Energy* **2017**, *103*, 401–417. [CrossRef]

26. Kaldellis, J.; Efstratiou, C.; Nomikos, G.; Kondili, E. Wave Energy Exploitation in the North Aegean Sea: Spatial Planning of Potential Wave Power Stations. In Proceedings of the 15th International Conference on Environmental Science and Technology, Rhodes, Greece, 31 August–2 September 2017.

27. Soukissian, T.; Denaxa, D.; Karathanasi, F.; Prospathopoulos, A.; Sarantakos, K.; Iona, A.; Georgantas, K.; Mavrakos, S. Marine Renewable Energy in the Mediterranean Sea: Status and Perspectives. *Energies* **2017**, *10*, 1512. [CrossRef]

28. Bai, W.; Lee, D.; Lee, K.Y. Stochastic Dynamic AC Optimal Power Flow Based on a Multivariate Short-Term Wind Power Scenario Forecasting Model. *Energies* **2017**, *10*, 2138. [CrossRef]

29. Lavidas, G.; Venugopal, V.; Friedrich, D. Investigating the opportunities for wave energy in the Aegean Sea. In Proceedings of the 7th International Science Conference Energy Climate Change, Athens, Greece, 8–10 October 2014.

30. Fusco, F.; Nolan, G.; Ringwood, J.V. Variability reduction through optimal combination of wind/wave resources-An Irish case study. *Energy* **2010**, *35*, 314–325. [CrossRef]

31. Cradden, L.; Mouslim, H.; Duperray, O.; Ingram, D. Joint Exploitation of Wave and Offshore Wind Power. Available online: https://www.research.ed.ac.uk/portal/files/21760167/221.pdf (accessed on 2 February 2015).

32. Saha, S.; Moorthi, S.; Pan, H.L.; Wu, X.; Wang, J.; Nadiga, S.; Tripp, P.; Kistler, R.; Woollen, J.; Behringer, D.; et al. The NCEP climate forecast system reanalysis. *Bull. Am. Meteorol. Soc.* **2010**, *91*, 1015–1057. [CrossRef]

33. Lavidas, G.; Agarwal, A.; Venugopal, V. Long-Term Evaluation of the Wave Climate and Energy Potential in the Mediterranean Sea. In proceedings of the 4th IAHR Europe Congress, Liege, Belgium, 27–29 July 2016.

34. Vestas. Vestas Offshore V112-3.3 MW Product IEC IB. Technical Report. 2016. Available online: https://www.nhsec.nh.gov/projects/2013-02/documents/131212appendix_15.pdf (accessed on 11 November 2018).

35. Babarit, A. A database of capture width ratio of wave energy converters. *Renew. Energy* **2015**, *80*, 610–628. [CrossRef]

36. Rusu, E.; Onea, F. Estimation of the wave energy conversion efficiency in the Atlantic Ocean close to the European islands. *Renew. Energy* **2016**, *85*, 687–703. [CrossRef]

37. Babarit, A.; Hals, J.; Muliawan, M.; Kurniawan, A.; Moan, T.; Krokstad, J. Numerical benchmarking study of a selection of wave energy converters. *Renew. Energy* **2012**, *41*, 44–63. [CrossRef]

38. Astariz, S.; Iglesias, G. Output power smoothing and reduced downtime period by combined wind and wave energy farms. *Energy* **2016**, *97*, 69–81. [CrossRef]

39. Manwell, J.; McGowan, J.; Rogers, A. *Wind Energy Explained: Theory, Design and Application*, 2nd ed.; John Wiley & Sons Ltd.: Hoboken, NJ, USA, 2009.

40. Zhang, X.; Yang, J. Power capture performance of an oscillating-body WEC with nonlinear snap through PTO systems in irregular waves. *Appl. Ocean Res.* **2015**, *52*, 261–273. [CrossRef]

41. Martins, J.; Goulart, M.; Gomes, M.N.; Souza, J.; Rocha, L.; Isoldi, L.; dos Santos, E. Geometric evaluation of the main operational principle of an overtopping wave energy converter by means of Constructal Design. *Renew. Energy* **2018**, *118*, 727–741. [CrossRef]

42. WaveStar. 2015. Available online: http://wavestarenergy.com/ (accessed on 5 January 2015).

43. Bozzi, S.; Miquel, A.M.; Antonini, A.; Passoni, G.; Archetti, R. Modeling of a point absorber for energy conversion in Italian seas. *Energies* **2013**, *6*, 3033–3051. [CrossRef]

44. Bozzi, S.; Archetti, R.; Passoni, G. Wave electricity production in Italian offshore: A preliminary investigation. *Renew. Energy* **2014**, *62*, 407–416. [CrossRef]

energies

MDPI

Article

Assessing the Macro-Economic Benefit of Installing a Farm of Oscillating Water Columns in Scotland and Portugal

Samuel Draycott [1,*], **Iwona Szadkowska** [1], **Marta Silva** [2] and **David M. Ingram** [1]

[1] School of Engineering, Institute for Energy Systems, The University of Edinburgh, Edinburgh EH9 3DW, UK; s1454311@sms.ed.ac.uk (I.S.); David.Ingram@ed.ac.uk (D.M.I.)
[2] WavEC Offshore Renewables, 1400-119 Lisboa, Portugal; marta@wavec.org
* Correspondence: S.Draycott@ed.ac.uk; Tel.: +44-(0)131-651-3556

Received: 14 August 2018 ; Accepted: 16 October 2018; Published: 19 October 2018

Abstract: The nascent wave energy sector has the potential to contribute significantly to global renewables targets, yet at present there are no proven commercially viable technologies. Macro-economic assessment is seldom used to assess wave energy projects, yet can provide insightful information on the wider economic benefits and can be used in conjunction with techno-economic analysis to inform policy makers, investors and funding bodies. Herein, we present a coupled techno–macro-economic model, which is used to assess the macro-economic benefit of installing a 5.25 MW farm of oscillating water column wave energy devices at two locations: Orkney in Scotland and Leixoes in Portugal. Through an input-output analysis, the wide-reaching macro-economic benefit of the prospective projects is highlighted; evidenced by the finding that all 29 industry sectors considered are either directly or indirectly stimulated by the project for both locations. Peak annual employment is expected to be 420 and 190 jobs in Portugal and Scotland respectively during the combined installation and manufacturing stage, with an associated peak annual GVA of over €16.6 m and €12.8 m. The discrepancies between the two locations is concluded to largely be a result of the site-specific attributes of the farm locations: specifically, increased water depth and distance to shore for the Portuguese site, resulting in higher costs associated with mooring and electrical cables and vessels. The insights gained through the presented results demonstrate the merit of macro-economic analysis for understanding the wider economic benefit of wave energy projects, while providing an understanding over key physical factors which will dominate estimated effects.

Keywords: input-output modeling; macro-economic assessment; wave energy; oscillating water column

---

1. Introduction

Wave energy has the potential to provide copious amounts of clean, safe and reliable renewable energy [1], yet at present remains a largely under-explored resource. The oceans span 71% of the surface of the Earth [2], and are associated with a total global wave energy potential of around 2.11 TW [3]. At present, however, the development of Wave Energy Converter (WEC) devices to exploit this resource are still in the early stages and are not currently commercially viable. It has been suggested that almost a thousand WEC prototypes have been invented [4], yet at present most full-scale devices are still in the prototype stage and, as such, currently represent a significant risk to investors. By comparison, the wind power sector is now a mature industry built on decades of cumulative research efforts and optimization; resulting in reliable devices which attract consistent investment [5]. Although the rapid deployment of wind farms is helping countries approach their renewables targets, recent studies suggest that many European countries will still fall short of meeting them [6]. In future, the contribution from other technologies, such as WECs, will be important for both meeting these targets and ensuring diversification of intermittent renewable sources supplying the grid.

A possible solution to improving the invest-ability of wave energy is through thorough economic analysis of potential farms of WEC devices, enabling device design to be driven from predicted economic performance. One of the most widely used and effective methods is the techno-economic model [7]. Techno-economic analysis is a cost-benefit technique used to evaluate the cost and performance of a technology. It combines process modelling and engineering design with economic evaluation, which provides a way to assess the impact of different configurations and research breakthroughs on the economic viability of the system under analysis. For wave energy, this will include the estimated energy production for a given deployment location, while accounting for expenditure associated with materials and manufacture, statistically expected repairs and maintenance, and man-hours. This method allows for a comprehensive understanding to be gained of the areas most critical to obtaining good economic performance, and can be used to inform future design changes. Additionally if a techno-economic model indicates good Return On Invested Capital (ROIC) then it may be used to attract investment. Good examples highlighting the power of techno-economic assessment of wave energy can be found in [8,9], whereby multiple WECs are assessed in a variety of locations. This type of analysis can hence be used to identify the the most favourable device–location combination, along with providing a breakdown of where the costs lie, and hence, key areas for improvement. Other examples of techno-economic analysis of wave energy projects can be found in [7,10], while an informative review of the state-of-the-art in economic and socio-economic assessment of offshore renewables can be found in [11].

The partial limitation of the techno-economic model is the fact that the investment is assessed in its own context, separate from the rest of the market. This economic assessment, made in isolation from the industries affected, means that the wider-reaching effects are ignored. This negligence is in part a result of decades of scorn towards macro-economics which begun in the 1970s and caused a shift among the academic and business sectors in favour of micro-economics [12]. Yet, this self-imposed constraint removes a useful argument to be made to investors and policy makers, especially those closely working with governments, international organizations and local markets. These organizations are interested in part on the possible impact on the populace in making of financial decisions—which can be provided by a macro-economic assessment [11]. It incorporates the socio-economic influences on the studied economies based on the underlying market relations e.g., outsourcing tendencies, workforce available and ongoing internal structures [13]. Key outputs of macro-economic models include include total jobs created or Gross Value Added (GVA) associated with the proposed project [14]. For assessing wave energy farms, macro-economic methods may be particularly favourable as the total environmental and economic benefit can be used to inform decision making. This may be particularly relevant for policy makers when deciding the appropriate level of subsidy to stimulate the sector [15].

Several macro-economic studies have been carried out on the effect of renewables including those focusing on the impact of wind energy (e.g., [16,17]) and marine energy on regional economies. Work done by SQW Consulting [18] assesses the macro-economics of the Aquamarine Power's Oyster device focusing on implications on employment in Orkney. In [19], the macro-economics associated with a farm of heaving buoys is assessed, using a reverse Levelized Cost of Energy (LCOE) approach with device characteristics and project spend derived from literature. In both [20] and [13] the macro-economic impact of large-scale deployment of marine renewables in Scotland is assessed (3 GW and 1.6 GW respectively) using Computable General Equilibrium (GCE) and Input-Output (IO) models [14] to provide insightful results. The inputs to drive the models described in the aforementioned publications, as is typical for macro-economic assessment, tend not to be based on a detailed techno-economic analysis. This, in addition to introducing additional uncertainty in the model results, omits some of the complexity of the expenditure and the associated consequences on the regional economic sectors.

This paper details the development of a coupled techno-macro-economic model, which is subsequently used to assess the macro-economic influence of a farm of Oscillating Water Column (OWC) devices if installed in chosen locations in Scotland and Portugal. This techno-economic aspect of the presented model includes detailed analysis of location specific expenditure such as device installation and maintenance, resource use, along with the nature of the labour required for completion. This data is directly connected to the expenditure on human resources, agreements (like rental of equipment) and supply chain management (e.g., logistics), which is intertwined with local economic profiles. Local economies offer their unique enterprise range and employee variety. These factors impact the costs of hire and transport, as well as times required for manufacture and installation—local businesses are closest to the site, have an established regional network and provide a service with knowledge of external competition. The combination of a comprehensive techno-economic model and up-to-date macro-economic profile will result in an IO model that effectively incorporates both technological nuances as well as regional impacts.

This article will describe the aforementioned IO modelling approach, carried out as part of Wave Energy Transition to Future by Evolution of Engineering and Technology (WETFEET) project [21], which is a part of European Union's Framework Horizon 2020 (H2020). This project is detailed further in Section 2.1. The remaining sections are laid out as follows. First, the methodology and theoretical overview is given with emphasis on the information flow and assumptions made in Section 2. Results of the IO modelling work are presented for both Scotland and Portugal in Section 3. Additional discussion is carried out in Section 4, prior to offering concluding remarks in Section 5.

2. Methodology

This section provides information about the WETFEET project (Section 2.1), the farm of OWC devices modelled (Section 2.2) and the specific locations chosen for analysis (Section 2.3), before detailing the Input-Output modelling approach and implementation in Sections 2.4 and 2.5 respectively.

2.1. WETFEET H2020 Project

WETFEET is a project participant from May 2015 under a grant number 641,334, funded by H2020 EU.3.3.5. programme for New Knowledge and Technologies established in 2013 [22]. It addresses the issues found with implementation of wave energy technologies through an analysis of identified features of critical impact: referred as 'breakthrough features' in project's chosen terminology. Its main goal is to provide information that will assist in improving the wave energy technology performance in hopes of accelerating progress in the sector. The potential design breakthroughs identified for the OWC device are as follows [21]:

1. Survivability: Assessed by device submergence under bad weather conditions at sea.
2. Operation and Maintenance: Assessed by continuous submergence and adjustment of elements and strategies.
3. Power Take-Off (PTO): Assessed by the evaluation of new PTO options and their development via dielectric membrane, opposed to standard electromechanical approach.
4. Array: Assessed by distributing the connections and seabed attachments between multiple devices.
5. Performance: Assessed by the functionality of an experiment involving negative spring (NS) for OWC.

These features were refined by participating member organizations. In this paper the most promising breakthrough design variant, when assessed in terms of techno-economics, was chosen for macro-economic assessment. This breakthrough, the "Negative Spring" (NS) variant—addressing the performance improvement—is used for modelling the farm of OWC devices in both Scotland and Portugal.

2.2. The Farm of Oscillating Water Column WECs

A detailed description of the OWC device and NS design variant is available in [23]. Specifics of the farm of devices chosen for the modelling work is provided here.

2.2.1. Farm Design

The Oscillating Water Column Device

Oscillating water column devices, first reported in 1978 [24], extract energy by exploiting the moving air-water interface introduced by wave action. Air turbines are subsequently used to extract the energy. Many OWC designs are fixed structures integrated into the natural coastline or breakwaters; however, the OWC modelled in this work is floating: categorized as a floating spar buoy OWC device [23]. One of the main advantages of a floating device is the ability to be deployed offshore in areas with larger wave resource, with the added benefit of reducing the visual impact of the farm; and hence likely opposition to its installation.

Farm Size

Consistent with the WETFEET project, a fixed farm size has been used throughout the analysis: 5.25 MW. For the farm of 150 kW OWC WECs this means arrays of 35 devices are considered. A multiple of 5 devices was chosen in order to allow a comparison with other design variants assessed as part of the WETFEET project; specifically the "shared moorings" breakthrough which couples five devices in a compact array.

Moorings

The mooring set-up for the devices is depicted in Figure 1. A three-point catenary mooring is used using a combination of chain and synthetic rope with floats and clump weights.

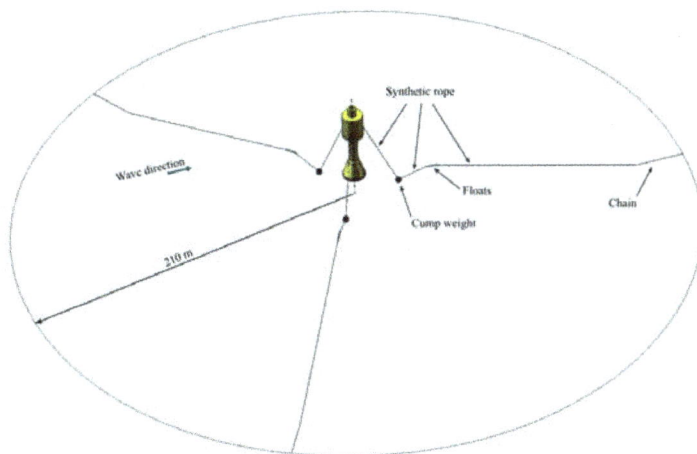

Figure 1. Diagram depicting the three point catenary mooring used for the model.

Array Layout and Configuration

A variety of electrical configurations and array layouts are considered in the techno-economic model. For the macro-economic assessment a single array layout and electrical configuration was considered. For the farm layout, effectively 2 rows of devices is considered, with the distance between devices set as just over 13 device diameters. For the electrical configuration, a star array was chosen as it provided a lower LCOE than a string configuration, and has advantages in terms of redundancy of the system. In this configuration, each device has its own umbilical cable, and 'stars' of devices are grouped prior to connection to the offshore substation. This is illustrated in Figure 2, whereby one star of five devices is depicted. Six other static cables connect the remaining 30 devices to the offshore substation.

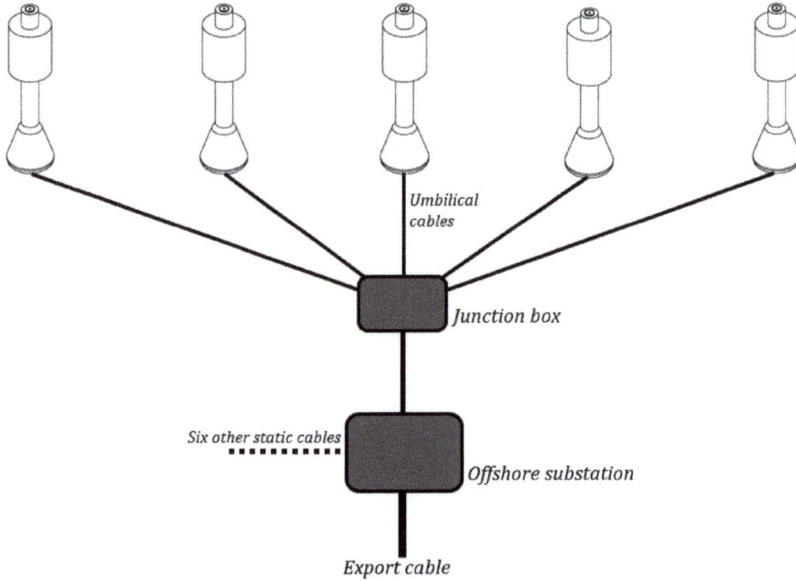

Figure 2. Diagram depicting the farm of OWC devices (5 of 35 devices shown). The sub-sea cable configuration is shown, using a star configuration for umbilical cables.

2.2.2. Farm Lifetime

The operational lifetime of the farm is assumed to be 20 years. Prior to the operational phase, three years of manufacturing is assumed, with two years of installation commencing after the first year of manufacturing. A one year decommissioning phase is incorporated after the 20 year lifetime of the farm.

2.3. The Locations

Precise locations are required for completion of the modelling work. For the techno-economics the deployment site is critical as the associated wave climate dominates the expected power output of the farm. The distances to small and large ports, water depth, and wave climate also influence the cost of installation along with the O&M costs over the project lifetime. It is therefore necessary to identify two prospective sites. These locations are detailed below, and have been chosen due to their associated wave climate and potential suitability as WEC farm deployment locations:

1. European Marine Energy Centre (EMEC) [25], Orkney, Scotland, UK
   Grid-connected test facility for wave and tidal energy devices
2. Leixoes, Portugal
   Major port in the north of Portugal, located in Matosinhos near the city of Porto.

These two locations vary greatly on values of distances to shore, largest port and nearest O&M ports, as well as water depth, as shown in Table 1. In case of Leixoes, the distances from site and ports are almost two times larger compared to the EMEC site, and the distance from site to shore differs in a factor of 7. In terms of depth, the Leixões is at 80 m, and EMEC at 50 m. This suggests that the costs associated with water transport (especially in installation stage), electrical cables and mooring components (textiles and fabricated metal) will be significantly higher in case of Leixoes than EMEC.

Table 1. Site-specific distances and water depths.

	Leixoes	EMEC
Distance from nearest large port to site (km)	25	13
Distance from nearest small O&M port to site (km)	25	13
Distance from site to shore (km)	26	3.7
Distance from shore to substation/grid (km)	2	0.25
Water depth at central farm location (km)	80	50

Contours describing the relative abundance of significant wave height, H_{m0}, and energy period, T_E, for the two sites are shown in Figure 3, based on data from [26,27]. This demonstrates the differing nature of the two sites in terms of spread of likely sea state conditions. The mean values, however, are comparable, with the Leixoes site having mean values of H_{m0} and T_E of 2.0 m and 8.6 s, and EMEC 1.8 m and 8.8 s.

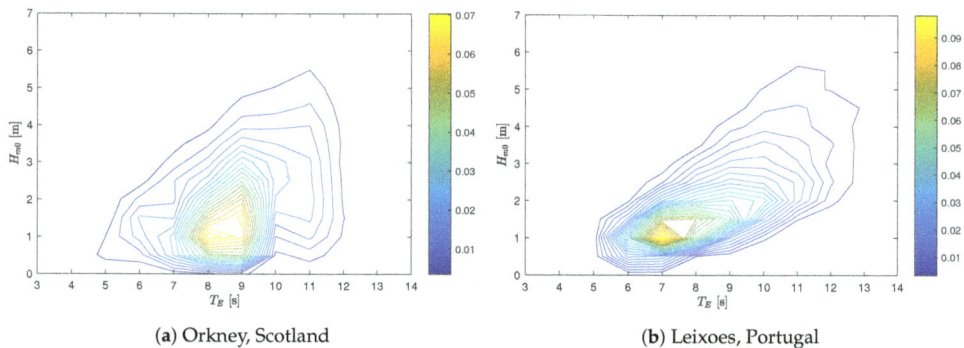

(a) Orkney, Scotland   (b) Leixoes, Portugal

Figure 3. Scatter diagrams of significant wave height and energy period for the locations in Scotland and Portugal. Use of colour and associated values denote the probability of occurrence of a sea state in a given bin for the site. Bin sizes are 0.5 m and 1 s.

2.4. Input-Output Modelling

IO modelling is a quantitative method of macro-economic analysis, considering interdependencies between different branches of the economy. This modelling approach enables the wider economic benefit to a specified region to be assessed, based on knowledge of direct sectoral spend along with relevant multipliers accounting for the inter-relationships between economic sectors. Estimates can be obtained for the number of created jobs and the total Gross Added Value (GVA) associated with the proposed project. For this work, IO modelling is used to quantify and understand the effects of installing a farm of oscillating water column WECs on Scottish and Portuguese economies.

To complete classical IO modelling it is necessary to make several simplifying assumptions. The key assumptions required are as follows [11,19]:

1. The supply side of the regional economy is passive, and does not itself influence demand
2. The intervention with the model takes all supply regardless of external demand
3. Fixed coefficients can be used to describe the interdependencies between sectors i.e., sectors inputs respond linearly to changes in output.

The implications of these assumptions are discussed in Section 4.2.2. Underlying all IO models are Industry by Industry (IxI) matrices which describe the total inter-sectoral spend for all Standard

Industrial Classes (SIC). These are published by countries, along with certain autonomous regions, and are normalised to compute multiplier effect from the known interdependencies. To compute macro-economic effects it is therefore required to have up-to-date IxI matrices, and to allocate all project spend to SICs.

2.5. Model Implementation

To complete the input-output computations, reliable estimates of project expenditure are needed. The IO model is therefore coupled to a techno-economic model, which is described along with the model considerations and options in Section 2.5.1. The procedure to carry out the IO modelling is described separately Section 2.5.2.

2.5.1. Coupled Techno-economic–IO Model

The techno-economic–IO model considers a wide variety of parameters, enabling the assessment of various locations, device types (and associated failure rates), farm layouts, materials, vessels and other variables. The details of the logistic model used to compute the techno-economic analysis can be found in [28,29], with techno-economic model outputs for the WETFEET project in [30]. Detailed CAPEX and OPEX entries are then passed to the IO model, along with other key variables, such as location, which directly influence the modelling work. The key inputs and considerations in the model is depicted in Figure 4. Outputs of the IO model for all of the breakthrough cases can be found in [31].

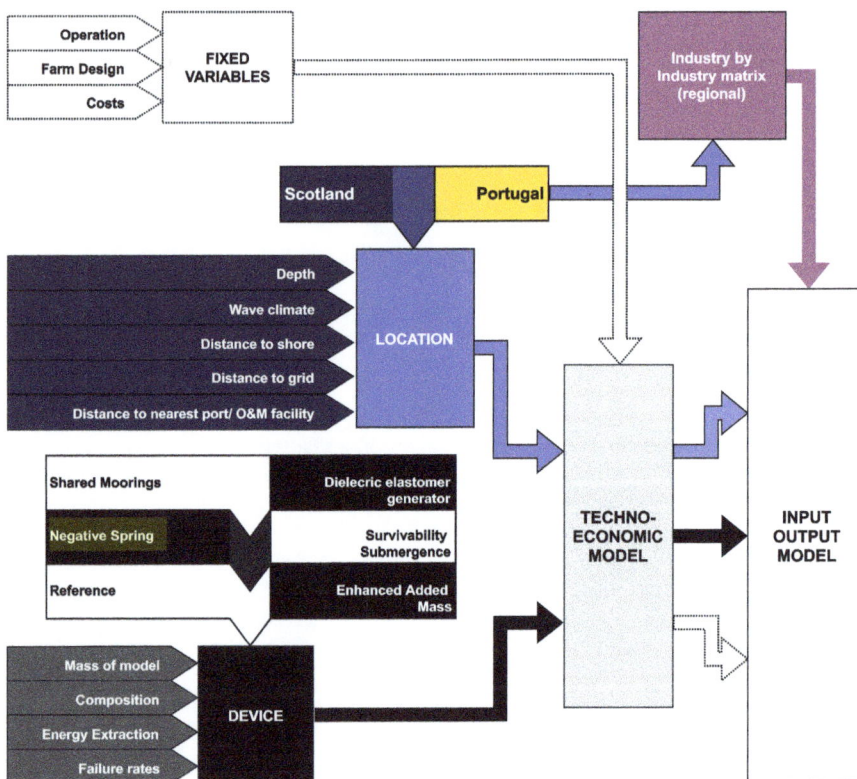

Figure 4. Diagram showing the main inputs and information flow for the coupled techno-economic–IO model. The key model choices (location and breakthrough type) are displayed above the blocks, while the attributes associated with those choices are shown from the left.

It is demonstrated in Figure 4 that there are two main categories of inputs to the model: those related to the device specification and those associated with the deployment location, which are accompanied by several additional variables required to complete the analysis. These catagories are expanded upon below:

Device

Specifics of the mechanical, electrical, and electro-mechanical components of the device are required to compute the CAPEX associated with the devices themselves. Associated device failure rates and power matrices enable the O&M costs and electricity sales to be considered in the overall economic analysis.

Location

Detailed understanding of the deployment location is required (as described in Section 2.3) to ensure distances are properly accounted for to obtain reasonable estimates for mooring and electrical cable lengths, along with vessel journey times for installation, maintenance and decommissioning. A basic understanding of the wave climate is also required, in conjunction to the power matrix of the device, to predict annual energy production. This is key to Levelized Cost of Energy (LCOE) calculations; however, is not an input to the IO model.

Fixed Variables

In addition to key inputs associated with the device and location, additional fixed variables are required to complete the modelling work. Crucially, it is required to use a database of costs (for e.g., wage and material costs) to convert from technical and logistical aspects of the model to equivalent CAPEX and OPEX entries. Although easy to change in the model, for the purpose of the analysis the farm operation and design have been considered as fixed variables; details of which are provided in Section 2.2.

2.5.2. IO Modelling

The outputs of the techno-economic model, along with the IxI matrices are used as inputs to the IO model. The procedure used to carry out the IO modelling work is as follows:

1. Allocate CAPEX and OPEX expenditure entries to SICs
2. Create grouped, simplified IxI matrices
3. Apply Ready Reckoners (RR) and simulate time-series of expenditure for each SIC class
4. Compute IO model: obtain direct and type II output, jobs and GVA

The methodology for completing this procedure is described below, with the process depicted in Figure 5.

Allocate CAPEX and OPEX Expenditure Entries to SICs

As the matrices which describe sector inter-dependency use standard classes, it is required that all project expenditure be allocated to these classes. To achieve this, each CAPEX and OPEX entry of the techno-economic model has been separated into the differing associated materials and services, and costs allocated to the most appropriate classes.

The attribution to classes for CAPEX and OPEX has been done by detailed assessment identifying the industry most influenced by the cost entry. In some cases, this means attributing costs between multiple industries by the expected relative influence. Summing total expenditure in each class provides indication as to the key sectors being shocked, and hence which ones should be kept as separate classes, and which can be aggregated to simplify the analysis and presentation of results.

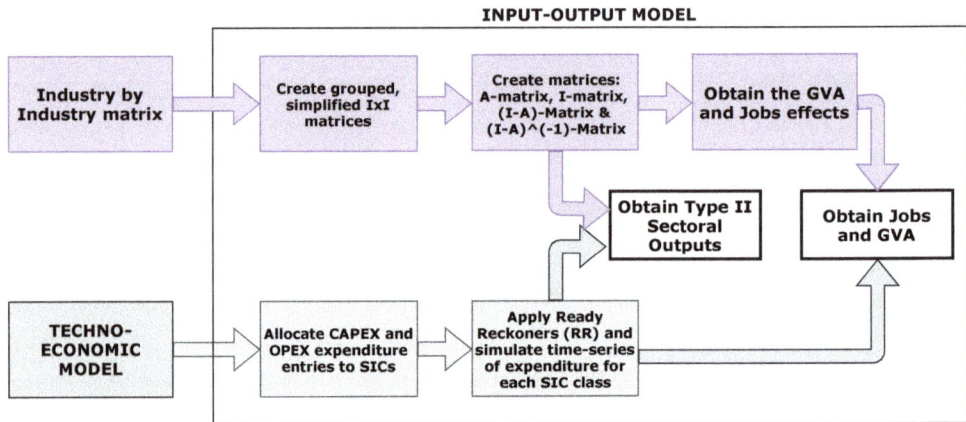

Figure 5. Diagram depicting the process and interdependencies of completing the IO model.

Create Grouped, Simplified IxI Matrices

Once key industrial sectors have been identified in the classification procedure, a process of aggregation can be carried out on those sectors of reduced interest. For this model the process resulted in 29 groups, clustered from the original SIC (2007) list of 98 separate industries. This has been carried out by identifying common characteristics e.g., the aggregated category "Food and Drink processing" encompasses industries such as dairy, meat and wine. The resulting aggregated groups are presented in Table 2, where sectors without aggregation are interpreted as those of most interest to the study.

Once the aggregated groups have been formulated, it is necessary to use these new classes to create grouped IxI matrices, describing the inter-dependency between defined aggregated groups. The same methodology needs to be applied to compute aggregated expenditure, and updated values of multipliers corresponding to the new classes.

Apply Ready Reckoners (RR) and Simulate Time-Series of Expenditure for Each SIC Class

Ready reckoners are required to compute the net spend in each of the grouped cost centres for the area of interest. These additionalities contain information that try to account for the extent the project is directly responsible for the influence on the economy. To estimate the total net spend in each of the classes Equation (1) is used, where e.g., [32] can be used for definitions of the different ready reckoners.

$$Y = GI(1 - L)(1 - D_p)(1 - S)(1 - D_E) \tag{1}$$

where Y is the net sector demand and GI is gross impact. The ready reckoners are Leakage (L), Displacement (D_p), Deadweight (D_e) and Substitution (S).

For implementation in the model, the ready reckoners presented in Table 3 have been used for both Scotland and Portugal. These are effectively 'best guess' values based on the country profiles and nature of the sectors. Only sectors directly 'shocked' have associated RR values and are presented in Table 3. The sensitivity to these assumptions are discussed further in Section 4.2.

Table 2. Aggregated SIC classes used for IO modelling.

SIC	Grouped Sector Names
38.–39.	Waste, remediation & management
01.–03.	Agriculture, forestry and fishing
09.	Other mining and quarrying
10.–11.	Food and drinks processing
14.	Clothing
19.–20.	Chemicals
15.–18.	Metal and non-metal goods
50.	Water transport
21.–24., 31.–32.	Other manufacturing
36.–37.	Water
41.–43., 81	Construction
49., 51.–52.	Distribution and other transport
61.–64., 66.–68., 82.	Communications, finance and business
53.–60., 73.–75., 78.–80., 84.–97.	Education, public and other services
05.–08.	Coal, Oil and Gas extraction
12.	Tobacco
35.	Gas and Electricity
45.–47.	Wholesale and Retail
13.	Textiles
25.	Fabricated metal
26.–27.	Electrical equipment
28.	Machinery and equipment
29.	Motor Vehicles
30.	Other transport equipment
33.	Repair and maintenance
65.	Insurance and pensions
69.	Legal activities
70.–72.	Architectural services etc
77.	Rental and leasing services

Table 3. Ready Reckoners implemented in the IO model.

Textiles	Deadweight	Leakage	Displacement	Substitution
Cement lime and plaster	0.00	0.50	0.00	0.00
Fabricated metal	0.00	0.50	0.00	0.00
Electrical equipment	0.00	0.50	0.00	0.00
Machinery and equipment	0.00	0.50	0.00	0.00
Motor Vehicles	0.00	0.50	0.00	0.00
Other transport equipment	0.00	0.80	0.25	0.25
Repair and maintenance	0.00	0.00	0.00	0.00
Construction	0.00	0.00	0.00	0.00
Water transport	0.00	0.00	0.25	0.50
Insurance and pensions	0.00	0.00	0.00	0.20
Legal activities	0.00	0.00	0.00	0.00
Architectural services etc	0.00	0.00	0.00	0.00
Rental and leasing services	0.00	0.00	0.00	0.00

Once the total demand values, Y, are estimated via ready reckoners and gross impact (allocated to SICs) a time-series of expenditure for each grouped class can be created. The classified CAPEX/OPEX entries are allocated appropriately to the 4 phases described in Section 2.2.2, providing a time-series of the annual demand for each directly shocked sectors.

Compute IO Model: Obtain Direct and Type II Output, Jobs and GVA

The methodology described provides the final demand, Y, of the aggregated sectors, j, as appropriate for the region of interest. The IO model enables the wider effect of this spend to be assessed considering the multiplier effect resulting from sector interdependency. These multiplier effects can be split in two categories:

1. Supply linked—due to companies' supply chain. Sometimes referred as indirect multiplier.
2. Income linked—due to expenditure from people whose income is supplied from the project. Sometimes called induced multiplier.

Type II incorporates both effects, while Type I only incorporates indirect multiplier effects. Type II multiplier effects are considered in this work to fully account for the macro-economic benefit of the wave farms; incorporating direct, indirect and induced effects on sector output, jobs and GVA.

Type II Sector Output

The basic principle of computing IO models is that developed by Leontief [33], in that sectoral outputs can be linked to final demand via the well-known matrix equation:

$$X = [I - A_I]^{-1} Y \tag{2}$$

where X is the sectoral outputs and Y is the demand. I is an identity matrix. The A_I matrix (Type I) is essentially the normalised equivalent of the IxI matrix developed for the aggregated groups. For Type I multipliers $A_{i,j}$ describes the relative amount of sector i required to create one unit of output for sector j.

For Type II, the effects of households also need to be considered, which can be formally described as:

$$\begin{bmatrix} A_I & A_{IH} \\ A_{HI} & 0 \end{bmatrix} \tag{3}$$

where A_I is the Type I matrix, A_{IH} is the amount of industry i required per unit of household income, and A_{HI} is the compensation of employees per unit of output of sector i. Type II sectoral outputs can then be calculated by:

$$X = [I - A_{II}]^{-1} Y \tag{4}$$

where $L = [I - A_{II}]^{-1}$ is commonly referred to as the Type II inverse Leontief matrix.

Type II Employment and GVA

Type II GVA and employment can be computed using the following equations [34]:

$$J_j = Y_j E_j \tag{5}$$

$$GVA_j = Y_j G_j \tag{6}$$

$$E_j = \sum_i w_i L_{i,j} \tag{7}$$

$$G_j = \sum_i g_i L_{i,j} \tag{8}$$

where w_i is the Full Time Equivalent (FTE) employment for industry i divided by the column total of total output at basic prices, and g_i is the GVA for industry i divided by the column total. E_j represents the total impact on employment throughout the economy resulting from a unit change in final demand of industry j, and G_j the GVA equivalent. E_j and G_j are commonly referred to as the employment effect and GVA effect. These must be calculated for the aggregated groups defined in Table 2.

3. Results

3.1. Levelised Cost of Energy (LCOE)

Although the focus of this article is on the macro-economic outputs, the techno-economic component of the coupled model enables key performance metrics to be calculated including the Levelized Cost of Energy (LCOE), computed using Equation (9):

$$LCOE = \frac{\sum_{t=0}^{n} \frac{(Investment_t + O\&M_t + Fuel_t + Carbon_t + Decomissioning_t)}{(1+r)^t}}{\sum_{t=0}^{n} \frac{AEP_t}{(1+r)^t}} \tag{9}$$

where $LCOE$ is the levelized cost of energy, AEP_t is the annual electricity production at year t, r is the discount rate and n is the system lifetime. The numerator represents the sum of expenditure in year t. Further details on the inputs to the calculation can be found in [30].

The LCOE computed for the OWC farms installed in EMEC, Scotland and Leixoes, Portugal are 105 c€/kWh and 130 c€/kWh respectively, with the corresponding breakdown of the LCOE into major cost centers is depicted in Figure 6a,b. The values for LCOE are notably higher than conventional electricity generation, which as detailed in [35] tend to range between 5 c€/kWh and 20 c€/kWh. It is worth noting, however, that there is no learning rate assumed in the LCOE calculation and the farm is relatively small and hence does not take advantage of economies of scale. It is also true (and evident in Figure 6) that there is high ratio of manufacturing cost, and material, to the rated power of the OWC. This suggests that the size of the OWC device may be sub-optimal, and a larger device may represent a more material efficient machine. This aspect, however, is not the focus of the presented study.

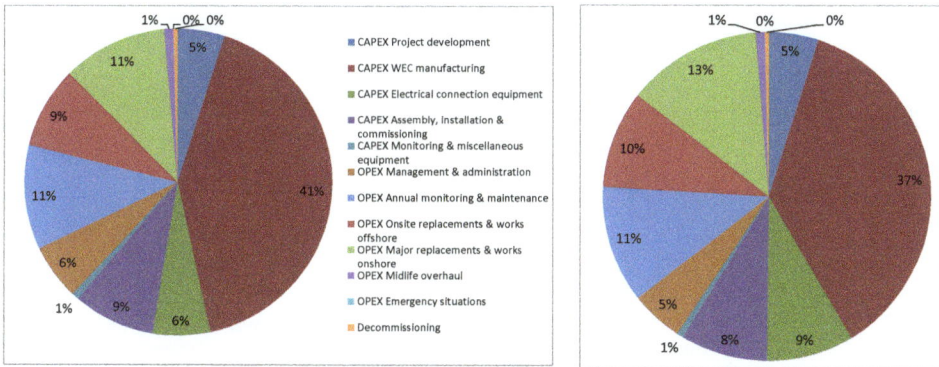

(a) Orkney, Scotland (b) Leixoes, Portugal

Figure 6. LCOE breakdown for OWC farms installed in in Scotland and Portugal.

Assessing the LCOE breakdowns for the OWC farm installed in Scotland and Portugal, it is evident that for the Leixoes site a higher proportion of LCOE is attributed to installation, and works both onshore and offshore. This is attributed to the greater distances to port and to shore for the Leixoes site, increasing the costs associated with mooring and electrical cables, and vessels for installation and maintenance. WEC manufacturing is associated with a smaller relative value of LCOE compared with the EMEC site, despite the cost of the device being slightly higher for the Leixões case (due to higher mooring costs). This relative reduction, however, reflects that the expenditure in other cost centres has increased, rather than manufacturing costs decreasing for this site.

3.2. Direct and Type II Sectoral Outputs

The type II outputs for the 29 aggregated sectors is shown here for the projects deployed in Scotland and Portugal. Results are shown relative to the direct demand associated with the sectors for the prospective projects. This comparison is concisely presented in Figure 7, enabling comparisons to be made between type II outputs and demand, along with the relative values for the two locations studied.

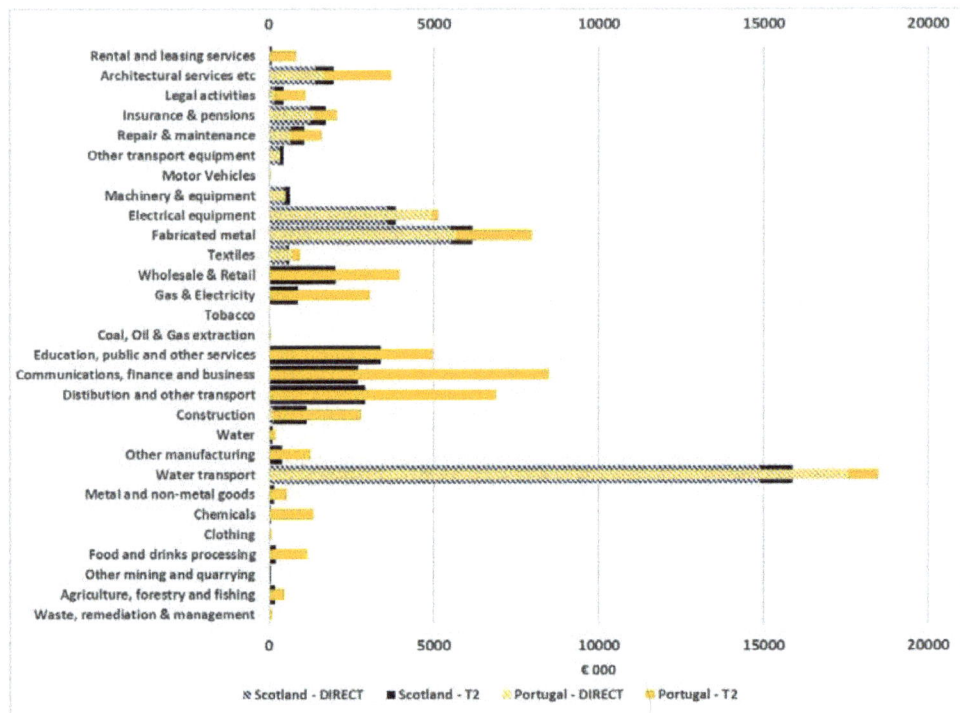

Figure 7. Total direct and Type II output for the 29 aggregated sectors, shown for the OWC farm installed in Portugal and Scotland.

From Figure 7, several findings can be inferred. Comparing the direct to the type II outputs indicates the extent of the multiplier effect introduced from inter-sectoral dependencies in the studied economies. It is clear that, although only 13 sectors are directly shocked, practically all sectors are affected by the project indirectly. Some of these indirect and induced outputs are very significant (e.g., "communications, finance and business") demonstrating the significant and wide-reaching effects resulting from project developments of this type. Also evident is the extent of the multiplier effects for those sectors which are directly stimulated, for example. "Architectural services", where the total type II output is double the demand for the Portuguese location.

Comparing between the two locations it is clear that the output associated with installing the farm of OWC devices in Portugal is significantly higher than that when installed in Scotland. Assessing the location profiles describes in Table 1 in conjunction with Figure 7 the reasons for this become apparent. Water depths along with distances to ports and shore are significantly larger for the Leixoes site. The greater water depth results in increased costs due to larger mooring and electrical cable requirements. The larger distance to shore also results in longer electrical cables, while the distance to port means the vessel costs (categorized as water transport) for installation, O&M and decommissioning are also increased. The increased demand for Portugal consequently increases the type II outputs for dependent sectors, and had a positive effect on macro-economics. This highlights an apparent trade-off between positive techno-economics and macro-economics, in that higher project costs are associated with more favourable macro-economics. However, this apparent trade-off is under the crude assumption that a project is able to sustain itself at arbitrarily high costs, which is clearly not sustainable.

3.3. Jobs Creation

The total jobs associated with the four project phases are presented in Figure 8 for both Scotland and Portugal. In this figure the job years are presented, which is the sum over the project phase of the total (type II) number of jobs supported in each year, across all sectors. Peak employment occurs during the combined installation and manufacturing phase, where over 420 jobs are supported each year for Portugal and over 190 for Scotland. This large difference is owed to the larger CAPEX required for the Portugese development, due to the additional mooring and cabling requirements for the Leixoes site compared with EMEC. Similarly, the increase in jobs supported for the operational, installation and decommissioning phase is largely due to the cost associated with increased water transport for O&M tasks and the associated multiplier effects through the economies.

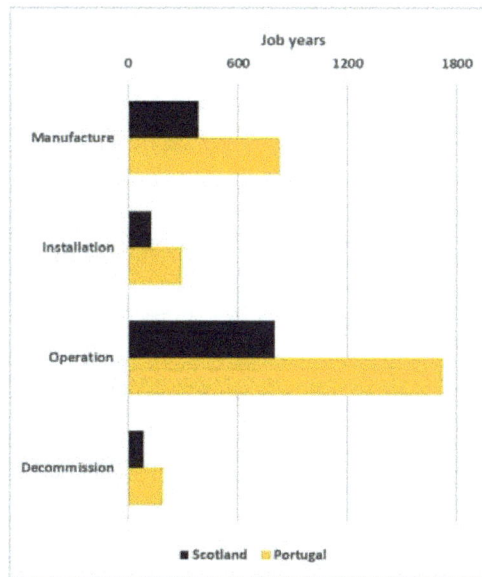

Figure 8. Total job years per project phase, shown for the farm of OWC device installed in both Portugal and Scotland.

For comparison, the number of job years associated with a hypothetical 200 MW deployment of Aquamarine's Oyster devices in Orkney was calculated to be 8503 [18]; equalling 45.5 job years/MW. For the current study, 60 and 133 job years/MW are computed for the farm of OWC devices deployed in Scotland and Portugal respectively; indicating similar but larger numbers for the OWC deployed in Orkney then for the Oyster device. This suggests that the the expenditure (per MW) associated with the OWC device is greater (assuming distribution of expenditure across economic sectors is approximately equal). The values computed in this analysis also compare favourably to other forms of electricity generation. For example, a 2015 study attributes only around 2 and 17 job years/MW for gas and coal power plants respectively [36].

3.4. GVA

The total GVA associated with each of the project phases is presented in Figure 9. Peak annual GVA exceeds €16.5 m for the Portugal development, and €12.8 m for the Scotland equivelent. As expected there is a strong correlation between number of jobs and GVA for each project phase. However, it is evident that there is a larger proportion of total GVA associated with the operational phase, than the proportion of total jobs attributed to this phase. This is likely due to the nature of employment associated with the two phases; suggesting that a greater number of personnel is required per output of manufacture work than operations, which may be attributed to the large proportion of costs associated with vessel hire.

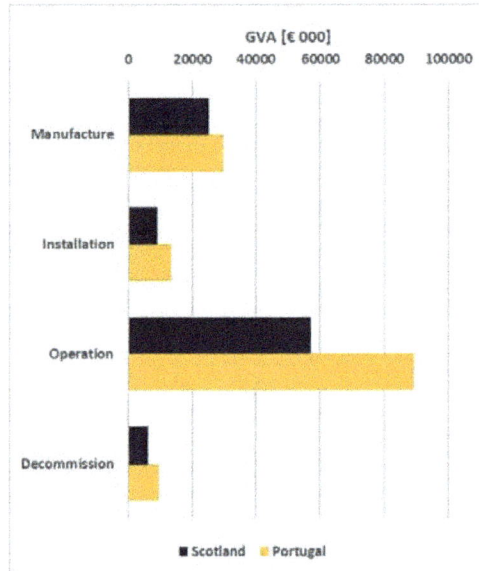

Figure 9. Total GVA per project phase, shown for the farm of OWC device installed in both Portugal and Scotland.

For the farm of OWC devices deployed in Scottish and Portuguese waters the normalized GVA was found to be €4.14 m/MW and €5.83 m/MW respectively. This (under normal exchange rates) is significantly larger than the hypothetical deployment of Aquamarine's Oyster devices in Orkney with a value of £1.44 m/MW [18]. This is predominantly attributed to greater expenditure for the OWC device per MW than the Oyster; indicative of poorer techno-economic performance yet seemingly more beneficial macro-economics.

4. Discussion

4.1. Potential for Integrated Analysis Approach

Techno-economic models can be used to assess the LCOE of a given project, the potential cost reduction of a given technology, to aid in the design process of a technology and for preliminary feasibility studies for a project. By coupling a techno-economic model with a macro-economic model, the macro-economic benefits can be assessed at the same time, and inform in the decision making process.

As noted throughout the results section (Section 3), there is an apparent trade-off between techno-economic performance and macro-economic benefit of projects. The coupled model presented demonstrates the ability to integrate the analysis of both macro and techno-economics, enabling a consideration of this trade-off and a new (potentially subjective) optimum found for the technical design of the project. The most desirable outcome of the coupled analysis would depend on the perspective of the individual assessor. For example, project developers may prioritize reduction in

LCOE and simply present the resulting expected number of jobs and GVA to funding bodies and local councils, whereas government bodies may prioritize jobs created associated with a project providing a threshold LCOE value is satisfied.

This type of integrated analysis approach also lends itself to the assessment of the both macro and techno-economic benefits of a circular economy implementation (see e.g., [37,38]), when assessed for a specific project. It has been suggested that the transition to a circular economy should have a broadly positive effect on the macro-economics (or at least not very negative) [39]. The project-specific effect of, for example: re-use, refurbishment and recycling of components in terms of micro-economics can be identified in the techno-economic model. A coupled model, similar to the one presented, also enables the possibility to assess the macro-economic implications associated with the change in external activities as a result of the circular economy system implementation.

4.2. Uncertainty and Model Sensitivity

As with all modeling work, there are several sources of uncertainty, stemming from the datasets used, assumptions implemented, and the modeling approach used. These are expanded upon in this section and are separated into those associated with the techno-economic model, and macro-economic model on Sections 4.2.1 and 4.2.2 respectively.

4.2.1. Techno-Economic Component

Techno-economic analysis presupposes a certain number of assumptions. At early stage technology development this is especially true, as many of the inputs are yet to be determined or decided, and it is common to use this type of analysis for design choice considerations.

In the case of the WETFEET project, the goal was to do a preliminary analysis under the Multi-disciplinary Assessment for Large-scale Deployment workpackage, in order to assess the viability of each breakthrough in relation to a reference case with no breakthroughs. While the analysis assessed the LCOE, the results presented were of a normalized LCOE, in reference to the reference case. The assumptions for this analysis were equal across breakthroughs and the reference case.

For this analysis, the power matrices used for energy output calculation were produced through numerical modelling, with no experimental validation. While the energy output has no impact on the macro-economic analysis, it has an inverse power relationship with the LCOE, meaning that a small deviation of capacity factor will not have a big impact, but a capacity factor 25% lower can represent a LCOE increase of 33%.

The costing of the device and its variants was done on the basis of the materials required, and typical values for fabrication of steel devices. Likewise, PTO costs were derived from cost curves for OWC devices. The structural costs are one of the major drivers of the CAPEX, and any variation will have a strong impact on the final LCOE (a change of 10% in structural costs represents a 3–4% change in LCOE).

For the logistic operations of installation and maintenance, the algorithms developed for the DTOcean project [29] have been adapted within the model, making use of the vessel database developed in the same project. This database in by no means exhaustive, and while all the unfeasible vessels (in terms of lifting capability, bollard pull and deck space) have been filtered out for this analysis, over dimensioned vessels were still included. The choice of vessel was made on the basis of minimizing the LCOE; however, for the same class of vessel, there can be a variability of cost of over 2x, that can represent up to an increase of 8% on the LCOE.

Furthermore, for the calculation of weather windows and storm conditions, statistical data derived from 20-year time series was used. While this is a rather simplistic approach, for the purpose of the project, a more detailed analysis was deemed unnecessary.

4.2.2. Macro-Economic Component

As the techno-economic outputs are used as inputs to the macro-economic model, any uncertainty in the CAPEX and OPEX values propagate into the macro-economic outputs. As the IO model is linear, as is the manner in which errors propagate. The extent these will affect the total macro-economic assessment will depend on magnitude of the errors in each sector, along with their associated multipliers.

In addition to uncertainty resulting from the techno-economic modeling outputs there are number of notable areas of uncertainty pertinent to the macro-economic modeling work. One of these is the method of classifying CAPEX and OPEX expenditure entries to SICs. As discussed in Section 2.5.2, sometimes this relied on judgment for the allocation, and the occasional individual entry was split across multiple SICs. Although the overall errors associated with mis-classification are likely to be minimal, they are somewhat difficult to quantify, with their relative influence determined by the precise nature of the disputed sectors.

The use of estimated ready reckoner values (see Section 2.5.2) enables a more conservative and realistic figure to be used for the macro-economic modeling work. The model is highly sensitive to these estimated values, as they essentially reduce the expenditure to account only for the amount which influences the economic region of interest, and as such it is a significant area of uncertainty. Again, resulting from the linear nature of the model a change in ready reckoner for a given directly stimulated sector will induce a proportional alteration to total jobs, GVA and output associated with that SIC class. Specifically how this affects indirectly stimulated sectors will depend on the inter-sector relationships, and how the total (all class) values vary will depend on the relative expenditure associated with the class and its respective type II multipliers.

Lastly, it is worth discussing the implications of using an IO model to assess the macro-economic implications of the proposed wave farm. As mentioned (Section 2.4) this approach makes several simplifying assumptions, including that the supply side is completely passive. It is suggested in [13] that this has the effect of over-estimating the immediate effects and ignoring longer-term "legacy" effects when compared to more advanced Computable General Equilibrium (CGE) methods. Hence in reality the peak value of jobs, GVA and sectoral outputs are likely to be reduced when compared to the presented IO results, yet will be more smoothed out over time and will extend beyond the end of the project lifetime.

5. Conclusions

A coupled techno-economic–IO model was developed as part of the WETFEET project, and used to assess the macro-economic benefit of installing a 5.25 MW farm of oscillating water column wave energy devices in chosen locations in Scotland and Portugal. The analysis indicates that the development will directly stimulate several sectors associated with the construction, installation and operation of the devices, along with those associated with insurance and project management. Peak employment of 420 jobs and 190 jobs, for Portugal and Scotland respectively, is predicted during the combined installation and manufacturing phase, which is accompanied by an associated GVA for the same year of over €16.5 m and €12.8 m respectively. The differences for these project phases, along with the operation and decommissioning phases, are attributed to the increased distance to shore/port and the increased water depth at the Leixoes site in Portugal. This results in larger costs for the mooring and electrical cables, along with the expenditure on vessels for installation and maintenance of the farm.

The model outputs highlight the wide-reaching macro-economic benefit of projects of this type, and when used in combination with techno-economic analysis will provide additional information to investors, policy makers and funding bodies. The apparent trade-off between techno and macro-economics highlights the benefits of considering both and taking a holistic approach to project assessment.

Energies **2018**, *11*, 2824

Author Contributions: S.D. and I.S. jointly created the Input-Output model and led on the subsequent analysis and write up. M.S. developed and provided the techno-economic aspects of the model, while D.M.I. led the study. All authors read, edited and approved the manuscript.

Funding: This research was funded by the European Commission though the WETFEET project. Grant no. 641334.

Acknowledgments: The authors are grateful to the European Commission for funding the WETFEET project as part of the Horizon 2020 framework. A special thanks is owed to Grant Allan and David Crooks for their early guidance in the model development.

Conflicts of Interest: The authors declare no conflict of interest.

References

1. Evans, D.V. A theory for wave-power absorption by oscillating bodies. *J. Fluid Mech.* **1976**, *77*, 1–25, doi:10.1017/S0022112076001109. [CrossRef]
2. Cheng, E. Area of Earth's Oceans—The Physics Factbook. 1997. Available online: https://hypertextbook.com/facts/1997/EricCheng.shtml (accessed on 11 September 2018).
3. Gunn, K.; Stock-Williams, C. Quantifying the global wave power resource. *Renew. Energy* **2012**, *44*, 296–304, doi:10.1016/j.renene.2012.01.101. [CrossRef]
4. Mustapa, M.; Yaakob, O.; Ahmed, Y.M.; Rheem, C.K.; Koh, K.; Adnan, F.A. Wave energy device and breakwater integration: A review. *Renew. Sustain. Energy Rev.* **2017**, *77*, 43–58, doi:10.1016/j.rser.2017.03.110. [CrossRef]
5. ECONorthwest. *Economic Impact Analysis of Wave Energy: Phase One*; Technical report; ECONorthwest: Portland, OR, USA, 2009.
6. Cucchiella, F. Future Trajectories of Renewable Energy Consumption in the European Union. *Resources* **2018**, *7*, 10, doi:10.3390/resources7010010. [CrossRef]
7. De Andres, A.; Maillet, J.; Todalshaug, J.H.; Möller, P.; Bould, D.; Jeffrey, H. Techno-Economic Related Metrics for a Wave Energy Converters Feasibility Assessment. *Sustainability* **2016**, *8*, 1109, doi:10.3390/su8111109. [CrossRef]
8. Quitoras, M.R.D.; Abundo, M.L.S.; Danao, L.A.M. A techno-economic assessment of wave energy resources in the Philippines. *Renew. Sustain. Energy Rev.* **2018**, *88*, 68–81, doi:10.1016/j.rser.2018.02.016. [CrossRef]
9. O'Connor, M.; Lewis, T.; Dalton, G. Techno-economic performance of the Pelamis P1 and Wavestar at different ratings and various locations in Europe. *Renew. Energy* **2013**, *50*, 889–900, doi:10.1016/j.renene.2012.08.009. [CrossRef]
10. Costello, R.; Teillant, B.; Weber, J.; Ringwood, J.V. Techno-Economic Optimisation for Wave Energy Converters. In Proceedings of the 4h International Conference on Ocean Energy, Dublin, Ireland, 17–19 October 2012.
11. Dalton, G.; Allan, G.; Beaumont, N.; Georgakaki, A.; Hacking, N.; Hooper, T.; Kerr, S.; O'Hagan, A.M.; Reilly, K.; Ricci, P.; et al. Economic and socio-economic assessment methods for ocean renewable energy: Public and private perspectives. *Renew. Sustain. Energy Rev.* **2015**, *45*, 850–878, doi:10.1016/j.rser.2015.01.068. [CrossRef]
12. Bårdsen, G.; Nymoen, R. *The Econometrics of Macroeconomic Modelling*; Oxford Univeristy Press: Oxford, UK, 2004.
13. Allan, G.J.; Lecca, P.; McGregor, P.G.; Swales, J.K. The economic impacts of marine energy developments: A case study from Scotland. *Mar. Policy* **2014**, *43*, 122–131, doi:10.1016/j.marpol.2013.05.003. [CrossRef]
14. *Calculating the Economic Contribution of Beatrice Offshore Windfarm Limited*; Beatrice Offshore Windfarm Ltd.: Moray Firth, UK, 2017. Available online: https://www.google.com.tw/url?sa=t&rct=j&q=&esrc=s&source=web&cd=1&ved=2ahUKEwi3toX9yZHeAhVOPHAKHad9D18QFjAAegQICRAC&url=http%3A%2F%2Fsse.com%2Fmedia%2F475205%2FBOWL-methodology-document-FINAL.pdf&usg=AOvVaw34RcX2spXhUXMh2RXwCjQG (accessed on 11 September 2018)
15. Fraser of Allander Institute. *Impact Study into the Development of the UK Offshore Renewable Energy Industry to 2020*; Project Report; Fraser of Allander Institute: Glasgow, UK, 2014; pp. 1–24.
16. Ernst & Young. *Analysis of the Value Creation Potential of Wind Energy Policies*; Ernst & Young: London, UK, 2012.

17. Okkonen, L.; Lehtonen, O. Socio-economic impacts of community wind power projects in Northern Scotland. *Renew. Energy* **2016**, *85*, 826–833, doi:10.1016/j.renene.2015.07.047. [CrossRef]
18. SQW Consulting. *Socio-Economic Impact Assessment of Aquamarine Power's Oyster Projects: Report to Aquamarine Power*; Technical Report; SQW Consulting: Cambridge, UK, 2009.
19. Crooks, D.; De Andres, A.; Medina-Lopez, E.; Jeffrey, H. Demonstration of a Socio-economic Cost of Energy Analysis of a Wave Energy Converter Array. In Proceedings of the 12th European Wave and Tidal Energy Conference, Cork, Ireland, 27 August–1 September 2017; pp. 1–11.
20. Allan, G.J.; Bryden, I.; McGregor, P.G.; Stallard, T.; Kim Swales, J.; Turner, K.; Wallace, R. Concurrent and legacy economic and environmental impacts from establishing a marine energy sector in Scotland. *Energy Policy* **2008**, *36*, 2734–2753.10.1016/j.enpol.2008.02.020. [CrossRef]
21. WETFEET. WETFEET Project. 2018. Available online: http://www.wetfeet.eu/wetfeet-project/ (accessed on 11 September 2018).
22. CORDIS. *Wave Energy Transition to Future by Evolution of Engineering and Technology*; H2020; CORDIS, 2015. Available online: https://cordis.europa.eu/project/rcn/193803_en.html (accessed on 11 September 2018).
23. Teillant, B.; Debruyne, Y.; Sarmento, A.; Silva, M.; Simas, T.; Gomes, R.P.; Henriques, J.C.; Philippe, M.; Combourieu, A.; Fontana, M. *D2.1—Designs and Specifications of an OWC Able to Integrate The Negative Spring*; Technical Report 641334; WETFEET: Lisboa, Portugal, 2016.
24. Evans, D.V. The oscillating water column wave energy device. *IMA J. Appl. Math.* **1978**, *22*, 423–433.
25. EMEC. Facilities, 2015. Available online: http://www.emec.org.uk/facilities/ (accessed on 11 September 2018).
26. Pontes, M.T.; Rebêlo, L.; Silva, P.; Pata, C. Database of wave energy potential in Portugal. In Proceedings of the ASME 2005 24th International Conference on Offshore Mechanics and Arctic Engineering, Halkidiki, Greece, 12–17 June 2005; pp. 803–809.
27. Wave Energy Scotland. *WES O&M Model: P2*; Wave Energy Scotland: Inverness, UK, 2017.
28. Teillant, B.; Raventos, A.; Chainho, P.; Victor, L.; Goormachtigh, J.; Nava, V.; Ruiz, P.; Minguela, R.J. *Deliverable 5.1: Methodology Report and Logistic Model Flow Charts*; DTOcean, 2014. Available online: https://www.google.com.tw/search?newwindow=1&ei=GlPJW4z7ApH1wAPoz6XADQ&q=Deliverable+5.1%3A+Methodology+Report+and+Logistic+Model+Flow+Charts&oq=Deliverable+5.1%3A+Methodology+Report+and+Logistic+Model+Flow+Charts&gs_l=psy-ab.3...0.0.0.28320.0.0.0.0.0.0.0.0.0....0...1..64.psy-ab..0.0.0....0.zPkarVc4m_Q (accessed on 11 September 2018).
29. Teillant, B.; Chainho, P.; Vrousos, C.; Vicente, P.; Charbonier, K.; Ybert, S.; Monbet, P.; Giebhardt, J. *Report on Logistical Model for Ocean Energy Array and Considerations*; Project Deliverable D5.6; DTOcean, 2016. Available online: http://www.dtocean.eu/Deliverables/Deliverables/Logistical-model (accessed on 11 September 2018).
30. Silva, M.; Afferni, L.; Sebastian, R. *D7.3 Techno-Economic Assessment of the Proposed Breakthroughs For Large Scale Deployment*; Technical Report; WETFEET: Lisboa, Portugal, 2018.
31. Draycott, S.; Szankowska, I.; Chaperon, C.; Yerzhanov, A. *D7.5—LCA and Socio-Economic Implications of Large Scale Deployment of the Proposed Breakthroughs*; Technical Report; WETFEET: Lisboa, Portugal, 2018.
32. Scottish Enterprise. *Additionality & Economic Impact Assessment Guidance Note*; Scottish Enterprise: Glasgow, UK, 2008; pp. 1–28.
33. Leontief, W. *Input-Ouput Economics*; Oxford University Press: Oxford, UK, 1986; p. 396.
34. Crossdale, S.; Campbell, G.; O'Neill, K. *Input-Output Methodology Guide*; The Scottish Government: Edinburgh, UK, 2015.
35. None, N. *Annual Energy Outlook 2016 With Projections to 2040*; Technical Report; USDOE Energy Information Administration (EI): Washington, DC, USA, 2016.
36. Rutovitz, J.; Dominish, E.; Downes, J. *Calculating Global Energy Sector Jobs: 2015 Methodology Update*; Institute for Sustainable Futures: University of Technology, Sydney, Australia, 2015.
37. Prieto-Sandoval, V.; Jaca, C.; Ormazabal, M. Towards a consensus on the circular economy. *J. Clean. Prod.* **2018**, *179*, 605–615, doi:10.1016/j.jclepro.2017.12.224. [CrossRef]

38. Kirchherr, J.; Piscicelli, L.; Bour, R.; Kostense-Smit, E.; Muller, J.; Huibrechtse-Truijens, A.; Hekkert, M. Barriers to the Circular Economy: Evidence From the European Union (EU). *Ecol. Econ.* **2018**, *150*, 264–272, doi:10.1016/j.ecolecon.2018.04.028. [CrossRef]

39. McCarthy, A.; Dellink, R.; Bibas, R. *The Macroeconomics of the Circular Economy Transition: A Critical Review of Modelling Approaches*; Technical Report; OECD Publishing: Paris, France, 2018.

energies

MDPI

Article

A Study about Performance and Robustness of Model Predictive Controllers in a WEC System

Rafael Guardeño *, Agustín Consegliere and Manuel J. López

Escuela Superior de Ingeniería, Universidad de Cádiz, 11519 Puerto Real, Spain;
agustin.consegliere@uca.es (A.C.); manueljesus.lopez@uca.es (M.J.L.)
* Correspondence: rafael.guardeno@uca.es; Tel.: +34-638-88-70-86

Received: 27 September 2018; Accepted: 19 October 2018; Published: 22 October 2018

Abstract: This work is located in a growing sector within the field of renewable energies, wave energy converters (WECs). Specifically, it focuses on one of the point absorber waves (PAWs) of the hybrid platform W2POWER. With the aim of maximizing the mechanical power extracted from the waves by these WECs and reducing their mechanical fatigue, the design of five different model predictive controllers (MPCs) with hard and soft constraints has been carried out. As a contribution of this paper, two of the MPCs have been designed with the addition of an embedded integrator. In order to analyze and compare the MPCs with conventional PI type control, an exhaustive study about performance and robustness is realized through the computer simulations carried out, in which uncertainties in the WEC dynamics and JONSWAP spectrum are considered. The results obtained show how the MPCs with embedded integrator improve power production of the WEC system studied in this work.

Keywords: wave energy converter; model predictive control; robustness analysis; embedded integrator; mathematical model; system identification; JONSWAP spectrum

1. Introduction

Nowadays, the main motivation for the research and development of wave energy converters (WECs) is the advantages offered by waves: a clean and abundant source of energy. As evidence, the authors in [1] compared a global study of net wave power (estimated at about 3 TW) with the electrical power consumed globally in 2008 (equivalent to an average power of 2.3 TW). However, it should be noted that currently, there is not a clear line of development, but a great diversity of systems based on different approaches to extract energy from the waves. In particular, the ocean energy systems collaboration program [2] classifies three kinds of WEC systems: oscillating water columns, overtopping and wave-activated bodies (WABs). This work focuses on a type of WAB system, a point absorber wave (PAW) energy converter from the W2POWER (Wind and Wave Power) platform [3]. These systems are characterized because their extension is significantly smaller than the predominant wavelengths. In addition, PAWs extract the maximum mechanical power from the sea when they are in resonance with the excitation force caused by the waves [4]. In order to favor this situation, it is necessary to enlarge the bandwidth of these devices. For this reason, several control systems are used, and the most common are: passive loading control, reactive loading control and latching control [5,6]. Although, since the last decade, more complex controllers like MPC (model predictive control) are being employed. The interest in implementing MPCs in WEC systems is motivated by the need to increase the productive/economic viability of these systems; due to the fact that these controllers allow them to minimize mechanical fatigue in their structures (limiting the operating ranges) and to focus the control strategy on the maximization of the extracted power directly.

Actually, several authors have developed predictive controllers for WECs [7–17]. These MPCs can be grouped according to the characteristics of the cost function optimized. On the one hand, the authors in [8,10–16] used a cost function in which the extraction of mechanical power was directly maximized. Whereas, on the other hand, in [9,17], the authors proposed a cost function to maximize power extraction by minimizing the error between the speed of oscillation of the system and a setpoint for it. In addition to the above classification, MPCs can be distinguished according to the mathematical model used for their design. On one side, in [9–11,13,14], a reduced model was used for the design, which did not consider the dynamics of the radiation force. Meanwhile, in [7,8,12,13,15–17], such dynamics were considered in the design model. Moreover, the previous works did not consider the dynamics of the power take-off system (PTO), neither in the evaluation, nor in the design of the MPCs. In this aspect, the authors in [11,14], although they did not consider the PTO dynamics, optimized the cost function for the increment of the control signal, thus limiting the slew rate of the actuator. Finally, the treatment carried out in the case of non-feasibility when solving the cost functions with restrictions should be highlighted. In this aspect, the works in [9,10,13] considered a more complete approach by adding soft constrains in the case of non-feasibility.

This paper analyses the main approaches of MPCs applied to PAW systems [8–17]. In addition, a new design is proposed: MPC based on a model with an embedded integrator for controllers that follow a setpoint for the speed of oscillation of the PAW. This approach is recommended in the theory of predictive control in the space of states [18,19], and it is proposed as an alternative method to the one carried out in [9,17] for PAW systems. Moreover, all predictive controllers of this work consider soft and hard constraints. On the other hand, the PTO dynamics is taken into account to validate the MPC controllers, as well as in the design of some of these controllers. After an exhaustive fine-tuning for all MPCs, the main contribution of this work is obtained, and an in-depth study about performance and robustness of all MPCs through simulations is carried out. The results obtained for a sea state defined by the JONSWAP spectrum [15,16] are compared with conventional controllers: I-P control and resistive damping (RD). Furthermore, the mathematical model of the WEC system is obtained using a software simulation based on the boundary element method, such as openWEC [20].

The rest of the article is organized as follows: Section 2 presents the generic mathematical model of a PAW, the standard identification methodology applied in this paper and the treatment applied to the model identified for its later use in the design of the MPCs. Section 3 details the five MPC controllers designed with hard and soft restrictions. Section 4 presents a performance and robustness comparison analysis for the MPCs and conventional (P, PI) controllers. Finally, in Section 5, the main conclusions are indicated.

2. Mathematical Model

Given the importance of mathematical modeling in the design of MPC controllers, this section begins by describing the generic model of a PAW. This is followed by the standard identification process realized in this paper for the study system, one of the WEC systems of the platform W2POWER [3]. Later, the treatment of the model is detailed for its later use in the design of MPC controllers.

2.1. Generic Mathematical Model for Point Absorber Wave

The modeling of the forces affecting a PAW that extracts power from the waves using a single degree of freedom (heave) is widely used in the literature [4,7–9,21–27]; see Figure 1. The main dynamic interactions between the buoy and the waves are collected by:

$$m\ddot{z} = F_e + F_r + F_s + F_u \tag{1}$$

where m is the mass of the system, z the vertical displacement of the buoy, F_e the excitation force caused by the wave, F_r the radiation force, F_s the hydrostatic restoring force and F_u the control force realized by the PTO system.

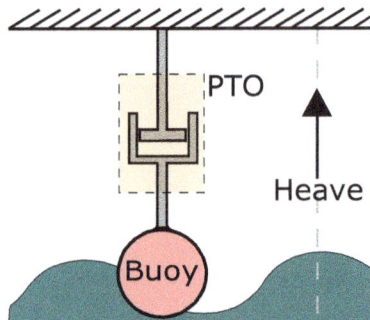

Figure 1. Illustrative scheme: point absorber wave (PAW) system of the W2POWER platform.

The force of radiation is due to the effect produced on the system by the waves it radiates when oscillating. It is modeled by the Cummins equation [28] as (2), where the radiation force F_r is composed of two terms: Fr_{m_∞} and Fr_{K_r}. The first is a function of the acceleration of the system and the mass of water added m_∞; while the second defines a radiation force as a function of the speed of oscillation as an integral of convolution.

$$F_r = -\underbrace{m_\infty \dot{w}(t)}_{Fr_{m_\infty}} - \underbrace{\int_\infty^t h_r(t-\tau)w(\tau)d\tau}_{Fr_{K_r}} \tag{2}$$

The convolution term represents the impulse response that relates the speed of the oscillation system with the force of radiation, where $h_r(t)$ represents the radiation impulse response function (RIRF). To avoid convolution calculations [21,24,26], an approximation is made based on ordinary differential equations (ODE) or as a transfer function in the Laplace domain (3).

$$\frac{Fr_{K_r}(s)}{W(s)} = K_r(s) = \frac{a_n s^n + a_{n-1}s^{n-1} + \ldots + a_0}{b_n s^n + b_{n-1}s^{n-1} + \ldots + b_0} \tag{3}$$

The hydrostatic restoring force represents the effect of Archimedes and gravity on the buoy (4). By linearizing Equation (4), the hydrostatic force is approximated as (5).

$$F_{res} = -(V_{desp}(z)\rho g - mg) \tag{4}$$

$$F_{res} = -k_{res}z(t)$$
$$k_{res} = \rho g A_W \tag{5}$$

where V_{desp} is the volume of water displaced, ρ is the density of seawater, g is the gravity constant, k_{res} is the hydrostatic restoring coefficient and A_W is the area of the buoy on its waterline.

For the excitation force caused by the waves, the components with the highest frequency are negligible. For this reason, authors such as [7,8,11,12,15,16,24–27] have defined the excitation force as a low pass filter of first/second order at wave height (6). This modeling is used later in the performance study.

$$G_{Fe}(s) = \frac{Fe(s)}{\eta(s)} = \frac{K_\tau}{s^2 + 2\zeta_\tau \omega_{n\tau}s + \omega_{n\tau}^2}. \tag{6}$$

On the other hand, in order to model the force on a buoy, the Morison equation [29] can be used, which is habitually employed to estimate the wave loads in the design of offshore structures. By linearizing the Morison equation for a point absorber wave in heave, the excitation force is defined according to Equation (7). This modeling is used later in the robustness study.

$$F_e(t) = m_\infty \ddot{\eta}(t) + B_{aprox}\dot{\eta}(t) + k_{res}\eta(t) \tag{7}$$

where η is the height of the incoming wave and B_{aprox} represents the damping of the system, which can be obtained as the stationary gain of the transfer function (3).

The PTO dynamics can be approximated as a linear second order system with a force limitation. The linear relation between the demanded force by the control system, (F_u), and the force applied to the buoy, (F_{pto}), is given by (8).

$$G_{pto}(s) = \frac{F_{pto}(s)}{F_u(s)} = \frac{d_a \omega_n{}^2}{s^2 + 2\zeta \omega_n s + \omega_n{}^2}, \tag{8}$$

2.2. Forces Identification Using Simulation Based on BEM

A common approach to determine the hydrodynamic forces of interaction between the wave and buoy is to use the linear wave theory, which assumes that waves are the sum of incident, radiated and diffracted wave components [21]. These components can be modeled using linear coefficients obtained from the frequency-domain potential flow BEM (boundary element method) solver Nemoh [30]. The BEM solutions are obtained by solving the Laplace equation for the velocity potential, which assumes that the flow is inviscid, incompressible and irrotational. In this work, openWEC software is employed. It is an open-source tool to simulate the hydrodynamic behavior and energy yield from single-body wave energy converters [20]. Two software packages are coupled in openWEC, a frequency domain solver Nemoh and a time domain solver. In particular, this numerical tool is applied to solve the fluid equation for a submerged body. The equations are solved for the following effects in the heave direction on a buoy: excitation force F_e, radiation force F_r and restoring force F_{res}. Thus, the frequency domain modeling is performed with the Nemoh BEM solver. For each panel of a mesh (see Figure 2a), the hydrodynamic parameters are calculated for a frequency range. The pressures caused by the incoming wave are integrated into resulting forces on the buoy; see Figure 2b. From these data in the frequency domain, a filter is established to convert a wave into an exciting force (9). A magnitude diagram of the resulting filter is also shown in Figure 2b. The magnitude response resembles a second order filter with a cut-off frequency of 0.9 rad/s.

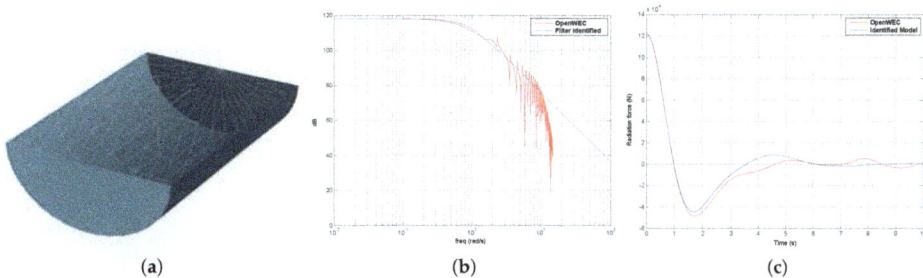

(a)　　　　　　　　　　(b)　　　　　　　　　　(c)

Figure 2. (a) Mesh defined by openWEC for a cylindrical buoy (11 m long and 7.5 m in diameter). Comparison of the identified transfer functions: (b) Bode diagram for excitation force $F_e(s)$; (c) impulse response for radiation force $K_r(s)$.

An advanced function of Nemoh can be enabled to calculate the impulse response function (IRF). A comparison between the IRF identified and that provided by openWEC is shown in Figure 2c. To avoid convolution calculations in (2), it is replaced as ordinary differential equations (ODE) or as a transfer function in the Laplace domain (3). This can be performed using Prony's method [21,31,32], which is implemented in MATLAB 2016b [33]. Different orders of ODEs have been tested to fit the impulse response. By comparing these responses, it was observed that for orders above eighth, the improvement in the system response was not significant.

$$\frac{F_e(s)}{\eta(s)} = \frac{6.5 \times 10^5}{(s+0.9)^2} \quad \left[\frac{N}{m}\right] \tag{9}$$

$$K_r(s) = \frac{F_r(s)}{W(s)} = \frac{a_{K_8}s^8 + a_{K_7}s^7 + a_{K_6}s^6 + a_{K_5}s^5 + a_{K_4}s^4 + a_{K_3}s^3 + a_{K_2}s^2 + a_{K_1}s + a_{K_0}}{s^8 + b_{K_7}s^7 + b_{K_6}s^6 + b_{K_5}s^5 + b_{K_4}s^4 + b_{K_3}s^3 + b_{K_2}s^2 + b_{K_1}s + b_{K_0}} \quad \left[\frac{N}{m/s}\right] \tag{10}$$

where the coefficients a_K and b_K are listed in Table 1.

Then, regrouping terms according to Equation (1), the transfer function (11) that relates external forces with the position of the buoy (z) is obtained.

$$G_{WEC}(s) = \frac{Z(s)}{F_{ext}(s)} = \frac{a_8 s^8 + a_7 s^7 + a_6 s^6 + a_5 s^5 + a_4 s^4 + a_3 s^3 + a_2 s^2 + a_1 s + a_0}{s^{10} + b_9 s^9 + b_8 s^8 + b_7 s^7 + b_6 s^6 + b_5 s^5 + b_4 s^4 + b_3 s^3 + b_2 s^2 + b_1 s + b_0} \quad \left[\frac{m}{N}\right] \tag{11}$$

where the coefficients a_n and b_n are listed in Table 1.

Table 1. Model coefficients identified for the WEC system.

Coefficient	Value	Coefficient	Value	Coefficient	Value	Coefficient	Value
—	—	b_9	8.8×10^1	—	—	—	—
a_8	2.4×10^{-6}	b_8	1.5×10^5	a_{K_8}	1.2×10^3	—	—
a_7	2.2×10^{-4}	b_7	6.7×10^6	a_{K_7}	2.3×10^5	b_{K_7}	8.9×10^1
a_6	3.6×10^{-1}	b_6	4.5×10^9	a_{K_6}	1.9×10^8	b_{K_6}	1.5×10^5
a_5	1.6×10^1	b_5	1.3×10^{10}	a_{K_5}	2.6×10^{10}	b_{K_5}	6.7×10^6
a_4	1.1×10^4	b_4	6.4×10^{10}	a_{K_4}	6.3×10^{12}	b_{K_4}	4.5×10^9
a_3	3.3×10^4	b_3	8.6×10^{10}	a_{K_3}	5.6×10^{14}	b_{K_3}	1.3×10^{10}
a_2	1.3×10^5	b_2	1.8×10^{11}	a_{K_2}	1.7×10^{15}	b_{K_2}	5.3×10^{10}
a_1	1.4×10^5	b_1	1.1×10^{11}	a_{K_1}	4.7×10^{15}	b_{K_1}	5.5×10^{10}
a_0	1.6×10^5	b_0	1.3×10^{11}	a_{K_0}	1.4×10^{15}	b_{K_0}	6.5×10^{10}

2.3. Treatment of the Mathematical Model for the Design of MPCs

In order to ensure safe behavior and reduce mechanical fatigue in PAW systems, a model that allows them to constrain the oscillation speed and the position of the buoy is necessary. For this reason, a brief study of the identified model (11) is realized. By representing it in the state space, it can be verified that the system obtained is not completely controllable, and therefore, it is not a minimal realization [34]. Furthermore, it is not enough to reduce the model (11) to its minimum order. This is because, except for the system output (z), the other state variables lack physical sense, and this does not allow it to impose speed constraints (w) directly. As a solution, we study the transfer function (10) that defines the dynamics of the radiation force. Representing (10) in the state space, it can be seen how the model obtained is not of minimum order, so the system is minimized to get (12). Note that the state variables of this model (x_r) will not be controlled, so it is not necessary that they have physical sense.

$$\dot{x}_r(t) = A_r x_r(t) + B_r w(t)$$
$$Fr_{K_r}(t) = C_r x_r(t) + D_r w(t) \tag{12}$$

where the matrices A_r, B_r, C_r and D_r are given by (13).

$$A_r = \begin{pmatrix} -3.0044 & -1.4736 & -0.3820 & -0.2258 \\ 8.1656 & 0.0180 & 0.0041 & 0.0025 \\ 0.0015 & 4.0015 & 0.0004 & 0.0002 \\ -0.0000 & -0.0000 & 2.0000 & -0.0000 \end{pmatrix}, \quad B_r = \begin{pmatrix} 121.3816 \\ -1.3375 \\ -0.1201 \\ 0.0005 \end{pmatrix} \tag{13}$$

$$C_r = 10^3 \begin{pmatrix} 1.0083 & 0.3683 & 0.2624 & 0.0377 \end{pmatrix}, \quad D_r = 1218.70$$

A complete model must consider the dynamics of the power take-off system. Given the similarity between the Wavestar system and the WEC studied in this work, the PTO model proposed in [24] is used, where the PTO dynamics are modeled according to (8). By representing this model in the state space, with the parameters indicated in [24], the matrices (14) have been obtained. For this purpose, the observable canonical form has been chosen. Thus, one of the state variables corresponds to the output of the PTO system, so MPCs may impose restrictions on the output of the actuator.

$$A_{pto} = \begin{pmatrix} -8.7965 & 1.0000 \\ -157.9137 & 0 \end{pmatrix}, \quad B_{pto} = \begin{pmatrix} 0 \\ 157.9137 \end{pmatrix}, \quad C_{pto} = \begin{pmatrix} 1 & 0 \end{pmatrix}, \quad D_{pto} = 0 \quad (14)$$

After that, in this paper, we propose (15) as one of the design models, and the MPCs can impose restrictions on the following state variables: force applied by the PTO (x_{pto_1}), oscillation speed (w) and buoy position (z). This model is similar to the one proposed in [8,12], but also considers the PTO dynamics.

$$\underbrace{\begin{pmatrix} \dot{z} \\ \dot{w} \\ \dot{x}_r \\ \dot{x}_{pto} \end{pmatrix}}_{\dot{x}} = \underbrace{\begin{pmatrix} 0 & 1 & 0 & 0 \\ -\dfrac{k_{res}}{m_T} & -\dfrac{1}{m_T}D_r & -\dfrac{1}{m_T}C_r & \dfrac{1}{m_T}C_{pto} \\ 0 & B_r & A_r & 0 \\ 0 & 0 & 0 & A_{pto} \end{pmatrix}}_{A_{WEC}} \underbrace{\begin{pmatrix} z \\ w \\ x_r \\ x_{pto} \end{pmatrix}}_{x} + \underbrace{\begin{pmatrix} 0 \\ 0 \\ 0 \\ B_{pto} \end{pmatrix}}_{B_{WEC_u}} F_u + \underbrace{\begin{pmatrix} 0 \\ \dfrac{1}{m_T} \\ 0 \\ 0 \end{pmatrix}}_{B_{WEC_{F_e}}} F_e \quad (15)$$

$$\underbrace{\begin{pmatrix} z \\ w \end{pmatrix}}_{y} = \underbrace{\begin{pmatrix} 1 & 0 & 0 & 0 \\ 0 & 1 & 0 & 0 \end{pmatrix}}_{C_{WEC}} \underbrace{\begin{pmatrix} z \\ w \\ x_r \\ x_{pto} \end{pmatrix}}_{x}$$

where $m_T = m + m_\infty$ and the matrices I (unit matrix) and zero have the required size according to their location, and the parameters not yet presented are listed in Table 2.

Table 2. Parameters of the models (15) and (16) for the WEC system.

Symbol	Description	Value
k_{res}	Hydrostatic restoring coefficient	809,325 N/m
B_{aprox}	Stationary approximation to system damping	21,497 N s/m
m	Mass of water displaced by the buoy at rest	241,601.9 kg
m_∞	Mass of water added	167,700 kg

Simplified Model for the Design

In this paper, as in others works [9–11,13,14], a simplified model (16) is used for the design of some MPCs. It differs from the previous model (15) in that it does not take into account the dynamic of the radiation force or the PTO dynamics. Figure 3 shows a comparison of the outputs of the simplified model and the complete model.

$$
\underbrace{\begin{pmatrix} \dot{z} \\ \dot{w} \end{pmatrix}}_{\dot{x}} = \underbrace{\begin{pmatrix} 0 & 1 \\ -\dfrac{k_{res}}{m_T} & -\dfrac{B_{aprox}}{m_T} \end{pmatrix}}_{A_{WECr}} \underbrace{\begin{pmatrix} z \\ w \end{pmatrix}}_{x} + \underbrace{\begin{pmatrix} 0 \\ \dfrac{1}{m_T} \end{pmatrix}}_{B_{WECr}} (F_u + F_e)
$$

$$
\underbrace{\begin{pmatrix} z \\ w \end{pmatrix}}_{y} = \underbrace{\begin{pmatrix} 1 & 0 \\ 0 & 1 \end{pmatrix}}_{C_{WECr}} \underbrace{\begin{pmatrix} z \\ w \end{pmatrix}}_{x}
$$

(16)

where $m_T = m + m_\infty$, and the parameter values are listed in Table 2.

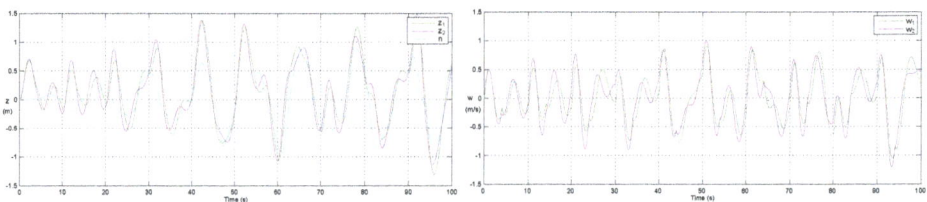

Figure 3. Comparison of models: complete ($z1$, $w1$) vs. simplified ($z2$, $w2$). The height of the wave is n.

3. Model Predictive Control for the Point Absorber WEC

This section details the design of the commonly-used MPCs for PAWs. Moreover, in this work, two MPCs are proposed based on the addition of an embedded integrator. In order to make a complete design, all the MPCs take into account constraints and the possibility of relaxing them, in the case of non-feasibility, in their cost functions. In particular, these constraints are applied to the control force of the PTO system, the position of the buoy and its oscillation speed. However, in the case of non-feasibility, only soft-constraints are applied to the position and oscillation speed of the system; due to the PTO being an actuator whose physical limit cannot be exceeded. Finally, for each controller, a sampling period of T_m is set, which is used to discretize the mathematical model using the zero order hold (ZOH) approximation.

3.1. MPC_1

The cost function that minimizes this controller considers the maximization of extracted power directly; as the design model is used (16), for which the state vector estimated for a prediction horizon M and a control horizon N is defined according to Equation (17) [18,19].

$$
X = \underbrace{\begin{bmatrix} A \\ A^2 \\ A^3 \\ \vdots \\ A^M \end{bmatrix}}_{Jx_{(Mn \times n)}} x_k + \underbrace{\begin{bmatrix} B & 0 & 0 & \cdots & 0 \\ AB & B & 0 & \cdots & 0 \\ A^2B & AB & B & \cdots & 0 \\ \vdots & \vdots & \vdots & \ddots & \vdots \\ A^{M-1}B & A^{M-2}B & A^{M-3}B & \cdots & A^{M-N}B \end{bmatrix}}_{Ju_{(Mn \times N)}} (F_{pto} + F_e)
$$

(17)

where X represents the estimated state vector for a prediction horizon M, x_k represents the state vector at the current instant, the matrices A and B are obtained from the model (16) discretized (ZOH), n is the order of the model and F_{pto} and F_e are vectors that contain the force applied by the PTO and the excitation force for whole control horizon N, respectively.

In a more compact form, the above equation can be expressed as:

$$X = J_x x_k + J_u(F_{pto} + F_e) \tag{18}$$

On the other side, as defined in [10,12,26], the mechanical power generated is given by (19). The expression of the power generated for the whole prediction horizon is obtained (20).

$$P_{gen}(t) = -w(t)F_{pto}(t) \tag{19}$$

$$P_{gen} = -W^T F_{pto} \tag{20}$$

where W and F_{pto} are vectors with length M, which represent the oscillation speed and control force for the whole prediction horizon, respectively.

By replacing (18) in (20),

$$P_{gen} = -(S_w X)^T F_{pto}$$
$$P_{gen} = -(S_w(J_x x_k + J_u F_{pto} + J_u F_e))^T F_{pto} \tag{21}$$

where S_w is a selector matrix for speed w (size $M \times Mn$).

Developing (21) and grouping in terms of least squares, the cost function is obtained:

$$J(F_{pto}) = \frac{1}{2} F_{pto}{}^T \underbrace{\left(J_u{}^T S_w{}^T + R\right)}_{H} F_{pto} + \frac{1}{2} \underbrace{\left(F_e{}^T J_u{}^T S_w{}^T + x_k{}^T J_x{}^T S_w{}^T\right)}_{b} F_{pto} \tag{22}$$

where the matrix R weights the control effort.

In addition, constraints are imposed to: force demanded for the PTO position and oscillation speed of the buoy (24). Therefore, the cost function (22) with constraints is defined as:

$$J(F_{pto}) = \frac{1}{2} F_{pto}{}^T H F_{pto} + \frac{1}{2} b F_{pto}$$
$$A_g F_{pto} \le B_g \tag{23}$$

where $A_g = [A_1 \quad A_2 \quad A_3]^T$ and $B_g = [B_1 \quad B_2 \quad B_3]^T$; see Equation (24).

$$\underbrace{\begin{bmatrix} I \\ -I \end{bmatrix}}_{A_1} F_{pto} \le \underbrace{\begin{bmatrix} F_{PTO_{max}} \\ -F_{PTO_{min}} \end{bmatrix}}_{B_1}, \quad \underbrace{\begin{bmatrix} S_z J_u \\ -S_z J_u \end{bmatrix}}_{A_2} F_{pto} \le \underbrace{\begin{bmatrix} Z_{n_{max}} - S_z(J_x x_k + J_u F_e) \\ -Z_{n_{min}} + S_z(J_x x_k + J_u F_e) \end{bmatrix}}_{B_2}$$

$$\underbrace{\begin{bmatrix} S_w J_u \\ -S_w J_u \end{bmatrix}}_{A_3} F_{pto} \le \underbrace{\begin{bmatrix} W_{n_{max}} - S_w(J_x x_k + J_u F_e) \\ -W_{n_{min}} + S_w(J_x x_k + J_u F_e) \end{bmatrix}}_{B_3} \tag{24}$$

where S_z is a selector matrix for position (size $M \times Mn$), the vectors $F_{PTO_{max}}$ and $F_{PTO_{min}}$ (size $N \times 1$) define the nominal limits of the force applied by the PTO and the vectors $W_{n_{max}}$, $W_{n_{min}}$, $Z_{n_{max}}$ and $Z_{n_{min}}$ (size $M \times 1$) define the nominal limits of the buoy position and oscillation speed, respectively. In addition, the matrices I (unit matrix) and 0 have the required size according to their location.

In the case of non-feasibility, soft constraints to the position and oscillation speed of the system are applied. Thus, the cost function (22) would be defined as:

$$J(F_{pto}, \varepsilon_z, \varepsilon_w) = \frac{1}{2} F_{pto}{}^T H F_{pto} + b F_{pto} + \varepsilon_z{}^T W_{\varepsilon_z} \varepsilon_z + \varepsilon_w{}^T W_{\varepsilon_w} \varepsilon_w \tag{25}$$

where ε_z and ε_w represent the relaxation applied to position and speed along the prediction horizon M and the matrices W_{ε_z} and W_{ε_w} (size $M \times M$) weight these slacks.

By regrouping terms and adding soft restrictions, the cost function (25) can be expressed as:

$$J(\beta) = \frac{1}{2}\beta^T \begin{bmatrix} H & 0 & 0 \\ 0 & W_{\varepsilon z} & 0 \\ 0 & 0 & W_{\varepsilon w} \end{bmatrix} \beta + \begin{bmatrix} b & 0 & 0 \end{bmatrix} \beta \tag{26}$$

$$A_g \beta \leq B_g$$

where $\beta = \begin{bmatrix} F_{pto} & \varepsilon_z & \varepsilon_w \end{bmatrix}^T$, with size $(N + 2M) \times 1$, $A_g = \begin{bmatrix} A_1 & A_2 & A_3 & A_4 & A_5 \end{bmatrix}^T$ and $B_g = \begin{bmatrix} B_1 & B_2 & B_3 & B_4 & B_5 \end{bmatrix}^T$; see Equation (27). Matrices 0 have the required size according to their location.

$$\underbrace{\begin{bmatrix} S_z J_u & -I & 0 \\ -S_z J_u & -I & 0 \end{bmatrix}}_{A_2} \beta \leq \underbrace{\begin{bmatrix} Z_{n_{max}} - S_z(J_x x_k + J_u F_e) \\ -Z_{n_{min}} + S_z(J_x x_k + J_u F_e) \end{bmatrix}}_{B_2}, \quad \underbrace{\begin{bmatrix} S_w J_u & 0 & I \\ -S_w J_u & 0 & I \end{bmatrix}}_{A_3} \beta \leq \underbrace{\begin{bmatrix} W_{n_{max}} - S_w(J_x x_k + J_u F_e) \\ -W_{n_{min}} + S_w(J_x x_k + J_u F_e) \end{bmatrix}}_{B_3}$$

$$\underbrace{\begin{bmatrix} I & 0 & 0 \\ -I & 0 & 0 \end{bmatrix}}_{A_1} \beta \leq \underbrace{\begin{bmatrix} F_{PTO_{max}} \\ -F_{PTO_{min}} \end{bmatrix}}_{B_1}, \quad \underbrace{\begin{bmatrix} 0 & I & 0 \\ 0 & -I & 0 \end{bmatrix}}_{A_4} \beta \leq \underbrace{\begin{bmatrix} \kappa_z \\ 0 \end{bmatrix}}_{B_4}, \quad \underbrace{\begin{bmatrix} 0 & 0 & I \\ 0 & 0 & -I \end{bmatrix}}_{A_5} \beta \leq \underbrace{\begin{bmatrix} \kappa_w \\ 0 \end{bmatrix}}_{B_5} \tag{27}$$

where κ_z and κ_w are vectors (size $M \times 1$) that represent the maximum slack allowed for position and oscillation speed, respectively. The matrices I (unit matrix) and 0 have the required size according to their location.

3.2. MPC$_2$

This controller uses the simplified model (16). Its cost function is based on maximizing the extracted power by tracking a setpoint for the oscillation speed (w_{ref}) along a prediction horizon M,

$$J = (\tilde{w} - w_{ref})^T Q(\tilde{w} - w_{ref}) + F_{pto}{}^T R F_{pto} \tag{28}$$

where Q and R are diagonal matrices of size $(M \times M)$ and $(N \times N)$ that weight the tracking error and the control effort, respectively.

Substituting (18) in (28), the cost function for this controller can be written as (29).

$$J(F_{pto}) = (\underbrace{J_x x_k + J_u F_{pto} + J_u F_e}_{f} - w_{ref})^T Q(\underbrace{J_x x_k + J_u F_{pto} + J_u F_e}_{f} - w_{ref}) + F_{pto}{}^T R F_{pto} \tag{29}$$

By developing the cost function (29) and grouping terms, the following expression is obtained:

$$J(F_{pto}) = F_{pto}{}^T \underbrace{(J_u^T \delta J_u + R)}_{H} F_{pto} + 2\underbrace{(f - w_{ref})^T Q J_u}_{b} F_{pto} + \underbrace{(f - w_{ref})^T Q(f - w_{ref})}_{l} \tag{30}$$

Note that the term l can be ignored when the cost function is minimized, because it does not depend on the variable to be optimized (F_{pto}),

$$J(F_{pto}) = \frac{1}{2}F_{pto}{}^T \underbrace{(J_u^T \delta J_u + R)}_{H} F_{pto} + \underbrace{(J_x x_k + J_u F_{pto} + J_u F_e - w_{ref})^T Q J_u}_{b} F_{pto} \tag{31}$$

The constraints imposed on this cost function can be expressed in the same way as in the MPC_1 controller; using (23) for hard constraints and (26) for soft constraints. On the other hand, the reference trajectory, or setpoint for the oscillation speed, is defined by the approach proposed in [4], $w_{ref} = F_e/2B_{aprox}$.

3.3. MPC$_3$

This approach is a contribution made in this work. This controller uses the simplified model (16) to which an embedded integrator has been added according to the theory of predictive controllers in the state space [18,19]. Its cost function is based on the maximization of the extracted power through the tracking of a setpoint for the oscillation speed. To add the integrator, it is necessary to multiply the model (16) discretized by the operator $\triangle = 1 - z^{-1}$. Regrouping terms, an extended state vector is defined as:

$$
\underbrace{\begin{bmatrix} \triangle x(t+1) \\ y(t+1) \end{bmatrix}}_{x_e(t+1)} = \underbrace{\begin{bmatrix} A & 0 \\ CA & I \end{bmatrix}}_{A_e} \underbrace{\begin{bmatrix} \triangle x(t) \\ y(t) \end{bmatrix}}_{x_e(t)} + \underbrace{\begin{bmatrix} B \\ CB \end{bmatrix}}_{B_e} (\triangle F_{pto}(t) + \triangle F_e(t))
$$

$$
y(t) = \underbrace{\begin{bmatrix} 0 & I \end{bmatrix}}_{C_e} \underbrace{\begin{bmatrix} \triangle x(t) \\ y(t) \end{bmatrix}}_{x_e(t)}
$$

(32)

where the output vector $y(t)$ is formed by the position and oscillation speed of the WEC system, $\triangle x(t)$ represents the state vector increment, the matrices A, B and C are from the model (16) discretized (ZOH) and the matrices 0 have the required size according to their location.

Using the extended model (32), the prediction of the outputs (z, w) is defined for a prediction horizon M and a control horizon N according to:

$$
Y = \underbrace{\begin{bmatrix} CA \\ CA^2 \\ CA^3 \\ \vdots \\ CA^M \end{bmatrix}}_{F_{(2M \times n_e)}} X_e + \underbrace{\begin{bmatrix} CB & 0 & 0 & \cdots & 0 \\ CAB & CB & 0 & \cdots & 0 \\ CA^2B & CAB & CB & \cdots & 0 \\ \vdots & \vdots & \vdots & \ddots & \vdots \\ CA^{M-1}B & CA^{M-2}B & CA^{M-3}B & \cdots & CA^{M-N}B \end{bmatrix}}_{G_{(2M \times N)}} (\triangle F_{pto} + \triangle F_e)
$$

(33)

where $n_e = n + j$ represents the order of the extended model and j its number of outputs. In the matrices A, B and C, the sub-index e has been omitted to get a clearer notation.

By adding the embedded integrator, this controller minimizes a cost function that gets the optimal increase in control force (F_{pto}) for the full control horizon N,

$$
J = (\tilde{w} - w_{ref})^T Q (\tilde{w} - w_{ref}) + \triangle F_{pto}{}^T R \triangle F_{pto}
$$

(34)

where Q and R are diagonal matrices of size $(M \times M)$ and $(N \times N)$ that weigh the tracking error and the control effort, respectively.

Replacing the output prediction (33) in (34) and obviating the independent term of $\triangle F_{pto}$:

$$
J(\triangle F_{pto}) = \frac{1}{2} \triangle F_{pto}{}^T \underbrace{(G^T \delta G + R)}_{H} \triangle F_{pto} + \underbrace{(F x_{e_k} + G \triangle F_e - w_{ref})^T Q G}_{b} \triangle F_{pto}
$$

(35)

In addition, constraints are added to the demanded force on the PTO, position and oscillation speed of the system. Therefore, the cost function (35) subject to the restrictions is defined as:

$$
J(\triangle F_{pto}) = \frac{1}{2} \triangle F_{pto}{}^T H \triangle F_{pto} + b \triangle F_{pto}
$$

$$
A_g \triangle F_{pto} \le B_g
$$

(36)

where $A_g = [A_1 \quad A_2 \quad A_3]^T$ and $B_g = [B_1 \quad B_2 \quad B_3]^T$; see Equation (37).

$$\underbrace{\begin{bmatrix} T \\ -T \end{bmatrix}}_{A_1} \Delta F_{pto} \le \underbrace{\begin{bmatrix} F_{PTO_{max}} \\ -F_{PTO_{min}} \end{bmatrix}}_{B_1}, \quad \underbrace{\begin{bmatrix} S_z G \\ -S_z G \end{bmatrix}}_{A_2} \Delta F_{pto} \le \underbrace{\begin{bmatrix} Z_{n_{max}} - S_z(Fx_{e_k} + G\Delta F_e) \\ -Z_{n_{min}} + S_z(Fx_{e_k} + G\Delta F_e) \end{bmatrix}}_{B_2}$$

$$\underbrace{\begin{bmatrix} S_w G \\ -S_w G \end{bmatrix}}_{A_3} \Delta F_{pto} \le \underbrace{\begin{bmatrix} W_{n_{max}} - S_w(Fx_{e_k} + G\Delta F_e) \\ -W_{n_{min}} + S_w(Fx_{e_k} + G\Delta F_e) \end{bmatrix}}_{B_3}$$

$$(37)$$

where S_z and S_w are selector matrices for z and w (size $M \times Mn$), T is a lower triangular matrix (size $M \times N$), the vectors $F_{PTO_{max}}$ and $F_{PTO_{min}}$ (size $N \times 1$) define the nominal limits of the force applied by the PTO and the vectors $W_{n_{max}}$, $W_{n_{min}}$, $Z_{n_{max}}$ and $Z_{n_{min}}$ (size $M \times 1$) define the nominal limits of the buoy position and oscillation speed, respectively. The matrices I (unit matrix) and 0 have the required size according to their location.

In the case of non-feasibility in the cost function (36), soft constraints are applied,

$$J(\beta) = \frac{1}{2}\beta^T \begin{bmatrix} H & 0 & 0 \\ 0 & W_{\varepsilon_z} & 0 \\ 0 & 0 & W_{\varepsilon_w} \end{bmatrix} \beta + \begin{bmatrix} b & 0 & 0 \end{bmatrix} \beta$$

$$(38)$$

$$A_g \beta \le B_g$$

where $A_g = [A_1 \quad A_2 \quad A_3 \quad A_4 \quad A_5]^T$ and $B_g = [B_1 \quad B_2 \quad B_3 \quad B_4 \quad B_5]^T$; see Equation (38).

$$\underbrace{\begin{bmatrix} T & 0 & 0 \\ -T & 0 & 0 \end{bmatrix}}_{A_1} \beta \le \underbrace{\begin{bmatrix} F_{PTO_{max}} \\ -F_{PTO_{min}} \end{bmatrix}}_{B_1}$$

$$\underbrace{\begin{bmatrix} S_z G & -I & 0 \\ -S_z G & -I & 0 \end{bmatrix}}_{A_2} \beta \le \underbrace{\begin{bmatrix} Z_{n_{max}} - S_z(Fx_{e_k} + G\Delta F_e) \\ -Z_{n_{min}} + S_z(Fx_{e_k} + G\Delta F_e) \end{bmatrix}}_{B_2}, \quad \underbrace{\begin{bmatrix} 0 & I & 0 \\ 0 & -I & 0 \end{bmatrix}}_{A_4} \beta \le \underbrace{\begin{bmatrix} \kappa_z \\ 0 \end{bmatrix}}_{B_4}$$

$$(39)$$

$$\underbrace{\begin{bmatrix} S_w G & 0 & I \\ -S_w G & 0 & I \end{bmatrix}}_{A_3} \beta \le \underbrace{\begin{bmatrix} W_{n_{max}} - S_w(Fx_{e_k} + G\Delta F_e) \\ -W_{n_{min}} + S_w(Fx_{e_k} + G\Delta F_e) \end{bmatrix}}_{B_3}, \quad \underbrace{\begin{bmatrix} 0 & 0 & I \\ 0 & 0 & -I \end{bmatrix}}_{A_5} \beta \le \underbrace{\begin{bmatrix} \kappa_w \\ 0 \end{bmatrix}}_{B_5}$$

where κ_z and κ_w are vectors (size $M \times 1$) that represent the maximum slack allowed for z and w, respectively. The matrices I (unit matrix) and 0 have the required size according to their location.

3.4. MPC$_4$

This controller is made using the model (15). The cost function that minimizes this controller is focused on the maximization of extracted power directly. The matrix development needed to express this controller as a least squares problem is analogous to that presented for controller MPC_1. Although, when the PTO dynamics are taken into account, Equation (18) should be redefined as:

$$X = J_x x_k + J_u F_{pto} + J_f F_e \tag{40}$$

where J_x is a matrix already defined in Equation (17), while the matrices J_u and J_f are given by:

$$J_u = \begin{bmatrix} B_u & 0 & 0 & \cdots & 0 \\ AB_u & B_u & 0 & \cdots & 0 \\ A^2 B_u & AB_u & B_u & \cdots & 0 \\ \vdots & \vdots & \vdots & \ddots & \vdots \\ A^{M-1} B_u & A^{M-2} B_u & A^{M-3} B_u & \cdots & A^{M-N} B_u \end{bmatrix}$$

$$J_f = \begin{bmatrix} B_{F_e} & 0 & 0 & \cdots & 0 \\ AB_{F_e} & B_{F_e} & 0 & \cdots & 0 \\ A^2 B_{F_e} & AB_{F_e} & B_{F_e} & \cdots & 0 \\ \vdots & \vdots & \vdots & \ddots & \vdots \\ A^{M-1} B_{F_e} & A^{M-2} B_{F_e} & A^{M-3} B_{F_e} & \cdots & A^{M-N} B_{F_e} \end{bmatrix}$$

(41)

where A, B_u and B_{F_e} are obtained by discretizing (ZOH) A_{WEC}, B_{WEC_u} and $B_{WEC_{F_e}}$ of the model (15).

The cost function to be minimized by this controller can be expressed according to (42). Its development is analogous to that carried out for the controller MPC_1.

$$J(F_{pto}) = \frac{1}{2} F_{pto}^T \underbrace{(J_u^T S_w^T + R)}_{H} F_{pto} + \frac{1}{2} \underbrace{(F_e^T J_f^T S_w^T + x_k^T J_x^T S_w^T)}_{b} F_{pto}$$

(42)

In addition, the nominal constraints imposed on the system must be added, which are defined in the same way as in the controller MPC_1, Equation (24). However, in this case, it must be considered that the state vector prediction is given by Equation (40). Therefore, the cost function (42) subject to the constraints is defined as:

$$J(F_{pto}) = \frac{1}{2} F_{pto}^T H F_{pto} + \frac{1}{2} b F_{pto}$$
$$A_g F_{pto} \leq B_g$$

(43)

where $A_g = [A_1 \quad A_2 \quad A_3]^T$ and $B_g = [B_1 \quad B_2 \quad B_3]^T$; see Equation (24).

Finally, in the case of non-feasibility in the function (48), soft constraints will be applied to the system. These can be expressed in a similar way to the development shown for MPC_1, Equation (27). However, in this case, it must be considered that the prediction of the state vector is given by Equation (40). Therefore, the cost function of this controller subject to soft constraints is given by:

$$J(\beta) = \frac{1}{2} \beta^T \begin{bmatrix} H & 0 & 0 \\ 0 & W_{\epsilon z} & 0 \\ 0 & 0 & W_{\epsilon w} \end{bmatrix} \beta + \begin{bmatrix} b & 0 & 0 \end{bmatrix} \beta$$
$$A_g \beta \leq B_g$$

(44)

where $A_g = [A_1 \quad A_2 \quad A_3 \quad A_4 \quad A_5]^T$ and $B_g = [B_1 \quad B_2 \quad B_3 \quad B_4 \quad B_5]^T$; see Equation (27).

3.5. MPC_5

This controller is another contribution of this work. It uses the model (15), to which an embedded integrator has been added according to the theory of predictive controllers in the state space [18,19]. Its cost function to minimize is based on the maximization of the extracted power through the tracking of a setpoint for the oscillation speed of the system w_{ref}. As in the MPC_3, the state vector is extended by adding an embedded integrator (32). Despite taking into account the PTO dynamics, the prediction of the output vector for the full prediction horizon M is defined by:

$$Y = G_u \triangle F_{pto} + \underbrace{F x_{e_k} + G_f \triangle F_e}_{f} \tag{45}$$

where F is a matrix already defined in Equation (33), and the matrices G_u and G_f are given by:

$$G_u = \begin{bmatrix} C_e B_{e_u} & 0 & 0 & \cdots & 0 \\ C_e A_e B_{e_u} & C_e B_{e_u} & 0 & \cdots & 0 \\ C_e A_e{}^2 B_{e_u} & C_e A_e B_{e_u} & C_e B_{e_u} & \cdots & 0 \\ \vdots & \vdots & \vdots & \ddots & \vdots \\ C_e A_e{}^{M-1} B_{e_u} & C_e A_e{}^{M-2} B_{e_u} & C_e A_e{}^{M-3} B_{e_u} & \cdots & C_e A_e{}^{M-N} B_{e_u} \end{bmatrix}$$

$$\tag{46}$$

$$G_f = \begin{bmatrix} C_e B_{e_{F_e}} & 0 & 0 & \cdots & 0 \\ C_e A_e B_{e_{F_e}} & C_e B_{e_{F_e}} & 0 & \cdots & 0 \\ C_e A_e{}^2 B_{e_{F_e}} & C_e A_e B_{e_{F_e}} & C_e B_{e_{F_e}} & \cdots & 0 \\ \vdots & \vdots & \vdots & \ddots & \vdots \\ C_e A_e{}^{M-1} B_{e_{F_e}} & C_e A_e{}^{M-2} B_{e_{F_e}} & C_e A_e{}^{M-3} B_{e_{F_e}} & \cdots & C_e A_e{}^{M-N} B_{e_{F_e}} \end{bmatrix}$$

where B_{e_u} and $B_{e_{F_e}}$ are obtained by discretizing the matrices that define the inputs of (15) extended.

Analogous to controller MPC_3, the cost function to be minimized can be expressed according to:

$$J(\triangle F_{pto}) = \frac{1}{2} \triangle F_{pto}{}^T \underbrace{\left(G_u{}^T \delta G_u + R \right)}_{H} \triangle F_{pto} + \underbrace{\left(F x_{e_k} + G_f \triangle F_e - w_{ref} \right)^T Q G_u}_{b} \triangle F_{pto} \tag{47}$$

In addition, it is necessary to add nominal constraints to the system, which are defined as in the controller MPC_3. However, in this case, it must be considered that the prediction of the system outputs is given by (45). Thus, the cost function (47) subject to the constraint is defined as:

$$J(\triangle F_{pto}) = \frac{1}{2} \triangle F_{pto}{}^T H \triangle F_{pto} + \frac{1}{2} b \triangle F_{pto}$$
$$A_g \triangle F_{pto} \leq B_g \tag{48}$$

where $A_g = [A_1 \quad A_2 \quad A_3]^T$ and $B_g = [B_1 \quad B_2 \quad B_3]^T$; see Equation (37).

Furthermore, in the case of the non-feasibility in the function (47), soft constraints are used. The soft constraints are expressed analogously to the development made for controller MPC_3, Equation (39). However, in this case, the prediction of the system output is given by (45). Therefore, the cost function of this controller subject to soft constraints is defined as:

$$J(\beta) = \frac{1}{2} \beta^T \begin{bmatrix} H & 0 & 0 \\ 0 & W_{\varepsilon_z} & 0 \\ 0 & 0 & W_{\varepsilon_w} \end{bmatrix} \beta + \begin{bmatrix} b & 0 & 0 \end{bmatrix} \beta \tag{49}$$
$$A_g \beta \leq B_g$$

where $A_g = [A_1 \quad A_2 \quad A_3 \quad A_4 \quad A_5]^T$ and $B_g = [B_1 \quad B_2 \quad B_3 \quad B_4 \quad B_5]^T$; see Equation (39).

3.6. Conventional Controllers

In order to make a comparative analysis of the proposed predictive controllers, this section presents the design of two of the controllers most commonly used in WEC systems [5,6]. Firstly, a resistive damping (RD) controller (or proportional control) has been tuned for the PTO system, whose control law is given by:

$$F_{pto}(k) = -K_{RD} w(k) \tag{50}$$

where k is the current sampling time and K_{RD} is a proportional gain (K_P), which must be tuned to maximize the power generated.

On the other hand, an I-P control has been designed whose control law is defined by Equation (51). A compromise between generated power and keeping the system within its nominal limits of operation is maintained by tuning the gains K_P and K_I.

$$
\begin{aligned}
F_{pto}(k) &= u_P(k) + u_I(k) \\
u_P(k) &= -K_P w(k) \\
u_I(k) &= K_I T_m(w_{ref}(k) - w(k)) + u_I(k-1)
\end{aligned}
\tag{51}
$$

where k is the current moment. The backward Euler method has been used to discretize the controller for a sampling period T_m.

4. Study about Performances and Robustness

This section presents the results obtained from an in-depth study of the performance and robustness of the five MPCs designs. This study assumes that the system state vector and the excitation force (F_e) for the whole prediction horizon (M) are known. The computer simulations have been carried out using Simulink, employing a fourth order Runge–Kutta integration method with a fixed step of one millisecond. In order to solve the optimization problems with constraints, associated with MPCs' design, the MATLAB quadprog function has been employed [35]. Note that although the quadprog function provides F_{pto} for the whole control horizon N, only the setpoint obtained for the current instant k is applied [18,19].

Due to the fact that the WEC system of the W2POWER platform has not yet been built, its operating limits and physical limits are not available. Therefore, these limits have been chosen on the basis of [12,15,16,24], whose WEC systems are similar to the system studied in this paper; see Table 3.

Table 3. Hard and soft constraints for the WEC system.

Symbol	Description	Value
$F_{PTO_{max}}$	Maximum stationary force for the power take-off system	450 KN
$F_{PTO_{min}}$	Minimum stationary force for the power take-off system	$-450\ KN$
$z_{n_{max}}$	Maximum nominal limit for the buoy position	1.25 m
$z_{n_{min}}$	Minimum nominal limit for the buoy position	-1.25 m
$w_{n_{max}}$	Maximum nominal limit for oscillation speed	1 m/s
$w_{n_{min}}$	Minimum nominal limit for oscillation speed	-1 m/s
$z_{f_{max}}$	Maximum physical limit for the buoy position	1.7 m
$z_{f_{min}}$	Minimum physical limit for the buoy position	-1.7 m
$w_{f_{max}}$	Maximum physical limit for oscillation speed	1.3 m/s
$w_{f_{min}}$	Minimum physical limit for oscillation speed	-1.3 m/s
κ_z	Maximum slack applied to the position nominal limit	0.45 m
κ_w	Maximum slack applied to the speed nominal limit	0.3 m/s

4.1. Performance Comparison

This section shows a comparison between the seven controllers designed. The controllers' performances are compared in terms of: average power generated, reduction of instantaneous power peaks, overshoot of nominal limits and control effort. A sea state defined by the JONSWAP spectrum [15,16], with a significant wave height of three meters and a peak period of 11 s, has been chosen as a realistic scenario for evaluating these characteristics. In order to obtain a truthful performance comparison, all predictive controllers must have the same information about the incoming wave, which is recorded in the prediction time t_f. To set the prediction horizon, two factors have been taken into account. Firstly, in [7,8,10], the authors used realistic sea states, and they set the prediction times between two and four seconds. Furthermore, in [8], the authors checked that t_f could also be reduced to three or four seconds without significant reduction of the harvested energy. Secondly, in [36], the authors showed how short-term wave forecasting models maintain good performance up to five seconds of prediction for wide wave spectra. Therefore, in this work, the prediction time chosen was three seconds, the same as the one set in [10]. With this t_f, a prediction horizon $M = 75$, a control horizon $N = 75$ and a sampling period $T_m = 0.04$ s has been set for all MPCs. The RD and I-P controllers have the same sampling period. On the other hand, as a result of a fine-tuning for this realistic sea state, the parameters of the controllers are listed in Table 4.

Figure 4 shows a comparison of the instantaneous mechanical powers generated by applying the seven controllers to the mathematical model (15). In this comparison, it can be seen how MPCs with an embedded integrator (MPC$_3$ and MPC$_5$) achieve more regular power than the MPCs most used for WEC systems, those whose optimization criteria maximize the extracted power directly (MPC$_1$ and MPC$_4$). In addition, Figure 5 shows how MPCs with embedded integrators achieve less overshoot in the control force applied by the PTO system than all other MPCs (reducing actuator overstress).

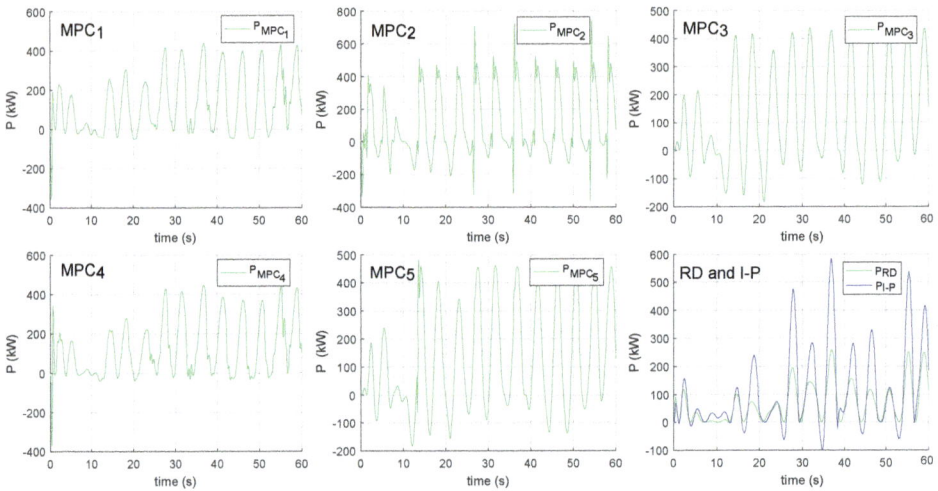

Figure 4. Simulation of the instantaneous mechanical powers generated by the WEC system when applying the seven controllers to the mathematical model (15).

Table 4. Control parameter set for the sea state defined by the JONSWAP spectrum (3 m of significant wave height and 11 s of peak period).

Controller	R	Q	W_{ε_z}	W_{ε_w}	K_P	K_I
MPC$_1$	4.500×10^{-7}	—	1.000×10^{10}	1.000×10^7	—	—
MPC$_2$	1.000×10^{-7}	9.150×10^4	1.000×10^7	2.000×10^9	—	—
MPC$_3$	5.000×10^{-5}	4.575×10^4	5.000×10^{10}	1.000×10^5	—	—
MPC$_4$	4.500×10^{-7}	—	1.000×10^8	1.000×10^4	—	—
MPC$_5$	5.000×10^{-5}	4.000×10^4	1.000×10^7	1.000×10^9	—	—
RD	—	—	—	—	7.902×10^5	—
I-P	—	—	—	—	7.050×10^5	2.228×10^4

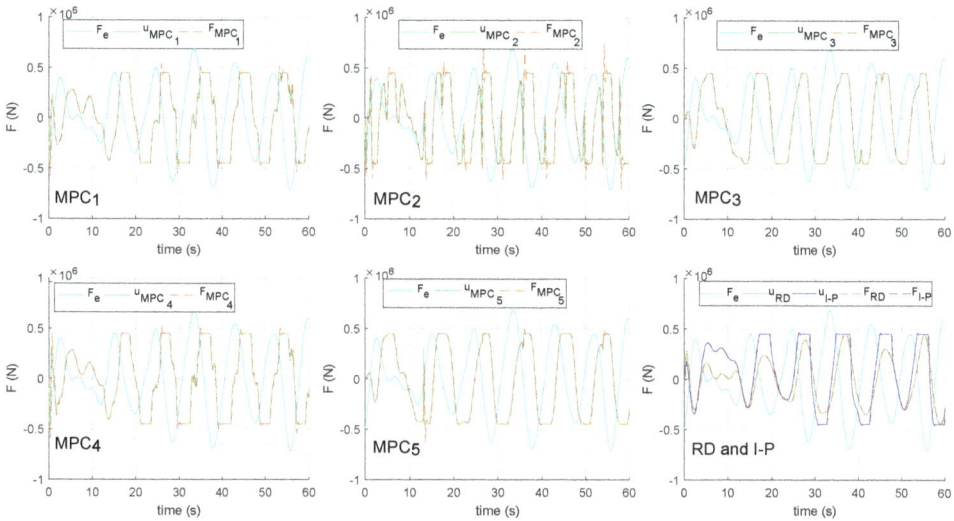

Figure 5. Simulation of: wave force, force setpoint demanded for the PTO and real force produced by the PTO, when applying the seven controllers to the mathematical model (15).

By applying the MPC$_2$ to the system, it gives the most irregular power; while the power generated with the MPC$_3$, designed from the same model and following the same optimization criteria, is much cleaner, with fewer occasional peaks. This is due to the fact that the MPC$_1$ is continuously applying soft constraints to the oscillation speed, because it does not carry out a good control of the force that the PTO system exerts on the WEC during the time (see Figure 5). On the other hand, Table 5 shows how the addition of the embedded integrator (MPC$_3$) considerably improves the behavior of the system with respect to that obtained by applying the MPC$_2$. Since, the MPC$_3$ does not exceed nominal limits of the position and oscillation speed on any occasion. Furthermore, the MPC$_3$ controller generates a clearer control signal than the MPC$_2$ (see Figure 5), and as a consequence, the underdamped response of the PTO system decreases greatly.

Table 5. Results obtained in the application of the seven controllers to the WEC system. The powers are expressed in kW. Note that $ONLP$ indicates overshoot of nominal limits for position and $ONLS$ indicates overshoot of nominal limits for speed.

Controller	$\bar{P}_{gen}$	$ONLP$	$ONLS$	$P_{gen_{Max}}$	$P_{gen_{Min}}$
MPC$_1$	127.60	0.0000	0.0023	437.48	−356.33
MPC$_2$	110.72	0.0000	0.0281	744.06	−364.97
MPC$_3$	125.53	0.0000	0.0000	444.24	−183.62
MPC$_4$	127.50	0.0000	0.0107	452.75	−369.27
MPC$_5$	129.01	0.0000	0.0413	481.53	−182.85
RD	67.02	0.0000	0.0000	258.88	−1.05
I-P	109.90	0.0073	0.0511	583.73	−107.34

Table 5 records the most significant quality indicators of the control performed by each controller. First, this table shows the average mechanical powers generated by the WEC system when applying each controller. In this aspect, the MPC$_5$ controller is the one that generates more power, followed very closely by the MPC$_1$, MPC$_4$ and, with a bit more distance, the MPC$_3$. On the other side, the indicators $ONLP$ and $ONLS$ quantify the area of overshoot from the nominal limits of the position and oscillation speed, respectively. In this sense, the MPC$_3$ (together with the RD control) provides the best behavior, since it does not apply slack to the nominal limits on any occasion, then it is followed by MPC$_1$.

This can be verified in Figures 6 and 7, which show a comparison between the positions and oscillation speeds obtained by applying the designed controllers to the mathematical model (15). As can be seen, all controllers keep the WEC system within its physical operating limits. Finally, the last two columns of Table 5 show the maximum and minimum power peaks obtained when applying each controller. In this aspect, ignoring the resistive damping control, MPCs with embedded integrators are once again the best performers. Note that these power peaks will cause an oversizing of: electrical machines, power electronics, accumulators, etc.

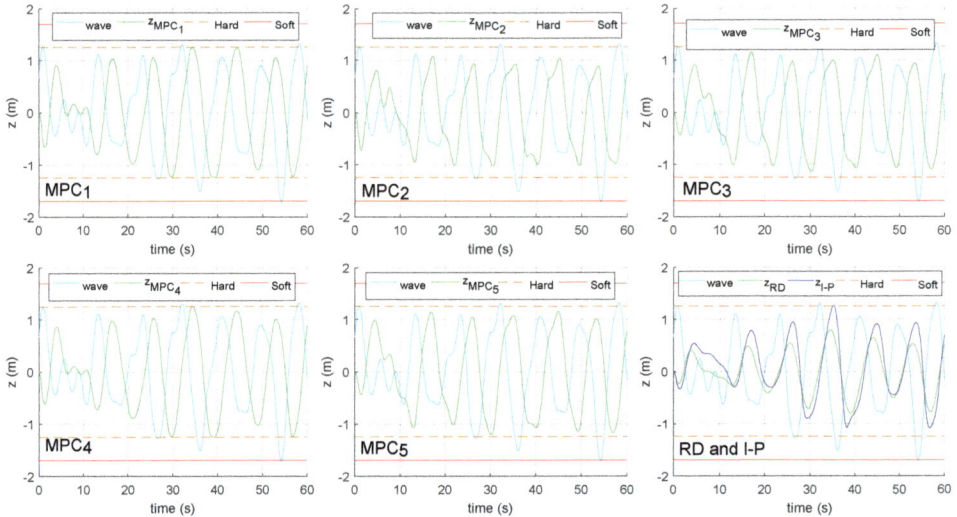

Figure 6. Simulation of the positions obtained in the WEC system when applying the seven controllers to the mathematical model (15).

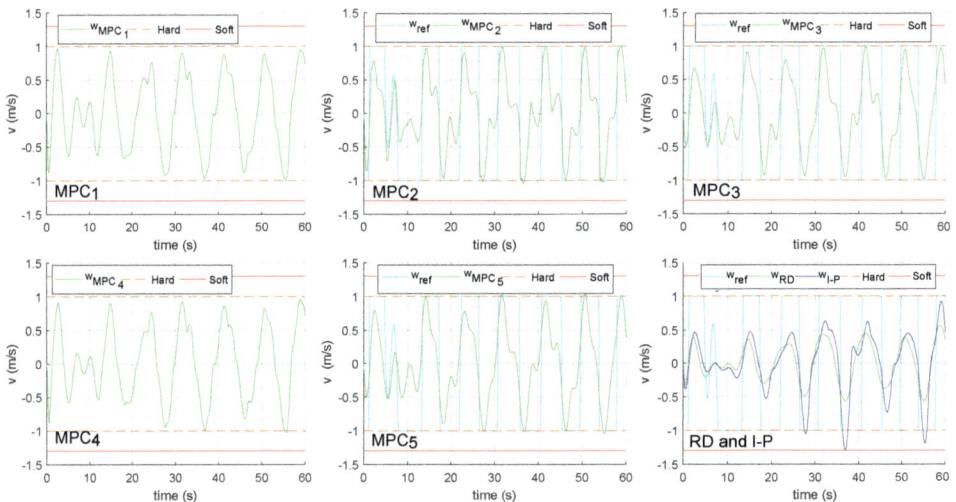

Figure 7. Simulation of the oscillation speeds obtained in the WEC system when applying the seven controllers to the mathematical model (15).

With respect to the two optimization criteria compared in this paper, it can be concluded that the MPCs that employ an optimization criterion based on the minimization of the error between w and w_{ref} should only be used if they are designed from a model with an embedded integrator. Thus, this approach achieves better performance (in terms of diminution of instantaneous power peaks and reduction of mechanical fatigue due to exceeding nominal limits) than the standard optimization criteria based on maximizing the extracted power directly. On the other hand, with respect to the use of a complete model or a simplified model, it can be concluded that (in terms of instantaneous power generated and reduction of mechanical fatigue) the use of a complete model does not improve the performance of MPCs for this WEC significantly. This is because, the increase in the average power generated is minimal, and only in the case of the MPC_1 and MPC_4, the use of a complete model reduces the overshoot of the PTO system a bit. Finally, the variation of the extracted mechanical power as a function of the prediction time t_f is studied. In particular, Figure 8 shows a comparison between MPC_3 and MPC_1. As can be seen, in both cases, after a prediction time of 3 s, the power generation does not improve significantly, while the computation effort increases. In addition, it should be pointed that MPC_3 does not have a monotonously increasing behavior that relates t_f to the power generated, as would be expected. Nevertheless, in this aspect, the MPC_3 is better than the MPC_1, because it can generate more power with less information of the future excitation force (up to 2.5 s).

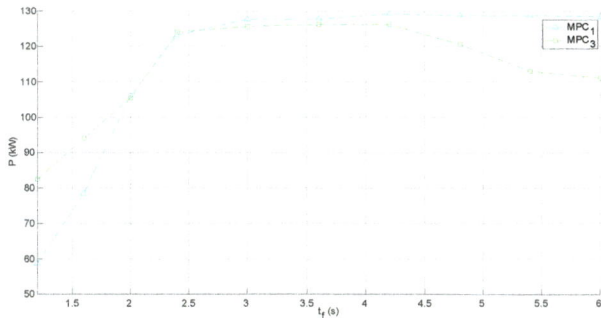

Figure 8. Variation of the extracted mechanical power as a function of the prediction time for MPC_1 and MPC_3 (control parameters listed in Table 4).

To conclude the performance study, by comparing the previous figures and the values recorded in Table 5, it can be seen how the MPCs almost double the average power generated by the WEC system with respect to that obtained by tuning an adequate resistive damping for the PTO system (RD control). In addition, like the RD control, the MPC_3 keeps the system within its nominal operating limits. On the other hand, it is also verified that the MPCs are superior in performances than the conventional I-P controller, in terms of: average mechanical powers generated, diminution of instantaneous power peaks and reduction of mechanical fatigue (due to exceeding nominal limits).

4.2. Robustness Comparison

Another contribution of this work, searching for the greatest realism in the comparative of the designed controllers, is that uncertainty is added to the complete system model (15) to the most significant identified parameters (52): added mass and dynamics of the radiation force, which have been obtained through the openWEC software, and the hydrostatic restoring coefficient K_{res} (a nonlinear parameter that has been linearized during the modeling of the WEC system).

$$F_{res}(t) = -k_{res}(1 + \triangle_{k_{res}})z(t), \quad F_{rm_{\infty}}(t) = m_{\infty}(1 + \triangle_{m_{\infty}})\ddot{w}(t), \quad \frac{Fr_{K_r}(s)}{W(s)} = (1 + \triangle_B)K_r(s) \quad (52)$$

where $\triangle$ represents the added uncertainty in each parameter.

Note that when modifying the physical parameters of the WEC system, the frequency response of the filter (9) is affected. Therefore, looking for a truthful comparison, it would be necessary to identify a new filter (6) for each added uncertainty. Given the high number of simulations required, applying different levels of uncertainty to each of the parameters, this is not feasible. For this reason, in this work, the Morison model (7) is used to define F_e, allowing one to modify the excitation force caused by the wave as a function of the added uncertainty more easily. Thus, when modifying the physical parameters of the system, the external force that the wave causes on the system also varies. Therefore, Equation (7) is redefined for this analysis as:

$$F_e(t) = m_\infty(1 + \triangle_{m_\infty})\ddot{\eta}(t) + B(1 + \triangle_B)\dot{\eta}(t) + k_{res}(1 + \triangle_{k_{res}})\eta(t) \tag{53}$$

It should be noted that Equation (53) uses a first and a second derivative of wave height. As a consequence, if the Morison model is used for the excitation force in a realistic sea state, it will amplify the high-frequency harmonics that are part of the wave. This is the opposite of the bandwidth of the WEC system provided by the openWEC software (see Figure 2b). For this reason, in this part of the paper, the simulations are performed in an irregular sea state formed by the fifteen sinusoidal components listed in Table 6.

Table 6. Sinusoidal components used for sea-state (values expressed in international units).

Component	Amplitude	Period	Phase	Component	Amplitude	Period	Phase
s_1	0.420	13.00	$-\pi$	s_9	0.200	9.00	0.00
s_2	0.520	12.50	1.5π	s_{10}	0.180	8.50	π
s_3	0.420	12.25	0.40	s_{11}	0.200	7.50	0.10
s_4	0.520	11.50	0.20	s_{12}	0.150	6.50	-0.77
s_5	0.450	11.25	0.11π	s_{13}	0.100	5.50	0.5π
s_6	0.300	10.50	-1.50	s_{14}	0.075	5.00	0.00
s_7	0.500	10.00	-0.33	s_{15}	0.020	3.70	0.12
s_8	0.210	9.50	0.78	—	—	—	—

In order to create an unfavorable test scenario for the MPCs, two factors have been adjusted. In the first place, the sea state recorded in Table 6 is not favorable to the controllers, because the wave force becomes more than twice the stationary force that the PTO system can apply (recorded in Table 3). Moreover, the prediction time t_f has been limited to a maximum of 1.5 s. After this, a fine-tuning has been made to all the MPCs looking for a balance between the generated power and the overshoot of the nominal limits of the WEC system. The result of this fine-tuning is listed in Table 7.

Table 7. Control parameters set for the sea state recorded in Table 7.

Controller	T_m [s]	T_f [s]	R	Q	W_{ε_z}	W_{ε_w}
MPC_1	0.05	1.5	1.1×10^{-7}	—	1.0×10^{10}	1.0×10^7
MPC_2	0.04	1.2	1.0×10^{-7}	2.15×10^4	1.0×10^7	2.0×10^9
MPC_3	0.04	1.2	5.0×10^{-5}	1.475×10^4	5.0×10^{10}	1.0×10^5
MPC_4	0.05	1.5	4.5×10^{-7}	—	1.0×10^8	1.0×10^4
MPC_5	0.02	0.6	5.0×10^{-5}	1.75×10^4	1.0×10^7	1.0×10^9

Note that for the same wave height, the excitation force can increase or decrease according to the uncertainty added in each parameter. Therefore, there will be situations where, for the sea state defined in Table 6, the controller cannot keep the system within its physical limits. This is because, the actuation force of the PTO system will be much lower than the excitation force. In this paper, this non-feasibility situation will be considered as the robustness limit that the controller can support. This limit is defined for the uncertainty added in each of the parameters (52). This non-feasibility situation with soft constraints does not mean that the closed-loop system becomes unstable, but that the controller cannot keep the WEC system within the physical limits defined in Table 3. Furthermore, in order to obtain a more complete analysis of how this uncertainty affects the closed-loop system, the average powers generated for each value of added uncertainty to each parameter are recorded. A large number of simulations has been carried out for this purpose; all of them have a duration of 120 s and use fourth order Runge–Kutta (RK4) integration method with an integration step of 1 ms.

Once the robustness study has been defined, Figure 9 shows the results obtained as a function of the added uncertainties; feasible limits obtained for the five MPCs and the variation of their mean power generated. With respect to the added uncertainty in m_∞, the MPC$_2$ is the least robust. Meanwhile, the MPC$_1$ offers the best features in a power-robustness ratio. However, it should be noted that when considering more reasonable added uncertainty values (interval $[-50, 50]$%), the MPC$_5$ extracts significantly more power than the others. It should also be noted that the controllers that directly maximize power in their cost function (MPC$_1$ and MPC$_4$) have the most predictable behavior with respect to the added uncertainty in m_∞. On the other hand, the MPC$_5$ offers the best features with respect to the uncertainty added to the dynamics of the radiation force (up to 400%). In contrast, the MPC$_2$ gives very bad results in this respect. The MPC$_1$ also gets good results, because it achieves a practically constant power production despite variations of $\triangle_B$.

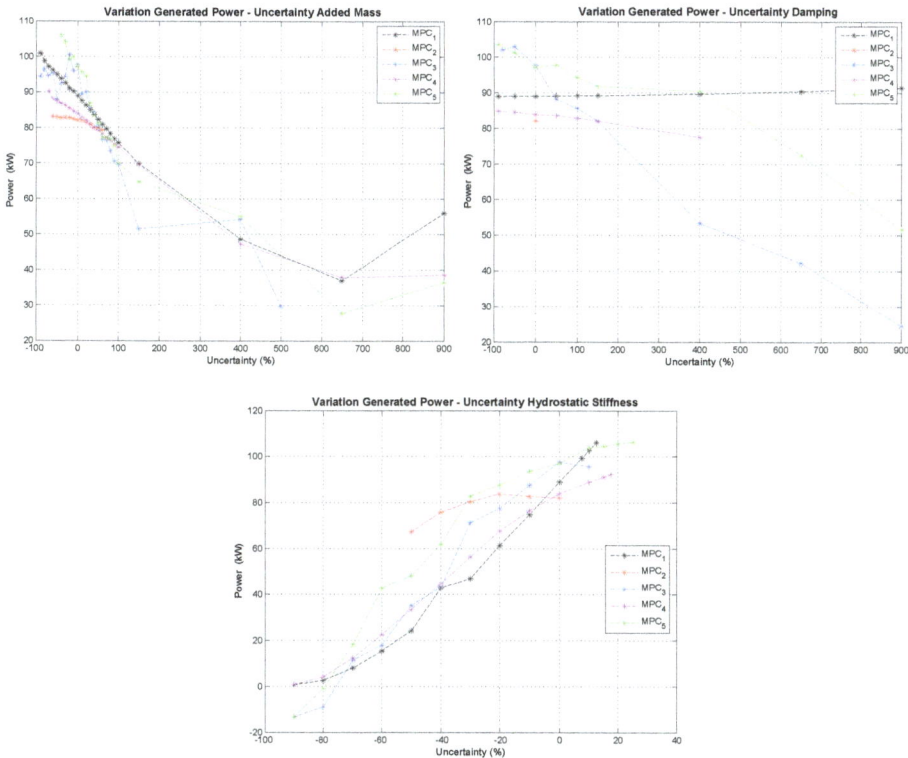

Figure 9. Variations of the average mechanical power generated as a function of the added uncertainty to: added mass, damping coefficient and hydrostatic restoring coefficient.

Finally, Figure 9 shows how the power generated by the different controllers varies according to the uncertainty added to the hydrostatic restoring coefficient of the system. In this aspect, it can be appreciated how the robustness of all the controllers is more limited. If the value of the coefficient k_{res} increases, the excitation force that the wave exerts on the system (53) increases proportionally.

Therefore, the margin of action of the PTO system decreases noticeably. Even so, the MPC$_5$ supports an added uncertainty of 25%, again being the one that provides the best robustness results even with the smallest prediction time t_f. Note that, for such added uncertainty, the excitation force becomes more than three-times the force that the PTO system can apply to the buoy; see Figure 10. After the MPC$_5$, the MPC$_4$ and MPC$_1$ get the best results, in this order.

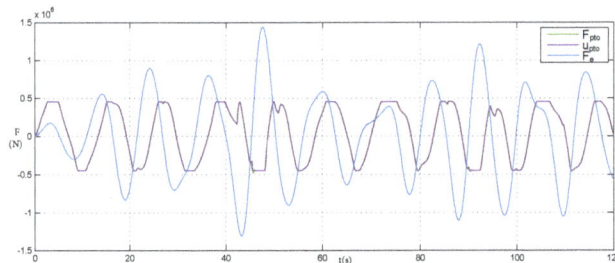

Figure 10. Wave force, force setpoint demanded and real force produced by the PTO, by applying the MPC$_5$ to the model (15), which has an added uncertainty to the hydrostatic restoring coefficient of 25%.

5. Conclusions

The interest in implementing MPCs in WEC systems is motivated by the need to increase the productive/economic viability of these systems. For this reason, in this work, five different predictive controllers have been designed. All these controllers allow minimizing mechanical fatigue by limiting the operating range of the WEC system by means of hard and soft constraints. The main contribution of this work is the study of performance and robustness carried out for the five MPCs designed. This study demonstrates how the addition of an embedded integrator to the design model improves the performance of the WEC device referring to: average power generated, diminution of instantaneous power peaks, reduction of mechanical fatigue and robustness of the closed-loop system, in comparison with the other MPCs. In addition, the MPCs have been compared with two conventional controllers for WEC systems (resistive damping and I-P control) in a realistic sea state defined by the JONSWAP spectrum. Predictive controllers have proven that, compared to these conventional controllers, they minimize mechanical fatigue due to overshoot of nominal limits and increase the mechanical power generated by this type of WEC system.

Author Contributions: R.G. was responsible for designing the predictive controllers, studying them and writing most of the paper. A.C. and M.J.L. performed the modeling of the WEC system, proposed the identification methodology for PAW systems and the comparative robustness, reviewed all the work done and wrote part of the document. Furthermore, M.J.L. and A.C. proposed the control problems to solve using the model predictive control strategy and participated in the design and analysis of the controllers.

Funding: This project is funded by the Spanish Ministry of Economy, Industry and Competitiveness within the framework of the State Plan for Scientific and Technical Research and Innovation 2013-2016/State Programme for R+D+I aimed at the challenges of society.

Acknowledgments: Thanks to all ORPHEO (Optimización de la Rentabilidad de Plataformas Híbridas de energía Eólica y de las Olas) project partners for giving us the opportunity and funds to research in this promising field.

Conflicts of Interest: The authors declare no conflict of interest.

References

1. Pecher, A.; Kofoed, J.P. *Handbook of Ocean Wave Energy*; Sprinder Open: Boca Raton, FL, USA, 2016; ISBN 978-3-319-39888-4.
2. The Ocean Energy Systems Technology Collaboration Programme. Available online: www.ocean-energy-systems.org/index.php (accessed on 4 May 2017).
3. Pelagic Power. Available online: www.pelagicpower.no/about.html (accessed on 6 May 2017).

4. Falnes, J. *Ocean Waves and Oscillating Systems*; Cambridge University Press: New York, NY, USA, 2002; ISBN 0-521-78211-2.

5. Drew, B.; Plummer, A.R.; Sahinkaya, M.N. A review of wave energy converter technology. *J. Power Energy* **2009**, *223*, 887–902. [CrossRef]

6. Valério, D.; Beirao, P.; Mendes, M.J.G.C.; da Costa, J.S. Comparison of control strategies performance for a Wave Energy Converter. In Proceedings of the 16th Mediterranean Conference on Control and Automation, Ajaccio, France, 25–27 June 2008; pp. 773–778.

7. Li, G. Predictive control of a wave energy converter with wave prediction using differential flatness. In Proceedings of the 54th IEEE Conference on Decision and Control, Osaka, Japan, 15–18 December 2015; pp. 3230–3235.

8. Andersen, P.; Pedersen, T.S.; Nielsen, K.M.; Vidal, E. Model Predictive Control of a Wave Energy Converter. In Proceedings of the 2015 IEEE Conference on Control Applications (CCA 2015), Sydney, Australia, 21–23 September 2015; pp. 1540–1545.

9. Brekken, T.K. On Model Predictive Control for a point absorber Wave Energy Converter. In Proceedings of the IEEE Trondheim PowerTech, Trondheim, Norway, 19–23 June 2011; pp. 1–8.

10. Richter, M.; Magana, M.E.; Sawodny, O.; Brekken, T.K.A. Nonlinear Model Predictive Control of a Point Absorber Wave Energy Converter. *IEEE Trans. Sustain. Energy* **2013**, *4*, 118–126. [CrossRef]

11. Li, G.; Belmont, M.R. Model predictive control of sea wave energy converters—Part I: A convex approach for the case of a single device. *Renew. Energy* **2014**, *69*, 453–463. [CrossRef]

12. Cavaglieri, D.; Bewley, T.R.; Previsic, M. Model Predictive Control leveraging Ensemble Kalman forecasting for optimal power take-off in wave energy conversion systems. In Proceedings of the American Control Conference, Chicago, IL, USA, 1–3 July 2015; Volume 2015, pp. 5224–5230.

13. Oetinger, D.; Magaña, M.E.; Member, S.; Sawodny, O. Decentralized Model Predictive Control for Wave Energy Converter Arrays. *IEEE Trans. Sustain. Energy* **2014**, *5*, 1099–1107. [CrossRef]

14. Li, G.; Belmont, M.R. Model predictive control of a sea wave energy converter: A convex approach. In Proceedings of the International Federation of Automatic Control (IFAC 2014), Cape Town, South Africa, 24–29 August 2014; Volume 19, pp. 11987–11992.

15. Soltani, M.N.; Sichani, M.T.; Mirzaei, M. Model Predictive Control of Buoy Type Wave Energy Converter. In Proceedings of the International Federation of Automatic Control (IFAC 2014), Cape Town, South Africa, 24–29 August 2014; Volume 47, pp. 11159–11164.

16. Starrett, M.; So, R.; Brekken, T.K.A.; McCall, A. Increasing power capture from multibody wave energy conversion systems using model predictive control. In Proceedings of the 2015 IEEE Conference on Technologies for Sustainability (SusTech), Ogden, UT, USA, 30 July–1 August 2015; pp. 20–26.

17. Lagoun, M.S.; Benalia, A.; Benbouzid, M.E.H. A predictive power control of Doubly fed induction generator for wave energy converter in irregular waves. In Proceedings of the 1st International Conference on Green Energy, Sfax, Tunisia, 25–27 March 2014; pp. 26–31.

18. Camacho, E.F.; Bordons, C. *Model Predictive Control*; Sprinder: Sevilla, Spain, 1998; ISBN 978-1-85233-694-3.

19. Wang, L. *Model Predictive Control System Desing and Implementation Using MATLAB*; Sprinder: Melbourne, Australia, 2009; ISBN 978-1-84882-330-3.

20. Openore. Available online: https://openore.org/2016/04/28/openwec/ (accessed on 14 May 2017).

21. Ricci, P. Time-Domain Models. In *Numerical Modelling of Wave Energy Converters*; Elsevier Inc.: New York, NY, USA, 2016; pp. 31–66, ISBN 978-0-12-803210-7.

22. Bozzi, S.; Miquel, A.M.; Antonini, A.; Passoni, G.; Archetti R. Modeling of a point absorber for energy conversion in Italian seas. *Energies* **2013**, *6*, 3033–3051. [CrossRef]

23. Hong, Y.; Eriksson, M.; Boström, C.; Waters, R. Impact of generator stroke length on energy production for a direct drivewave energy converter. *Energies* **2016**, *9*, 730. [CrossRef]

24. Hansen, R.H.; Kramer, M.M. Modelling and Control of the Wavestar Prototype. In Proceedings of the 9th European Wave and Tidal Energy Conference, Southampton, UK, 5–9 September 2011; pp. 1–10.

25. Kovaltchouk, T.; Multon, B.; BenAhmed, H.; Glumineau, A.; Aubry, J. Influence of control strategy on the global efficiency of a Direct Wave Energy Converter with electric Power Take-Off. In Proceedings of the Eighth International Conference and Exhibition on Ecological Vehicles and Renewable Energies, Monte Carlo, Monaco, 27–30 March 2013; pp. 1–10.

26. Fusco, F.; Ringwood, J.V. A study of the prediction requirements in real-time control of wave energy converters. *IEEE Trans. Sustain. Energy* **2012**, *3*, 176–184. [CrossRef]

27. Tedeschi, E.; Carraro, M.; Molinas, M.; Mattavelli, P. Effect of control strategies and power take-off efficiency on the power capture from sea waves. *IEEE Trans. Energy Convers.* **2011**, *26*, 1088–1098. [CrossRef]

28. Cummins, W. The Impulse Response Function and Ship Motions. *Schiffstechnick* **1962**, *9*, 101–109. [CrossRef]

29. Morison, J.R.; O'Brien, M.P.; Johnson, J.W.; Schaaf, S.A. The forces exerted by surface waves on piles. *Soc. Pet. Eng.* **1950**, *189*, 149–154. [CrossRef]

30. Openore. Available online: https://openore.org/2014/01/21/nemoh-open-source-bem/ (accessed on 13 October 2018).

31. Armesto, J.A.; Guanche, R.; Jesus, F.; Iturrioz, A.; Losada, I.J. Comparative analysis of the methods to compute the radiation term in Cummins' equation. *J. Ocean Eng. Mar. Energy* **2015**, *1*, 377–393. [CrossRef]

32. Duarte, T.; Sarmento, A.; Alves, M.; Jonkman, J. State-Space Realization of the Wave-Radiation Force within FAST. In Proceedings of the ASME 2013 32nd International Conference on Ocean, Offshore and Arctic Engineering, Nantes, France, 9–14 June 2013; pp. 1–12.

33. Mathworks. Available online: https://es.mathworks.com/help/signal/ref/prony.html (accessed on 2 February 2017).

34. Ogata, K. *Ingeniería de Control Moderna*; Pearson Educación S.A.: Madrid, Spain, 2010; ISBN 9788483226605.

35. Mathworks. Available online: https://es.mathworks.com/help/optim/ug/quadprog.html (accessed on 17 May 2017).

36. Fusco, F.; Ringwood, J.V. Short-Term Wave Forecasting for Real-Time Control of Wave Energy Converters. *IEEE Trans. Sustain. Energy* **2010**, *1*, 99–106. [CrossRef]

energies

MDPI

Article

Coupling Methodology for Studying the Far Field Effects of Wave Energy Converter Arrays over a Varying Bathymetry

Gael Verao Fernandez *, Philip Balitsky, Vasiliki Stratigaki and Peter Troch

Department of Civil Engineering, Ghent University, Technologierpark 904, B-9052 Zwijnaarde, Belgium;
philip.balitsky@ugent.be (P.B.); vasiliki.stratigaki@UGent.be (V.S.); peter.troch@ugent.be (P.T.)
* Correspondence: gael.veraofernandez@ugent.be; Tel.: +32-9-264-5489; Fax: +32-9-264-5837

Received: 1 October 2018; Accepted: 18 October 2018; Published: 25 October 2018

Abstract: For renewable wave energy to operate at grid scale, large arrays of Wave Energy Converters (WECs) need to be deployed in the ocean. Due to the hydrodynamic interactions between the individual WECs of an array, the overall power absorption and surrounding wave field will be affected, both close to the WECs (near field effects) and at large distances from their location (far field effects). Therefore, it is essential to model both the near field and far field effects of WEC arrays. It is difficult, however, to model both effects using a single numerical model that offers the desired accuracy at a reasonable computational time. The objective of this paper is to present a generic coupling methodology that will allow to model both effects accurately. The presented coupling methodology is exemplified using the mild slope wave propagation model MILDwave and the Boundary Elements Methods (BEM) solver NEMOH. NEMOH is used to model the near field effects while MILDwave is used to model the WEC array far field effects. The information between the two models is transferred using a one-way coupling. The results of the NEMOH-MILDwave coupled model are compared to the results from using only NEMOH for various test cases in uniform water depth. Additionally, the NEMOH-MILDwave coupled model is validated against available experimental wave data for a 9-WEC array. The coupling methodology proves to be a reliable numerical tool as the results demonstrate a difference between the numerical simulations results smaller than 5% and between the numerical simulations results and the experimental data ranging from 3% to 11%. The simulations are subsequently extended for a varying bathymetry, which will affect the far field effects. As a result, our coupled model proves to be a suitable numerical tool for simulating far field effects of WEC arrays for regular and irregular waves over a varying bathymetry.

Keywords: coupling; wave propagation model; MILDwave; BEM; NEMOH; array; farm; near field; far field; WECwakes project; experimental validation; WECWakes Hydralab IV project

1. Introduction

Wave energy is a renewable energy source with the potential to contribute to reduce the world's dependency on fossil fuels. Compared to other renewable energy sources, such as wind and solar, wave energy conversion lacks technologic and economic development. In order for wave energy to be economically viable, large farms of Wave Energy Converters (WECs) have to be deployed at the same location, which will enable wave energy to operate at grid scale and compete with wind farms. This is usually termed as WEC arrays or WEC farms in literature. For this study, the term WEC farm refers to a scale comparable to a wind farm, while a WEC array is a small group of WECs closely spaced within the farm. Due to the hydrodynamic interactions between the individual WECs in the array, the overall power absorption will be affected. The hydrodynamic problem of power absorption is characterized by two different problems: the diffraction problem and the radiation problem. The diffraction problem studies the change in direction of the incident wave field due to the presence of the WECs. Assuming the WECs to be stationary and depending on their geometry, a diffracted wave field around the WECs can be obtained. The radiation problem refers to the generation of a radiated wave field around the WECs due to the oscillations of the bodies caused by the incident wave field. The superposition of the diffracted and radiated wave fields using linear wave theory results in a complex perturbed wave field around the WECs. This is often described as the near field effects in literature (illustrated in Figure 1). In addition, the absorption and redistribution of the wave energy around the WECs will also cause a wake behind the WEC array, which is an area of reduced wave height in the lee of the WECs. Wake effects can have a positive or negative impact on the coastline and other sea users. This is often described as the far field effect of WECs.

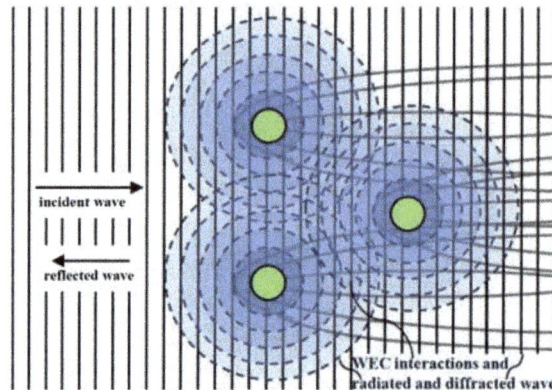

Figure 1. Visual representation of the near field effects between neighboring oscillating WECs (represented by solid circles) in a WEC array under incident waves [1].

Substantial numerical research has been carried out to study the near field effects and interaction factors of WECs, to optimize the array lay-out for maximizing power output. To date, various wave-structure interaction models have been used: numerical array models based on semi-analytical coefficient calculation [2,3], Boundary Elements Methods (BEM) based on potential flow theory [4,5] and Computational Fluid Dynamics (CFD) [6].

Far field effects are traditionally studied using wave propagation models. In [7–11], phase-averaging spectral models are used to obtain the wave field in the lee of a WEC array. The WEC arrays in these studies are simplified as obstacles with a fixed transmission coefficient. In the same way, Ref. [12] used a time-dependent mild-slope equation model and simplified each WEC as a wave power absorbing obstacle. To obtain the absorption coefficient for phase-averaging spectral models and the wave power absorbing obstacle coefficient for time-dependent mild-slope equation models, tank testing or numerical modelling is required. Therefore, the modelling of the hydrodynamic interactions is not taken into account resulting in a simplified WEC parametrization, which leads to low accuracy results.

As pointed out in [13], the currently available approaches either focus on modelling the near field effects at high fidelity but with high computational cost or the far field effects with low fidelity but low computational cost, in part due to the limitation of modelling both effects simultaneously using a single solver. On the one hand, wave-structure interaction solvers require a long computational time, which increases exponentially with the number of bodies of the WEC array and the size of the domain. Additionally, BEM solvers [14,15] are limited to a constant bathymetry, whilst other solvers like CFD solvers increase the computational time even more when considering irregular bathymetry. As a result, BEM solvers and CFD solvers are not suitable for studying far field effects of WEC arrays which require an even larger domain. On the other hand, wave propagation models offer a lower computation time for modelling large domains and study the WEC array impact at a regional scale.

Nonetheless, the simplification in modelling the WEC hydrodynamic problem can lead to erroneous model conclusions.

Various coupling methodologies to rectify these limitations have recently been advanced in [16–20]. These coupling methodologies are based on the work of [1], who first presented a coupling between a wave propagation model (MILDwave [21]) and a wave-structure interaction solver (WAMIT [14]). Pairing models with different resolutions and computational costs can enable the modeler to obtain results for different sub-domains of the problem while keeping the computational cost reasonable. This allows higher precision in the estimation of near field effects using wave-structure interaction solves. Subsequently, the resulting wave field of the wave-structure interaction solver is propagated using wave propagation models which solve wave propagation and transformation over large distances with varying bathymetry. Given the fact that the cost of installation of floating structures increases significantly in larger water depths, installation of floating structures in smaller depths where realistic bathymetries become significant could be a solution to this high cost. Therefore, modelling WEC array effects for irregular wave conditions and for realistic bathymetries can play an import role in further developments in the wave energy sector.

In this paper, a generic methodology for coupling a wave-structure interaction solver with a wave propagation model for any (floating) structure is presented and validated, with a novel application for irregular waves over a varying bathymetry. In Section 2, the details are presented of the generic coupling methodology between any wave-structure interaction solver and any wave propagation model. Section 3 illustrates the coupling methodology applied to a test case between the wave-structure solver, NEMOH, and the wave propagation model, MILD wave, for an array of nine floating WECs. Section 4.1 provides a verification of the results from the proposed coupling methodology against the wave fields simulated using the wave-structure interaction solver. Additionally, the coupling methodology is compared to an experimental data set for a 9-WEC array in Section 4.2. Section 4.3 advances an implementation of the coupling methodology with varying bathymetry. In Section 5 , the ability to simulate the far fields effects with high accuracy of the proposed coupling methodology is discussed. Finally, the conclusions of this work and future work are discussed in Section 6.

2. Generic Coupling Methodology

The proposed generic coupling methodology introduced in [1] and refined in [16] consists of four steps, as illustrated in Figure 2. Firstly (Step 1), the wave propagation model is used to obtain the incident wave field at the location of the structure(s) when the structure(s) is(are) not present. Secondly (Step 2), the obtained wave field is used as input for the wave-structure interaction solver at the location of the structure(s). Now, we can solve the motion of the structures(s) and obtain an accurate solution of the radiated and diffracted wave fields around the structure(s), namely the perturbed wave field. Thirdly (Step 3), the perturbed wave field is used as input in the wave propagation model and is propagated throughout a large domain. This is done by prescribing an internal wave generation boundary around the structure location. Finally (Step 4), the total wave field due to the presence of the structure(s) is obtained as the superposition of the incident wave field and the perturbed wave field in the wave propagation model.

The aforementioned coupling methodology can also be classified into a one-way coupling or two-way coupling depending on how Step 4 is implemented. Figure 3 shows the schematics of a one-way and a two-way coupling, respectively. The inner model domain corresponds to the location of the structure(s), where the near field effects are solved. The outer model domain corresponds to the area where the far field effects are evaluated. Both the inner model domain and the outer model domain are represented not to scale. In a one way-coupling, the wave field for each numerical problem is calculated independently. Thus, the main coupling mechanism in this example is the superposition of two different simulations obtained in the wave propagation model: an incident wave field calculated intrinsically and a perturbed wave field calculated using a wave generation boundary. In a two-way coupling, there is an exchange of information between the wave propagation model and the wave-structure interaction solver in each time step of the simulation and therefore Steps 2, 3 and 4 of the coupling methodology have to be re-calculated at each simulation time step.

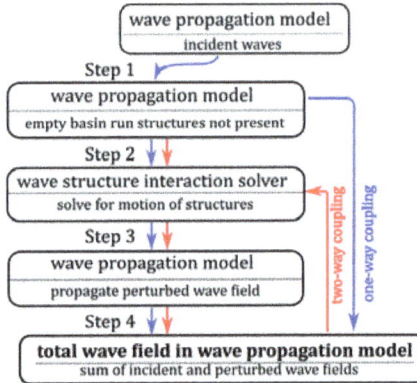

Figure 2. Flow chart of the generic coupling methodology between a wave-structure interaction solver and a wave propagation model.

Nevertheless, both solutions are able to obtain the far field effects of the WEC array at a reasonable computational cost and accuracy taking into account bathymetric effects and wave transformation processes, with an accurate description of the perturbed wave field around the structure/WEC.

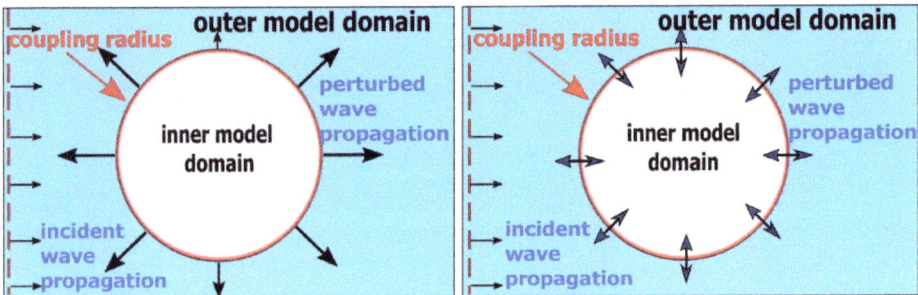

Figure 3. Schematic of a one-way coupling (**left**) and two-way coupling (**right**). In the inner model domain, the motions of the studied structure(s)/WEC(s) are solved.

The proposed coupling methodology is a generic tool that can be applied in the following cases:

1. Any wave-structure interaction solver that describes the perturbed wave field is suitable for obtaining the input parameters for the internal wave generation boundary. Models based on potential flow theory (e.g., BEM [17,22,23]) or analytical models based on analytical calculation of coefficients or numerical models based on resolving the Navier–Stokes equations (e.g., CFD [6] or SPH) are all suitable in obtaining the perturbed wave field around the WEC array [20].

2. Any wave propagation model can be used. A wave propagation boundary can be implemented in both phase-resolving and phase-averaging models.

3. The methodology applies to any kind of oscillating or floating structure. In this paper, a WEC array of heaving point absorber WECs is modelled using a phase-resolving model (in order to demonstrate this numerical coupling methodology). However, it can be applied to oscillating water column WECs, overtopping WECs, wave surge WECs, floating breakwaters or platforms.

3. Application of the Coupling Methodology between the Wave Propagation Model, MILDwave, and the BEM Solver, NEMOH

In this section, the generic coupling methodology presented in Section 2 will be demonstrated for obtaining the far field effects due to a WEC array. First, the applied wave theory is discussed. Secondly, a description of the two numerical models employed is provided. Thirdly, the application of the coupling methodology is described for regular and irregular waves. Finally, the coupling methodology is validated against WEC array experimental data from the WECwakes project [1,24,25] which has been co-ordinated by the co-authors of the present paper.

3.1. Numerical Background

3.1.1. Linear Potential Flow

Both models employed are based on linear potential flow theory [22] that allows the flow velocity, v, to be expressed as the gradient of the potential, Φ:

$$v = \nabla \Phi. \tag{1}$$

The assumptions underlying potential flow theory are the following:

1. The flow is inviscid.
2. The flow is irrotational.
3. The flow is incompressible.

The standard assumption of linear theory that the motion amplitudes of the bodies are much smaller than the wavelength also applies. Linear potential flow theory has hitherto been utilized in a majority of the investigations into WEC array modelling—for example, see [26]. Due to the principle of superposition, linear potential theory allows for the separation of the total wave field into the following components of the velocity potential :

$$\varphi_t(x,y,z) = \varphi_i + \varphi_{diff} + \sum_i^6 \varphi_{rad}, \tag{2}$$

where φ_t is the total velocity potential, φ_i is the incident wave velocity potential, φ_{diff} is the diffracted wave velocity potential and $\sum_i^6 \varphi_{rad}$ is the sum of the radiated wave velocity potentials for each degree of freedom of motion of the WEC(s). In our investigation, we also make use of the term "perturbed wave" to denote the wave resulting from the sum of the diffracted and radiated wave velocity potentials.

3.1.2. Wave Propagation Model MILDwave

The wave propagation model chosen for demonstrating the coupling methodology is the mild-slope wave propagation model MILDwave [21,27]. MILDwave is a phase-resolving model based on the depth-integrated mild-slope equations of Radder and Dingemans [28]. MILDwave allows for solving the shoaling and refraction of waves propagating above mild-slope varying bathymetries. Furthermore, MILDwave has been widely used in the modelling of WEC arrays [12,16,17,19,24,27,29,30]. The mild-slope equations (Equations (3) and (4)) are resolved using a finite difference scheme that consists of a two-step space-centered, time-staggered computational grid, as detailed in [31]:

$$\frac{\partial \eta}{\partial t} = \frac{\omega^2 - k^2 CC_g}{g} \varphi - \nabla \left(\frac{CC_g}{g} \nabla \varphi \right), \tag{3}$$

$$\frac{\partial \varphi}{\partial t} = -g\eta, \tag{4}$$

where η and φ are, respectively, the free water surface elevation and the wave velocity potential at the free water surface, g is the gravitational acceleration, C is the phase velocity and C_g is the group velocity for a wave with wave number k and angular frequency ω.

3.1.3. Wave-Structure Interaction Solver NEMOH

The wave-structure interaction solver chosen for demonstrating the coupling methodology and to solve the diffraction/radiation problem is the open-source potential flow BEM solver NEMOH. Given Equation (1), NEMOH solves the Laplace Equation (5) for the complex wave velocity potential, φ:

$$\Delta\varphi = 0, \tag{5}$$

given a set of boundary conditions on the wetted body surface, the free surface, sea bottom and the far field area. Equation (5) is solved by employing Green's functions [15]. An important restriction imposed by Green's functions is the assumption that the water depth h is constant throughout the BEM domain. The free surface elevation η is calculated by taking the real part of the complex potential $\overline{\eta}$ from the free surface boundary condition, presented by Equation (6). From the superposition principle, presented by Equation (2), free surface elevations can be obtained separately from the vertical motions of the WEC(s) due to the diffracted and radiated potentials:

$$\eta = -\frac{1}{g}\left(\frac{\partial\varphi}{\partial t}\right)_{z=0}. \tag{6}$$

3.1.4. Modelled WECs

The examined WEC array consists of nine heaving buoys. The buoy shape is a flat circular cylinder with a diameter of 10 m and a draft of 2 m. The shape was chosen to represent several promising WEC technologies that are being developed at the moment [32]. The Power Take-off (PTO) of each WEC is modeled as a resistive damper. The damping coefficient of the PTO, B_{PTO}, is kept constant during the investigation with $B_{PTO} = 3.6 \times 10^5$ kgs^{-2}. The chosen B_{PTO} leads to a resonance condition for a wave period of 8 s [18]. However, it has been found that a variation of B_{PTO} depending on the wave period does not have a significant impact on the WEC motion [16]. The WEC array layout is sketched in Figure 4. Here, $d_x = 30$ m and $d_y = 30$ m are the inter-array separation distances, $l_x = 30$ m and $l_y = 120$ m are the total array dimensions and the WEC array is located at the center of the domain with co-ordinates $x = 0$ m and $y = 0$ m. The layout is a staggered grid lay-out. The array layout is kept constant during the analysis.

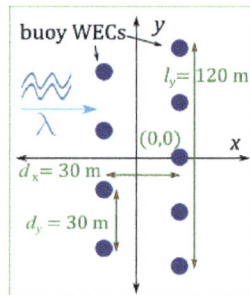

Figure 4. Plane (Top) view of the WEC array layout for nine heaving buoys. λ indicates the direction of wave propagation.

3.1.5. Wave Characteristics

The results presented here comprise two sets of regular and irregular waves included in Table 1. We aim to demonstrate the application of the coupling methodology and impact of the WEC array on the wave field for different wave periods and for a fixed wave direction $\theta = 0°$. The regular wave set consists of a wave height $H = 2$ m and a wave period $T = 6, 8$ and 10 s. The irregular wave set consists of a significant wave height $H_s = 2$ m and a peak period $T_p = 6, 8$ and 10 s. The three wave periods chosen range from 6 to 10 s as they range from values closer to the resonance period (T_r) of the WECs, for which the WEC motions are large, to values far from T_r where the motions of the WECs are reduced reproducing possible operational wave conditions for the WEC array.

Table 1. Incident wave data sets used for the coupling methodology test cases.

Regular Waves			
Case Name	T (s)	H (m)	$\theta = 0°$
A	6	2	0
B	8	2	0
C	10	2	0
Irregular Waves			
Case Name	T_p (s)	H_s (m)	$\theta = 0°$
D	6	2	0
E	8	2	0
F	10	2	0

For the regular wave cases (A–C), the results are presented in all points of the domain by calculating the $K_{D,r}$, defined as the ratio between the numerically calculated total wave, H_{tot}, and the incident wave height, H_i:

$$K_{D,r} = \frac{H_{tot}}{H_i} = \frac{\sqrt{8 \cdot \sum_t^{\Delta t} \eta(x,y)_t^2 \cdot \frac{dt}{\Delta t}}}{H_i}, \tag{7}$$

where $\eta(x,y)$ is the resulting surface elevation in each simulation time step dt and Δt is the time window over the K_D is computed.

For the irregular wave cases (D–F), a Pierson–Moskovitz spectrum is utilized to represent a realistic sea state [33]. This methodology for generating irregular waves has already been used by [34] when modelling WEC arrays.

The irregular wave spectrum is defined by Equation (8):

$$S(f) = \frac{B}{f^5} e^{-\frac{C}{f^4}}, \tag{8}$$

where

$$B = \frac{5}{16} \frac{H_s^2}{T_p^4}, \tag{9}$$

$$C = \frac{5}{4} \frac{1}{T_p^4}. \tag{10}$$

The wave spectrum is discretized in a total of $n = 20$ regular wave components. The wave height H_i and wave amplitude a_i corresponding to each regular wave component of wave frequency f_i is obtained by:

$$H_i = 2\sqrt{2S(f_i)\Delta f}, \tag{11}$$

$$a_i = \frac{H_i}{2} e^{i\varphi_i}, \tag{12}$$

where $\Delta f = 0.2$ corresponds with the wave frequency discretization of the spectrum and φ_i with a random number between $\check{}\pi$ and π corresponds to the phase angle of each wave frequency component.

The surface elevation η_i for each regular wave component is then obtained using Equation (13):

$$\eta_i(x,y) = a_i\eta_{i1}(x,y), \tag{13}$$

where $\eta_{i1}(x,y)$ corresponds to the surface elevation of unit wave amplitude for the ith regular wave component of the wave spectrum.

The irregular surface elevation $\eta_{ir}(x,y)$ at each point of the numerical domain is then obtained as a superposition of the N regular wave components of the spectrum using Equation (14):

$$\eta_{irr}(x,y) = \sum_{i}^{N} \eta_i(x,y). \tag{14}$$

As in the case of regular waves, the results are presented as the $K_{D,irr}$ for irregular waves, defined as the ratio between the numerically calculated significant wave height, $H_{s,tot}$, and the incident significant wave height, $H_{s,i}$:

$$K_{D,irr} = \frac{H_{s,tot}}{H_{s,i}} = \frac{4 \cdot \sqrt{\sum_{t}^{\Delta t} \eta_{irr}(x,y)_t^2 \cdot \frac{dt}{\Delta t}}}{H_{s,i}}. \tag{15}$$

It has to be noted that Equations (13) and (14) can also be used to obtain the diffracted and radiated wave fields around the WEC array.

3.2. Coupling Methodology Implementation

This section shows the application of the generic coupling methodology using the selected numerical models. The objective is to obtain the total wave field in the MILDwave domain due to the presence of the WEC array. This is performed by superimposing the incident wave field, the diffracted wave field and the radiated wave field generated in MILDwave. The first step (Step 1 in Figure 2) of the coupling methodology is to obtain the incident wave field in the wave propagation model, MILDwave, at the location of the WEC array (Figure 2). A numerical basin is set-up in MILDwave where the incident waves are generated along a linear wave generation boundary perpendicular to the wave propagation direction. In the MILDwave domain, both constant or varying bathymetries can be modelled. To minimize unwanted wave reflection absorption zones (implemented in MILDwave as sponge layers), are placed down-wave and up-wave the basin. From this simulation, the surface elevations at the WEC array location are obtained and used as the input value for NEMOH, which is Step 2 in Figure 2.

In the second step (Step 2 in Figure 2) of the coupling methodology, the radiated/diffracted wave field is obtained around the WEC array using the wave-structure interaction solver NEMOH. NEMOH resolves the wave frequency dependent wave radiation problem for each individual WEC and the diffraction over a predetermined numerical grid. The input values for NEMOH are the WEC array, the wave amplitude at the WEC array location obtained in Step 1, the wave period and the water depth at the WEC array location. As a result, NEMOH gives the complex radiated and diffracted wave fields described by Equations (16) and (17) respectively:

$$\eta_{rad} = \sum_{j}^{M} i\omega \tilde{X}(\omega) \left|\eta_{rad}\right| e^{i\varphi_{rad}}, \tag{16}$$

$$\eta_{diff} = \left|\eta_{diff}\right| e^{i\varphi_{diff}}, \tag{17}$$

where ω corresponds to the wave angular frequency (rad/s), i is the imaginary number part, φ_{rad} and φ_{diff} correspond to the radiated and diffracted wave phase angle, respectively, M is the total number of WECs and $\tilde{X}$ is the Response Amplitude Operator (RAO) of each WEC given by:

$$\tilde{X} = \frac{a_c \tilde{F}_{ex}}{-(M + M_A)\omega^2 - i\omega(B_{hyd} + B_{PTO}) + K_H}. \tag{18}$$

Here, a_c corresponds to the wave amplitude at the coupling region, $\tilde{F}_{ex}$ is the excitation force, M is the WEC mass, M_A is the added mass, B_{hyd} is the hydrodynamic damping coefficient, B_{PTO} is the Power Take-Off damping coefficient and K_H is the hydrodynamic stiffness.

The radiated and diffracted wave fields are then summed up in the frequency domain to obtain the perturbed wave field:

$$\eta_{pert} = \eta_{diff} + \eta_{rad}. \tag{19}$$

In the third step (Step 3 in Figure 2), the perturbed wave field is then transformed from the frequency domain to the time domain and propagated into MILDwave using a circular wave generation boundary (Figure 5). Waves are forced away from the circular wave generation boundary by imposing the free surface elevation along the circle, $\eta_{circ}(x, y, t)$:

$$\eta_{circ}(x, y, t) = |\eta_{pert,i}| \cos(\varphi_{pert,c} - \omega t). \tag{20}$$

To avoid wave reflection, absorption layers (implemented in MILDwave as sponge layers) are placed up-wave, down-wave and also along the sides of the MILDwave numerical domain. In the fourth step (Step 4 in 2), the incident wave field obtained in Step 1 and the perturbed wave field obtained in Step 3 are combined to obtain the total wave field due to the presence of the WECs array in the MILDwave domain:

$$\eta_{tot}(x, y, t) = \eta_{inc}(x, y, t) + \eta_{pert}(x, y, t). \tag{21}$$

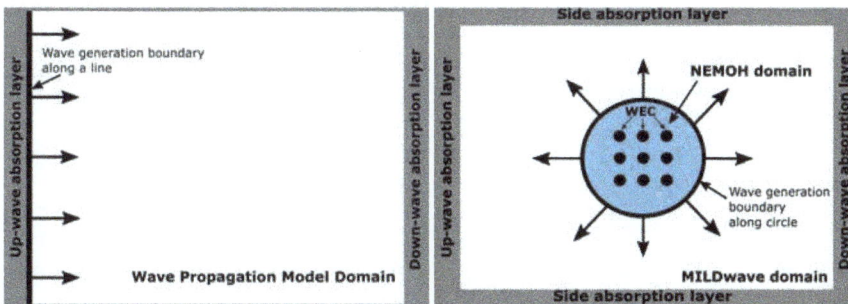

Figure 5. Sketch of the incident wave propagation (**left**) and perturbed wave propagation (**right**) in MILDwave. The black line corresponds to the wave generation line, the black circle corresponds to the circular wave generation boundary and the grey areas correspond to absorption zones (sponge layers) down-wave, up-wave and along the sides of the numerical domain.

3.3. Experimental Data-Set Used for Numerical Validation Purposes

This section gives a description of the experimental data-set used to validate the coupling methodology described in this study. The experimental data-set from the WECwakes project [1,24,25] conducted in the Shallow Water Wave Basin of DHI, Hørsholm, Denmark is used. In the WECwakes project, WEC arrays up to 25 devices were tested to study near field and far field effects of heaving point absorber WECs. First, the characteristics of an individual WEC are described. Secondly, a description of the wave basin is provided. Thirdly, a general description of the WECwakes experiment is included. Finally, the numerical model implementation of the experimental set-up is discussed.

Each of the 25 WEC heaving point absorbers tested consists of a buoy heaving through a metallic vertical shaft mounted on a metal base installed at the bottom of the wave basin (Figure 6). Each buoy consists of a cylindrical body and a spherical bottom with a diameter of 0.315 m. The draft of the WEC is 0.323 m. The water depth is fixed to 0.700 m. The PTO system is composed of Teflon (PTFE)-blocks at the top, which causes energy dissipation through friction damping of the WEC heave motion. Additionally, the presence of the vertical shaft through the buoy causes additional frictional forces on the WEC buoy.

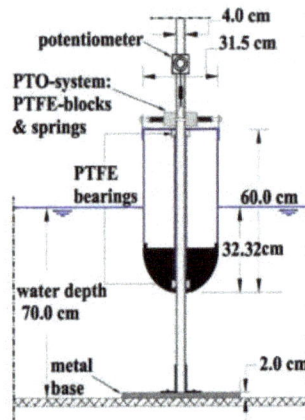

Figure 6. Definition sketch of the WECwakes WEC unit. Adopted from [1,24,25].

The DHI wave basin experimental domain is 22 m wide and 25 m long and overall depth of 0.8 m. Forty-four piston type wave paddles, each of width 0.5 m generate waves at one end of the wave basin. During the WECwakes project, arrays of up to 25 heaving point absorber WECs (see Figure 7) have been tested using different geometric WEC array configurations. By testing different WEC array configurations under a wide range of sea states a large experimental data-set has been generated and is publicly available for numerical validation purposes. The wave field around the WECs is recorded using 41 resistive wave gauges (WGs) distributed in the basin as shown in Figure 7. A potentiometer is installed at the top of each WEC unit to measure its heave displacement. Furthermore, two load cells were installed in the five WECs located on the central line of the array to measure surge forces.

The WECwakes project has led to a database of 591 test focusing on different array geometrical configurations and wave characteristics. For this validation, an array of nine WECs arranged in a 3 × 3 WEC layout has been selected (see Figure 8). A total of 15 wave gauges located in the front, leeward and sides of the array are identified to compare the free surface elevations between the NEMOH-MILDwave coupled model and the experimental data-set. The separating distance between the different WEC units is equal to 1.575 m. The incident regular wave conditions used are a wave height of $H = 0.074$ m and a wave period of $T = 1.26$ s.

The effect of the WEC's PTO system is included in the numerical simulation by adding an external damping coefficient, B_{PTO}, to the equation of motion (Equation (18)). The value for B_{PTO} is calculated empirically to account for (i) the PTO system itself which mimics a coulomb damper, (ii) viscous damping of the WEC's motion due to the presence of water between the vertical supporting axis and the shaft through the WEC unit [35] and (iii) the wave-induced surge forces pushing the WEC against its vertical supporting axis [36]. Therefore, a single WEC has been modelled in NEMOH, similar to the experimental set-up but without the shaft bearing, and regular waves are generated. When the difference between numerical and experimental results of the free surface elevations of the total wave field was smaller than 5%, the applied external damping coefficient is considered sufficient. This methodology resulted in a value of $B_{PTO} = 28.5$ kg/s.

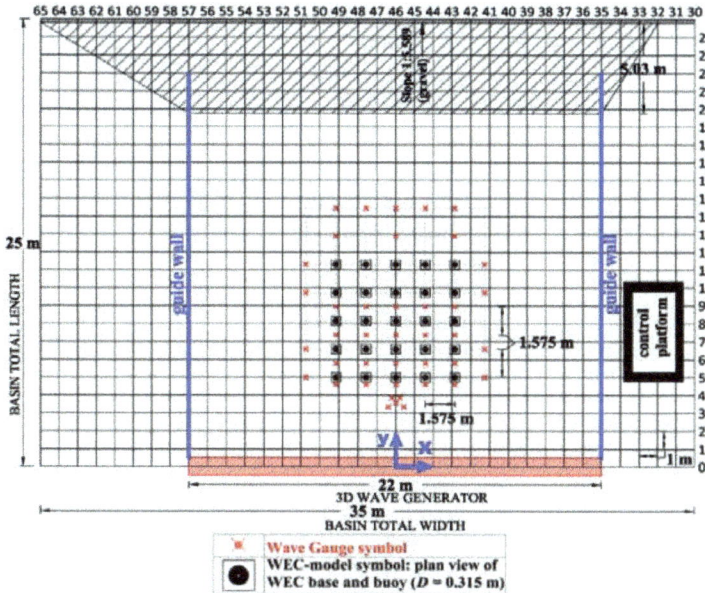

Figure 7. Plan view of the WECwakes experimental set-up in the DHI wave basin as a 5 × 5 rectilinear array. The red crosses indicate the position of all the wave gauges installed in the DHI wave basin during the experiments and the black circles indicate the location of the different WEC units. The wave paddles are denoted by the red hatched area at the bottom of the figure while the black hatched area at the top of the figure represents the installed absorbing beach. Two guiding walls were installed at the sides of the basin, denoted in blue lines [1].

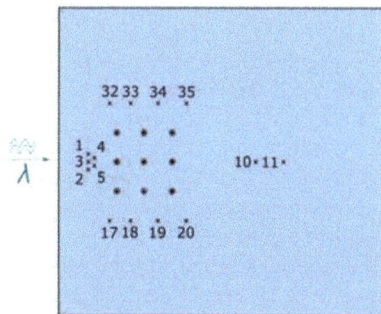

Figure 8. Set up of the WEC array layout used for the comparison between the coupled model and the experimental tests. WECs are represented by • and wave gauges (WGs) by x. The WG are numbered as they appear in the WECwakes experimental data set. The direction of wave propagation, indicated by λ, is from left to right.

3.4. Test Program

3.4.1. Coupling Methodology Implementation for Constant Bottom Bathymetry

The first objective of this research is to show the accuracy of the NEMOH-MILDwave coupled model in obtaining the total wave field around a WEC array. A comparison between NEMOH and the NEMOH-MILDwave coupled model will be done using the numerical results from NEMOH as a benchmark for the NEMOH-MILDwave coupled model. To quantify the difference in the K_D between

NEMOH and the NEMOH-MILDwave coupled model, the relative K_D difference between NEMOH and the NEMOH-MILDwave coupled model is determined:

$$RD = \frac{(K_{D,NEMOH} - K_{D,coupled})}{K_{D,NEMOH}} \cdot 100. \tag{22}$$

A test program, Table 2, has been designed based on the regular waves and irregular waves cases presented in Table 1. The total wave field around the 9-WEC array illustrated in Figure 4 is simulated in deep water. The direction of the wave propagation, indicated by λ in Figure 4, is from left to right. The basin water depth is set to a constant depth of $d = 40$ m to provide deep water conditions.

Table 2. Test program for the coupling methodology implementation in constant bathymetry.

Test Number	Numerical Models	Wave Type	*H* (m)	*T* (s)	Water Depth *d* (m)
1	NEMOH	REG	2	6	40
2	NEMOH	REG	2	8	40
3	NEMOH	REG	2	10	40
4	NEMOH-MILDwave	REG	2	6	40
5	NEMOH-MILDwave	REG	2	8	40
6	NEMOH-MILDwave	REG	2	10	40
7	NEMOH	IRREG	2	6	40
8	NEMOH	IRREG	2	8	40
9	NEMOH	IRREG	2	10	40
10	NEMOH-MILDwave	IRREG	2	6	40
11	NEMOH-MILDwave	IRREG	2	8	40
12	NEMOH-MILDwave	IRREG	2	10	40

Tests 1–3 and tests 7–9 are performed using the BEM code NEMOH. NEMOH has important constraints as noted in [15] but accurately solves array interactions within linear theory. To maximize the accuracy of the results, a numerical basin of 800 m × 800 m is used for each simulation with a grid size of 2 m with an equal spacing in *x*- and *y*-directions. It has to be noted that the results from NEMOH are used as the input values for the NEMOH-MILDwave coupled model. NEMOH gives the total wave field individually for each body; however, the 9-WEC array is implemented directly in the NEMOH-MILDwave coupled model.

Tests 4–6 and 10–11 are performed using the NEMOH-MILDwave coupled model. The coupling methodology is implemented along an internal circular wave generation boundary of coupling radius $r_c = 120$ m. According to [16], a wave generation circle with a minimum rc distance equal to the distance from the center of the coupling region to the most further WEC + 2· r_{WEC} offers an accurate solution. The grid size for the simulation is set to 2 m with an equal spacing in the *x*- and *y*-directions and an effective domain of 2800 m × 1600 m. Absorbing sponge layers are placed in the edges of the numerical basin to avoid wave reflection. Each simulation is run for 2000 s to obtain a fully developed sea state.

3.4.2. Coupling Methodology Validation for Constant Bottom Bathymetry against Experimental Data

The second objective of this research is to validate the NEMOH-MILDwave numerical model against existing experimental data. For this purpose, the experimental set-up presented in Section 3.3 will be simulated using the coupling methodology presented in Section 3. The 3 × 3 WEC array show in Figure 8 is implemented in the MILDwave domain at a distance of 6.575 m from the wave generation line within an internal circular wave generation boundary of coupling radius $r_c = 2.5$ m. The wave height is $H = 0.074$ m and the wave period is $T = 1.26$ s. The numerical basin depth is set to a constant depth of $d = 0.7$ m. The grid cell size for the simulation is set to $d_x = d_y = 0.04$ m with an equal spacing in the x- and y-directions over a MILDwave effective domain of 25 m $\times$ 22 m. Each simulation is run for 100 s to obtain a completely developed wave field.

The comparison between the results of the NEMOH-MILDwave model and the experimental WECwakes data is based on the difference of the free surface elevations recorded at the 15 resistive wave gauges (WG) shown in Figure 8. To have a quantitative estimation of the extent of the differences in the free surface elevation, the Root Mean Square Error (RMSE) over a time series of 10 s has been obtained and normalized between the experimental data and the NEMOH-MILDwave coupled model:

$$RMSE = \frac{\sqrt{\frac{1}{N} \sum_i^n (\eta_{e,i} - \eta_{n,i})^2}}{\eta_{e,max} - \eta_{e,min}}. \tag{23}$$

3.4.3. Coupling Methodology Implementation for Varying Bathymetry

The third objective of this research is to illustrate the capabilities of the developed NEMOH-MILDwave coupled model to model wave transformations over a large domain. Subsequently, a varying bathymetry (as sketched in Figure 9) is applied to the numerical wave basin by modifying the tests 4–6 and tests 10–12 bathymetrical input (see Table 3). The 9-WEC array (see Figure 4) is placed in the center of the domain at a constant water depth of 20 m. The numerical simulations are performed using identical parameters to those reported in Section 3.1. The latter will provide a benchmark to assess the capability of the model for propagating the wave field over a varying bathymetry.

Table 3. Test program for the coupling methodology implementation in varying bathymetry.

Test Number	Numerical Models	Wave Type	H (m)	T (s)	Water Depth d (m)
13	NEMOH-MILDwave	REG	2	6	VAR
14	NEMOH-MILDwave	REG	2	8	VAR
15	NEMOH-MILDwave	REG	2	10	VAR
16	NEMOH-MILDwave	IRREG	2	6	VAR
17	NEMOH-MILDwave	IRREG	2	8	VAR
18	NEMOH-MILDwave	IRREG	2	10	VAR

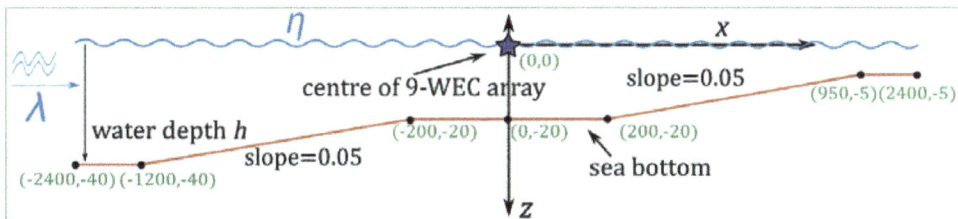

Figure 9. Depth view showing the location of the 9-WEC array. x–z plane (side) profile. The coastline is located at the right side of the figure. The 9-WEC array is located at the center of the domain.

4. Results

4.1. Coupling Methodology Implementation for Constant Bottom Bathymetry

In this section, the accuracy of the NEMOH-MILDwave coupled model using the presented coupling methodology is discussed. First, the benchmark for the coupling methodology comparison is obtained calculating the total wave field around the 9-WEC array for the BEM model NEMOH. Then, the same simulation is performed using the NEMOH-MILDwave coupled model. Finally, a comparison study between NEMOH and the NEMOH-MILDwave coupled model results is performed by comparing the K_D.

4.1.1. NEMOH Wave Field

The K_D for NEMOH is illustrated in Figure 10. A diffracted-radiated pattern of waves interacting with the WEC array is observed. In front of the WEC array, there is a wave reflection pattern, while, in the lee of the WEC array, there is a wake effect with reduced values of K_D. The diffracted wave pattern does not differ substantially for the three different wave periods both for regular waves and irregular waves. This is due to the fact that the size of the WEC modelled (WEC radius = 5 m) is smaller relative to the incoming wave length. In contrast, the magnitude of the diffracted wave over the radiated wave is increased when the wave period is reduced. As a result, there is almost no wave reflection observed for the case of $T = 10$ s and $T_p = 10$ s, while for $T = 6$ s and 8 s and $T_p = 6$ s and 8 s the wave reflection pattern is enhanced close to the center of the array, where more WECs are present.

Finally, a comparison between regular and irregular waves shows that less wave reflection occurs in front of the WEC array and reduced wake effect in the lee of the WEC array for smaller wave periods and irregular waves. This decreasing effect possibly originates from the superposition of 20 different frequencies resulting in the total wave field. The superposition of high frequency components will have a major contribution in modifying the incident wave. Due to the type of WEC modelled with a fixed B_{PTO}, a larger effect of wave diffraction over wave radiation is observed for higher frequencies. On the contrary, low frequency components will contribute to reduce the wave diffraction and wave radiation effect due to irregular waves as they barely affect the incident wave field. This results in a reduction of both effects when modelling irregular waves leading to smaller values of K_D compared to regular waves.

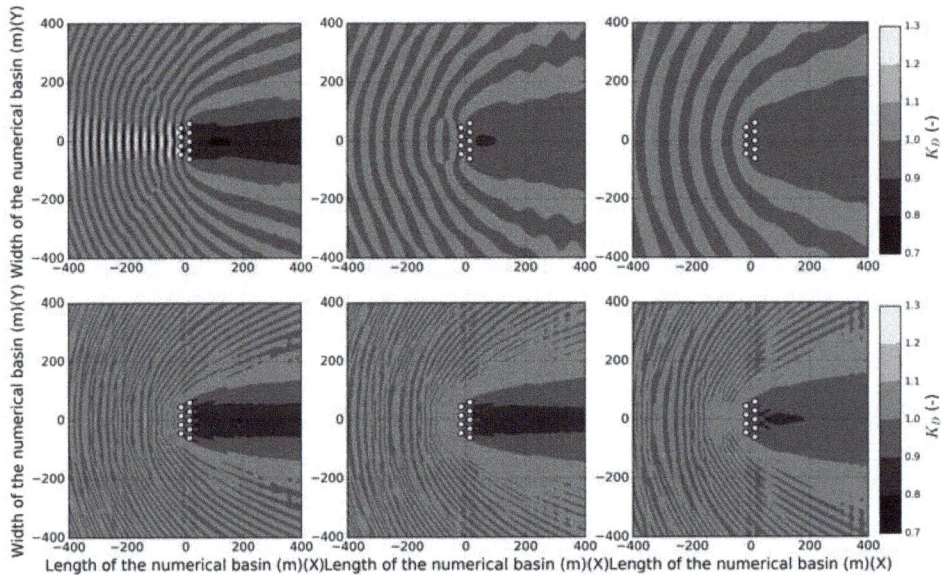

Figure 10. NEMOH results of K_D for a 9-WEC array for regular waves (**top**) with $T = 6$ s (**left**), $T = 8$ s (**middle**) and $T = 10$ s (**right**) and for irregular waves (**bottom**) with $T_p = 6$ s (**left**), $T_p = 8$ s (**middle**) and $T_p = 10$ s (**right**). Contour levels are set at an interval 0.04 m. The white solid circles indicate the location of the WECs

4.1.2. NEMOH-MILDwave Coupled Model Wave Field

The K_D values for the NEMOH-MILDwave coupled model are illustrated in Figure 11. As reported before, there is a reflection pattern in front of the 9-WEC array while there is a reduction of K_D values in the lee of the WEC array. In contrast to the NEMOH simulations, the numerical domain in MILDwave has now been increased considerably. As a result, it is possible to study the extent of these wave reflection and wake effects in a larger domain overcoming the size limitations of the wave-structure interaction solver. Qualitatively, the wave reflection and wake effects are larger for regular waves than for irregular waves.

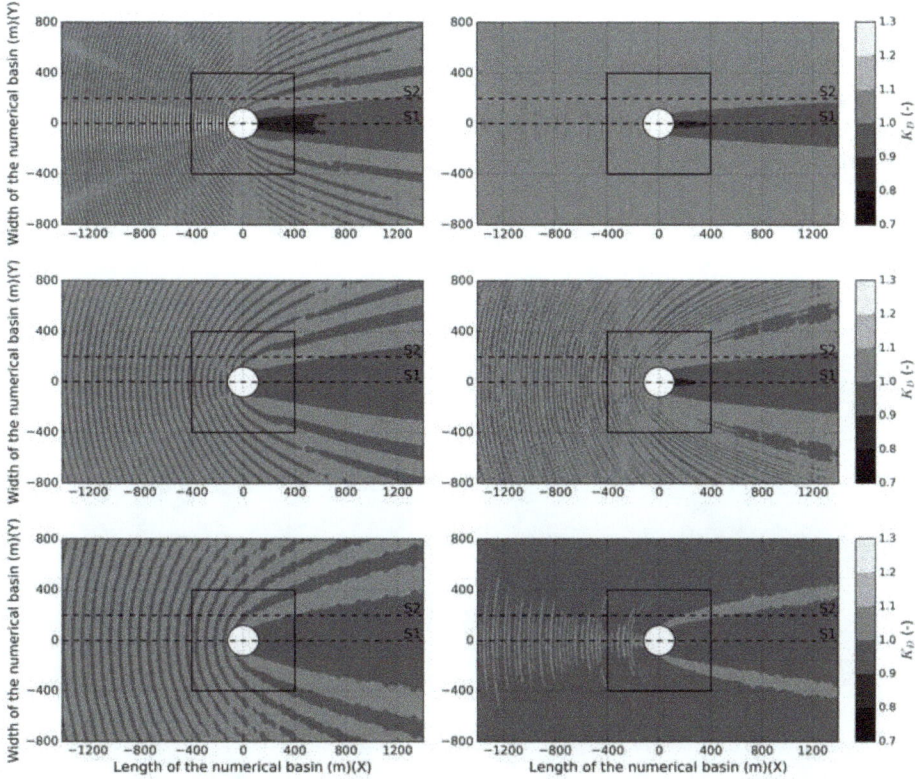

Figure 11. NEMOH-MILDwave couple model K_D values of a 9-WEC array for regular waves with $T = 6$ s, $T = 8$ s and $T = 10$ s and for irregular waves with $T_p = 6$ s, $T_p = 8$ s and $T_p = 10$'s. Contour levels are set at an interval 0.04 m. The water depth is 40 m. Waves are propagating from left to right. The coupling region is masked out using a white circle and includes the WECs. The NEMOH numerical domain is limited by black square.

4.1.3. Comparison of the Total Wave Field Generated by NEMOH and the NEMOH-MILDwave Coupled Model

The K_D for NEMOH and the NEMOH-MILDwave coupled modes results are depicted in Figures 10 and 11, respectively. The observed K_D close to the internal coupling generation line is slightly different in Figures 10 and 11. In front of the WEC array, there is a positive difference in K_D of 0.02 between NEMOH and the NEMOH-MILDwave coupled model, while, in the lee of the array, there is a negative difference in K_D of -0.01. Despite these small differences, on a region of 800 m $\times$ 800 m around the 9-WEC array, the obtained correspondence of the results is very good.

In Figure 12, six plots show the relative % difference in K_D between the two models for the wave periods studied for regular and irregular waves. The coupling region is masked out in a white circle as the total wave field in this region corresponds to the NEMOH results.

The largest observed relative % difference in K_D between the NEMOH-MILDwave coupled model and the NEMOH model is located in the region in front of the array for all cases. These relative % differences in K_D never exceed 7% for all cases. Moreover, the difference in the observed total wave field is higher for regular waves than for irregular waves, being maximized for smaller periods. Thereafter, there is an overestimation in the NEMOH-MILDwave model when propagating the internal prescribed perturbed wave. This overestimation is concentrated locally close to the coupling region up to a maximum value of 3% and is reduced when moving away from it. Subsequently, an assumption can be made that the NEMOH-MILDwave coupled model is accurate when propagating the wave field on a large domain.

To have a better look at the agreement between the NEMOH-MILDwave coupled model and NEMOH, longitudinal cross-sections are taken at two different locations: S1 at the center of the domain, $y = 0$ m (Figure 13), and S2 with an off-set of 200 m along the y-axis, $y = 200$ m (Figure 14). The cross-sections shown on the left side of the figures correspond to regular waves, while the cross sections shown on the right side of the figures correspond to irregular waves. The near field region that is modelled with NEMOH and is not part of the coupling methodology analysis is masked out in gray between two black vertical lines. Both coupled models show a good correlation as the values of K_D obtained are close to each other. As it has already been mentioned, for irregular waves, the wake behind the WEC array is increased in both magnitude and distance but is narrower, in contrast to the higher wave reflection values in front of the WEC array given for regular waves.

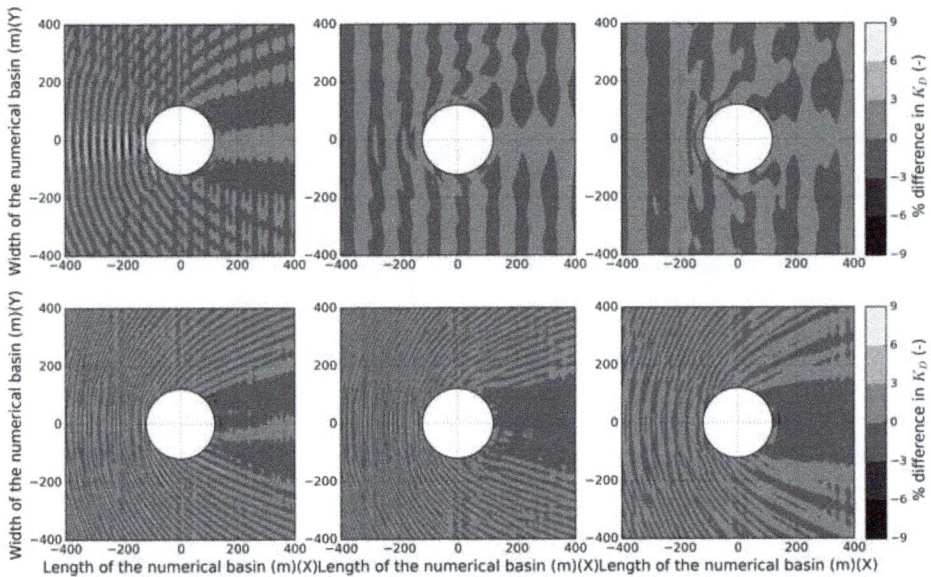

Figure 12. Relative difference (%) in K_D between NEMOH-MILDwave coupled model and NEMOH. Regular waves (**top**) and Irregular waves (**bottom**). For regular waves with $T = 6$ s, $T = 8$ s and $T = 10$ s and for irregular waves with $T_p = 6$ s, $T_p = 8$ s and $T_p = 10$ s. The coupling region is masked out using a white circle.

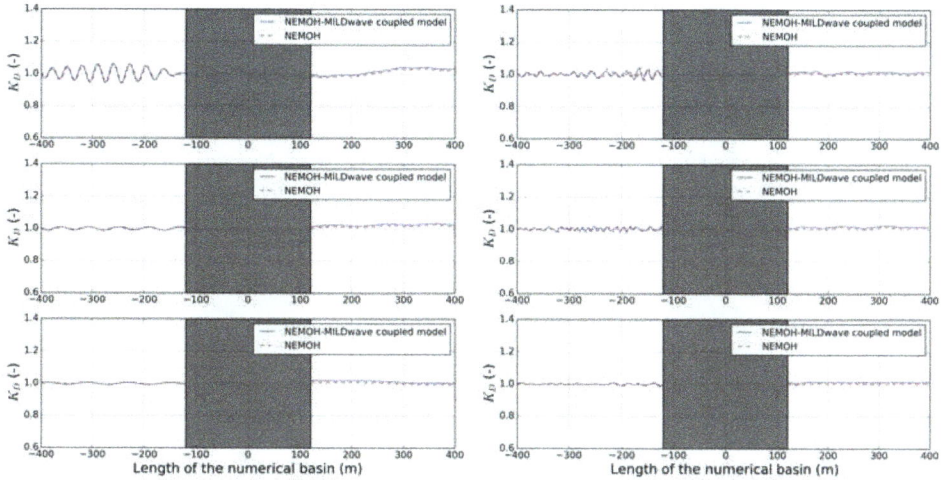

Figure 13. Cross-section S1 of the K_D for a 9-WEC array at $y = 0$ m for regular waves (**left**) and irregular waves (**right**) for regular waves with $T = 6$ s, $T = 8$ s and $T = 10$ s and for irregular waves with $T_p = 6$ s, $T_p = 8$ s and $T_p = 10$ s. The coupling region is masked out in gray between two vertical black lines.

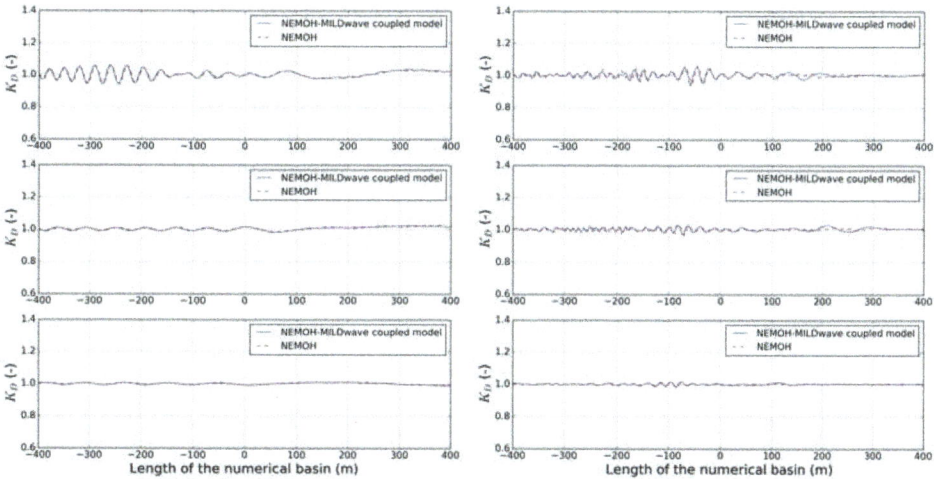

Figure 14. Cross-section S2 of the K_D for a 9-WEC array at $y = 200$ m for regular waves (**left**) and irregular waves (**right**) for regular waves with $T = 6$ s, $T = 8$ s and $T = 10$ s and for irregular waves with $T_p = 6$ s, $T_p = 8$ s and $T_p = 10$ s.

4.2. Coupling Methodology Validation for Constant Bottom Bathymetry against Experimental Data

The comparison between the NEMOH-MILDwave coupled model and the experimental data is shown in Figure 15. Overall, there is a good agreement between the NEMOH-MILDwave coupled model and the experimental data as it can be seen in Figure 15. However, it can be noticed that wave gauges 1, 10, 20, 32 and 33 show small differences on the surface elevation pattern.

The error of the free surface elevation for each wave gauge is included in Figure 16. It can be seen that the best agreement is obtained in front of the WEC array (WG 1-5) and in the wake of the array (WG 11) with an error ranging from 3–5%. The biggest difference is also obtained in the wake of the array (WG10) close to the WEC array with an error of 11%. While evaluating this biggest difference, it has to be considered that experimental data is intrinsically nonlinear [24]. Nonlinear effects such as viscosity or the friction between the shaft an the WECs cannot be modeled with the coupling methodology employed as it is based on linear wave theory. This cannot be modeled with the coupling methodology employed as it is based on linear wave theory. When the wave propagates further from the WEC, these nonlinearities are reduced and therefore the agreement between experimental and numerical data is better. Finally, on the sides of the coupling region the error ranges from 6 to 9% showing that the numerical model is not accurately representing the wave diffraction around the WEC array.

It has to be noted, however, that, within this numerical validation, a linear model is compared with experimental data that has nonlinear effects. Therefore, it can be seen that further away from the location of the WEC array, a better agreement in the free surface elevation is obtained.

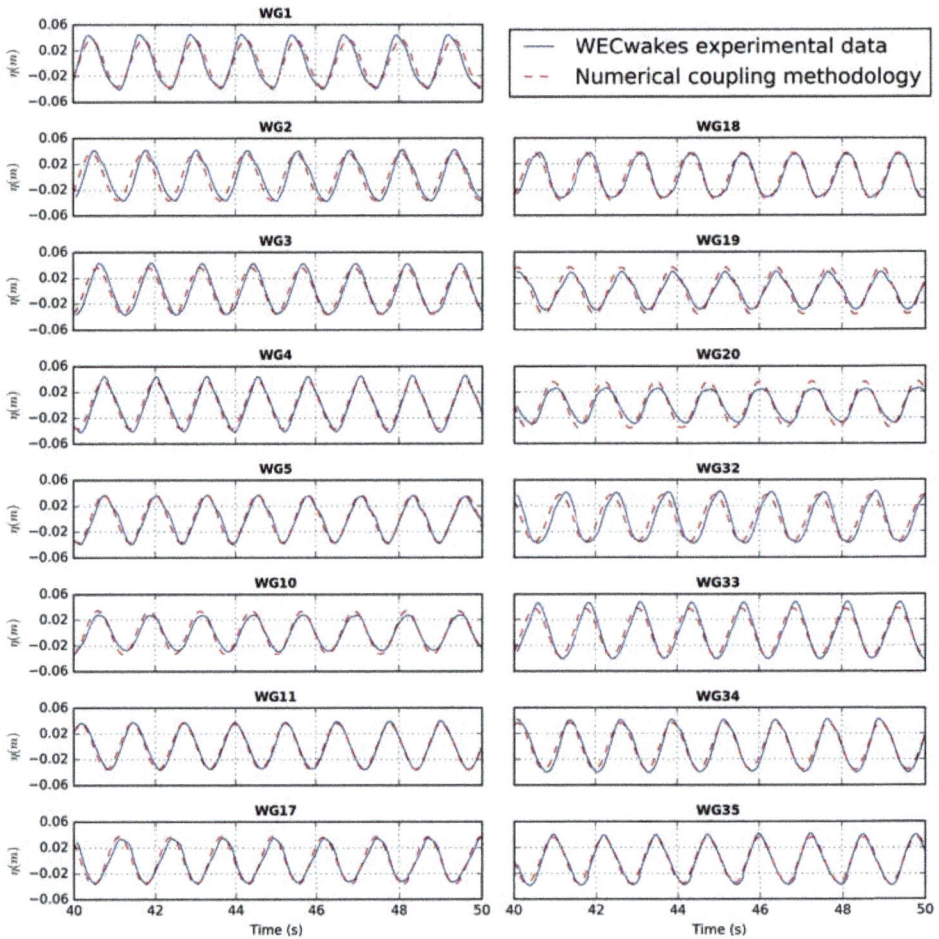

Figure 15. Surface elevations η for the NEMOH-MILDwave coupled model and the WECwakes experimental data for a total of 15 wave gauges shown in Figure 8.

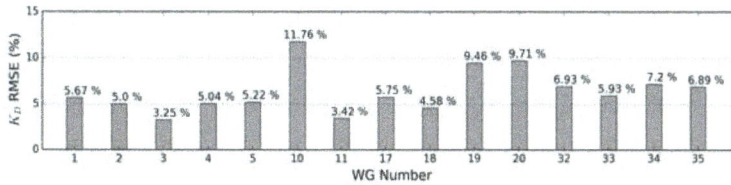

Figure 16. Root Mean Square Error (RMSE) values for the free surface elevation η for the 15 wave gauges analyzed from the data set (see Figure 8).

4.3. Coupling Implementation for Varying Bathymetry

The K_D are given in Figure 17 for both regular and irregular waves. Again, a wave reflection pattern is formed in front of the WEC array and a reduction in K_D values appears in the lee of the WEC array. However, the presence of the bathymetry is changing the magnitude of the wave radiation and wave diffraction effects. As mentioned in Section 4.2, there is a reduction in the extent of the wake effects for higher periods and for irregular waves. In contrast to the constant bathymetry case, the extension of the wake effect tends to disappear with an increasing wave period for a varying bathymetry. Furthermore, there is a reduction in the wave reflection pattern up-wave. This is due to the shoaling effect that is expected to be bigger on waves with a larger wave length and consequently has a higher impact in higher wave periods than the generated wake behind the WEC array which is lower. Only for $T = 6$ s does the array total wave field modification outweigh the shoaling effects and thus the WEC array has an impact both up-wave and down-wave. For the other two wave periods modelled, the effect of the WEC array is practically nil in the region down-wave the WEC array where the shoaling effects are much greater than the wake effects.

To have a closer look at the effects of shoaling over the wave field generated around the WEC array, a longitudinal cross-section S1 is taken at the center of the domain, $y = 0$ m (Figure 18), for the constant and irregular bathymetry test cases. The cross-sections shown on the left side of the figures correspond to regular waves, while the cross sections shown on the right side of the figures correspond to irregular waves. The coupling region is masked out in gray between two vertical black lines, as the results between NEMOH and the NEMOH-MILDwave coupled model cannot be compared. Consistently, the influence of shoaling over wave reflection and wave diffraction effect of the WEC array is reduced with a reduction in the wave period. It can be observed that the changes in the wave reflection magnitude are due to the reduction of the incident wave caused by bottom induced propagation effects in the wave propagation model MILDwave. On the contrary, the reduction of the extent of the wake effects is generated by the increase of the wave height as the shoaling effect takes over.

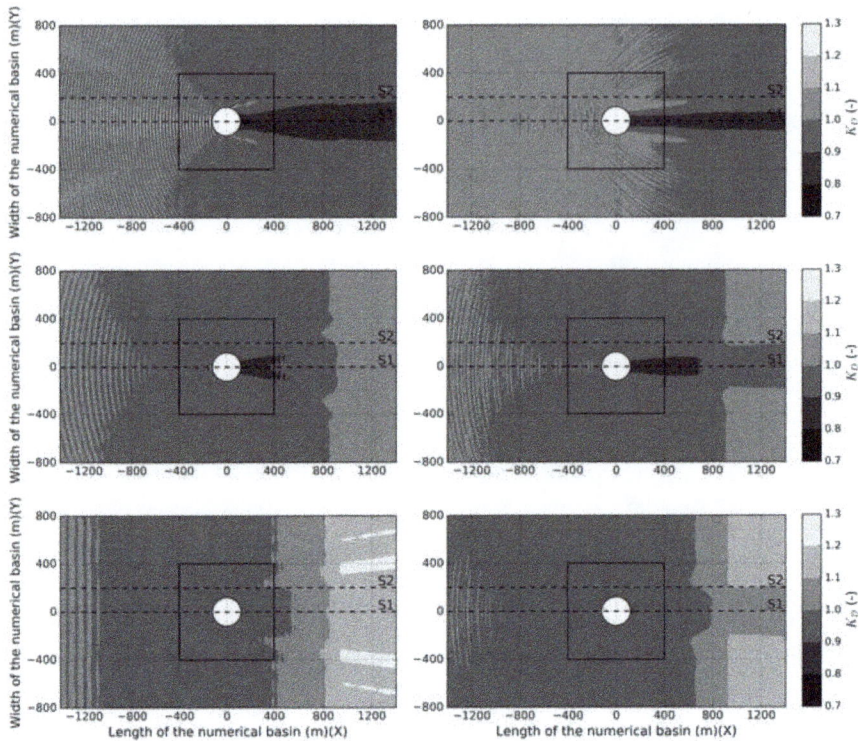

Figure 17. NEMOH-MILDwave couple model K_D values of a 9-WEC array for regular waves (**left**) with $T = 6$ s, $T = 8$ s and $T = 10$ s and for irregular waves (**right**) with $T_p = 6$ s, $T_p = 8$ s and $T_p = 10$ s with a slopping bathymetry. Contour levels are set at an interval 0.04 m. Waves are propagating from left to right. The coupling region is masked out with a white circle.

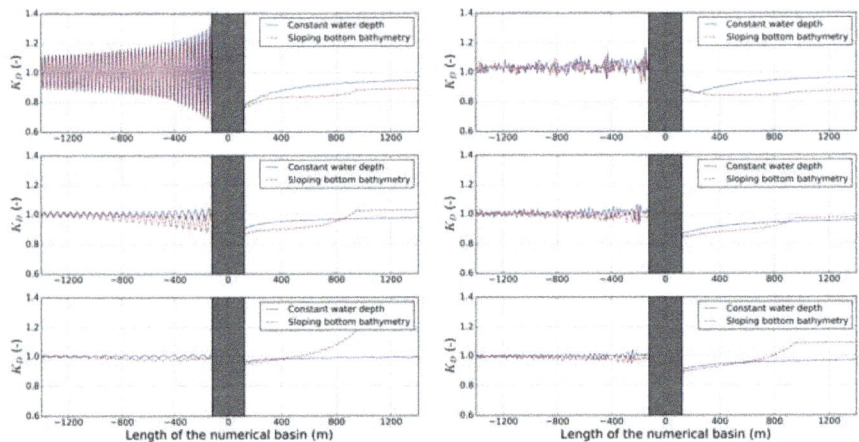

Figure 18. Cross-section S1 of the K_D for a 9-WEC array at $y = 0$ m for regular waves (**left**) with $T = 6$ s, $T = 8$ s and $T = 10$ s and for irregular waves (**right**) with $T_p = 6$ s, $T_p = 8$ s and $T_p = 10$ s at a constant depth and a sloping bathymetry. The coupling region is masked out in gray between two vertical black lines.

5. Discussion

In Section 4.1, it can be seen that the NEMOH-MILDwave coupled model can accurately propagate the total wave field around a WEC array according to linear wave theory. Nevertheless, there are some discrepancies for K_D between NEMOH and the NEMOH-MILDwave coupled model close to the coupling region. These relative differences for K_D remain under 3%. Moreover, when moving away from the coupling region these relative differences in K_D are reduced. These reductions in the error are showing that the complexity of the hydrodynamic interactions when modelling the far field effects is not that influential, as wave diffraction and wave radiation effects diminish with distance. As a result, it can be concluded that the NEMOH-MILDwave coupled model is able to replicate the numerical NEMOH simulation and extend it to a larger domain. Furthermore, the simulations in Section 4.1 have been extended including a varying bathymetry in Section 4.3. Those simulations have given good results, showing the effect of bottom induced propagation effects over the wave reflection and wake effects caused by the WEC array in the incident wave. Subsequently, it has been demonstrated that the NEMOH-MILDwave coupled model can be used to analyze the impacts of a WEC array in a wave field over a varying bathymetry over a large coastal zone.

However, coupling NEMOH and MILDwave has some limitations. Firstly, despite the fact that the time spent to compute a small WEC array in this study is not very long, it can increase considerably when increasing the number of WECs present. The NEMOH simulations are performed for nine WECs with one degree of freedom (DOF) and one direction. For an array of J WECs with six DOFs, the computational time of a BEM model increases as σ^{6J}, with increased simulation time in larger domains. Consequently, the maximum number of WECs that can be used is limited. Secondly, NEMOH calculations can only be performed on a constant bathymetry. This assumption is valid for closely spaced WECs in deep water. However, when the WECs are placed in intermediate or shallow waters, the effects of the bathymetry are significant as shown in Section 4.2 and according to [18]. It is necessary to study the influence of the bathymetry in closely space WEC arrays in order to define the extent of the WEC array that can be modelled. Thirdly, irregular waves are calculated as a superposition of regular waves. To increase the accuracy of the results, an increase in the number of regular wave components is required, resulting in more computational time. Finally, a heaving buoy modelled with a passive resistive PTO has an effect on the surrounding areas that is highly influenced by wave diffraction and not so much by wave radiation. This clearly shows the importance of the frequency distribution in the WEC array effects and the need for a realistic modelling of the WEC PTO to maximize the power output and thus quantify its effect on the surrounding wave field [37].

In terms of limitations of the proposed coupling methodology, these depend each time on the type of models that are coupled. Specifically for coupling between two linear models such as NEMOH and MILDwave which are used here, the resulting coupled model will provide conservative results in study cases when nonlinear phenomena are dominating. Moreover, MILDwave is applied for mild-slope bathymetries. On the other hand, the above limitations can be overcome when applying the proposed coupling methodology, if, for instance, nonlinear models are coupled (which however often introduce computational instability and high computational cost).

6. Conclusions

In this paper, a generic coupling methodology has been presented to calculate far field effects of floating structures. The wave propagation model MILDwave and the wave structure interaction solver NEMOH are combined to model the wave field propagating around a WEC array under regular and irregular waves over a constant or varying bathymetry. Pairing models with different resolutions enables the modeler to obtain accurate results within a reasonable computational time. It has been demonstrated that there is a good agreement between the NEMOH-MILDwave coupled model and the NEMOH solution in the close proximity of the coupling region for both regular and irregular waves. As a result, it can be assumed that the NEMOH-MILDwave coupled model can be extended to larger domains. Furthermore, a good agreement is obtained between numerical and experimental results. Even though there are some discrepancies between the numerical and experimental results, these discrepancies are mainly caused due to non-linear effects during the experiments. The numerical results have been extended to a varying bathymetry. It is seen that the bathymetry highly influences the total wave field around the WEC array and thus the coupling methodology presented provides itself as a useful numerical tool for wave energy studies assessing the coastal impact of WECs arrays within linear wave theory. The numerical results have also shown some limitations regarding computational time, the number of WECs modelled and the accuracy of PTO modelling. Regardless of this limitations, the NEMOH-MILDwave coupled model introduced has proven to be a reliable tool that can be applied in a fast and efficient way to calculate far field effects of WEC arrays. The next steps in our modelling work is to extend the methodology to real bathymetries in the MILDwave domain to study the effect of WEC arrays on a real case scenario. As MILDwave correctly models coastal transformation processes, this will allow for calculating the impact of a wave farm on any particular coastal area given any type of WEC. Different array and PTO configurations will be tested in order to accurately represent the wave absorption of commercially viable WECs.

Author Contributions: G.V.F. set up the numerical experiments in collaboration with P.B.; G.V.F. performed the numerical experiments; G.V.F. compared the numerical experiments with experimental data with the assistance of P.B.; V.S. and P.T. have offered the WECwakes project experimental data-set; V.S. and P.T. provided the fundamentals of the coupling methodology; V.S. and P.T. proofread the text and helped in structuring the publication.

Funding: This research received no external funding.

Conflicts of Interest: The authors declare no conflict of interest.

Abbreviations

The following abbreviations are used in this manuscript:

WEC	Wave Energy Converter
BEM	Boundary Element Method
CFD	Computer Fluid Dynamics
SPH	Smoothed Particle Hydrodynamics
PTO	Power Take-Off
RAO	Response Amplitude Operator
DHI	Danish Hydraulic Institute
WG	Wave Gauge
RMSE	Root-Mean-Square-Error

References

1. Stratigaki, V. Experimental Study and Numerical Modelling of Intra-Array Interactions and Extra-Array Effects of Wave Energy Converter Arrays. Ph.D. Thesis, Ghent University, Gent, Belgium, 2014.
2. Child, B.M.F.; Venugopal, V. Optimal Configurations of wave energy devices. *Ocean Eng.* **2010**, *37*, 1402–1417. [CrossRef]
3. Garcia-Rosa, P.B.; Bacelli, G.; Ringwood, J. Control-Informed Optimal Array Layout for Wave Farms. *IEEE Trans. Sustain. Energy* **2015**, *6*, 575–582. [CrossRef]
4. Babarit, A. On the park effect in arrays of oscillating wave energy converters. *Renew. Energy* **2013**, *58*, 68–78. [CrossRef]
5. Borgarino, B.; Babarit, A.; Ferrant, P. Impact of wave interaction effects on energy absorbtion in large arrays of Wave Energy Converters. *Ocean Eng.* **2012**, *41*, 79–88. [CrossRef]
6. Devolder, B.; Rauwoens, P.; Troch, P. Towards the numerical simulation of 5 Floating Point Absorber Wave Energy Converters installed in a line array using OpenFOAM. In Proceedings of the 12th European Wave and Tidal Energy Conference (EWTEC 2017), Cork, Ireland, 27 August–1 September 2017; pp. 739–749.
7. Abanades, J.; Greaves, D.; Iglesias, G. Wave farm impact on the beach profile: A case study. *Coast. Eng.* **2014**, *86*, 36–44. [CrossRef]

8. Iglesias, G.; Carballo, R. Wave farm impact: The role of farm-to-coast distance. *Renew. Energy* **2014**, *69*, 375–385. [CrossRef]

9. Millar, D.L.; Smith, H.C.M.; Reeve, D.E. Modelling analysis of the sensitivity of shoreline change to a wave farm. *Ocean Eng.* **2007**, *34*, 884–901. [CrossRef]

10. Venugopal, V.; Smith, G. Wave Climate Investigation for an Array of Wave Power Devices. In Proceedings of the 7th European Wave and Tidal Energy Conference, Porto, Portugal, 11–13 September 2007; p. 10.

11. Smith, H.C.M.; Millar, D.L.; Reeve, D.E. Generalisation of wave farm impact assessment on inshore wave climate. In Proceedings of the 7th European Wave and Tidal Energy Conference, Porto, Portugal, 11–13 September 2007.

12. Beels, C.; Troch, P.; De Backer, G.; Vantorre, M.; De Rouck, J. Numerical implementation and sensitivity analysis of a wave energy converter in a time-dependent mild-slope equation model. *Coast. Eng.* **2010**, *57*, 471–492. [CrossRef]

13. Folley, M.; Babarit, A.; Child, B.; Forehand, D.; O'Boyle, L.; Silverthorne, K.; Spinneken, J.; Stratigaki, V.; Troch, P. A Review of Numerical Modelling of Wave Energy Converter Arrays. In Proceedings of the ASME 2012 31st International Conference on Ocean, Offshore and Arctic Engineering (OMAE), Rio de Janeiro, Brazil, 1–6 July 2012.

14. Lee, C.H.; Newman, J.N. *WAMIT User Manual, Versions 6.4, 6.4 PC, 6.3, 6.3S-PC*; WAMIT, Inc.: Chestnut Hill, MA, USA, 2006.

15. Babarit, A.; Delhommeau, G. Theoretical and numerical aspects of the open source BEM solver NEMOH. In Proceedings of the 11th European Wave and Tidal Energy Conference, Nantes, France, 6–11 September 2015.

16. Balitsky, P.; Verao Fernandez, G.; Stratigaki, V.; Troch, P. Coupling methodology for modelling the near-field and far-field effects of a Wave Energy Converter. In Proceedings of the ASME 36th International Conference on Ocean, Offshore and Arctic Engineering (OMAE2017), Trondheim, Norway, 25–30 June 2017.

17. Verbrugghe, T.; Stratigaki, V.; Troch, P.; Rabussier, R.; Kortenhaus, A. A comparison study of a generic coupling methodology for modeling wake effects of wave energy converter arrays. *Energies* **2017**, *10*, 1697. [CrossRef]

18. Charrayre, F.; Peyrard, C.; Benoit, M.; Babarit, A. A Coupled Methodology for Wave-Body Interactions at the Scale of a Farm of Wave Energy Converters Including Irregular Bathymetry. In Proceedings of the ASME 2014 33rd International Conference on Ocean, Offshore and Arctic Engineering, San Francisco, CA, USA, 8–13 June 2014.

19. Tomey-Bozo, N.; Murphy, J.; Lewis, T.; Troch, P.; Thomas, G. Flap type wave energy converter modelling into a time-dependent mild-slope equation model. In Proceedings of the 2nd International Conference on Offshore Renewable Energies, Lisbon, Portugal, 24–26 October 2016; pp. 277–284.

20. Verbrugghe, T.; Devolder, B.; Dominguez, J.; Kortenhaus, A.; Troch, P. Feasibility study of applying SPH in a coupled simulation tool for wave energy converter arrays. In Proceedings of the 12th European Wave and Tidal Energy Conference (EWTEC 2017), Cork, Ireland, 27 August–1 September 2017; pp. 679–689.

21. Troch, P. *MILDwave—A Numerical Model for Propagation and Transformation of Linear Water Waves*; Technical Report; Department of Civil Engineering, Ghent University: Ghent, Belgium, 1998.

22. Alves, M. Wave Energy Converter modelling techniques based on linear hydrodynamic theory. *Numer. Model. Wave Energy Convert.* **2016**, *1*, 11–65.

23. Verbrugghe, T.; Troch, P.; Kortenhaus, A.; Stratigaki, V.; Engsig-Karup, A.P. Development of a numerical modelling tool for combined near field and far field wave transformations using a coupling of potential flow solvers. In Proceedings of the 2nd International Conference on Renewable Energies Offshore, Lisbon, Portugal, 24–26 October 2016.

24. Stratigaki, V.; Troch, P.; Stallard, T.; Forehand, D.; Kofoed, J.P.; Folley, M.; Benoit, M.; Babarit, A.; Kirkegaard, J. Wave Basin Experiments with Large Wave Energy Converter Arrays to Study Interactions between the Converters and Effects on Other Users. *Energies* **2014**, *7*, 701–734. [CrossRef]

25. Stratigaki, V.; Troch, P.; Stallard, T.; Forehand, D.; Folley, M.; Kofoed, J.P.; Benoit, M.; Babarit, A.; Vantorre, M.; Kirkegaard, J. Sea-state modification and heaving float interaction factors from physical modelling of arrays of wave energy converters. *J. Renew. Sustain. Energy* **2015**, *7*, 061705. [CrossRef]

26. Penalba, M.; Touzón, I.; Lopez-Mendia, J.; Nava, V. A numerical study on the hydrodynamic impact of device slenderness and array size in wave energy farms in realistic wave climates. *Ocean Eng.* **2017**, *142*, 224–232. [CrossRef]

27. Troch, P.; Stratigaki, V. Phase-Resolving Wave Propagation Array Models. *Numer. Model. Wave Energy Convert.* **2016**, *10*, 191–216.

28. Radder, A.C.; Dingemans, M.W. Canonical equations for almost periodic, weakly nonlinear gravity waves. *Wave Motion* **1985**, *7*, 473–485. [CrossRef]

29. Beels, C.; Troch, P.; Kofoed, J.P.; Frigaard, P.; Kringelum, J.V.; Kromann, P.C.; Donovan, M.H.; De Rouck, J.; De Backer, G. A methodology for production and cost assessment of a farm of wave energy converters. *Renew. Energy* **2011**, *36*, 3402–3416. [CrossRef]

30. Stratigaki, V.; Troch, P.; Baelus, L.; Keppens, Y.U. Introducing wave generation by wind in a mild-slope wave propagation model MILDwave, to investigate the wake effects in the lee of a farm of wave energy converters. In Proceedings of the ASME 2011 30th International Conference on Ocean, Offshore and Arctic Engineering, Rotterdam, The Netherlands, 19–24 June 2011,

31. Brorsen, M.; Helm-Petersen, J. On the Reflection of Short-Crested Waves in Numerical Models. In Proceedings of the 26th International Conference on Coastal Engineering, Copenhagen, Denmark, 22–26 June 1998; pp. 394–407.

32. National Renewable Energy Laboratory. Marine and Hydrokinetic Technology Database. Available online: https://openei.org/wiki/Marine_and_Hydrokinetic_Technology_Database_759 (accessed on 1 June 2018).

33. Faltinsen, O.M. *Sea Loads on Ships and Offshore Structures*; Cambridge University Press: Cambridge, UK, 1990; p. 328.

34. Sismani, G.; Babarit, A.; Loukogeorgaki, E. Impact of Fixed Bottom Offshore Wind Farms on the Surrounding Wave Field. *Int. J. Offshore Polar Eng.* **2017**, *27*, 357–365. [CrossRef]

35. Devolder, B.; Rauwoens, P.; Troch, P. Numerical simulation of a single floating point absorber wave energy converter using OpenFOAM. In Proceedings of the 2nd International Conference on Renewable energies Offshore, Lisbon, Portugal, 24–26 October 2016; pp. 197–205.

36. Devolder, B.; Stratigaki, V.; Troch, P.; Rauwoens, P. CFD simulations of floating point absorber wave energy converter arrays subjected to regular waves. *Energies* **2018**, *11*, 641. [CrossRef]

37. Child, B.M.F. On the Configuration of Arrays of Floating Wave Energy Converters. Ph.D. Thesis, The University of Edinburgh, Edinburgh, UK, 2011.

![energies logo] **energies**

Article

Adaptive Sliding Mode Control for a Double Fed Induction Generator Used in an Oscillating Water Column System

Oscar Barambones *[iD], Jose M. Gonzalez de Durana and Isidro Calvo

Faculty of Engineering of Vitoria-Gasteiz, University of the Basque Country (UPV/EHU), Nieves Cano 12, 01006 Vitoria-Gasteiz, Spain; josemaria.gonzalezdedurana@ehu.eus (J.M.G.d.D.); isidro.calvo@ehu.eus (I.C.)
* Correspondence: oscar.barambones@ehu.es; Tel.: +34-945-013-235; Fax: +34-945-013-270

Received: 23 September 2018; Accepted: 25 October 2018; Published: 27 October 2018

Abstract: Wave power conversion systems are nonlinear dynamical systems that must endure strong uncertainties. Efficiency is a key issue for these systems, and the application of robust control algorithms can improve it considerably. Wave power generation plants are typically built using variable speed generators, such as the doubly fed induction generator (DFIG). These generators, compared with fixed speed generators, are very versatile since the turbine speed may be adjusted to improve the efficiency of the whole system. Nevertheless, a suitable speed controller is required for these systems, which must be able to avoid the stalling phenomenon and track the optimal reference for the turbine. This paper proposes a sliding mode control scheme aimed at oscillating water column (OWC) generation plants using Wells turbines and DFIGs. The contributions of the paper are (1) an adaptive sliding mode control scheme that does not require calculating the bounds of the system uncertainties, (2) a Lyapunov analysis of stability for the control algorithm against system uncertainties and disturbances, and (3) a validation of the proposed control scheme through several simulation examples with the Matlab/Simulink suite. The performance results, obtained by means of simulations, for a wave power generation plant (1) evidence that this control scheme improves the power generation of the system and (2) prove that this control scheme is robust in the presence of disturbances.

Keywords: renewable energy; wave energy; adaptive SMC; DFIG; lyapunov methods

---

1. Introduction

During the last decade, the use of renewable energy sources (mainly wind and solar) have gained increasing attention for several reasons such as the reduction of pollution caused by traditional sources of energy (mainly fossil fuels) and the removal of dependence on exhausted resources. Nowadays, the scientific community is focusing on wave energy due to its potential for supplying a considerable part of the electricity demand in some countries [1–5].

The wind over the ocean surface is a major cause of the waves. In many areas of the world, the wind generates continuous waves that may be converted into power in a consistent way. Following diverse approaches, several devices are being designed to harness the massive energy of the waves, either by extracting energy from the waves motion over the ocean surface or by benefiting from the pressure fluctuations below the surface [6–8]. In this work, a Wells turbine-based oscillating water column (OWC) device is used for converting the energy of the waves into mechanical energy [9]. The obtained mechanical energy depends on the wave characteristics—mainly height, speed, wavelength, and water density [10,11].

The doubly fed induction generator (DFIG) is used at many renewable power generation plants (e.g., wind turbines and wave energy plants) since they are able to adapt to sudden changes [12–16].

In these machines, the stator is connected directly to the grid, but the rotor is connected to the grid by means of a variable frequency converter (VFC). This configuration benefits from the fact that the VFC only manages a fraction, around (25–30%), of the nominal power for controlling the generator. Typically, one voltage source of the VFC is located at the grid side, the so-called grid-side converter (GSC), another is located at the rotor side, the so-called rotor-side converter (RSC), and a capacitor is used to connect back to back both converters [17–19].

The most frequent strategy to control these systems is a combination of vector control with cascaded PI control for current and power [20]. However, the nonlinear nature of these systems and their uncertainties suggest that the use of a more robust controller could improve the system performance.

One alternative is considering the sliding mode control (SMC) technique. SMC has proven to be suitable at systems in which a precise model is known. In addition, it reduces the sensitivity to variation parameters and rejects external disturbances [21,22]. During the last decade, the SMC techniques have been applied to control several types of electrical devices obtaining good results [23–26].

Traditionally, in OWC systems, the rotational speed of the turbine is adjusted taking into account the maximum pneumatic energy level available to the turbine [27–29]. However, in the control scheme proposed in this work, a new reference for the turbine speed, based on the airflow velocity, is proposed.

This work presents an SMC scheme aimed at improving the generation of power in an OWC system that incorporates a Wells turbine in order to transform the pneumatic energy into mechanical energy. The proposed approach is aimed at tracking the speed of the turbine in order to maximize power extraction. SMC theory and DFIG rotor current regulation were used to control the turbine speed. By regulating the speed of the turbine, the flow coefficients are optimized in order to maximize power generation even when there are uncertainties in the model or variations in the wave power.

The design of the traditional SMC scheme requires calculating an upper bound for the system uncertainties since the value of the sliding gain used must be higher. This value is directly related to the control effort, so a high sliding gain value implies a high control effort possibly causing the chattering problem. This problem arises because the sliding control signal is discontinuous across the sliding surfaces, so this kind of signal involves high control activity and may also excite the high frequency dynamic of the system. Therefore, the system uncertainties should be bounded accurately. In order to avoid this problem, the proposed control scheme includes an adaptive law to calculate the sliding gain.

The closed loop stability of the proposed control scheme was proven using the Lyapunov stability theory. The performance of this control scheme is validated by means of the Matlab/Simulink software. The conditions of operation, as well as the simulation results, are discussed in detail.

This paper is organized as follows: Section 2 presents a model for the plant used, the OWC. Section 3 introduces the design of the adaptive sliding mode control scheme. Section 4 presents the simulation results obtained. Section 5 presents the conclusions.

2. OWC Plant Model

In this work, an OWC system with a Wells turbine and a DFIG is considered.

Figure 1 shows how the OWC device uses the hydraulic energy of the waves to create an oscillating air flow that moves the turbine.

In an OWC system, the chamber collects the flow of air generated by the ocean waves. At the top of this chamber, there is a turbine, which is connected to a generator by means of a gear box. The airflow in the chamber is bidirectional since it depends on whether the waves hit or reflect. However, the Wells turbine rotates always in one direction, independently of the air flow direction, which has proven very useful to be used in maritime energy. This turbine was designed by Prof. Alan Arthur Wells of Queen's University Belfast at the end of the 1970s.

Energies **2018**, *11*, 2939

The power that can be extracted from the airflow in the chamber of the OWC is determined by [30]

$$P_{in} = \left(dP + \frac{\rho v_x^2}{2} \right) v_x a \qquad (1)$$

where $v_x(m/s)$ is the speed of the airflow, $dP(Pa)$ is the pressure drop at the turbine, and $a(m^2)$ is the area of the turbine duct.

It should be noted that the equations for the OWC system are quite similar to the equations obtained for the wind turbine systems. The term $\rho v_x^3 a/2$, that also appears in the wind turbines, takes into account the kinetic energy. The other term $dPv_x a$ takes into account the air pressure in the chamber and appears owing to an OWC chamber.

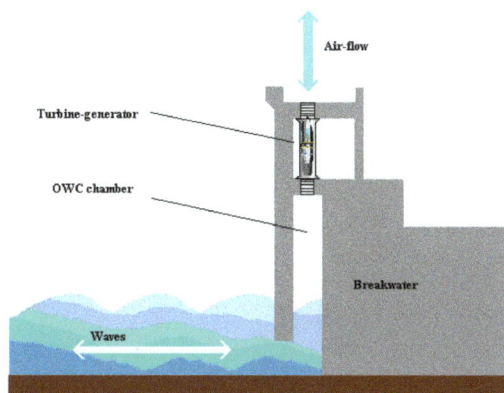

Figure 1. The oscillating water column (OWC) wave power system.

In this work, the turbogenerator module of the OWC is used with a Wells turbine, which is connected to a DFIG by means of a gearbox.

The DFIG is an induction machine typically used in renewable electric generation devices, such as wind turbines. A VFC is used to connect the rotor circuit of the DFIG to the grid, whereas the stator is directly connected. In order to deliver active power to the grid, regarding the frequency and voltage stability requirements of the grid, the flow between rotor and grid must be controlled in magnitude and direction. However, these machines are designed to operate at an extended range of rotational conditions, from subsynchronous to supersynchronous speed.

The use of this configuration allows that the power electronic converters deliver electrical power to the grid by managing a fraction of the nominal power, typically between 25 and 30%. The power generated at the stator goes directly to the grid, whereas the power obtained at the rotor is delivered by means of a VFC. As a result, the power converters needed can be relatively smaller when compared with other configurations. Moreover, this configuration provides additional benefits, such as improving the performance and reducing the cost of the equipment.

Typically, at this type of systems an inertia wheel drive is included for smoothing the curve of the output power delivered by the generator. This component facilitates the application of SMC schemes at these systems, since some phenomena that may appear, such as the chattering problem, can be absorbed by the inertia wheel.

This work assumes the use of Wells turbines for extracting the energy of the airflow. These turbines are designed to always turn in the same direction, independently of the sense of the airflow that moves the turbine, providing a robust behavior. The key at these turbines is the design of the blade, which is simple and symmetrical. The mathematical model for these turbines is given by the following equations [31]:

$$dP = C_a k_t \frac{1}{a} \left[v_x^2 + (r w)^2 \right] \tag{2}$$

$$T_t = C_t k_t r \left[v_x^2 + (r w)^2 \right] \tag{3}$$

$$T_t = \frac{C_t r a}{C_a} dP \tag{4}$$

$$k_t = \rho b n \frac{l}{2} \tag{5}$$

$$\phi = \frac{v_x}{r w} \tag{6}$$

$$q = v_x a \tag{7}$$

$$\eta = \frac{T_t w}{q \, dP} \tag{8}$$

where $T_t(Nm)$ is the generated torque at the Wells turbine, ϕ is the flow coefficient, C_a is the power coefficient, C_t is the torque coefficient, $k_t(Kg \cdot m)$ is a turbine specific constant, $r(m)$ is the radius or the turbine, $v_x(m/s)$ is the speed of the airflow, $w(rad/s)$ is the rotational speed of the turbine, $b(m)$ and $l(m)$ are constants related to the blades—respectively, the blade height and the blade chord length—n is the number of blades, $q(m^3/s)$ is the flow rate, $a(m^2)$ is the cross-sectional area of the turbine, and η is the efficiency of the turbine.

According to the equations of the presented model, both the torque and power generated by a Wells turbine can be obtained from the torque and power coefficients, respectively. The characteristic curves of the Wells turbine show the relationship between the torque and power coefficients versus the flow coefficient.

The performance of a Wells turbine decreases severely when the speed of the airflow goes beyond a critical value, which depends on the rotational speed of the turbine [31]. This phenomenon is known as stalling, since the Wells turbine stalls when the relative angle formed by the axial speed of the input airflow and the tangential velocity of the turbine surpasses a specific value. This value is typically close to $14°$.

Figure 2 depicts the torque versus flow coefficients ($\phi = \frac{v_x}{r w}$) for a specific Wells turbine. It can be seen in the figure that the value of C_t, the torque coefficient of the turbine, decreases severely when the turbine stalls (due to the stalling phenomenon). Even though this figure was obtained for a specific turbine and in general the behavior depends on some constructive parameters, the characteristic curves for diverse turbines present similar features.

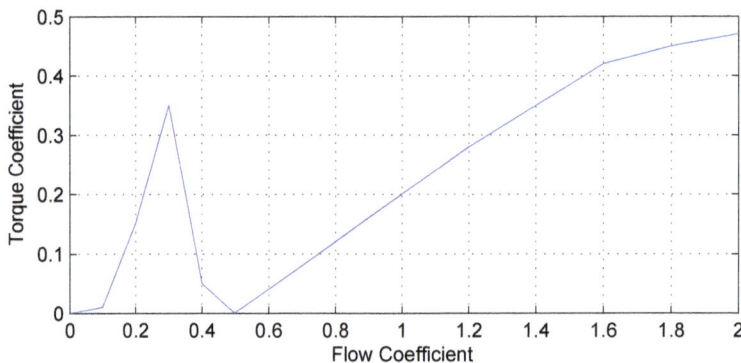

Figure 2. Torque coefficient versus flow coefficient.

For this specific turbine, Figure 2 illustrates that, when the flow coefficient reaches the value of $\phi = 0.3$, stalling occurs. As remarked above, the value for the flow coefficient differs depending on the characteristic curve for a specific Wells turbine. It may be concluded from the picture that a value for the flow coefficient of around $\phi_{opt} = 0.29$ is a good alternative. On one side, this value avoids the phenomenon of stalling and maximizes the value for torque C_t.

Since the movement of the airflow is caused by the ocean waves, which have an oscillatory nature, the value of the flow coefficient is consequently oscillating between zero and a positive value. Thus, the optimal zone of operation for the turbine is located from zero to the point at which the stalling phenomenon appears.

Taking into account the previous considerations and using Equation (6), the reference value for the turbine speed that maximizes the power extraction from the ocean waves can be obtained from

$$w^* = \frac{v_x}{r \cdot \phi_{opt}}. \tag{9}$$

Therefore, the flux coefficient in the Wells turbine may be optimized using an adequate turbine speed regulation that follows the reference value given by Equation (9).

3. Adaptive Sliding Mode Control Scheme for an OWC

In a DFIG-based OWC wave energy converter, the extracted power may be maximized by regulating the rotational speed of the Wells turbine. Thus, the value for the flux coefficient can be optimized while avoiding the stalling phenomenon. This objective may be achieved by means of an adaptive SMC scheme that controls the rotational velocity at the DFIG, which may be achieved using the quadrature component of the current at the rotor.

In the discussion about the model of the OWC system, it was pointed out that the shaft speed for the turbogenerator should be adjusted below the stalling phenomenon. Thus, the flow coefficient ϕ value provides an optimum value that extracts the maximum power from the sea waves. This would produce the maximum value for the torque coefficient C_t while avoiding stalling.

For a given pressure drop input dP, there is only a unique value for the speed reference that satisfies the condition for finding the optimum flow coefficient that maximizes the extraction of energy from the waves. This value can be obtained from the characteristic curve supplied by the manufacturer of the turbine. Obviously, this procedure is valid for any Wells turbine since they always present a similar stalling phenomenon.

This approach for designing the control scheme is quite similar to other well-established DFIG-based applications. The control strategies aimed at extracting the maximum power in wind turbines follow a similar approach. In such applications, the controller should follow the tracking reference in order to control the speed of the turbine. The tracking reference depends on the maximum variation of the power with the angular velocity for the DFIG. Obviously, the reference should differ since the aerodynamic characteristics for every turbine are also different, provided that different characteristic curves form the power coefficient versus the tip speed ratio [32].

The design of the adaptive SMC is based on the dynamics of the turbogenerator. Consequently, a dynamic model for the turbogenerator must be obtained. The dynamics of the turbogenerator involves both mechanical and electrical areas.

The following equation models the dynamics for the mechanical part of the turbogenerator:

$$J\dot{w} + Bw = T_t - \gamma T_e \tag{10}$$

where T_t is the turbine torque, generated by the airflow induced by the waves, B is the viscosity coefficient, T_e is the torque generated by the generator, J is the moment of inertia, w is the rotational velocity for the turbine, and the gear ratio $\gamma = w_e/w$ yields the ratio between the rotational speeds for the turbine, w, and the generator rotor, w_e.

The use of vector control techniques simplifies the DFIG model, since all expressions are referred to the stator flux reference frame. The d-axis is aligned with the stator flux linkage vector ψ_s; thus, $\psi_{ds} = \psi_s$ and $\psi_{qs} = 0$. The following expressions are thus obtained [33]:

$$i_{qs} = -\frac{L_m i_{qr}}{L_s} \tag{11}$$

$$i_{ds} = \frac{L_m(i_{ms} - i_{dr})}{L_s} \tag{12}$$

$$i_{ms} = \frac{v_{qs} - r_s i_{qs}}{w_s L_m} \tag{13}$$

$$T_e = -\frac{3p}{4}\frac{L_m^2 i_{ms} i_{qr}}{L_s} \tag{14}$$

$$v_{qr} = r_r i_{qr} + \sigma L_r \frac{di_{qr}}{dt} + sw_s\left(\sigma L_r i_{dr} + \frac{L_m^2 i_{ms}}{L_s}\right) \tag{15}$$

$$v_{dr} = r_r i_{dr} + \sigma L_r \frac{di_{qr}}{dt} - sw_s \sigma L_r i_{qr} \tag{16}$$

where i_{qs} and i_{ds} are the q-d components for the current at the stator, i_{qr} and i_{dr} are the q-d components for the current at the rotor, L_s, L_r, and L_m are the stator inductance, rotor inductance, and mutual inductances, respectively, i_{ms} is the current for magnetizing the stator, w_s is the angular velocity of the synchronous reference, $sw_s = w_s - w_e$ is the slip frequency, w_e is the velocity for the rotor at the generator, v_{qr} and v_{dr} are the q-d components of the voltage at the rotor, $\sigma = 1 - \frac{L_m^2}{L_s L_r}$ and p is the pole numbers.

The current for magnetizing the stator (i_{ms}) may be taken as a constant value due to that the stator resistance has small influence and the stator is directly connected to the grid [17]. Accordingly, the electromagnetic torque can be calculated as follows:

$$T_e = -K_T i_{qr} \tag{17}$$

where the torque constant value K_T is calculated below:

$$K_T = \frac{3p}{4}\frac{L_m^2 i_{ms}}{L_s}. \tag{18}$$

Substituting Equation (17) into Equation (10), a new expression for the dynamics is obtained. This expression includes the uncertainty terms:

$$\dot{w} = -(c_1 + \triangle c_1)w + (f + \triangle f) - (c_2 + \triangle c_2)i_{qs} \tag{19}$$

where $c_1 = \frac{B}{J}$, $c_2 = \frac{\gamma K_T}{J}$, $f = \frac{T_l}{J}$, and the terms $\triangle c_1$, $\triangle c_2$, and $\triangle f$ represent the uncertainties of the terms c_1, c_2, and f, respectively.

The dynamic Equation (19) can then be rewritten as

$$\dot{w} = -c_1 w + f + -c_2 i_{qs} + \Delta(t) \tag{20}$$

where all terms related to the uncertainty have been represented as Δ:

$$\Delta(t) = -\triangle c_1\, w(t) + \triangle f(t) - \triangle c_2\, i_{qr}(t). \tag{21}$$

Considering the properties of the SMC scheme, the proposed adaptive sliding mode compensates the uncertainties of the system.

The tracking error for the rotational velocity is defined with the following expression:

$$e(t) = w(t) - w^*(t) \tag{22}$$

where w^* is the reference for the turbine speed that maximizes power extraction from ocean waves.

The time derivative of the previous expression is

$$\dot{e}(t) = \dot{w} - \dot{w}^* = -c_1 w(t) + f(t) - c_2 i_{qr}(t) - \dot{w}^*(t) + \Delta(t). \tag{23}$$

An adaptive SMC scheme aimed at tracking the reference for the turbine velocity is then proposed. This controller is able to extract the maximum power from the waves in spite of the uncertainties at the plant.

The sliding variable $S(t)$ may be set as

$$S(t) = e(t) + \int_0^t k\,e(\tau)\,d\tau \tag{24}$$

where k represents a positive constant value for the gain.

The sliding surface proposed is as follows:

$$S(t) = e(t) + \int_0^t k\,e(\tau)\,d\tau = 0. \tag{25}$$

As a consequence, the sliding mode controller for the velocity is obtained:

$$i_{qr}^*(t) = \frac{1}{c_2}\left[k\,e + \hat{\beta}\,\gamma\,\text{sgn}(S) - c_1\,w - \dot{w}^* + f\right] \tag{26}$$

where $\hat{\beta}$ is the estimated switching gain, k is the gain defined previously, S is the sliding variable defined in Equation (24), $\text{sgn}(\cdot)$ is the signum function, and γ is a positive constant.

It should be noted that the control Equation (26) includes some terms that compensate the known dynamics of the system ($c_1\,w - \dot{w}^* + f$) and other terms that compensate the uncertainties ($k\,e + \hat{\beta}\,\gamma\,\text{sgn}(S)$).

The updating law for the switching gain $\hat{\beta}$ is defined as

$$\dot{\hat{\beta}}(t) = \gamma\,|S(t)| \qquad \hat{\beta}(0) = 0 \tag{27}$$

where the adaptation speed for the switching gain can be selected using the positive parameter γ.

The following condition should be fulfilled to obtain the tracking signal for the speed reference:

($\mathcal{C}$1) There are a finite and positive switching gain, β, which verifies the following:

$$\beta > \Delta_{max}$$

where $\Delta_{max} \geq |\Delta(t)| \quad \forall\,t$.

Note that further knowledge is not required for the upper bounds for the uncertainties. This assumption implies that the system uncertainties are bounded.

The Lyapunov stability theory will be used for ensuring the closed loop stability of the proposed control scheme for the wave power plant that moves the DFIG. This scheme implements the control law of Equation (26) using the value of Equation (27) for obtaining the switching gain. The dynamics for the DFIG is provided by (19).

Proof. The Lyapunov function candidate may be declared as

$$V(t) = \frac{1}{2}S(t)S(t) + \frac{1}{2}\tilde{\beta}(t)\tilde{\beta}(t) \tag{28}$$

where $\tilde{\beta}(t) = \hat{\beta}(t) - \beta$ and $S(t)$ is the previously defined sliding variable. $\square$

The time derivative of the previous Lyapunov function candidate is

$$
\begin{aligned}
\dot{V}(t) &= S(t)\dot{S}(t) + \tilde{\beta}(t)\dot{\hat{\beta}}(t) \\
&= S \cdot [\dot{e} + ke] + \tilde{\beta}(t)\dot{\hat{\beta}}(t) \\
&= S \cdot [-c_1 w + f - c_2 i_{qr} - \dot{w}^* + \Delta + ke] + \tilde{\beta}\gamma|S| \\
&= S \cdot [-c_1 w + f - (ke + \hat{\beta}\gamma\operatorname{sgn}(S) - c_1 w - \dot{w}^* + f) - \dot{w}^* + \Delta + ke] \\
&\quad + (\hat{\beta} - \beta)\gamma|S| \\
&= S \cdot [\Delta - \hat{\beta}\gamma\operatorname{sgn}(S)] + \hat{\beta}\gamma|S| - \beta\gamma|S| \\
&= \Delta S - \hat{\beta}\gamma|S| + \hat{\beta}\gamma|S| - \beta\gamma|S| \tag{29} \\
&\leq |\Delta||S| - \beta\gamma|S| \\
&\leq |\Delta||S| - \Delta_{max}\gamma|S| \\
&\leq 0. \tag{30}
\end{aligned}
$$

It should be indicated that in this proof the condition ($\mathcal{C}1$) and the Equations (23), (24), (26), and (27) have been used.

Based on the previous results, it can be stated that $\dot{V}(t)$ is a negative definite function. Hence, $V(t)$ is a positive definite function, and, when $S(t)$ tends to infinity, $V(t)$ tends to infinity. Consequently, it can be concluded that the equilibrium at the origin $S(t) = 0$ is globally asymptotically stable by means of the stability theory of Lyapunov. Accordingly, when **time** tends to infinity, $S(t)$ goes to zero. This means that the trajectories that start out of the sliding surface will reach it and must remain in this sliding surface. This behavior is commonly called *sliding mode* [34].

When the sliding surface, Equation (25), is reached, both $S(t)$ and $\dot{S}(t)$ decrease to zero. At that time, the dynamic behavior of this system tracking problem can be represented by the following equation:

$$\dot{S}(t) = 0 \quad \Rightarrow \quad \dot{e}(t) = -k\,e(t). \tag{31}$$

Considering that k is a positive constant it can be concluded that $e(t)$ converges exponentially to zero.

Therefore, the presented sliding mode control scheme can be employed in order to regulate the speed of the Wells turbine for wave power generation plants. This control scheme has been designed in order to extract the maximum power from the ocean waves under some uncertainties in the parameters of the system and under unmodeled dynamics.

4. Simulation Results

In this section, the performance of the proposed adaptive SMC scheme designed to regulate the turbine speed in an OWC wave power generator system is analyzed by means of some simulation examples. The goal of this control scheme is to obtain the maximum electrical power using the mechanical power extracted from the waves. For that, the optimal turbine speed command (that yields the maximum mechanical power) should be tracked by the turbine speed. In the following simulations, the proposed optimum tracking control scheme (that tracks the oscillating dynamics of the air flow in order to optimize the flow coefficient value) is compared with a control scheme without this optimum tracking control (the controller does not track the air flow variations in order to optimize the flow coefficient value).

Figure 3 shows the block diagram of the proposed adaptive SMC scheme designed to regulate the speed of the turbine. In this figure, the block "Reference Generator" gives the reference for the turbine speed in order to obtain the optimum flow coefficient and accordingly the maximum power. This block is implemented by Equation (9). The block "SMC" is the proposed adaptive SMC that provides the rotor current control designed to track the reference speed. This block is implemented by Equation (26).

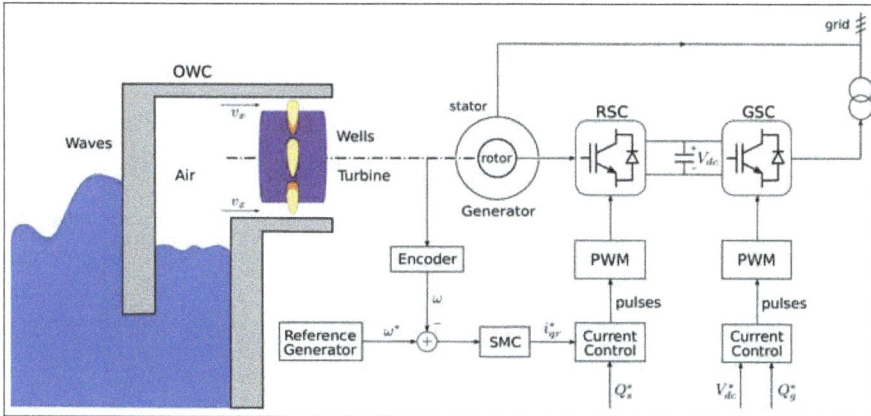

Figure 3. Block diagram of the proposed sliding mode control (SMC) scheme.

The Wells turbine used in the simulations has the following parameters:

- $n = 5$ (number of blades);
- $k = 0.7079$ Kg m (turbine torque coefficient);
- $r = 0.7285$ m (turbine radius);
- $a = 1.1763$ m^2 (turbine cross-sectional area);
- $b = 0.4$ m (turbine blade height);
- $l = 0.3$ m (turbine blade chord length).

The DFIG has the following parameters that are given in the per unit system:

- $P_{nom} = 27$ kW (nominal power);
- $p = 4$ (number of poles);
- $L_s = 0.18$ pu (stator inductance);
- $r_s = 0.023$ pu (stator resistance);
- $L_r = 0.16$ pu (rotor inductance);
- $r_r = 0.016$ pu (rotor resistance);
- $L_m = 2.9$ pu (mutual inductance);
- $H = 3.65$ s (inertia constant);
- $b = 0.01$ pu (friction factor).

Simulations were developed with the Matlab/Simulink software and the SimPowerSystems library [35].

In this control scheme validation, an uncertainty of 20% in the parameters of the system was considered. However, the simulation results show that the proposed control approach is able to cope with this uncertainty without deteriorating the system performance.

The next values for the parameters of the controller are selected: $k = 1.56$ and $\gamma = 2.34$. These values have been tuned experimentally considering their influence in the performance of the proposed control scheme. In this sense, the following rules should be considered in the selection of these parameters. An increase in the parameter k yields a decrease in the speed error convergence time. Unfortunately, this also increases the initial values of the control signal, because in the initial state the error is high, and this is undesirable in real applications. On the other hand, an increase in the parameter γ provides a faster adaptation dynamic for the sliding gain. Unfortunately, this also increments the final value of the sliding gain, which is not desirable in real applications.

In the first simulation example, it is considered that the ocean waves produce an oscillation in the pressure drop given by $dP = |7000\sin(0.1\pi t)|$ (Pa). This pressure variation is shown in Figure 4. This value for the pressure drop is considered low because it provides a flow coefficient values below the stalling behavior.

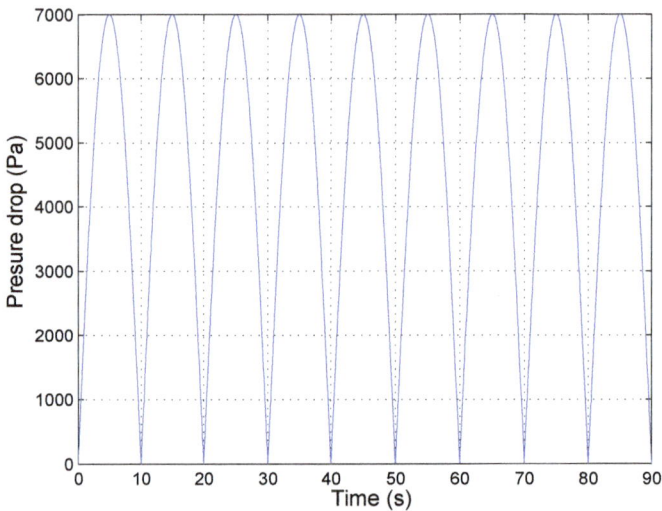

Figure 4. Low pressure drop for the first simulation example.

Figure 5 shows the turbine mechanical power produced by this pressure drop using the proposed adaptive sliding mode control in order to track the turbine speed command that provides the optimum flow coefficient value and hence the maximum power extraction. As can be observed, the average value of the mechanical power produced is 14.2 kW (green line). Figure 6 shows the electrical power generated whose average value is 10.7 kW (green line). Figure 7 shows the generator speed whose value is regulated using the proposed adaptive SMC that has been designed in order to optimize the flow coefficient and hence the mechanical power extraction from the Wells turbine system. Figure 8 shows the flow coefficient for this SMC. This figure shows that the speed regulation of the Wells turbine improves the flow coefficient values in order to maximize the mechanical power generation, because the flow coefficient is maintained, almost all of the time, close to the optimum value, $\phi_{opt} = 0.29$, which provides the maximum torque coefficient. It should be noted that the flow coefficient always have to decay to zero due to the oscillatory dynamics of the pressure drop that also decays to zero. However, the SMC tracking control increments the time in which the flow coefficient is maintained close to the optimum value. Figure 9 shows the time evolution of the sliding gain value, which is adapted online in order to overcome the system uncertainties. As can be seen in this figure, the sliding gain value starts from zero increasing its value until the sliding gain is high enough to compensate the uncertainties of the system. Figure 10 shows the sliding variable and Figure 11 shows the control signal. As can be observed, the control signal presents the chattering phenomenon that, as is well known, usually appears in the SMC schemes. However, this chattering will be filtered by the mechanical system inertia and it does not present a big problem in this case. Nevertheless, this chattering can be reduced by replacing the sing function with the saturation function [18].

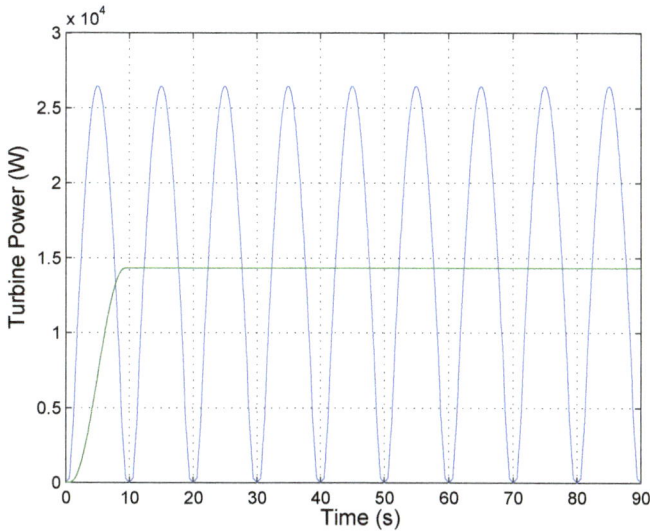

Figure 5. Turbine mechanical power for SMC tracking and low pressure drop.

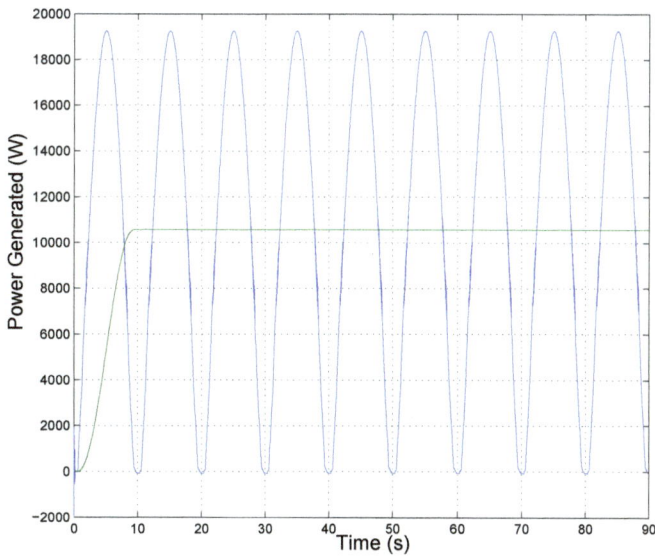

Figure 6. Electrical power generated for SMC tracking and low pressure drop.

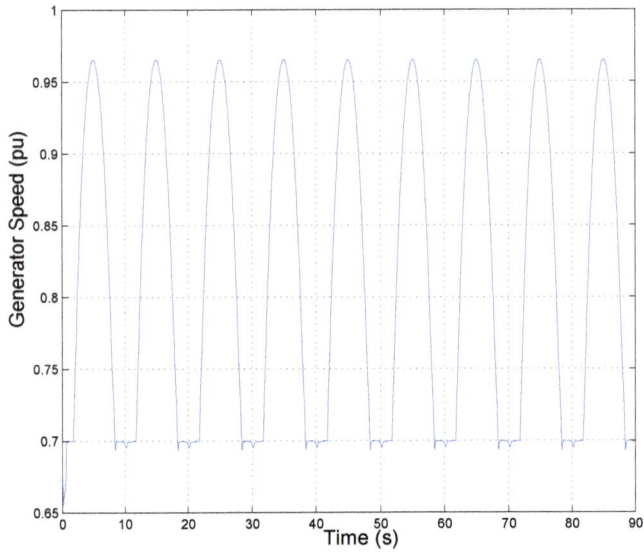

Figure 7. Generator speed for SMC tracking and low pressure drop.

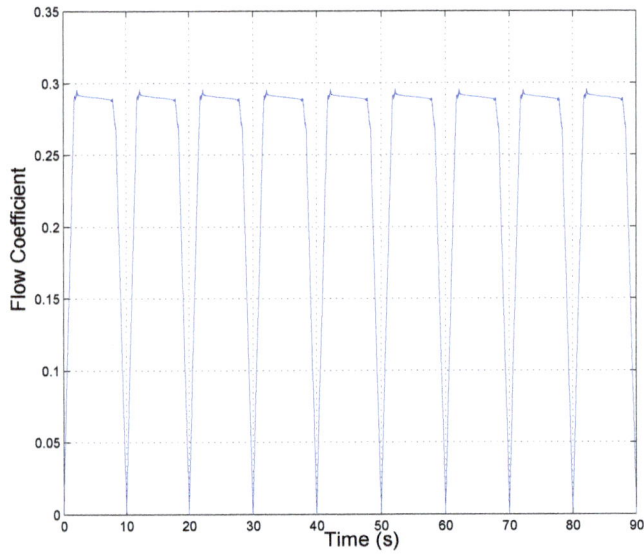

Figure 8. Flow coefficient for SMC tracking and low pressure drop.

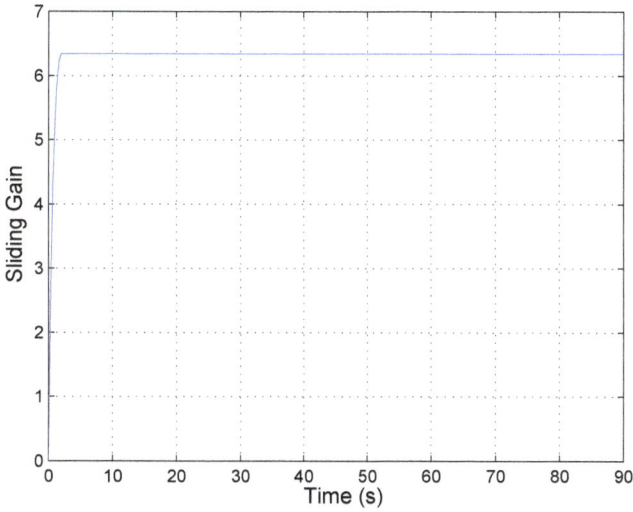

Figure 9. Sliding gain adaptation for SMC tracking and low pressure drop.

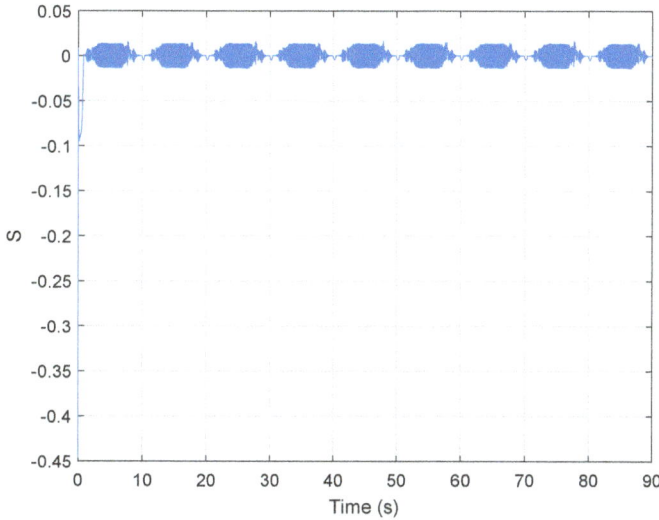

Figure 10. Sliding variable for SMC tracking and low pressure drop.

Figure 12 shows the turbine mechanical power produced by this pressure drop for this system without optimum tracking control; that is, in this case the turbine speed does not follow the reference value that provides the optimum flow coefficient. The average value of the mechanical power produced in this case is 10.8 kW (green line). Figure 13 shows the electrical power generated whose average value is 8.2 kW (green line). Figure 14 shows the flow coefficient obtained for this system without optimum tracking control. In this figure, it can be observed that the flow coefficient is not optimized in order to increment the mechanical power generation, and the flow coefficient is therefore not maintained close to the optimum value $\phi_{opt} = 0.29$. In this case, without optimum tracking control, the dynamics of the flow coefficient follows the oscillatory dynamics of the pressure drop.

Therefore, comparing Figures 6 and 13, it can be observed that the electrical power generated by an OWC-based wave power generation plat can be improved by controlling the generator speed in order to obtain the optimum flow coefficient value for the Wells turbine that produces the maximum wave energy extraction.

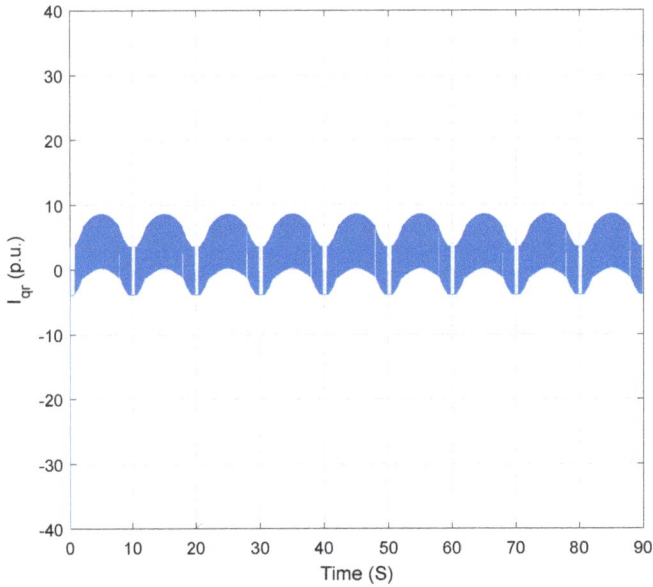

Figure 11. Control signal (i_{qr}) for SMC tracking and low pressure drop.

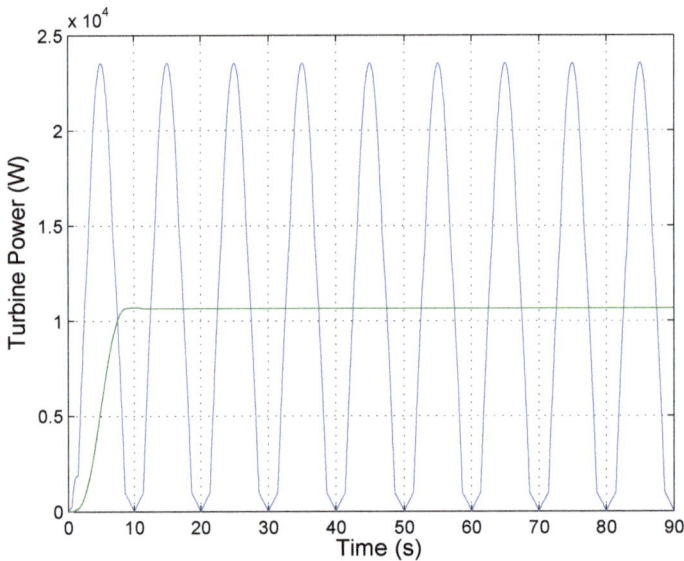

Figure 12. Turbine mechanical power without tracking and low pressure drop.

In the second simulation example, it is considered that the ocean waves produce an oscillation in the pressure drop given by $dP = |10,000\sin(0.1\pi t)|$ (Pa). This pressure variation is shown in Figure 15. This value for the pressure drop is considered high because it provides a flow coefficient that reaches the stalling behavior.

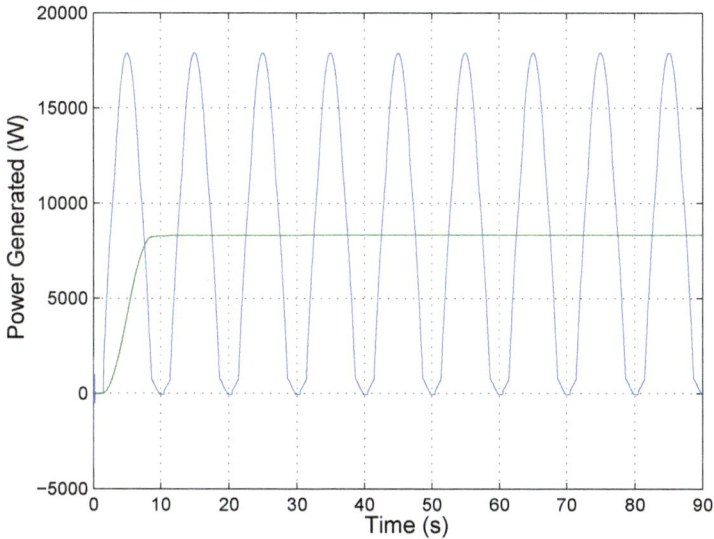

Figure 13. Electrical power generated without tracking and low pressure drop.

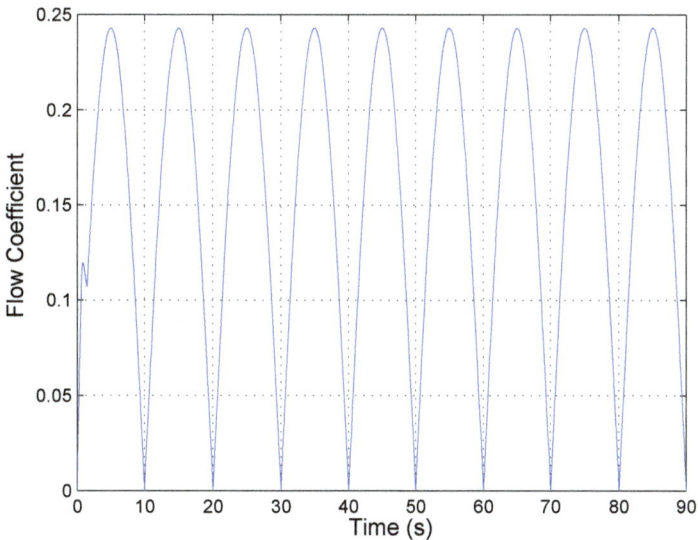

Figure 14. Flow coefficient without tracking and low pressure drop.

Figure 16 shows the turbine mechanical power produced by this pressure drop using this adaptive SMC. The average value of the mechanical power produced is 24.9 kW (green line). Figure 17 shows the electrical power generated whose average value is 18.3 kW (green line). Figure 18 shows the generator speed whose value is regulated by this adaptive sliding mode control in order to optimize the mechanical power extraction from the Wells turbine system.

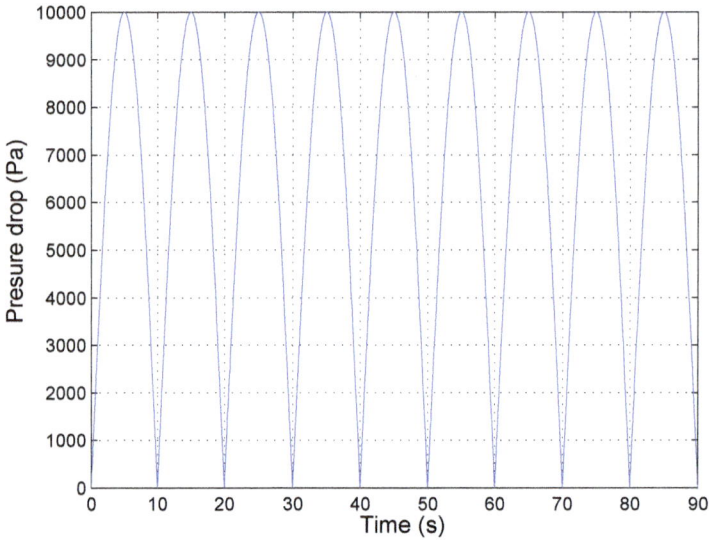

Figure 15. High pressure drop for the second simulation example.

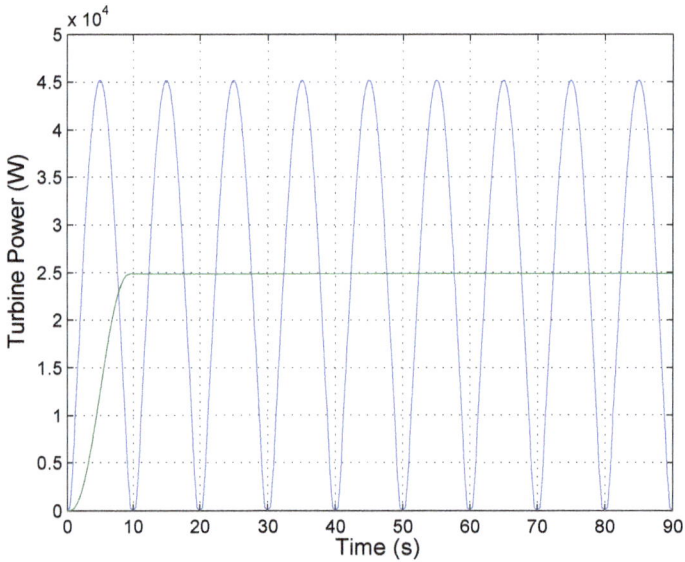

Figure 16. Turbine mechanical power for SMC tracking and high pressure drop.

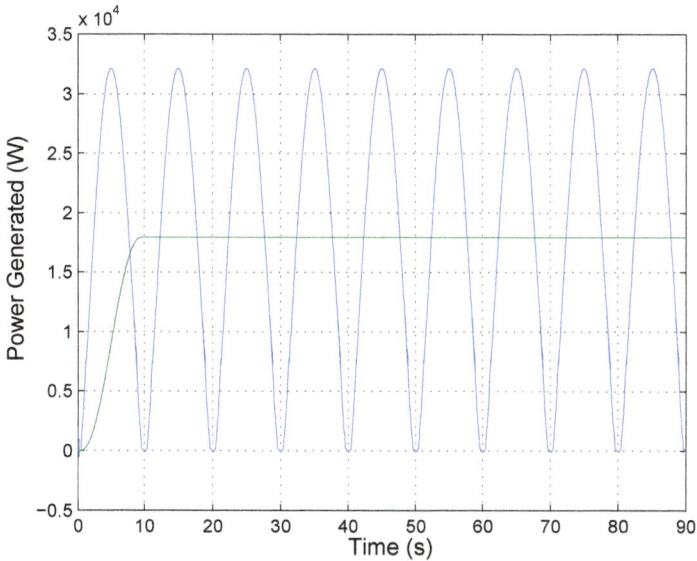

Figure 17. Electrical power generated for SMC tracking and high pressure drop.

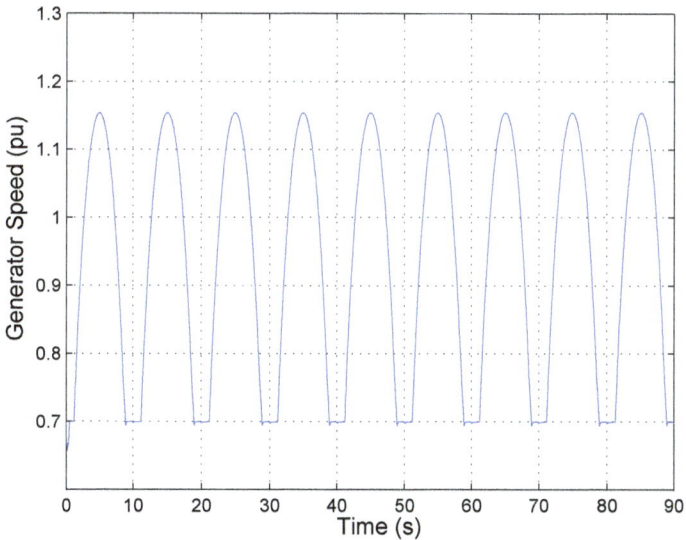

Figure 18. Generator speed for SMC tracking and high pressure drop.

Figure 19 shows the flow coefficient for the SMC case where the flow coefficient is maintained close to the optimum value. This figure shows that the speed regulation of the Wells turbine improves the flow coefficient values and hence optimizes the mechanical power generation. Moreover, the proposed speed regulation also avoids the stalling behavior that usually happens in the Wells turbine because the flow coefficient is maintained below the critical value $\phi = 0.3$. Figure 20 shows the sliding gain value that is adapted online in order to compensate the system uncertainties. As in the previous example, the sliding gain value is incremented until the value of the sliding gain can compensate the uncertainties of the system. Figure 21 shows the sliding variable and Figure 22 shows the control signal.

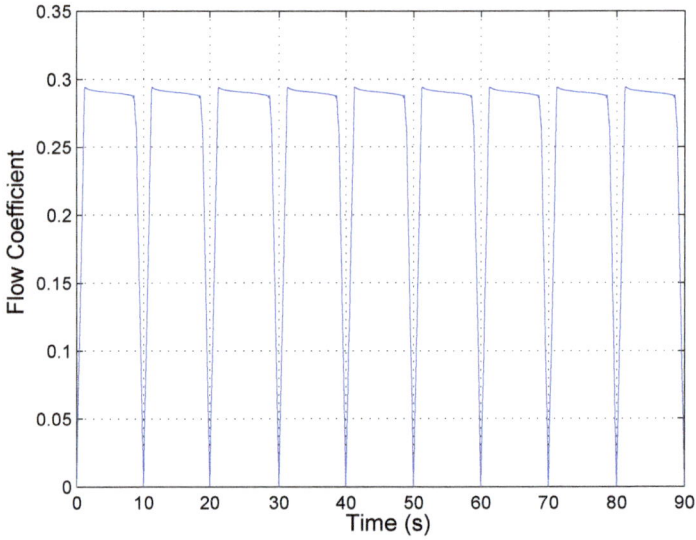

Figure 19. Flow coefficient for SMC tracking and high pressure drop.

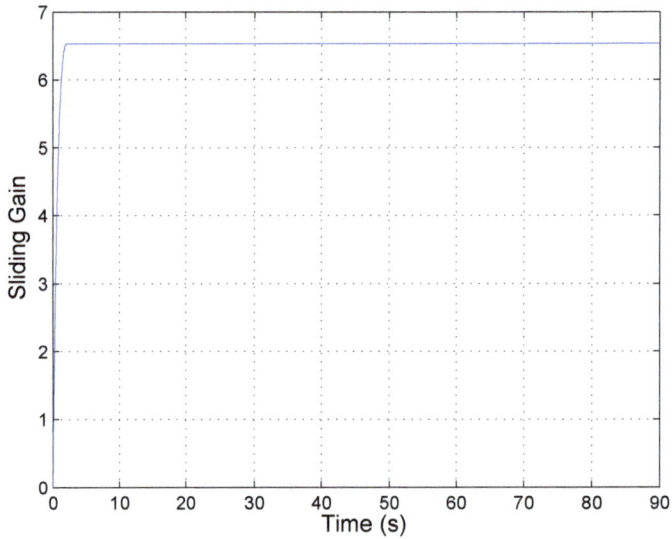

Figure 20. Sliding gain adaptation for SMC tracking and high pressure drop.

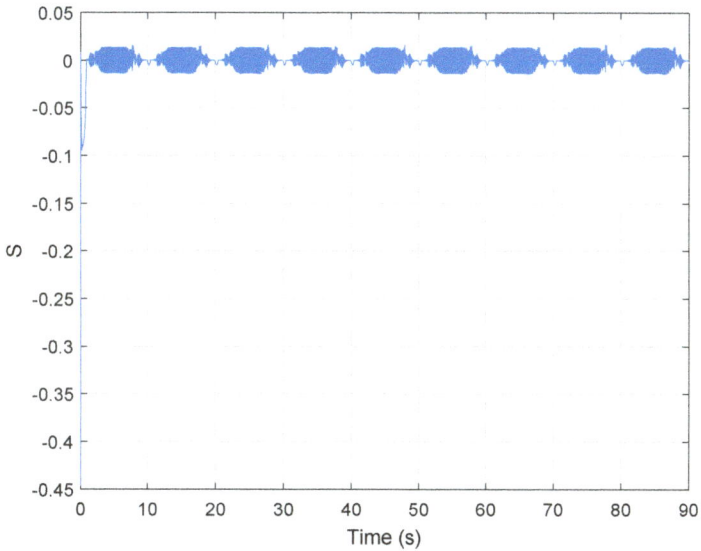

Figure 21. Sliding variable for SMC tracking and high pressure drop.

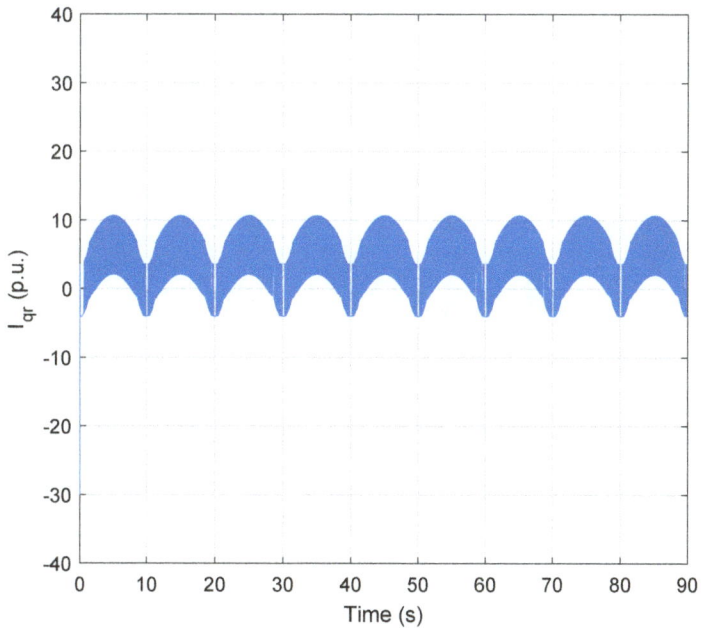

Figure 22. Control signal (i_{qr}) for SMC tracking and high pressure drop.

Figure 23 shows the turbine mechanical power produced by this pressure drop for this system without optimum tracking control. Unlike the previous optimum tracking SMC, the undesirable stalling behavior that produces power losses can be observed in this figure. The average value of the mechanical power produced in this case is 17.3 kW (green line). Figure 24 shows the electrical power generated whose average value is 12.9 kW (green line).

Figure 25 shows the flow coefficient for the case without optimum tracking control. In this figure, it can be seen that the flow coefficient exceeds the critical value $\phi = 0.3$, and the stalling behavior therefore appears in the dynamics of the Wells turbine because the flow coefficient is not optimized in order to increment the mechanical power generation.

Comparing Figures 17 and 24, it can be observed that the electrical power generated by an OWC system can be improved in two ways by means of the generator speed control. On the one hand, the optimum speed tracking improves the OWC system performance providing an optimum value of the flow coefficient for the Wells turbine that produces the maximum mechanical energy extraction from the ocean waves. On the other hand, the optimum speed tracking can also be used to avoid the stalling behavior in the Wells turbine dynamics, because the flow coefficient can be maintained below the critical value $\phi = 0.3$.

In the next simulation example, the proposed control scheme is evaluated under irregular waves. In this case, these irregular waves produce the pressure drop profile shown in Figure 26.

Figure 27 shows the turbine mechanical power produced by this pressure drop using the proposed control scheme. In this figure, the green line shows the average value of the produced mechanical power. Figure 28 shows the electrical power generated, and the average value of this power is also shown in the green line. Figure 29 shows the generator speed whose value is regulated by this adaptive SMC in order to optimize the mechanical power extraction from the Wells turbine system.

Figure 30 shows the flow coefficient for the SMC case under irregular waves scenario. This figure shows that the speed regulation of the Wells turbine improves the flow coefficient values and hence optimizes the mechanical power generation. Moreover, the proposed speed regulation also avoids the stalling behavior that usually happens in the Wells turbine because the flow coefficient is maintained below the critical value $\phi = 0.3$.

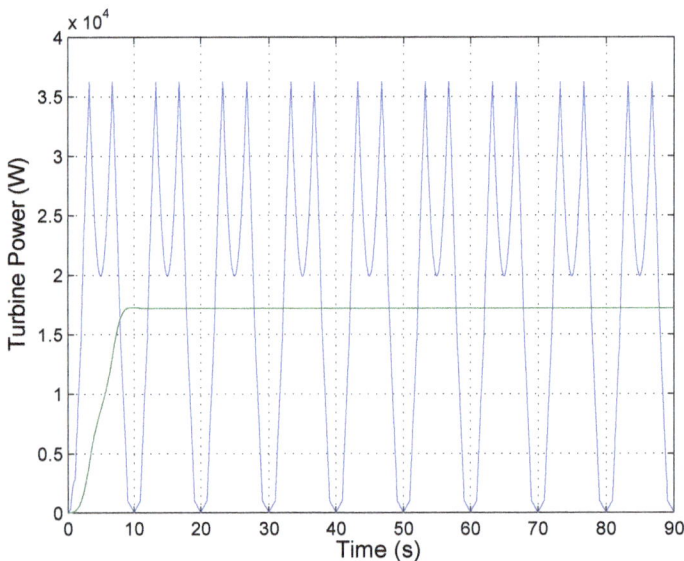

Figure 23. Turbine mechanical power without tracking and high pressure drop.

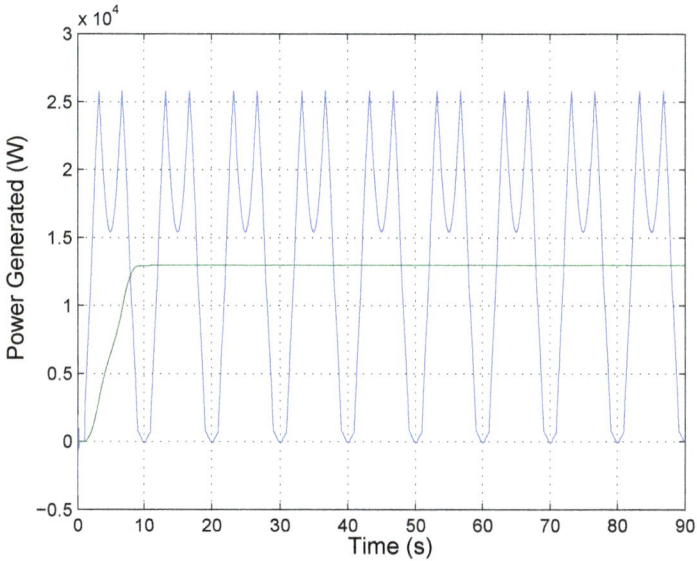

Figure 24. Electrical power generated without tracking and high pressure drop.

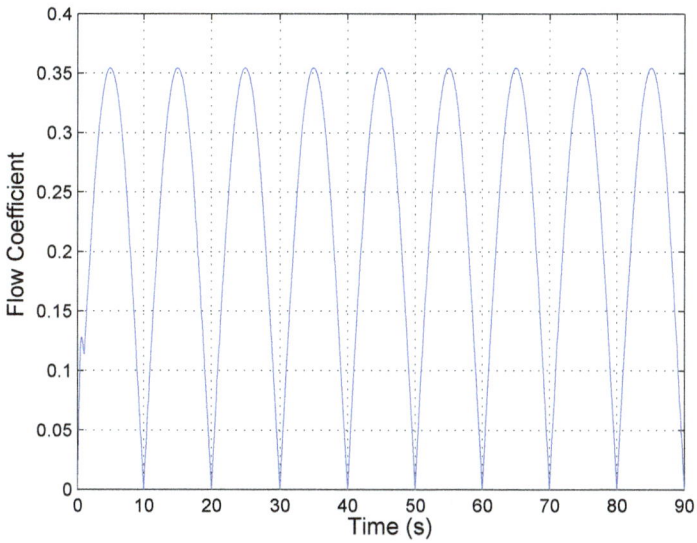

Figure 25. Flow coefficient without tracking and high pressure drop.

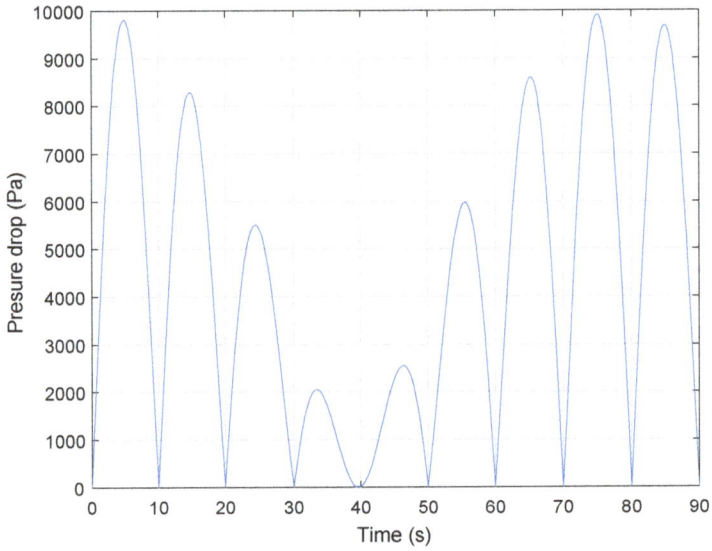

Figure 26. Pressure drop for the case of irregular waves.

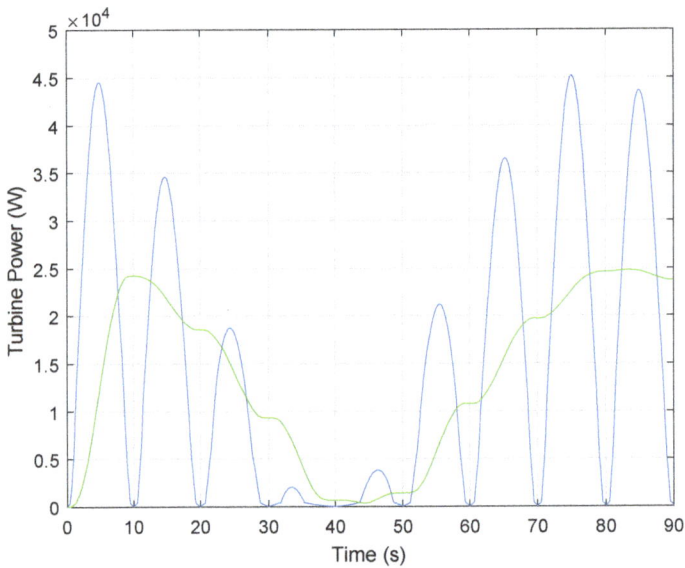

Figure 27. Turbine mechanical power for SMC tracking and irregular waves.

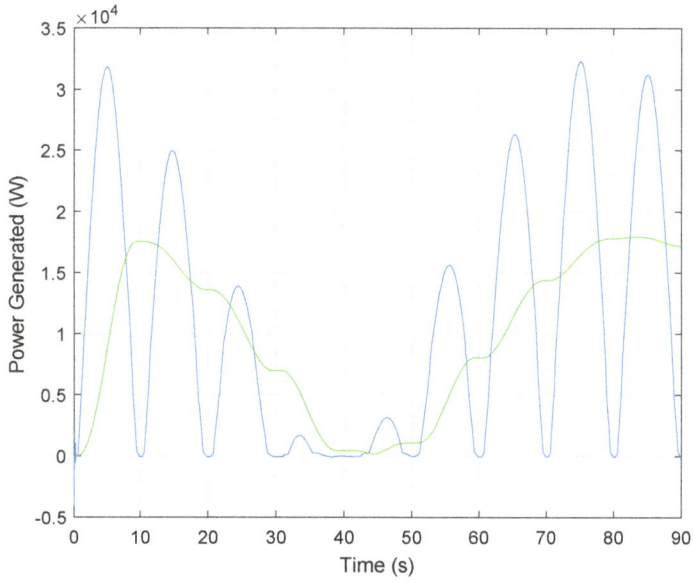

Figure 28. Electrical power generated for SMC tracking and irregular waves.

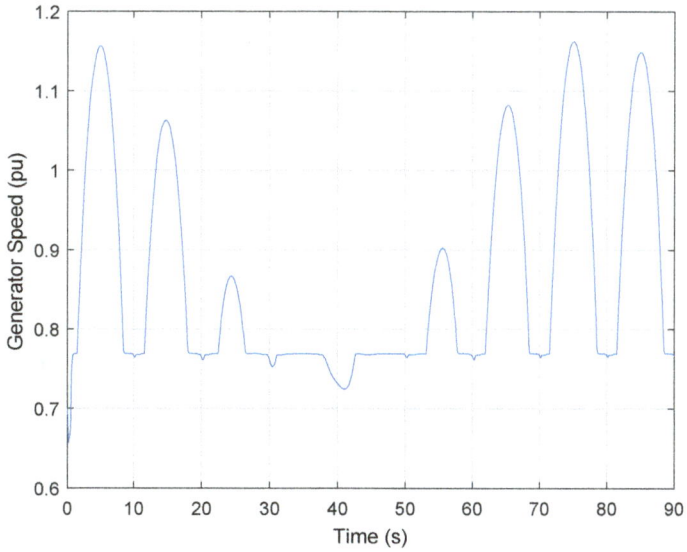

Figure 29. Generator speed for SMC tracking and irregular waves.

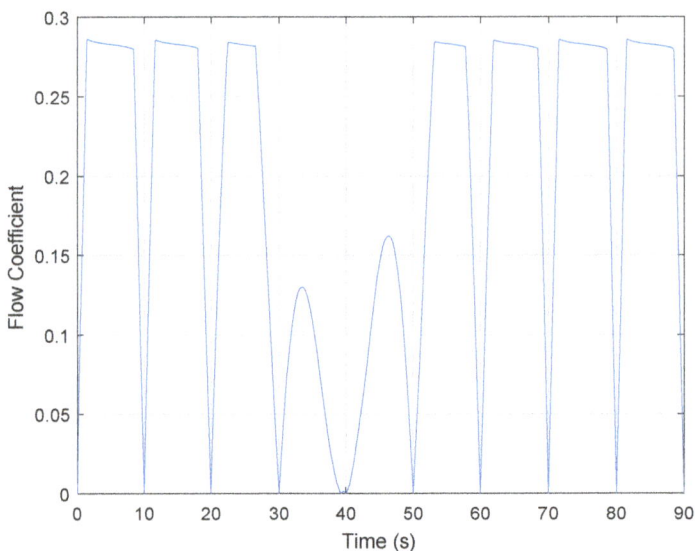

Figure 30. Flow coefficient for SMC tracking and irregular waves.

5. Conclusions

In this work, an adaptive SMC scheme for OWC wave power generation plants is proposed. It is assumed that the OWCs use Wells turbines and DFIGs. The SMC technique is inherently robust, so its application in such plants may reduce the effect of the uncertainties caused by the errors in the model of the plant as well as the disturbances of the system. However, in traditional SMC control schemes, it is necessary to calculate an upper limit for the system uncertainties in order to obtain an adequate value for the sliding gain. The SMC control scheme proposed in this work copes with this drawback by means of an adaptive switching gain value that can be calculated online. This work also analyzes the response and robustness of the proposed adaptive SMC, and it is concluded that the SMC is robust and presents a good dynamic response under system disturbances and modeling uncertainties.

Since the closed-loop stability of the proposed control scheme must be ensured, the Lyapunov stability theory was employed. The proposed design for the controller was successfully validated through several simulation examples with the Matlab/Simulink suite. Moreover, this work introduces a method to achieve the optimal flow coefficient value, while avoiding the stalling behavior. As a result, the Wells turbine is able to maximize the extraction of mechanical power from the ocean waves and, consequently, generate the maximum electrical power, improving the performance of the OWC-based wave power generation plans. The results of the simulations proved that the proposed adaptive SMC scheme adequately regulates the turbine speed command and therefore the optimum flow coefficient value; hence, the maximum power extraction can be obtained.

Author Contributions: This work was led by O.B. The theoretical development was performed by O.B. The modeling process was performed by O.B. and J.M.G.d.D. Simulation and analysis of the results was performed by O.B., J.M.G.d.D., and I.C. O.B., J.M.G.d.D., and I.C. participated in the writing, review, and editing of the manuscript. The final version was revised and has been approved by all the authors.

Funding: This research was partially funded by the Basque Government through the project ETORTEK KK-2017/00033 and by the UPV/EHU through the project PPGA18/04.

Acknowledgments: The authors wish to express their gratitude to the UPV/EHU and the Basque Government for supporting this work through the projects PPGA18/04 and ETORTEK KK-2017/00033.

Conflicts of Interest: The authors declare no conflict of interest.

Abbreviations

The following variables and symbols are used in this manuscript:

v_x	Airflow speed
dP	Pressure drop across the turbine
a	Area of turbine duct
ϕ	Flow coefficient
T_t	Torque generated by the Wells turbine
C_t	Torque coefficient
C_a	Power coefficient
k_t	Turbine constant
v_x	Air flow velocity
r	Turbine radius
w	Turbine angular velocity
n	Number of blades
b	Blade height
l	Blade chord length
a	Turbine cross-sectional area
q	Flow rate
η	Turbine efficiency
B	Viscous friction coefficient
T_e	Generator torque
J	Inertia moment
w	Angular velocity of the turbine shaft
w_e	Angular velocity of the generator rotor
i_{qs} and i_{ds}	q-d components of the stator current
i_{qr} and i_{dr}	q-d components of the rotor current
L_s, L_r, and L_m	Stator inductance, rotor inductance and mutual inductances
i_{ms}	Stator magnetizing current
w_s	Rotational speed of the synchronous reference frame
$sw_s = w_s - w_e$	Slip frequency
w_e	Generator rotor speed
v_{qr} and v_{dr}	q-d components of the rotor voltage
p	Pole numbers
$\hat{\beta}$	Estimated switching gain
S	Sliding variable
k	Positive constant (controller parameter)
γ	Positive constant (controller parameter)

References

1. Wan, Y.; Fan, C.; Zhang, J.; Meng, J.; Dai, Y.; Li, L.; Sun, W.; Zhou, P.; Wang, J.; Zhang, X. Wave Energy Resource Assessment off the Coast of China around the Zhoushan Islands. *Energies* **2017**, *10*, 1320. [CrossRef]
2. Wan, Y.; Fan, C.; Dai, Y.; Li, L.; Sun, W.; Zhou, P.; Qu, X. Assessment of the Joint Development Potential of Wave and Wind Energy in the South China Sea. *Energies* **2018**, *11*, 398. [CrossRef]
3. Wu, W.C.; Yang, Z.Q.; Wang, T.P. Wave resource characterization using an unstructured grid modeling approach. *Energies* **2018**, *11*, 605. [CrossRef]
4. Rusu, E. Numerical Modeling of theWave Energy Propagation in the Iberian Nearshore. *Energies* **2018**, *11*, 980. [CrossRef]
5. Stokes, C.; Conley, D. Modelling Offshore Wave farms for Coastal Process Impact Assessment: Waves, Beach Morphology, and Water Users. *Energies* **2018**, *11*, 2517. [CrossRef]
6. Son, D.; Yeung, R.W. Optimizing ocean-wave energy extraction of a dual coaxial-cylinder WEC using nonlinear model predictive control. *Appl. Energy* **2017**, *187*, 746–757. [CrossRef]
7. Aderinto, T.; Li, H. Ocean Wave Energy Converters: Status and Challenges. *Energies* **2018**, *11*, 1250. [CrossRef]

8. Zhou, Y.; Zhang, C.; Ning, D. Hydrodynamic Investigation of a Concentric Cylindrical OWC Wave Energy Converter. *Energies* **2018**, *11*, 985. [CrossRef]

9. Falcao, A.F.O.; Henriques, J.C.C. Oscillating-water-column wave energy converters and air turbines: A review. *Renew. Energy* **2016**, *85*, 1391–1424. [CrossRef]

10. Belibassakis, K.; Bonovas, M.; Rusu, E. A Novel Method for EstimatingWave Energy Converter Performance in Variable Bathymetry Regions and Applications. *Energies* **2018**, *11*, 2092. [CrossRef]

11. Rajapakse, G.; Jayasinghe, S.; Fleming, A.; Negnevitsky, M. Grid Integration and Power Smoothing of an OscillatingWater Column Wave Energy Converter. *Energies* **2018**, *11*, 1871. [CrossRef]

12. Kunwar, A.; Bansal, R.; Krause, O. Steady-state and transient voltage stability analysis of a weak distribution system with a remote doubly fed induction generator-based wind farm. *Energy Sci. Eng.* **2014**, *2*, 188–195. [CrossRef]

13. Barambones, O. Power Output maximization for Wave Power Generation Plants using an Adaptive Sliding Mode Control. In Proceedings of the 2013 International Conference on Renewable Energy Research and Applications (ICRERA), Madrid, Spain, 20–23 October 2013.

14. Rajapakse, A.; Jayasinghe, S.; Fleming, A.; Negnevitsky, M. A Novel DFIG Damping Control for Power System with High Wind Power Penetration. *Energies* **2016**, *9*, 521. [CrossRef]

15. Morshed, M.J.; Fekih, A. A new fault ride-through control for DFIG-based wind energy systems. *Electr. Power Syst. Res.* **2016**, *146*, 258–269. [CrossRef]

16. Morshed, M.J.; Fekih, A. A Fault-Tolerant Control Paradigm for Microgrid-Connected Wind Energy Systems. *IEEE Syst. J.* **2018**, *12*, 360–372. [CrossRef]

17. Pena, R.; Clare, J.C.; Asher, G.M. Doubly fed induction generator using back-to-back PWM converters and its application to variable-speed wind-energy generation. *Proc. Electr. Power Appl.* **1996**, *143*, 231–241. [CrossRef]

18. Barambones, O. Sliding Mode Control Strategy for Wind Turbine Power Maximization. *Energies* **2012**, *5*, 2310–2330. [CrossRef]

19. Xiong, L.; Li, Y.; Zhu, Y.; Yang, P.; Xu, Z. Coordinated Control Schemes of Super-Capacitor and Kinetic Energy of DFIG for System Frequency Support. *Energies* **2018**, *11*, 103. [CrossRef]

20. Taveiros, F.E.V.; Barros, L.S.; Costa, F.B. Back-to-back converter state-feedback control of DFIG (doubly-fed induction generator)-based wind turbines. *Energy* **2015**, *89*, 896–906. [CrossRef]

21. Polyakov, A.; Fridman, L. Stability notions and Lyapunov functions for sliding mode control systems. *J. Frankl. Inst.* **2014**, *351*, 1831–1865 [CrossRef]

22. Li, P.; Ma, J.; Zheng, Z. Robust adaptive sliding mode control for uncertain nonlinear MIMO system with guaranteed steady state tracking error bounds. *J. Frankl. Inst.* **2016**, *353*, 303–321 [CrossRef]

23. Barambones, O.; Alkorta, P.; de Durana, J.M.G. A real-time estimation and control scheme for induction motors based on sliding mode theory. *J. Frankl. Inst.* **2014**, *351*, 4251–4270 [CrossRef]

24. Merabet, A. Adaptive Sliding Mode Speed Control for Wind Energy Experimental System. *Energies* **2018**, *11*, 2238. [CrossRef]

25. Huang, X.; Zhang, C.; Lu, H.; Li, M. Adaptive reaching law based sliding mode control for electromagnetic formation flight with input saturation. *J. Frankl. Inst.* **2016**, *353*, 2398–2417 [CrossRef]

26. Farhat, M.; Barambones, O.; Sbita, L. A new maximum power point method based on a sliding mode approach for solar energy harvesting. *Appl. Energy* **2017**, *185*, 1185–1198. [CrossRef]

27. Garrido, I.; Garrido, Aitor J.; Alberdi, M.; Amundarain, M.; Barambones, O. Performance of an ocean energy conversion with DFIG sensorless control. *Math. Probl. Eng.* **2013**, 1–14. [CrossRef]

28. Falcão, A.F.O.; Henriques, J.C.C.; Gato, L.M.C. Rotational speed control and electrical rated power of an oscillating-water-column wave energy converter. *Energy* **2017**, *120*, 253–261. [CrossRef]

29. Mishra, S.K.; Purwar, S.; Kishor, N. Maximizing Output Power in Oscillating Water Column Wave Power Plants: An Optimization Based MPPT Algorithm. *Technologies* **2018**, *6*, 15. [CrossRef]

30. Alberdi, M.; Amundarain, M.; Maseda, F.J.; Barambones, O. Stalling behavior improvement by appropriately choosing the rotor resistance value in Wave Power Generation Plants. In Proceedings of the 2009 International Conference on Clean Electrical Power 2009, Capri, Italy, 9–11 June 2009; pp. 64–67.

31. Jayashankar, V.; Udayakumar, K.; Karthikeyan, B.; Manivannan, K.; Venkatraman, N.; Rangaprasad, S. Maximizing Power Output From A Wave Energy Plant. In Proceedings of the 2000 IEEE Power Engineering Society Winter Meeting. Conference Proceedings (Cat. No.00CH37077), Singapore, 23–27 January 2000; Volume 3, pp. 1796–1801.

32. Nagai, B.M.; Ameku, K.; Roy, J.N. Performance of a 3 kW wind turbine generator with variable pitch control system. *Appl. Energy* **2009**, *86*, 1774–1782. [CrossRef]

33. Barambones, O.; De La Sent, M.; Alkorta, P. A Robust Control of Double-feed Induction Generator for Wind Power Generation. In Proceedings of the Annual Conference of the IEEE Industrial Electronics Society, IECON 2009, Porto, Portugal, 3–5 November 2009; pp. 99–104.

34. Utkin, V.I. Sliding mode control design principles and applications to electric drives. *IEEE Trans. Ind. Electron.* **1993**, *40*, 26–36. [CrossRef]

35. Gilbert, S. and Patrice, B. *SimPowerSystems 5. User's Guide*; The MathWorks: Natick, MA, USA, 2005.

![energies logo] *energies*

MDPI

Article

A Study on Electrical Power for Multiple Linear Wave Energy Converter Considering the Interaction Effect

Qiao Li *[ID], **Motohiko Murai**[ID] **and Syu Kuwada**

Department of Environment and System Sciences, Yokohama National University, Yokohama,
Kanagawa 240-8501, Japan; murai-motohiko-pz@ynu.ac.jp (M.M.); kuwada-shu@jmuc.co.jp (S.K.)
* Correspondence: liqiao23@hotmail.com

Received: 20 September 2018 ; Accepted: 26 October 2018; Published: 1 November 2018

Abstract: A linear electrical generator can be used on wave energy converter for converting the kinetic energy of a floating structure to the electricity. A wave farm consists of multiple wave energy converters which equipped in a sea area. In the present paper, a numerical model is proposed considering not only the interference effect in the multiple floating structures, but also the controlling force of each linear electrical generator. In particular, the copper losses in the electrical generator is taken into account, when the electrical power is computed. In a case study, the heaving motions and electrical powers of the multiple wave energy converters are estimated in the straight arrangement and triangle arrangement. In addition, the average electrical power is analyzed in different distances of the floating structures. The aim of this paper is to clear the relationship between the interference effect and electric powers from wave energy converters. This will be useful for deciding the arrangement of multiple wave energy converters.

Keywords: wave energy converter; interaction effect; array condition

---

1. Introduction

Converting wave energy to electricity from ocean waves is one of the greatest attractions in ocean engineering. Till date, many different concepts have been proposed with the goal to convert the kinetic and potential energy of ocean waves [1,2]. Some significant types are given as follows: oscillating water column (OWC), over-topping device, hinged multi-module converter, point absorber etc. OWC is proposed based on the principle of wave induced air pressurization. Rezanejad, K. et al. [3] investigated the hydrodynamic performance of an OWC wave energy device using boundary integral equation method (BIEM) simulation and small scale model experiment. Viviano A. [4] discussed the scale effects of OWC by a small scale and a similar large scale device. Additionally, the feasibility study for a green touristic infrastructure by the installation of an OWC system in the port of Giardini Naxos [5]. An over-topping device captures sea water of incident waves in a reservoir above the sea level, then releases the water back to sea through turbines. An example of such a device is the Wave Dragon [6]. A hinged multi-module converter is made up of connected sections which flex and bend as waves pass, this motion is used to generate electricity. The effect of structural flexibility on the maximum wave energy conversion by two interconnected floaters is investigated [7]. A point absorber is a device that heave up and down on the surface of the water. Because of their small size, wave direction is not important for these devices [1]. A point absorber includes three technical parts: floater system which catches the wave energy, power take-off system and electrical energy generation system [8].

Floater system moves up and down on the water surface converting the wave energy into kinetic energy. Nagulan S. [8] provided the information on front end energy conversion of point absorber type wave energy converters. They indicated that the front end energy converter is the only responsible stage to capture as maximum energy as possible from the incoming wave and the natural frequency of the device should match with the wave frequency. Power take-off system converts the kinetic energy into electrical energy, and various control methods have been proposed. Method include resistive loading control (RL) [9], approximate complex-conjugate control (ACC) [10], model-predictive control (MPC) [11] and approximate complex-conjugate control considering generator copper losses (ACL) et al. Jorgen H. [12] and Dan-EI M.A. [13] compared some of these control methods performance by simulation. The experiment study of PTO systems was carried out by T. Taniguchi [14]. In his research, the motion of floating structure and electrical powers by different control systems have been investigated. Electrical energy system can be divided into linear generation, rotary generation and electro active polymer. Most of the PTO systems produce mechanical rotation but a few of them produce linear motion to energize linear generators [8]. Apart from these active control methods, nonlinear power capture mechanism such as bistable mechanism [15], tristable mechanism [16], snap through PTO system [17], negative stiffness [18] has also been researched as a hot spot in the recent years. These nonlinear power capture mechanism will assist the efficiency of the energy converter.

On the other hand, a wave farm includes multiple wave energy converters in a certain configuration. By using multiple wave energy converters, the wave conditions of them are varied because of the interference effect. The interaction phenomena is that the scattering waves from a floating body induces to the others. So, the interaction of waves effects to both in diffraction and radiation problem of the floating body. This means that the interference affects the performance of the effect generated electricity by the multiple WECs. In addition, the amount of the degree of the effect is changed by wave period and an arrangement of WECs. Several numerical methods have been proposed to analyze the response of arrays of wave energy converters to the incident wave climate and the resulting modification of wave conditions, particularly down-wave of such arrays [19]. Reviews of available modelling approaches and their applications are discussed in [20,21]. B. Borgarino [22] assessed the influence of distances between generic point-absorber WECs, and cleared that the electrical power increased in some distance compare to single one because of interference effects. His models include 9–25 cylinders arranged in squares, all of them have the same PTO characteristics. The experimental arrangement and the obtained database are presented by Vasiliki S. et al. [23]. They focus on the wave height in the arrangement.

In the present study, a point absorber WEC with a linear generator which proposed by "Linear-driving type Wave Energy Converter Project" of NEDO Japan is assumed. A numerical model is proposed to calculate the electrical power, considering both the controlling force and the interference effect in multiple floating structures. Especially, the copper loss in the linear electrical generator is taken into account. In the numerical model, the Three-dimensional Singularity Distribution Method (3D-SDM) and the dynamic equation of motions with controlling force are used. At first, the diffraction and radiation wave exciting forces from itself and other floating structures are calculated by 3D-SDM respectively. In addition, then, the motions of each floating structure are computed by the dynamic simultaneous equations of motions, here, the same and different PTO characteristics of each WEC will be discussed. Finally, the electrical power is computed by an absorbed power apart from a copper loss.

In past studies [22–24], the controlling force of each WEC in arrangement are the same to it in single condition. Thus, the interference effect is only considered in the transformation of wave energy to the motion of floating bodies, but the conversion from the motion to electric power is not optimized. In this simulation, each WEC in arrangement could be controlled in different controlling force following its position and wave condition. The total electric power of all WECs in arrangement will be discussed as an evaluation parameter. As an example of the proposed numerical model results, the electrical powers of the single model and three models in arrangement condition will be compared. In this paper, not only do we discuss the electrical powers, but we consider the controlling force coefficients, the heaving motions and absorbed powers as well.

We can take advantage of the numerical model to estimate the heaving motion and the expected electrical power of each wave energy converter in different relative positions and different controlling force. It can be used to find the best distance and relative position between floating structures. On the other hand, it can help us decide the range of controlling force in the design of the control system. The electrical power result computed by the present numerical model is an expected value, which is a target value that can be used to estimate different controlling methods.

2. Formulation and Solution Method

2.1. Hydrodynamic Forces

Three-dimensional Singularity Distribution Method (SDM) is a general method for analysis of hydrodynamic forces on offshore structure. The methodology of the SDM was introduced by W.D.Kim [25] and Garrison [26], and it has since been successfully applied to a variety of shapes by Faltinsen et al. [27], Oortmersen [28].

All the motions of floating body are assumed to be sinusoidal in time with circular frequency ω, and the velocity potential of the first-order incident wave progressing to x positive direction is expressed in the form

$$\Phi(x, y, z, t) = \text{Re}[\phi(x, y, z)e^{-i\omega t}] \tag{1}$$

The velocity potential which satisfies the governing equation and linearized boundary conditions, is summarized as follows:

$$[L] \quad \nabla^2 \phi = 0 \qquad \text{for } z \leq 0 \tag{2}$$

$$[F] \quad \frac{\partial \phi}{\partial z} - K\phi = 0 \quad \text{on } z = 0 \tag{3}$$

$$[B] \quad \frac{\partial \phi}{\partial z} = 0 \qquad \text{as } z = -h \tag{4}$$

$$[H] \begin{cases} \dfrac{\partial \phi_D}{\partial n} = 0 \\ \dfrac{\partial \phi}{\partial n} = \dot{\xi}_a n \end{cases} \quad \text{on } S_H \tag{5}$$

Here, n is the normal vector and ξ_a denotes the amplitude of incident wave. (2) is the Laplace equation in the fluid domain; (3) is the linear free-surface condition ($K = \omega^2/g$); (4) is a condition on the sea bottom and (5) is the conditions on the wetted body surface S_H. The diffraction potential ϕ_D in (5) is defined as the sum of the incident-wave potential ϕ_0 plus the scattering potential ϕ_d.

The velocity potentials which satisfy these boundary conditions can be calculated by free-surface Green function $G(P; Q)$ as

$$\phi(P) = \iint_{S_H} \sigma(Q) \cdot G(P; Q) ds \tag{6}$$

where $P = (x, y, z)$ is the field point, $Q(x', y', z')$ is the source point and $\sigma(Q)$ is the source density on the body surface. The Green function has been well studied and various expressions are known, and the source density could been solved by the SDM.

Once the velocity potentials ϕ on the body surface are obtained, it is straightforward to compute the radiation and diffraction forces. The radiation force F_R can be expressed as follows:

$$F_{Ri,j} = m_{i,j}\ddot{Z}_j + N_{i,j}\dot{Z}_j, \tag{7}$$

m_{ij} and N_{ij} is added mass and damping coefficient, which are given by

$$m_{ij} = -\rho Re[\sum_{m=1}^{N} \phi_{jm} \cdot n_{im} \cdot S_m] \quad (i,j = 1 \sim J) \tag{8}$$

$$N_{ij} = -\rho \omega Im[\sum_{m=1}^{N} \phi_{jm} \cdot n_{im} \cdot S_m] \quad (i,j = 1 \sim J) \tag{9}$$

i, j in the $F_{Ri,j}$, m_{ij} and N_{ij} is defined the coefficient of i-mode which is influenced by j-mode. n_{im} is the i-mode normal vector on m-th mesh, S_m is the mesh area of m-th. The floating bodies at wave move in 6 modes of motion: surge, sway, heave, roll, pitch and yaw, the mode numbers (i and j) are changed from 1 to 6 respectively, if the body number is one. J denotes the 6 modes times body number. For example, i or $j = 3$ is the heave motion of 1st body, 7 is the surge motion of 2nd body and 9 is the heave of 2nd body.

Using the incident wave amplitude(ξ_a), the diffraction force can be expressed as follows:

$$F_{Di} = f_{Di}\xi_a. \tag{10}$$

Here, the f_{Di} defined is the diffraction force coefficient as follows:

$$f_{Di} = i\omega\rho \sum_{m=1}^{N} (\phi_{0m} + \phi_{dm})n_{im}S_m \tag{11}$$

As a result, the coefficient of added mass, damping coefficient and diffraction force coefficient are established by the velocity potentials which are computed by SDM and can be applied to calculate the radiation and diffraction force.

2.2. Hydrodynamic Model of Multiple Floating Bodies

It is assumed that there are N floating bodies in array condition. All floating bodies can only move in the heaving motion ($i = 3, 9 \cdots 3 + (N-1) \times 6$) in Equations (7) and (10), the other degrees of freedom ideally restricting. Denoting the mass matrix and heaving acceleration of the n-th body with M_n and $\ddot{Z}_n$, and using (7) and (10) for the radiation and diffraction forces on the n-th body, the dynamic equation which describes the n-th floating body motion with a single degree of freedom in the time domain, oscillating in heave is:

$$M_n\ddot{Z}_n(t) = F_{D_n}(t) + \sum_{n'=1}^{N} F_{R_{nn'}}(t) + F_{S_n}(t) + F_{g_n}(t) \tag{12}$$

The aim of this article is to estimate the expected electrical power of wave energy converters. Therefore, the model in the frequency domain ω is used. Thus, Equation (12) results are adjusted applying the Fourier Transform as below:

$$M_n\ddot{Z}_n(\omega) = F_{D_n}(\omega) + \sum_{n'=1}^{N} F_{R_{nn'}}(\omega) + F_{S_n}(\omega) + F_{g_n}(\omega) \tag{13}$$

where $F_{D_n}(\omega)$ is the wave diffraction force which is the sum of pressure forces on n-th body surface due to incident and diffracted waves. It can be written as follows:

$$F_{D_n} = -i\omega\rho \iint_{SH_n} (\phi_0 + \phi_d)n_{(3,n)}dS\xi_a \tag{14}$$

$F_{Rnn'}(\omega)$ is the wave radiation force of n-th body in the heaving motion due to the radiated wave when the n'-th body moves, and can be obtained as follows:

$$F_{Rnn'} = -i\omega\rho \iint_{SH_n} \phi_{(3,n')}n_{(3,n)}dS\dot{Z}_{n'} \tag{15}$$

Here, ξ_a is the amplitude of incident wave; $\dot{Z}_{n'}$ means the speed of n'-th body and SH_n is the mesh number of n-th body. Assuming the inviscid and incompressible flow, the velocity potential ϕ is introduced. $\phi_{(3,n')}$ means the heaving radiation velocity potential in n'-th body; ϕ_0 is the incident-wave potential and ϕ_d is the scattering potential from n-th body. The velocity potentials are calculated by the Three-dimensional Singularity Distribution Method (3D-SDM) in the study. The normal vector n is defined as positive when directing into the fluid from the body surface, $n_{(3,n)}$ denotes the normal vector of n-th body in heave direction.

F_{S_n} is the restoring force, which is written as:

$$F_{S_n} = -\rho g A_{w_n} Z_n \tag{16}$$

where, A_{w_n} and Z_n is the water-plane area and heaving motion of n-th body; g is the gravitational acceleration.

F_{g_n} is the controlling force provided by the linear generator, which can be derived as

$$F_{g_n} = C_{g_n}\dot{Z}_n + K_{g_n}Z_n \tag{17}$$

Here, C_{g_n} and K_{g_n} are the coefficients for heaving velocity $\dot{Z}_n$ and heaving motion Z_n respectively, which could be produced by linear wave energy converter.

Using Euler's formula in complex number, heaving velocity $\dot{Z}_n$ and heaving acceleration $\ddot{Z}_n$ can be expressed as heaving motion

$$Z_n = z_n \dot{e}^{-i\omega t} \tag{18}$$

$$\dot{Z}_n = -i\omega z_n \cdot e^{-i\omega t} = -i\omega Z_n \tag{19}$$

$$\ddot{Z}_n = -\omega^2 z_n \cdot e^{-i\omega t} = -\omega^2 Z_n \tag{20}$$

As a result, the heaving motion Z_n of floating structures may be calculated by the dynamic equations of motions which are written as:

$$\begin{bmatrix} -\omega^2 M_1 + f_{R3,3} + \rho g A_{w,1} - (K_{g,1} - i\omega C_{g,1}) & f_{R3,9} & \cdots & f_{R3,3+(N-1)*6} \\ f_{R9,3} & -\omega^2 M_2 + f_{R9,9} + \rho g A_{w,2} - (K_{g,2} - i\omega C_{g,2}) & \cdots & f_{R6,3+(N-1)*6} \\ \vdots & \vdots & \ddots & \\ f_{R3+(N-1)*6,3} & f_{R3+(N-1)*6,9} & \cdots & -\omega^2 M_N + f_{R3+(N-1)*6,3+(N-1)*6} + \rho g A_{w,N} - (K_{g,N} - i\omega C_{g,N}) \end{bmatrix} \cdot \begin{bmatrix} Z_1 \\ Z_2 \\ \vdots \\ Z_N \end{bmatrix} = \begin{bmatrix} -f_{D3} \cdot \xi_a \\ -f_{D9} \cdot \xi_a \\ \vdots \\ -f_{D3+(N-1)*6} \cdot \xi_a \end{bmatrix} \tag{21}$$

where $f_{R_{nn'}}$ and f_{D_n} denote as

$$f_{D_n} = \iint_{SH_n} (\phi_0 + \phi_d) n_{(3,n)} dS \tag{22}$$

$$f_{R_{nn'}} = \iint_{SH_n} (\phi_{(3,n')}) n_{(3,n)} dS \tag{23}$$

2.3. Electric Power

The mechanism of the wave energy converter in this paper is based on the linear electrical generator system in which the floating body heaves up and down on the water surface and the dynamic energy is converted to electrical energy. There are two parts needed to consider in the energy conversion. One part is the absorbed power from the heave motion of the floating body, which is determined as:

$$P_n = F_{g_n} \times \dot{Z}_n \tag{24}$$

The other part is the copper loss in the electrical generator, which can be written as follows:

$$P_{C_n} = R \times (F_{g_n}/K_t)^2 \tag{25}$$

The copper loss is generated by heat produced in the linear electrical generator and decided by controlling force (F_g), R and K_t. R and K_t are winding resistance and force constant of linear electrical generator, which decide the performance of the linear electrical generator. In present study, we the assume that R is 0.6 Ω and K_t is 90 N/A for the model scale, which are the parameters of the linear electrical generator(GLM24-M-2530) made in THK CO., LTD. For the actual scale, R and K_t assume 0.3 Ω and 900 N/A as reasonable values.

The electrical power is denoted P'_n given below:

$$P'_n = P_n - P_{C_n} \tag{26}$$

Then, the absorbed power (P_n) and the copper loss (P_{C_n}) can be integrated in the wave period T,

$$\tilde{P}_n = \frac{1}{T}\int_0^T P_n dt = \frac{1}{T}\int_0^T (C_{g_n}\dot{Z}_n + K_{g_n}Z_n)\dot{Z}_n dt = \frac{1}{2}\omega^2 C_{g_n} Z_n^2 \tag{27}$$

$$\tilde{P_{C_n}} = \frac{1}{T}\int_0^T P_{C_n} dt = \frac{1}{T}\frac{R}{K_t^2}\int_0^T (C_{g_n}\dot{Z}_n + K_{g_n}Z_n)^2 dt = \frac{1}{2}\frac{R}{K_t^2}Z_n^2(\omega^2 C_{g_n}^2 + K_{g_n}^2) \tag{28}$$

$$\tilde{P'_n} = \tilde{P}_n - \tilde{P_{C_n}} = \frac{1}{2}\omega^2 C_{g_n} Z_n^2 - \frac{1}{2}\frac{R}{K_t^2}Z_n^2(\omega^2 C_{g_n}^2 + K_{g_n}^2) \tag{29}$$

which is an expected value within one wave period, and it is named Expected Electrical Power of n-th body. This expected value do not consider a specific kind of control strategy, because most classical control strategies can be expressed by controlling force coefficients C_g and K_g, such as the resistive loading control giving without K_g, the resonance control including C_g and K_g, and approximate complex-conjugate control (ACL control) considering the generator copper losses.

Furthermore, in order to discuss the relationship between interference effect and the electrical power, the average electrical power ($\tilde{P_{ave}}$) of multiple floating bodies are computed as

$$\tilde{P_{ave}} = \frac{\sum_{n=1}^{N}\tilde{P}'_n}{N} \tag{30}$$

3. Wave Energy Converter

3.1. Linear-Driving Type Wave Energy Converter

This study is a part of NEDO project which is named "Linear-driving type Wave Energy Converter". A point absorber device is proposed in the project as Figure 1. It generally consists of two separate parts: a spar part which is attached or moored to the seafloor, and a float part which oscillates with the waves. A linear generator is installed in the spar part and the resultant relative motion between two parts is used to generate electricity via a PTO system.

3.2. Simulation Model

The image picture and mesh picture of the WEC model is shown in Figure 2. Table 1 shows the principal particulars of the WEC model. The model scale is assumed 1/7 in the simulation in order to support the experiment.

3.3. Controlling Force Coefficients C_g and K_g for Single Model

To make clear the controlling force coefficients (C_g and K_g) effect to motion, the heaving motion of single model is calculated in wave period T from 0 s to 9.0 s. As shown in Figure 3, the resonance period of the floating structure is about 2.0 s without controlling force, and the Response Amplitude Operator (RAO) reached to 3.73.

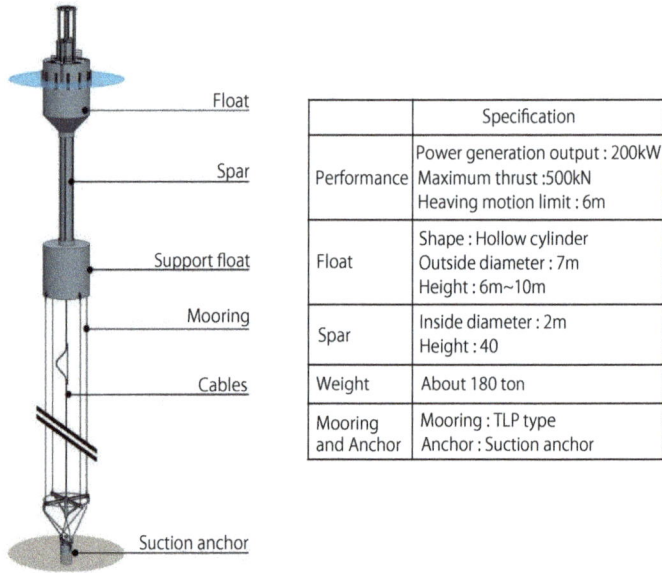

Figure 1. The image of a point absorber wave energy converter proposed in "Linear-driving type Wave Energy Converter Project".

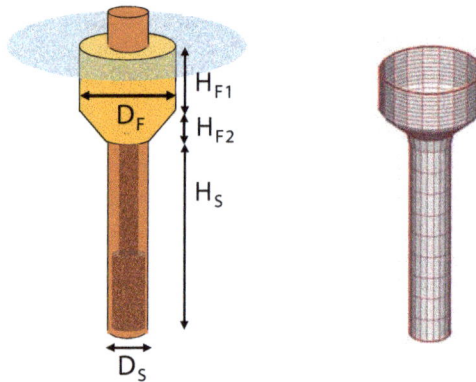

Figure 2. Simulation model.

Table 1. Principal particulars of WEC model.

Main Dimensions	Units	Model Scale	Actual Scale
Diameter of float (D_F)	m	1.00	7.00
Height-1 of float (H_{F1})	m	0.55	3.36
Height-2 of float (H_{F2})	m	0.30	2.10
Diameter of spar (D_s)	m	0.40	2.80
Height of spar (H_s)	m	2.70	17.67
Volume of float (Vol)	m^3	0.48	99.58
Water area of float (A_w)	m^2	0.66	29.59

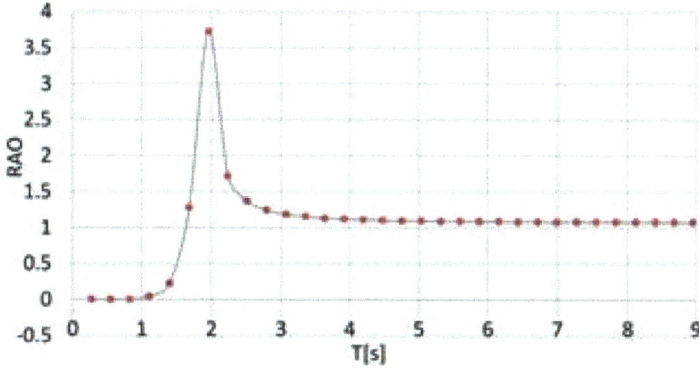

Figure 3. RAO of floating structure.

Next, we kept $K_g = 0$ N/m and changed C_g. Figure 4 shows the results of RAO in different C_g for each wave period. As shown in this figure, when the C_g in particular range, the RAO increase rapidly in the resonance period and the maximum value is 4.43, which is bigger than 3.73 (max RAO without controlling force). The resonance frequency could be changed by chose the appropriate C_g. On the other hand, we kept $C_g = 0$ Ns/m and changed K_g. As shown in Figure 5, the K_g can be used to change the resonance period of the floating body. By giving large K_g acting as reducing the restoring force as shown in Equation (21), the resonance period can be changed to large period, and the RAO became huge. The number of RAO is only a theoretical calculating number, the heaving motion limitation must be decided according to the mechanical structure of wave energy converter. According to the results, it could be understood that the function of control force coefficients (C_g and K_g) can change the resonance frequency and resonance period.

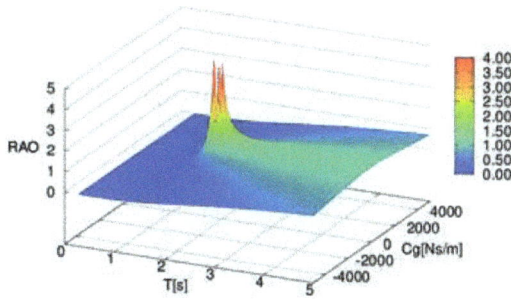

Figure 4. RAO of the floating structure in different Cg (Cg = −5000–5000 [Ns/m], Kg = 0 [N/m]).

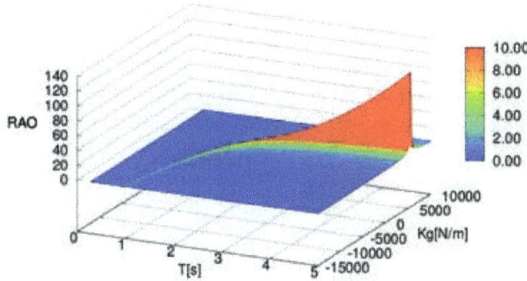

Figure 5. RAO of the floating structure in different Kg (Cg = 0 [Ns/m], Kg = −15,000–10,000 [N/m]).

4. The Results of Multiple WECs in Regular Wave

The results as the solution method will be shown and discussed in this section. The objective of the present study is shown as follows:

1. How to decide the controlling force to the linear generator in arrangement.
2. The relationship between the interference effect and electric power from WECs.
3. How to decide the arrangement of multiple WECs.

We cover these 3 objects to discuss the simulating results. At first, wave period, wave amplitude and WECs arrangement condition are changed as the parameter in the regular wave. Then, basis on the results in the regular wave, the estimation in the real sea wave condition are proceed. In the study, three kinds of arrangement will be discussed. They are single arrangement, straight arrangement and triangle arrangement as shown in Figure 6a–c, respectively. In a case study, three wave energy converter models are used in the straight and triangle arrangement. The limit of the heave motion is decided 0.2 m. As show in Figure 6, the models are set in the head-wave condition. The distances of each models are changed from 1.5 to 8.0 times diameter of the model in the straight and triangle arrangement. The reason of choosing this range of the distance is; the adjacent model may not contact each other in waves and it is not too far to appear clearly the hydrodynamic interaction effect among the models. In the simulation, the wave amplitude changes from 0.01 to 0.15 m, in every 0.01 m, and the wave period is from 0.6 to 5.0 s, in every 0.1 s. The depth of the water is assumed 4.5 m.

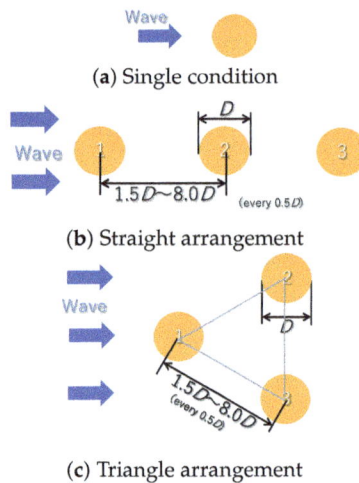

(a) Single condition

(b) Straight arrangement

(c) Triangle arrangement

Figure 6. Arrangement of multiple wave energy converters.

4.1. Investigation of Controlling Force in Arrange Condition

As the proposed method, the controlling force coefficients K_g and C_g are changed for wave periods, and the optimal values found by searching C_g in 0–2500 [Ns/m] and K_g in −5000–6000 [N/m] considering the controlling force limitation. We suppose that the controlling force coefficients of each floating body can be controlled commonly and independently. Controlled commonly means the K_g and C_g are the same in every WEC and decided only in wave condition. On the other hand, controlled independently purports that in addition to wave condition the interference effect are also considered to decide the controlling force coefficients. To compare these two method, we discuss the relationship between the controlling force of each WECs in arrangement and interference effect.

When the total expected electrical power $(P'_1 + P'_2 + P'_3)$ became maximum, the results of K_g, C_g, Z, P and P' of each floating body are exported. Figure 7 shows the results of 3 floating bodies in controlled commonly. Figure 8 shows the results in controlled independently. In addition, the results of the single body are also shown in the same figure, to compare with the variation. The wave amplitude is 0.1 m and the distances of each models are 3 times diameter (3 m) in straight arrangement.

At first, it will be easy to observe the difference of the two ways by comparing the curves of C_g in Figures 7b and 8b. It is imagined that this C_g's difference will directly affect the total power generation and the interference effect cannot be ignored in the certain range. We discuss the details of the interference effect by comparing with the two ways through Figures 7 and 8 as follows.

K_g of three models in two methods are similar to the single condition, as shown in Figures 7a and 8a. As the introduction in Section 3.3, K_g is used to change the resonance period of the floating body, and it is not changed belong to interference effect. However, the C_g of 1, 2 and 3 bodies in the straight arrangement are different to single's condition, from 1.2 s to 4.6 s. We can understand that the interference effect in the short and long wave period is not obvious. When the C_g controlled independently, heave motion Z of three WECs reach to the limitation, as shown in Figure 8c. On the other hand, when the C_g controlled commonly, a part of Z cannot reach to the limitation. Because of the difference of Z, the expected electrical power (P') of each WEC are different. As shown in Figures 7e and 8e, the generated electric power under controlling independently is better than that used the controlling parameters commonly in the array. Figures 7d and 8d show the absorbed power, and the copper loss could be calculated by generated electric power differenced absorbed power.

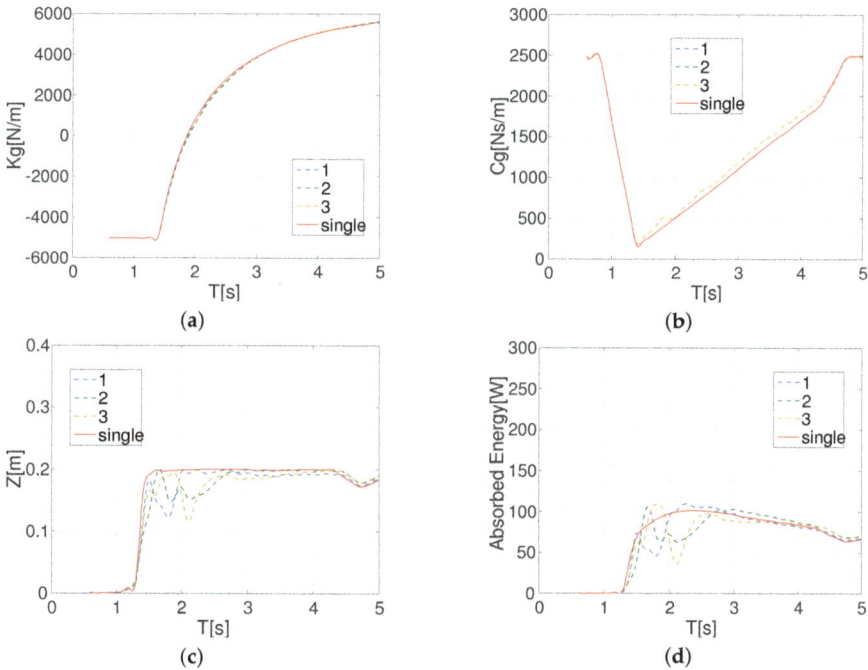

Figure 7. *Cont.*

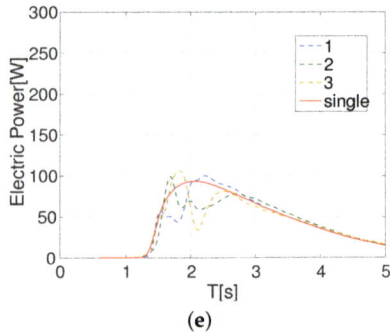

Figure 7. The results of multiple WECs in regular wave (3.0D, $Amp = 0.10$ [m], Use common parameters) (**a**: K_g, **b**: C_g, **c**: heave motion Z, **d**: the absorbed power, **e**: generated electric power).

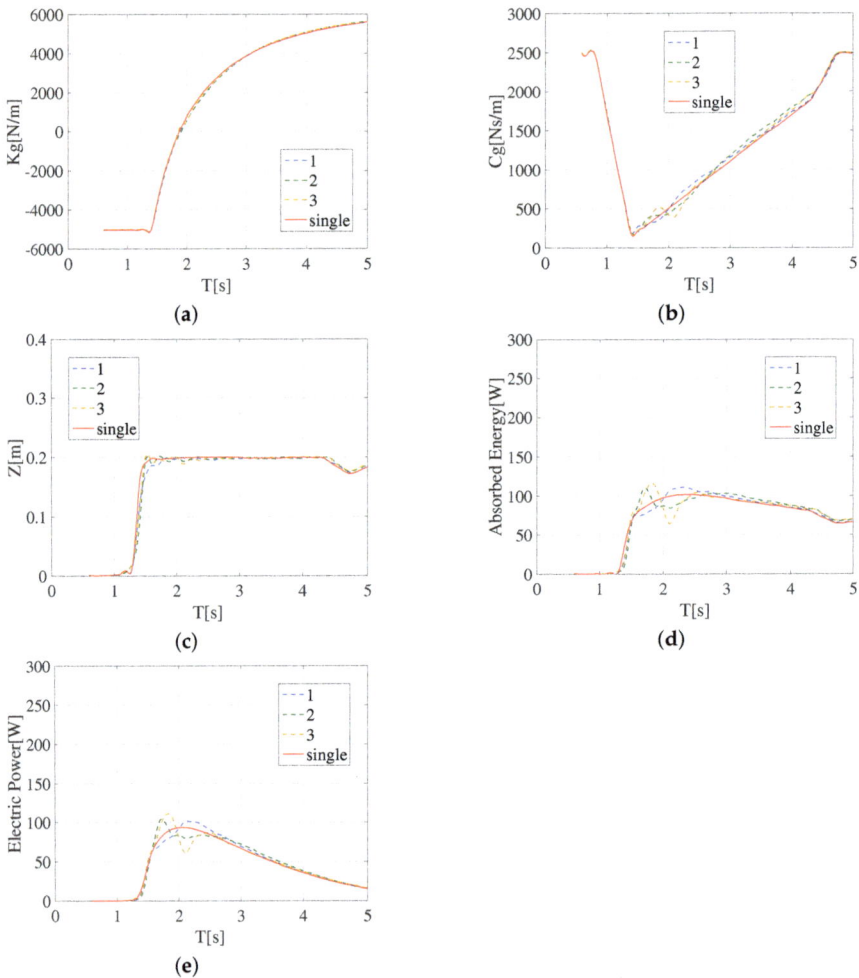

Figure 8. The results of multiple WECs in regular wave (3.0D, $Amp = 0.10$ [m], Use independent parameters) (**a**: K_g, **b**: C_g, **c**: heave motion Z, **d**: the absorbed power, **e**: generated electric power).

Next, in order to compared the electric power results in controlled by the common parameters with in controlled independently much more clearly, the rate of change E_{ave} are calculated as follows:

$$E_{ave} = \frac{\tilde{P}_{ave} \ (Controlled \ independently)}{\tilde{P}_{ave} \ (Controlled \ commonly)} \tag{31}$$

Figure 9 shows the E_{ave} as different color, the vertical line shows the distance between the WECs and the horizontal one the wave periods. As shown in this figure, around the resonance period (2.0 s) E_{ave} is bigger than "1", it is means the average electric power of three WECs in controlled independently is bigger than in controlled commonly. However, in the short and long wave period E_{ave} is "1".

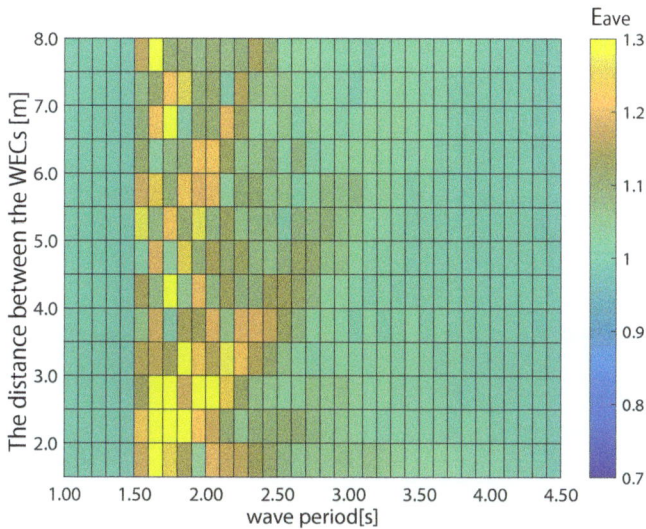

Figure 9. The rate of change of the two methods E_{ave}.

As a result, the C_g of each wave energy converter needs change independently following the wave period, especially around the resonance period. In addition, the average electric power in controlled independently is 10–20% bigger than in controlled commonly. On the other hand, K_g in two methods are similar to the single condition, and it is not changed belong to interference effect.

Therefore, the controlling force coefficients of each floating body controlled independently are used in the next calculation.

4.2. The Influence of Interference Effect to Electric Power

In preceding section, we cleared how to decide the controlling force of multiple WECs in arrangement. In this section, the electric power is focused on different wave periods, wave amplitude and distance between WECs considering the interference effect. As a case study, the expected electric power of each WECs are calculated in the single condition and in the straight arrangement (the distances of each bodies are 3 m). Here, besides the wave period, the wave amplitude is changed from 0.01 m to 0.15 m in every 0.01 m. In the calculation of changing wave amplitude, the wave forces are based on linear theory as Section 2.2, and the analysis mechanism of choosing the optimal values of C_g and K_g is comparison the expected electrical power on all C_g and K_g to find the best values within the allowable range. Therefore, the best controlling parameters are changed by not only a wave period but also an amplitude of waves.

Figure 10 shows the distribution map of single floating body in different the wave period and amplitude. *X*-axis shows the wave period, *Y*-axis shows the wave amplitude and *Z*-axis shows the number of electrical power. Moreover, the peak power and its happened period are also shown in the figure. It can be observed that the expected electrical power is increased following the wave amplitude, especially around the resonance period. However, there is no electricity power when the wave period is less than 1.2 s and the wave period is larger with small wave amplitude.

Figure 11 shows the power distribution map of each floating body when they are in the straight arrangement. The power distribution map and the peak power of each WECs in arrangement is different to the single condition. Therefore, the spacing between two absorbers are relatively small (for this case, L = 3D), the interference effect must be considered when the total electric power of multiple WECs in arrangement are calculated.

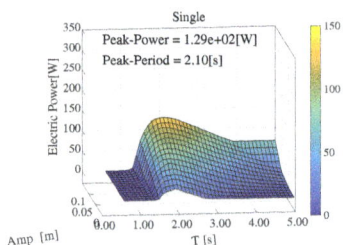

Figure 10. Electric power on single condition in different wave period and amplitude.

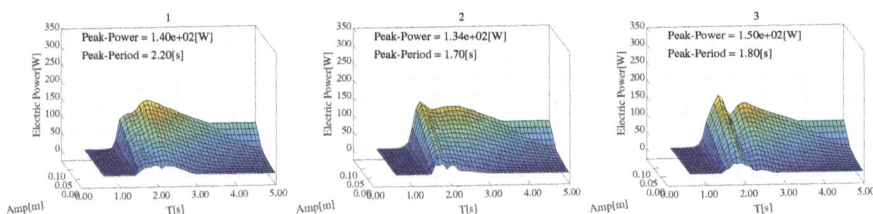

Figure 11. Electric power on straight condition in different wave period and amplitude.

Next, in order to estimate the interference effect, the changing ratio of electric power are defined as follows:

$$E_{ratio} = \frac{\tilde{P}'_{total}/N}{\tilde{P}'_{single}} \qquad (32)$$

When the changing ratio of electric power E_{ratio} greater than "1", the interference effect help the electric power increase. However, the E_{ratio} is a value which only use for estimating the interference effect, it does not mean that much more electric power can be obtained. Moreover, the triangle arrangement (Figure 6c) will be added to discuss.

The results of E_{ratio} in different non-dimensional parameter Wave-Length/Diameter (λ/D) are show in Figures 12–14. Figure 12 shows the results when the distance between WECs is 1.5D, Figure 13 is 3.0D and Figure 14 is 5.0D. For comparing the straight arrangement and triangle arrangement, the results in the same distance are shown in the same figure. We can confirm that the changing ratio of expected electrical power is increased and deceased in some area following the wave condition. In addition the increased and deceased area is changed for different distance. Therefore, the positive and negative influence of interference effect could be changed following the wave length (or period) and the distance between WECs.

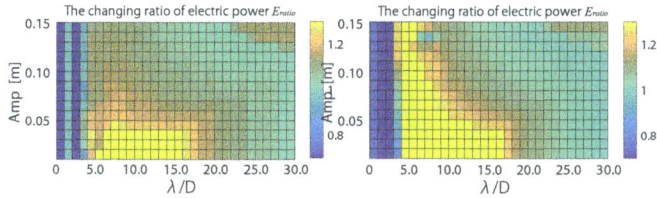

Figure 12. E_{ratio} in 1.5D (**Left**: Straight arrangement; **Right**: Triangle arrangement).

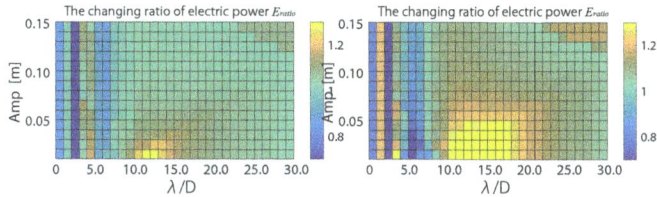

Figure 13. E_{ratio} in 3.0D (**Left**: Straight arrangement; **Right**: Triangle arrangement).

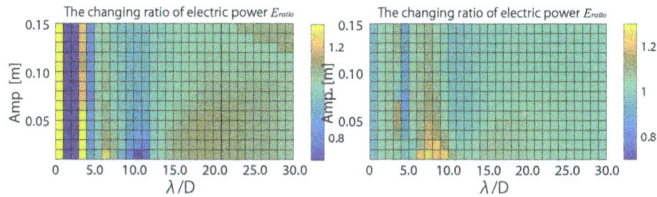

Figure 14. E_{ratio} in 5.0D (**Left**: Straight arrangement; **Right**: Triangle arrangement).

4.3. Decide the Distance between WECs

To evaluate the relationship between distance and electrical power, the changing ratio of electrical power E_{ratio} in straight arrangement and triangle arrangement is calculated to compare with the single one. Here, the distance between the adjacent bodies is changed from 1.5 m to 8.0 m every 0.5 m, the *Amp* kept in 0.1 m.

Figure 15a shows the E_{ratio} in straight arrangement. The relatively high electrical power performance range is shifted to the longer λ / D, when the distance becomes longer. The electrical power performance increase more than 10–15% in some wave condition, in different wave period as the different distance. Figure 15b shows the E_{ratio} in triangle arrangement. The trend is same to straight arrangement. The electrical power performance increment is larger than straight arrangement reached to more than 15%. Therefore, it can be said that the triangle arrangement is better than the straight arrangement for use interference effect in this case.

For showing the relationship between distance and E_{ratio} clearly, we organize that the vertical line shows the Distance/Wave-Length and the horizontal shows the E_{ratio}, as shown in Figure 16. The E_{ratio} fluctuate periodically as the Distance/Wave-Length. The peak in the graph means that high average electric power could be obtained by interference effect. Using this figure, the best distance could be choice following the wave condition characteristic of sea area.

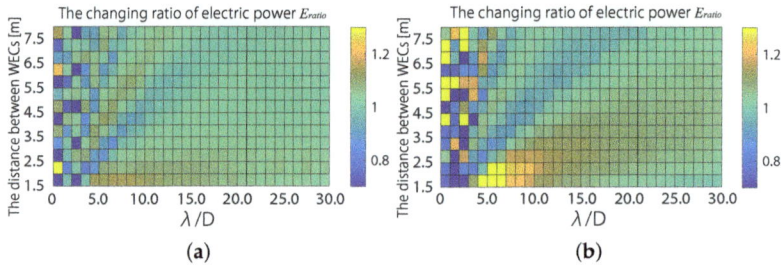

Figure 15. E_{ratio} in different distances in $Amp = 0.1$ m (**Left**: Straight arrangement; **Right**: Triangle arrangement).

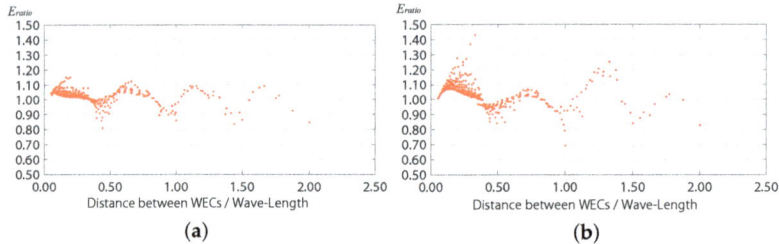

Figure 16. E_{ratio} as Distance/Wave-Length in $Amp = 0.1$ m (**Left**: Straight arrangement; **Right**: Triangle arrangement).

5. The Electrical Power Estimation of Multiple WECs in Real Sea

5.1. Formulation and Solution Method

We discussed multiple WECs in regular wave. However, a lot of different wave periods and amplitudes are fixed in the real sea, therefore the calculation in the irregular wave is necessary. In this section, the electric power of multiple WECs will be computed in the irregular wave and the interference effect in real sea will be discussed. To predict the response of floating body in the irregular wave, the response spectrum S_{qq} could be calculated by response in regular wave $H(\omega)$ and wave spectrum S_{QQ} as follows

$$S_{qq}(\omega) = H(\omega)^2 \cdot S_{QQ}(\omega) \tag{33}$$

which introduced in [29]. Here, the $H(\omega)$ is computed by the electric power divided by significant wave height, JONSWAP (Joint North Sea Wave Observation Project) spectrum [30] are used to show wave spectrum S_{QQ}.

To estimate the electric power in the real sea basic on the S_{qq}, the representative value of electric power is denoted σ given below:

$$\sigma = \sqrt{\int_0^\infty S_{qq}(\omega)\,d\omega} \tag{34}$$

Then, in order to discuss the electric power in certain sea area, the electric power distribution are calculated by using the H-T joint probability distribution P (Figure 17), as $P \times \sigma[W]$. Moreover, the summation of expected electric power in certain sea area could be calculated, which can be written as follows:

$$P_{Wsea} = \sum_{i=1}^I \sum_{j=1}^J P_{ij}\sigma_{ij} \tag{35}$$

Here, i, j is calculating point of wave period ($i = 1 \sim I$) and wave height ($j = 1 \sim J$), respectively. The representative value of electric power σ and the summation of expected electric power P_{Wsea} are used to discuss the WECs performance in the real sea area. Proposed method flow chart in real sea is presented in Figure 18.

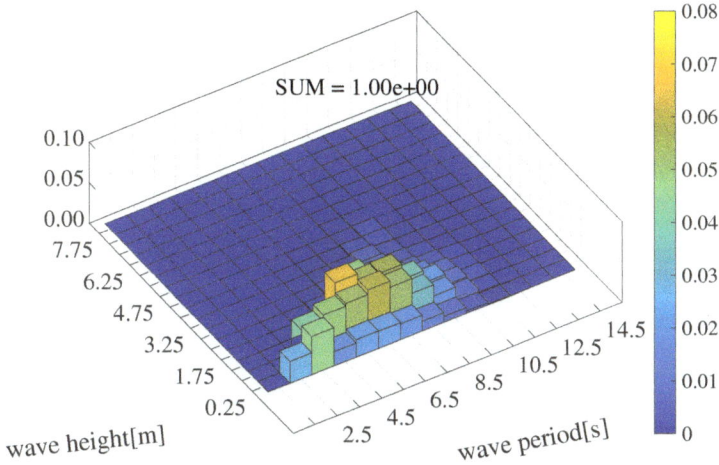

Figure 17. H-T joint probability distribution P in the bay of KAMAISHI.

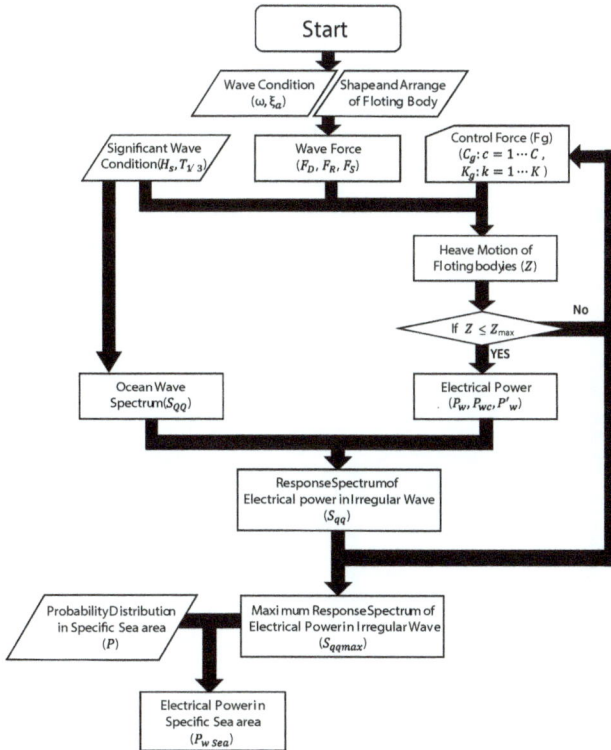

Figure 18. Simulation flow chart in the real sea.

Actual scale model which set in triangle arrangement and the distance 3.0D, is used in the real sea calculation. As an assumption sea area, the bay of Kamaishi where is located in northeast of Japan are used. The H-T joint probability distribution P of this sea area is shown in Figure 17, which is published in Web [31]. Here, the wave conditions shown by appearing frequency, so the totally frequency in the sea area is 1. The computational condition in real sea shown as Table 2.

Table 2. Computational condition in real sea.

Parameters	Units	Value
Wave period (T)	s	1.00–18.00 (every 0.25)
Significant wave height ($H_{1/3}$)	m	0.25–7.75 (every 0.25)
Significant period ($T_{1/3}$)	s	1.50–14.50 (every 1.00)
Controlling force coefficient C_g	Ns/m	0.000–500,000
Controlling force coefficient K_g	N/m	−250,000–300,000
Deep of water (H)	m	31.5
Motion limit of floating part (Z_{max})	m	2.00
Winding resistance (R)	ω	0.3
Force constant (K_t)	N/A	900

5.2. Compute Result

Figure 19 shows the result of electrical power distribution in single condition. It can be understood that the electrical power increases following the wave period and amplitude became bigger, because there is much more energy in a large wave's period and large amplitude. Whereas, the result of expected electric power P_{Wsea} is almost 0 when the wave condition is large wave period and large amplitude. The P_{Wsea} only can be obtained in the high appearing frequency of wave height and wave period.

Figure 20 shows the average result of 3 WECs in triangle arrangement. As the same to single condition, P_{Wsea} only can be obtained in the high appearing frequency of wave H-T. The totally average P_{Wsea} reach to 221 [kW], 7% bigger than 207 [kW] in single condition.

To compare the electric power of WEC in single condition and in arrangement condition from the viewpoint of wave period, Figure 21 shows the distribution map of them. By using the interference effect, the average electrical power in arrange condition increase from wave period 6.5 s, but decrease under 5.5 s compare with single condition. It could be said that, WECs in appropriately arranged conditions could obtain much more electrical power than single's, and the appropriate arranged conditions must be matched to the wave condition of the installed sea area.

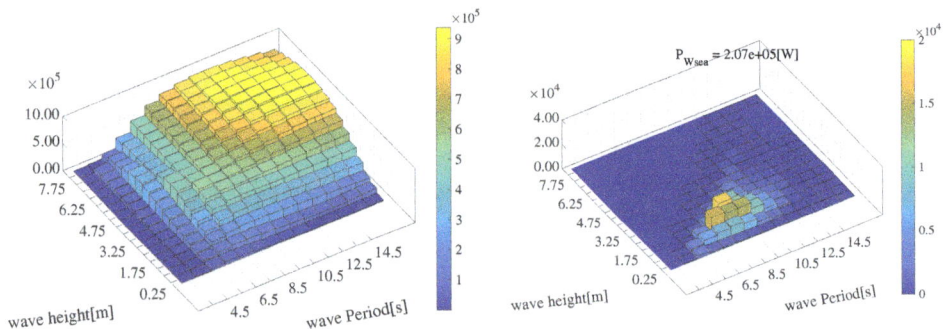

Figure 19. Single condition (Distribution map of electric power σ on the **left**, and distribution map of electric power in a sea area $P \times \sigma$ on the **right**.)

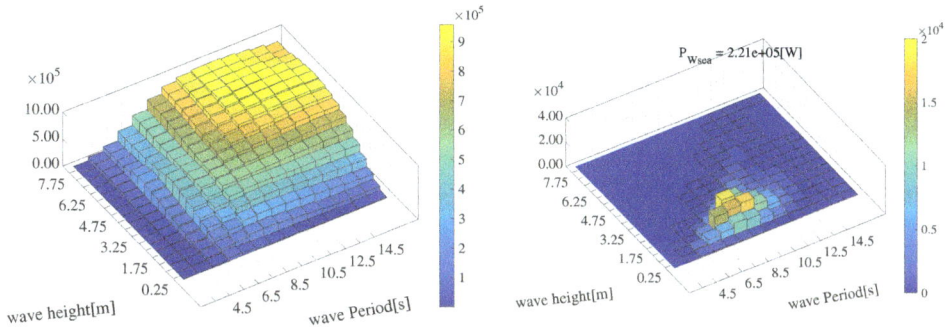

Figure 20. Arrange condition (Distribution map of electric power σ on the **left**, and distribution map of electric power in a sea area $P \times \sigma$ on the **right**.)

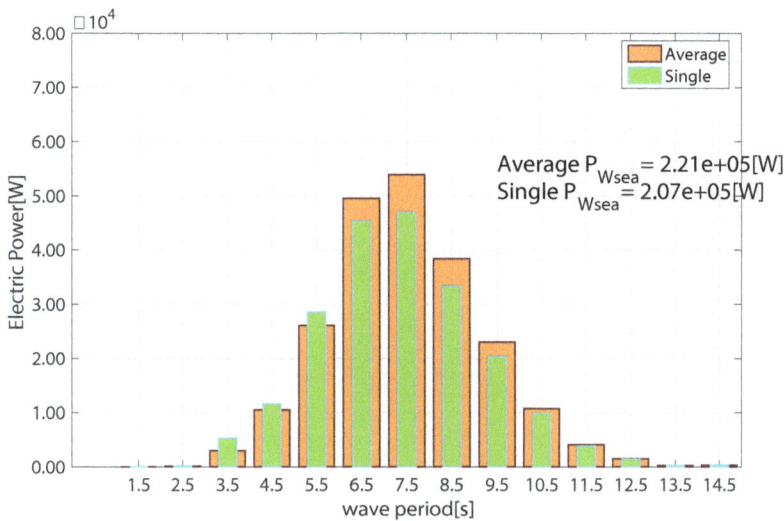

Figure 21. Electric power comparison in single condition and in arrange condition.

6. Conclusions

In general, the controlling force of WEC is optimized in single condition, and the same forces are used for each WEC in arrangement. Thus, the interference effect between the WECs is only considered in the transformation of wave energy to the motion of floating bodies, but the conversion from the motion to electric power is not optimized. In present paper, the interference effect of WECs is taken into account in motion calculation and motion-electric power conversion by appropriate arrangement for independently controlled WECs. The maximum electric power of multiple WECs is calculated and discussed in regular waves and real sea condition.

The following conclusions are obtained through the numerical consideration:

1. In arranged condition, the controlling force coefficient C_g needs change independently for each wave energy converter following the wave period.
2. The generated electric power under independent control is better than that used the controlling parameters commonly in the array.
3. In a case study of three WECs, the average electric power controlled independently is 10%–20% bigger than in controlled commonly.

4. The interference effect can be used to increase the electric power. In a case study of three WECs, the electrical power performance increases more than 10%–15% in straight arrangement and it reaches to more than 15% in triangle arrangement. The triangle arrangement is better than the straight arrangement for use interference effect.
5. Distance/Wave-Length determines the performance of WECs and the performance is changed cyclically along the parameter. It should be care that the interference effect appears even the distance of adjacent WEC is far.
6. WECs in appropriate arranging conditions could obtain much more electrical power than single's, and the appropriate arranging conditions must be match to the wave condition of the install sea area.

In the other our research it is suggested that the interference effect is increased by the number of consisted WECs. So, determining the parameters independently will be more important to the electric power maximization of huge number of WEC array. In that case, the knowledge obtained through this work will contribute much to the future work on it.

Author Contributions: Conceptualization, M.M.; Data curation, Q.L.; Formal analysis, Q.L.; Funding acquisition, M.M.; Investigation, Q.L.; Methodology, Q.L.; Project administration, M.M.; Software, Q.L.; Validation, Q.L. and S.K.; Visualization, S.K.; Writing—original draft, Q.L.; Writing—review & editing, Q.L.

Funding: This research received no external funding.

Acknowledgments: This work is part of Linear-driving type Wave Energy Converter project supported by the New Energy and Industrial Technology Development Organization (NEDO). Highest gratitude is expressed to NEDO and research partners.

Conflicts of Interest: The authors declare no conflict of interest.

References

1. Drew, B.; Plummer, A.R.; Sahinkaya, M.N. A review of wave energy converter technology. *Proc. Inst. Mech. Eng. Part A J. Power Energy* **2009**, *223*, 887–902. [CrossRef]
2. Day, A.H.; Babarit, A.; Fontaine, A.; He, Y.-P.; Kraskowski, M.; Murai, M.; Penesis, I.; Salvatore, F.; Shin, H.-K. Hydrodynamic modelling of marine renewable energy devices: A state of the art review. *Ocean Eng.* **2015**, *108*, 46–69. [CrossRef]
3. Rezanejad, K.; Soares, C.G.; Lopez, I.; Carballo, R. Experimental and numerical investigation of the hydrodynamic performance of an oscillating water column wave energy converter. *Renew. Energy* **2017**, *106*, 1–16. [CrossRef]
4. Viviano, A.; Naty, S.; Foti, E. Scale effects in physical modelling of a generalized OWC. *Ocean Eng.* **2018**, *162*, 248–258. [CrossRef]
5. Naty, S.; Viviano, A.; Foti, E. Feaseability study of a WEC integrated in the port of Giardini Naxos, Italy. In Proceedings of the Coastal Engineering Conference, Antalya, Turkey, 17–20 November 2016.
6. Tedd, J.; Kofoed, J.P. Measurements of overtopping flow time series on the Wave Dragon, wave energy converter. *Renew. Energy* **2009**, *34*, 711–717. [CrossRef]
7. Zhang, X.; Lu, D.; Guo, F.; Gao, Y.; Sun, Y. The maximum wave energy conversion by two interconnected floaters: Effects of structural flexibility. *Appl. Ocean Res.* **2018**, *71*, 34–47. [CrossRef]
8. Nagulan, S.; Venkatesan, B.; Arunachalam, A. A review on front end conversion in ocean wave energy converters. *Front. Energy* **2015**, *9*, 297–310.
9. Hals, J.; Bjarte-Larsson, T.; Falnes, J. Optimum Reactive Control and Control by Latching of a Wave-Absorbing Semisubmerged Heaving Sphere. In Proceedings of the International Conference on Offshore Mechanics and Arctic Engineering OMAE, Oslo, Norway, 23–28 June 2002; Volume 4, pp. 415–423.
10. Nebel, P. Maximizing the Efficiency of Wave-Energy Plants Using Complex Conjugate Control. *Proc. Inst. Mech. Eng. Part I J. Syst. Control. Eng.* **1992**, *206*, 225–236. [CrossRef]
11. Hals, J.; Falnes, J.; Moan, T. Constrained Optimal Control of a Heaving Buoy Wave-Energy Converter. *ASME J. Offshore Mech. Arct. Eng.* **2011**, *133*, 011401. [CrossRef]
12. Hals, J.; Johannes, F.; Torgeir, M. A comparison of selected strategies for adaptive control of wave energy converters. *J. Offshore Mech. Arct. Eng.* **2011**, *133*, 031101. [CrossRef]

13. Andrade, D.M.; Jaen, A.d.; Santana, A.G. Considering linear generator copper losses on model predictive control for a point absorber wave energy converter. *Energy Convers. Manag.* **2014**, *78*, 173–183. [CrossRef]

14. Taniguchi, T.; Umeda, J.; Fujiwara, T.; Goto, H.; Inoue, S. Experimental and Numerical study on Point Absorber Type Wave Energy Converter with Linear Generator. In Proceedings of the ASME 2017 36th International Conference on Ocean, Offshore and Arctic Engineering, Trondheim, Norway, 25–30 June 2017.

15. Zhang, X.; Tian, X.; Xiao, L.; Li, X.; Chen, L. Application of an adaptive bistable power capture mechanism to a point absorber wave energy converter. *Appl. Energy* **2018**, *228*, 450–467. [CrossRef]

16. Younesian, D.; Alam, M.R. Multi-stable mechanisms for high-efficiency and broadband ocean wave energy harvesting. *Appl. Energy* **2017**, *197*, 292–302. [CrossRef]

17. Zhang, X.; Yang, J. Power capture performance of an oscillating-body WEC with nonlinear snap through PTO systems in irregular waves. *Appl. Ocean Res.* **2015**, *52*, 261–273. [CrossRef]

18. Tetu, A.; Ferri, F.; Kramer, M.; Todalshaug, J. Physical and Mathematical Modeling of a Wave Energy Converter Equipped with a Negative Spring Mechanism for Phase Control. *Energies* **2018**, *11*, 2362. [CrossRef]

19. Troch, P.; Beels, C.; De Rouck, J.; De Backer, G. Wake Effects Behind a Farm of Wave Energy Converters for Irregular Long-Crested and Short-Crested Waves. In Proceedings of the International Conference on Coastal Engineering, Shanghai, China, 30 June–5 July 2010.

20. Folley, M.; Babarit, A.; O'f Boyle, L.; Child, B.; Forehand, D.; Silverthorne, K.; Spinneken, J.; Stratigaki, V.; Troch, P. A Review of Numerical Modeling of Wave Energy Converter Arrays. In Proceedings of the 31st International Conference on Offshore Mechanics & Arctic Engineering, Rio de Janeiro, Brazil, 10–15 June 2012.

21. Li, Y.; Yu, Y.-H. A synthesis of numerical methods for modeling wave energy converter-point absorbers. *Renew. Sustain. Energy Rev.* **2012**, *16*, 4352–4364. [CrossRef]

22. Borgarino, B.; Babarit, A.; Ferrant, P. Impact of wave interactions effects on energy absorption in large arrays of wave energy converters. *Ocean Eng.* **2012**, *41*, 79–88. [CrossRef]

23. Stratigaki, V.; Troch, P.; Stallard, T.; Forehand, D.; Kofoed, J.P.; Folley, M.; Benoit, M.; Babarit, A.; Kirkegaard, J. Wave Basin Experiments with Large Wave Energy Converter Arrays to Study Interactions between the Converters and Effects on Other Users in the Sea and the Coastal Area. *Energies* **2014**, *7*, 701–734. [CrossRef]

24. Leijon, M.; Bernhoff, H.; Agren, O.; Isberg, J.; Sundberg, J.; Berg, M.; Karlsson, K.E.; Wolfbrandt, A. Multiphysics Simulation of Wave Energy to Electric Energy Conversion by Permanent Magnet Linear Generator. *IEEE Trans. Energy Convers.* **2015**, *20*, 219–224 . [CrossRef]

25. Kim, W.D. On the Harmonic Oscillation of a Rigid Body on a Free Surface. *J. Fluid Mech.* **1965**, *21*, 427–451. [CrossRef]

26. Garrison, C.J.; Rao, V.S. Interaction of Waves with Submerged Objects. *J. Waterways Harb. Coast. Eng. Division* **1971**, *811*, 259–276.

27. Faltinsen, O.M.; Michelson, F.C. Motions of Large Structures in Waves at Zero Froude Number. In Proceedings of the International Symposium on the Dynamics of Marine Vehicles and Structures in Waves, London, UK, 1–5 April 1974; pp. 91–106.

28. Van Oortmerson, G. The motion of a ship in Shallow Water. *Ocean Eng.* **1976**, *3*, 221–255. [CrossRef]

29. Design of Ocean Systems-Ocean Wave Environment, MIT Open Course Ware. 7 March 2011. Available online: https://ocw.mit.edu/courses/mechanical-engineering/2--019-design-of-ocean-systems-spring-2011/lecture-notes/MIT_019S11_OWE.pdf (accessed on 23 October 2018).

30. Walter, H. Michl: Sea Spectra Revisited. *Mar. Technol.* **1999**, *36*, 211–227.

31. Statistical Database of Winds and Waves around Japan, National Maritime Research Institute. Available online: https://www.nmri.go.jp/wwjapan/namikaze_main_e.html (accessed on 23 October 2018).

![energies logo] *energies*

MDPI

Article

Fuzzy Supervision Based-Pitch Angle Control of a Tidal Stream Generator for a Disturbed Tidal Input

Khaoula Ghefiri [1,2,*](ID)**, Aitor J. Garrido** [1](ID)**, Eugen Rusu** [3](ID)**, Soufiene Bouallègue** [2](ID)**,**
Joseph Haggège [2] **and Izaskun Garrido** [1](ID)

[1]   Automatic Control Group—ACG, Department of Automatic Control and Systems Engineering,
     Engineering School of Bilbao, University of the Basque Country (UPV/EHU) , 48012 Bilbao, Spain;
     aitor.garrido@ehu.es (A.J.G.); izaskun.garrido@ehu.es (I.G.)
[2]   Laboratory of Research in Automatic Control—LA.R.A, National Engineering School of Tunis (ENIT),
     University of Tunis El Manar, 1002 Tunis, Tunisia; soufiene.bouallegue@issig.rnu.tn (S.B.);
     joseph.haggege@enit.rnu.tn (J.H.)
[3]   Department of Applied Mechanics, University Dunarea de Jos of Galati, Galati 800008, Romania;
     eugen.rusu@ugal.ro
*   Correspondence: kghefiri001@ikasle.ehu.eus or khaoulaghefiri@gmail.com; Tel.: +34-94-601-4443

Received: 30 September 2018; Accepted: 27 October 2018; Published: 1 November 2018

Abstract: Energy originating in tidal and ocean currents appears to be more intense and predictable than other renewables. In this area of research, the Tidal Stream Generator (TSG) power plant is one of the most recent forms of renewable energy to be developed. The main feature of this energy converter is related to the input resource which is the tidal current speed. Since its behaviour is variable and with disturbances, these systems must be able to maintain performance despite the input variations. This article deals with the design and control of a tidal stream converter system. The Fuzzy Gain Scheduling (FGS) technique is used to control the blade pitch angle of the turbine, in order to protect the plant in the case of a strong tidal range. Rotational speed control is investigated by means of the back-to-back power converters. The optimal speed is provided using the Maximum Power Point Tracking (MPPT) strategy to harness maximum power from the tidal speed. To verify the robustness of the developed methods, two scenarios of a disturbed tidal resource with regular and irregular conditions are considered. The performed results prove the output power optimization and adaptive change of the pitch angle control to maintain the plant within the tolerable limits.

Keywords: disturbed tidal resource; fuzzy gain scheduling; fuzzy supervisor; proportional integral derivative controller; pitch angle control; tidal energy; tidal stream generator

---

1. Introduction

Renewable energy consumption is predicted to grow in the range of 2.6% per year between 2012 and 2040 [1]. The increase in economic and structural changes will impact world energy consumption. Furthermore, with the development of countries and improvement of living conditions, the need for energy will increase rapidly [2,3]. The consumption of energy grew in the International Energy Outlook (IEO) 2016 Reference case [1]. The impact of fossil fuel dangers on the human environment and rising oil prices has prompted an expanded use of non-fossil renewable energy converters [4]. The worldwide energy demand is constantly increasing due to the evolution of modern society. Conventional energy sources, such as oil, gas, coal, and nuclear, are either at, or near the limits of their ability to grow in annual supply and will dwindle as the decades go forward [5]. The depletion of fossil fuel reserves, global warming due to CO2 emissions, the spread of health problems and increasing political tensions are some of the reasons why renewable energy should be promoted [6]. Research works have recently focused on renewable energy scavenging technologies which produce energy with small scale power. These technologies include triboelectric, nanogenerator and piezoelectric [7,8]. On large scale power, the switch to renewable energy sources should be done while fostering an evolution of personal, institutional and national values. These steps recognize the ultimate limits of the earth's carrying capacities which are presently being dramatically exceeded.

Tidal current energy, which harnesses the kinetic energy contained in tidal streams, is emerging as a great potential energy source [9,10]. It has a number of advantages compared to other renewable energies. The resource predictability, the minimal visual impact and land occupation, its high load factor and sustainability are some of the noteworthy features [11,12]. The benefits include reduced reliance on imported fuels, uninterrupted and affordable energy supplies, long-term price stability, decoupling hydrocarbon and resource risks, and environmental security [13]. However, realistic tidal locations are very perturbed with high range and disturbances are site-specific [14,15]. The swell considers the crucial phenomenon to be taken into account which affects the maritime structures [16]. The propagation of the submarine swell has the greatest influence on the marine current and the origin of the disturbance in small time scales for the tidal turbine. One can note that the harnessed output power will be affected in the case of a disturbed input. The turbulence must be estimated from field observations of the flow, which are inherently sparse and noisy [17].

Many studies concentrated on the optimization of the generated power in the case of high tidal speed using the angular position of the rotor's blades [18,19]. The pitch and stall angle controls have been developed in [20]. The work points out that the blade pitch angle control leads to more valuable responses concerning the energy yields than the stall regulated system. Some studies used the pitch angle control with several techniques [21]. Artificial intelligence has been used to handle renewable energy systems [22–24]. An artificial neural network is a designed method which is considered to solve many tasks of fitting applications. As detailed in [25], an artificial neural network has been conceived for the Tidal Stream Generator (TSG) to find the appropriate angle for each tidal speed variation. The study shows favourable results when compared with a conventional controller. The fuzzy reasoning approach is motivated by the flexibility in decision-making processes [26]. Interest in fuzzy logic has shown good results in the field of automatic control and the aim to extended it to renewable energy converters. This paper introduces a fuzzy rule-based scheme for gain scheduling of the pitch angle controller in power limitation mode. An adaptive fuzzy Proportional Integral Derivative (PID) controller with a gains scheduling mechanism is proposed. The fuzzy supervisor provides the gains to the controller in order to govern the blade pitch angle. The Maximum Power Point Tracking (MPPT) technique is used to generate the adequate trajectory to the rotational speed controller.

The remainder of this paper is structured as follows; Section 2 defines the realistic tidal site as a site evaluation tool for the tidal stream generator. Then, the design of the TSG system in a digital environment including the hydrodynamic, mechanical and electrical parts of the power plant is given in Section 3. Section 4 is devoted to the control objectives and strategies and presents the FGS-PID controller for the pitch angle control. Two study cases have been considered to test the investigated control approaches as presented in Section 5. Section 6 ends the paper with concluding remarks.

2. Alderney Race Tidal Site Profile

The Alderney Race is a straight located between the Channel Island of Alderney and Cap de la Hague on the West coast of France. The site is four meters wide and lies between Race rock (49°42′ N, 2°08′ W) and a rocky bank with a minimum depth of 17 m over it, which lies approximately 3.5 m from Cap de la Hague. The tides run in a northwesterly direction for a period of six hours starting at six hours before Dover High Water (DHW). After that, it switches direction to flow southeast for approximately six hours. The highest velocities are found on the east side. As an example, in the west of the La Foraine light buoy the spring current speed of the north going stream can reach 5 m/s and that of the south going stream is about 3.5 m/s [27].

This tidal site is an important profile for extracting marine energy because the density is large and the depths are suited for installing tidal stream turbines. One can note that the deployment of a TSG plant will have a huge load factor to generate electricity for a high time scale. Also, there are locations where the depth is about 30, 35 and 40 m which represents a suitable value for placing TSG plants. The local strength of the current is due to the acceleration of the tidal flow between the Alderney Island and La Hague cape (France). The average power density is around 5 kW/m^2 and depths varying between 30 and 60 m can be over a surface higher than 10 km^2 [27]. In this site, the data measurement of tidal velocities is provided by SHOM (French Navy Hydrographic and Oceanographic Service). As depicted in Figure 1, the propagation of tidal currents are spread over a wide range of values where high velocities can even exceed 4.5 m/s [28].

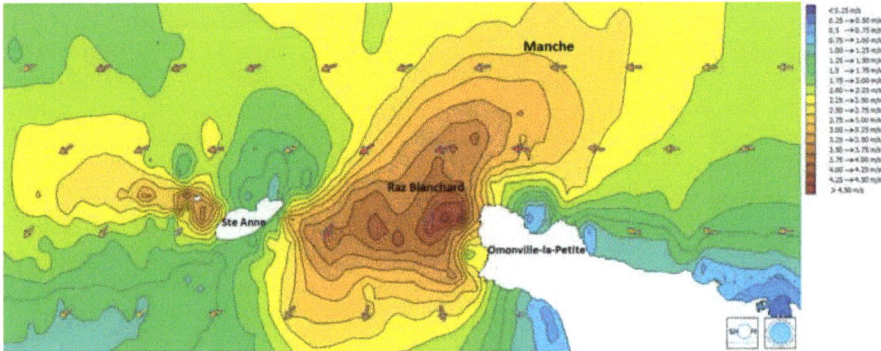

Figure 1. Tidal current speed in Alderney Race in the French western coast.

Fluctuation aspects of tidal power are based on two forms of energy disturbance: On a high time period corresponding to the neap and spring marine current changing each day, and on a small time period relating to swell effect phenomenon [29].

3. Model Statement

The development of high-efficiency tidal energy conversion systems requires multiple testings and continuous modifications to rapidly rectify and correct the behavior of the developed model. Therefore, it is better to perform these testings and rectifications in software in the loop framework. The structure of the TSG plant is illustrated in Figure 2. The tidal turbine is connected to the Doubly Fed Induction Generator (DFIG) via the drive train shaft. The hydrodynamic part is connected to the grid using the back-to-back power converters. In this sense, the dynamic modeling of the system requires the use of a computational tool including these nonlinear sub-models with a different timescale.

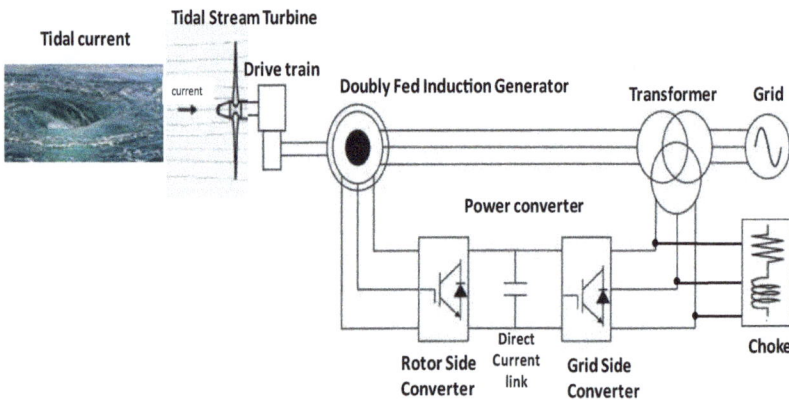

Figure 2. Scheme of the tidal stream generator system [30].

3.1. Tidal Turbine Model

The power generation from the marine current speed needs the hydrokinetic energy conversion to produce electrical power. It is described by the following equation [31]:

$$P_t = \frac{1}{2}C_p(\lambda, \beta)\rho\pi R^2 V^3 \tag{1}$$

where V is the tidal current speed in (m/s), P_t is the harnessed power from marine current (W), R is the rotor blade radius defined in (m), and ρ is the density of water (kg/m^3).

The kinetic power is corresponding to the speed of water V which passes through the channel section A as shown in Figure 3.

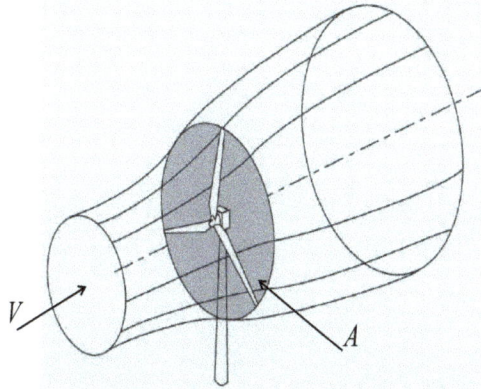

Figure 3. Tidal flow through the swept area of a rotor disk.

Bearing in mind that the TSG system can only extract a fraction of this available energy, so the power coefficient C_p characterizes the level of performance of the tidal stream turbine. Such a coefficient is defined as function of the pitch angle β in (deg) and the tip-speed ratio λ, given as [32,33]:

$$\lambda = \frac{\omega_t R}{V} \tag{2}$$

where ω_t is the rotor speed in (rad/s).

The hydrodynamic torque of the tidal turbine, defined in (Nm), is expressed as follows:

$$T_{tst} = \frac{P_t}{\omega_t} \tag{3}$$

3.2. Mechanical Shaft Model

The mechanical transmission is used to transform the low rotational speed at the rotor to high one at the generator side. The high rotational speed of the generator is necessary to apply compact constructed generators. The model of the shaft is chosen so as to regroup the hydrodynamic loads of the tidal turbine since they represent an important factor relating to the extracted output power. Therefore, the rotor shaft is assumed an important aspect of the Tidal Stream Turbine (TST) which has an impact on the power fluctuations. The two-mass model is used to describe the rotor shaft dynamics as follows [34]:

$$T_{tst} - T_t = 2H_t\frac{d\omega_t}{dt} \tag{4}$$

$$T_t = D_{sh}(\omega_t - \omega_g) + K_{sh}\int (\omega_t - \omega_g)dt \tag{5}$$

$$T_t - T_{em} = 2H_g \frac{d\omega_g}{dt} \tag{6}$$

where K_{sh} in (Nm/rad) and D_{sh} in (Nms/rad) are the stiffness and damping coefficients, respectively. T_t is the torque of the rotor shaft in (Nm), T_{em} is the electromagnetic torque in (Nm), and ω_g is the rotor speed in (rad/s). H_t and H_g are the inertia constants for the turbine and the generator in s, respectively.

3.3. Electrical Model

The hydrodynamic turbine should be able to operate over a wide range of tidal velocities in order to achieve optimum efficiency by tracking the optimal tip-speed ratio. Therefore, the DFIG system operates in both sub- and super-synchronous modes with a rotor speed range around the synchronous speed [35].

The model of the DFIG is given in the d–q synchronous frame using the Park's transformation as defined in [36]. The equations of the stator voltages and flux, in (V) and in (Wb) respectively, are written as follows:

$$\begin{cases} U_{sd} = R_s I_{sd} + \frac{d\varphi_{sd}}{dt} - \omega_s \varphi_{sq} \\ U_{sq} = R_s I_{sq} + \frac{d\varphi_{sq}}{dt} - \omega_s \varphi_{sd} \end{cases} \tag{7}$$

$$\begin{cases} \varphi_{sd} = L_s I_{sd} + L_m I_{rd} \\ \varphi_{sq} = L_s I_{sq} + L_m I_{rq} \end{cases} \tag{8}$$

The expressions of the rotor voltages and flux are given by the following equations:

$$\begin{cases} U_{rd} = R_r I_{rd} + \frac{d\varphi_{rd}}{dt} - \omega_r \varphi_{rq} \\ U_{rq} = R_r I_{rq} + \frac{d\varphi_{rq}}{dt} - \omega_r \varphi_{rd} \end{cases} \tag{9}$$

$$\begin{cases} \varphi_{rd} = L_r I_{rd} + L_m I_{sd} \\ \varphi_{rq} = L_r I_{rq} + L_m I_{sq} \end{cases} \tag{10}$$

The equation of the electromagnetic torque is defined as follows:

$$T_{em} = \frac{3}{2} p L_m \left(I_{sq} I_{rd} - I_{sd} I_{rq} \right) \tag{11}$$

where I_{sd}, I_{sq} are the stator currents given in (A), I_{rd}, I_{rq} are the rotor currents in (A), R_s and R_r are the resistances of the stator and rotor in (Ω), ω_s and ω_r are the pulsations of the stator and rotor in (rad/s), L_s and L_r are the inductances of the stator and rotor in (H), respectively, L_m is the magnetizing inductance in (H), and p is the number of the poles pairs.

3.4. Power Converters Model

Tidal stream converters aim to generate power and to guarantee cost reduction. For that reason, these systems use back-to-back power electronic converters since they ensure the connection with the grid [37]. These types of equipment ensure the conversion from a variable output frequency from the generator to a fixed one related to the grid [38]. The used back-to-back power converter includes the Rotor Side Converter (RSC) and the Grid Side Converter (GSC) which have been connected through the DC-link. This configuration has the advantage of applying the vector control method for both sides. The RSC is intended to control the operation of the generator. The aim of the GSC is to maintain constant voltage of the DC-link regardless of the magnitude and the direction of the rotor power.

The expressions of the active and reactive powers of the DFIG-based TST, in (W) and (VAR) respectively, are defined as:

$$P_g = \frac{3}{2} \left(U_{dg} I_{dg} - U_{qg} I_{qg} \right) \tag{12}$$

$$Q_g = \frac{3}{2}\left(U_{qg}I_{dg} - U_{dg}I_{qg}\right) \tag{13}$$

where U_{dg}, U_{qg} in (V) and I_{dg}, I_{qg} in (A) are the voltages and currents of the grid.

In order to achieve the voltage oriented control, the vectors of the d-axis and the grid voltage are aligned, $U_{dg} = U_g$ and $U_{qg} = 0$. So, the equations of the active and reactive powers are rewritten as:

$$P_g = \frac{3}{2}U_g I_{dg} \tag{14}$$

$$Q_g = -\frac{3}{2}U_g I_{qg} \tag{15}$$

The expression between the power stored in the DC-link and the power transferred to the grid is described as follows:

$$P_g = \frac{3}{2}U_g I_{dg} = U_{dc}I_{dc} \tag{16}$$

where U_{dc} and I_{dc} are the voltage and current of the DC-link.

4. Control Strategies

When the tidal stream generators are subjected to turbulent tidal current speed and strong swells, the pitch angle control is investigated to limit the generated power and maintain the system safe from overload. For that reason, it's important to study the system to optimize the extracted output power and to improve the efficiency. In this mode of operation, the pitch angle controller is set to regulate the pitch actuator when the marine current exceeds the threshold value, and thus maintain the generated power at its nominal condition. In each variable marine speed, the controller sends the adequate control signal in order to rotate the rotor blades to the desired angular position.

The power may be limited hydrodynamically using pitch control. The control scheme of the TSG power plant is depicted in Figure 4. Advanced control approaches are proposed in order to ensure better performances, especially to guarantee robustness under uncertainties. In this sense, the pitch angle control is investigated using the fuzzy logic approach in order to find the adaptive gains of the controller. Moreover, the rotational speed control is based on the MPPT strategy for which the maximum output power will be attained.

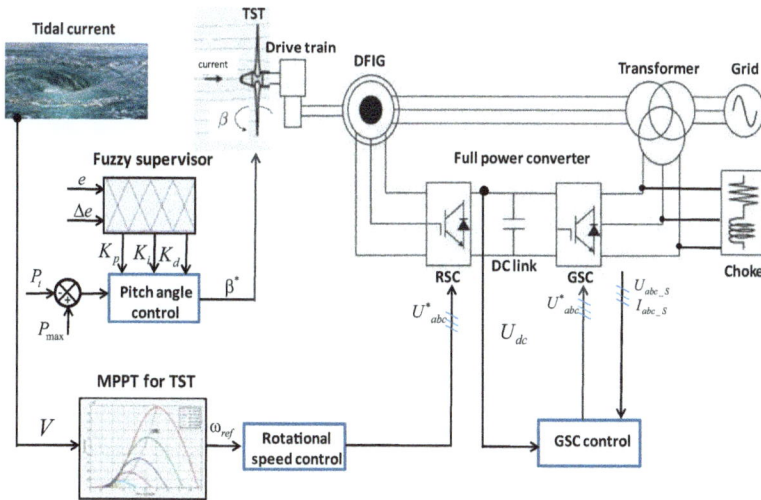

Figure 4. Tidal stream generator control scheme. TST, Doubly Fed Induction Generator (DFIG), Rotor Side Converter (RSC), Grid Side Converter (GSC), Maximum Power Point Tracking (MPPT).

4.1. Pitch Angle Control

The proposed control scheme is illustrated in Figure 5. The pitch angle control loop is designed using a fuzzy gain scheduling method because it represents a robust control technique regarding model uncertainties [39,40]. The investigated Fuzzy Gain Scheduling (FGS)-based PID control is used to generate and tune the gains in order to keep the required performance. The input of the PID controller is the error between the maximum power supported by the system which is 1.5 MW and the measured generated power.

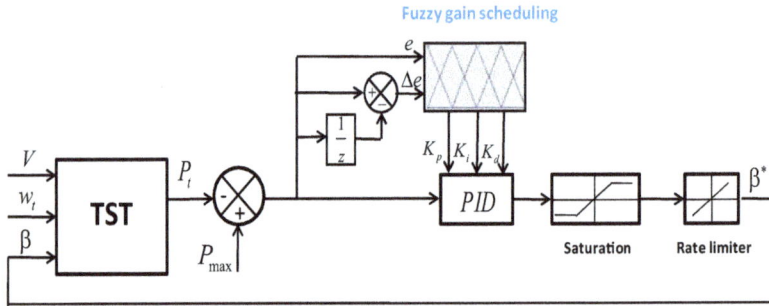

Figure 5. Fuzzy Gain Scheduling (FGS)-Proportional Integral Derivative (PID) based pitch angle control scheme.

The approach taken here is to exploit fuzzy rules and reasoning [41,42]. The variation of the studied tidal turbine under different values of the pitch angle β is depicted in Figure 6. One can note that as the angle β increases as the output power P_t decreases. The threshold value of the tidal velocity is calculated at 3.2 m/s. Over this value, the limitation mode will be used to protect the system.

The equation of the controller in the discrete-time domain is expressed as follows [39]:

$$u(k) = K_p \Delta e(k) + K_i T_s e(k) + K_d \Delta e(k) + u(k-1) \tag{17}$$

where $e(k)$ is the error between P_{max} and P_t, $\Delta e(k) = e(k) - e(k-1)$ is the change of the error, T_s is the sampling time and K_p, K_i and K_d are the PID controller parameters.

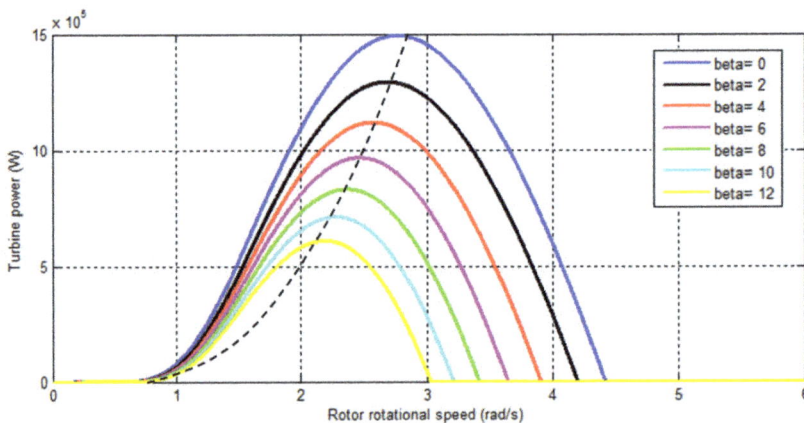

Figure 6. Output power versus the rotor speed for different blade pitch angles.

The gains K_p, K_i and K_d are normalized applying the linear transformation by the Equation (18) [43]:

$$
\begin{cases}
K'_p = (K_p - K_{p\,min}) / (K_{p\,max} - K_{p\,min}) \\
K'_i = (K_i - K_{i\,min}) / (K_{i\,max} - K_{i\,min}) \\
K'_d = (K_d - K_{d\,min}) / (K_{d\,max} - K_{d\,min})
\end{cases}
\tag{18}
$$

where $[K_{p\,min}, K_{p\,max}]$, $[K_{i\,min}, K_{i\,max}]$ and $[K_{d\,min}, K_{d\,max}]$ are the prescribed domains of the controller parameters.

The gain scheduling of the PID controller is calculated by means of the fuzzy rules given as follows:

$$
\begin{aligned}
&if \ \ e(k) \ \ is \ \ A_i \ \ and \ \ \Delta e(k) \ \ is \ \ B_i \\
&then \ \ K'_p \ \ is \ \ C_i \ \ and \ \ K'_i \ \ is \ \ D_i \ \ and \ \ K'_d \ \ is \ \ E_i
\end{aligned}
\tag{19}
$$

where A_i, B_i, C_i, D_i and E_i are the fuzzy sets on the relating linguistic variables where $i = 1, 2, ..., m$.

The types of membership functions used are triangular uniformly distributed and symmetrical in the universe of discourse. The corresponding linguistic levels are Negative Big (NB), Negative (N), Zero (Z), Positive (P) and Positive Big (PB) as shown by Figures 7 and 8.

The fuzzy rules proposed in this study are defined in Tables 1–3. The set of rules are proposed to fit the behavior of a PID conventionnel controller regarding the error and the error variation.

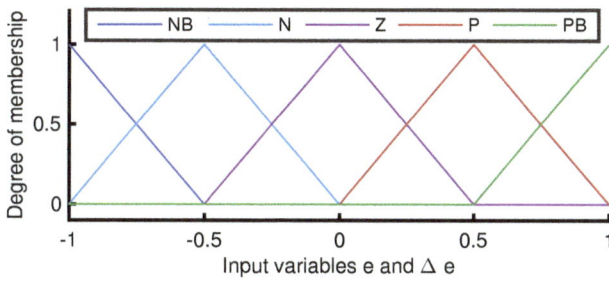

Figure 7. The inputs membership functions. Negative Big (NB), Negative (N), Zero (Z), Positive (P) and Positive Big (PB).

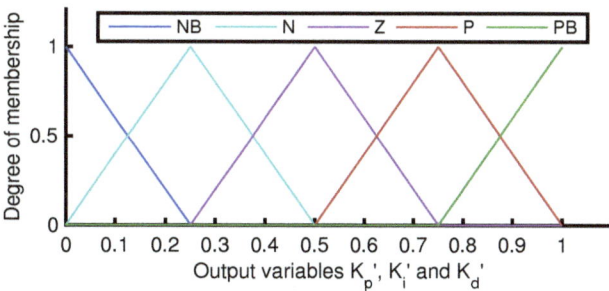

Figure 8. The outputs membership functions.

Table 1. Fuzzy rules for K_p gain [30].

$e(k)/\Delta e(k)$	NB	N	Z	P	PB
NB	NB	NB	NB	N	Z
N	NB	N	N	N	Z
Z	NB	N	Z	P	PB
P	Z	P	P	P	PB
PB	Z	P	PB	PB	PB

Table 2. Fuzzy rules for K_i gain [30].

$e(k)/\Delta e(k)$	NB	N	Z	P	PB
NB	PB	PB	PB	N	NB
N	PB	P	P	Z	NB
Z	P	P	Z	N	NB
P	Z	P	N	N	NB
PB	Z	N	NB	NB	NB

Table 3. Fuzzy rules for K_d gain.

$e(k)/\Delta e(k)$	NB	N	Z	P	PB
NB	NB	NB	NB	P	PB
N	N	N	N	Z	PB
Z	Z	N	Z	P	PB
P	Z	N	P	P	PB
PB	Z	P	PB	PB	PB

The equation of the defuzzification is described as follows:

$$
\begin{cases}
K'_p = \sum_{i=1}^{m} \mu_i \mu_{C_i} \\
K'_i = \sum_{i=1}^{m} \mu_i \mu_{D_i} \\
K'_d = \sum_{i=1}^{m} \mu_i \mu_{E_i}
\end{cases}
\tag{20}
$$

The decision-making output is calculated using a Max-Min fuzzy inference where the real outputs are calculated by the method of defuzzification center of gravity as:

$$
\begin{cases}
K_p = K_{p\,min} + (K_{p\,max} - K_{p\,min})K'_p \\
K_i = K_{i\,min} + (K_{i\,max} - K_{i\,min})K'_i \\
K_d = K_{d\,min} + (K_{d\,max} - K_{d\,min})K'_d
\end{cases}
\tag{21}
$$

By designing the fuzzy supervisor of the pitch controller based on the proposed fuzzy rules, the resulting fuzzy surfaces related to the gains K_p, K_i and K_d are illustrated in Figures 9–11, respectively.

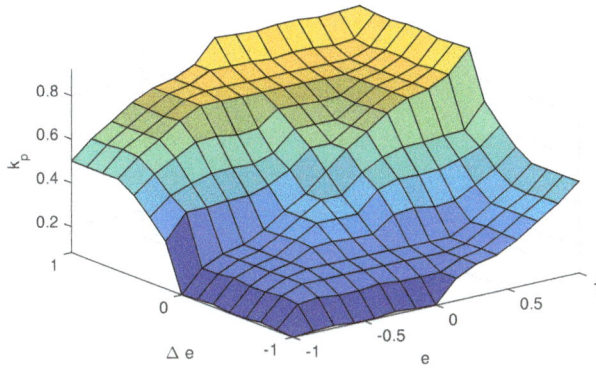

Figure 9. Fuzzy surface for K_p gain.

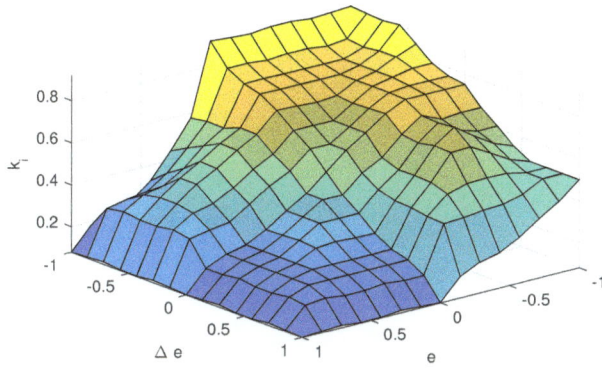

Figure 10. Fuzzy surface for K_i gain.

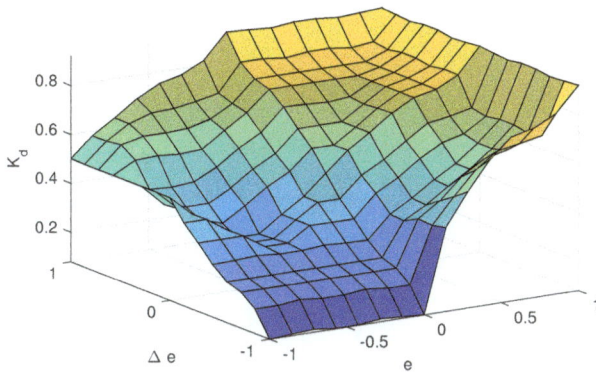

Figure 11. Fuzzy surface for K_d gain.

4.2. Rotational Speed Control

4.2.1. RSC Control Design

The control scheme related to the RSC component is illustrated in Figure 12. The stator flux oriented control is used in this study. The design of the control scheme includes one control loop to regulate the rotor speed and two control loops to regulate the currents.

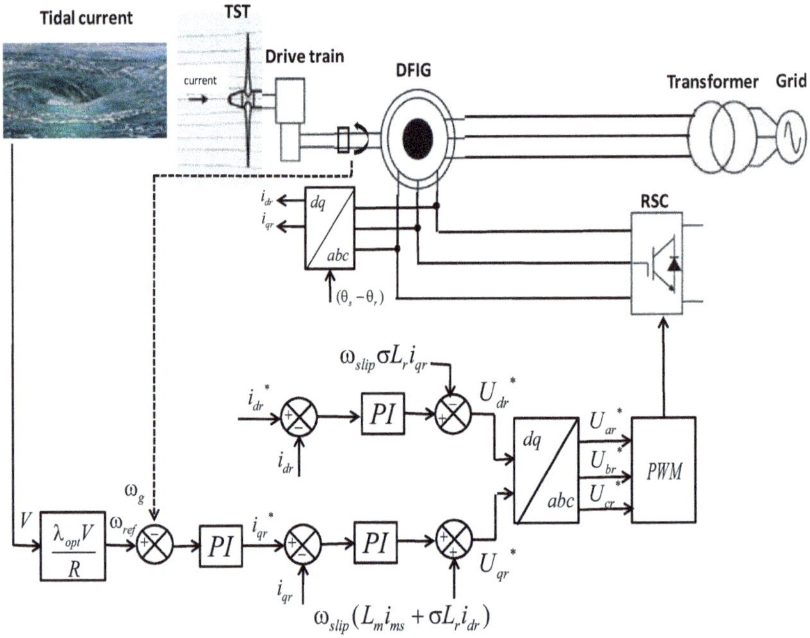

Figure 12. RSC control scheme design.

The MPPT generating the desired rotor speed to the outer loop is designed for the tidal turbine. It takes into account the characteristic curve shown in Figure 13 to follow the maximum power [44].

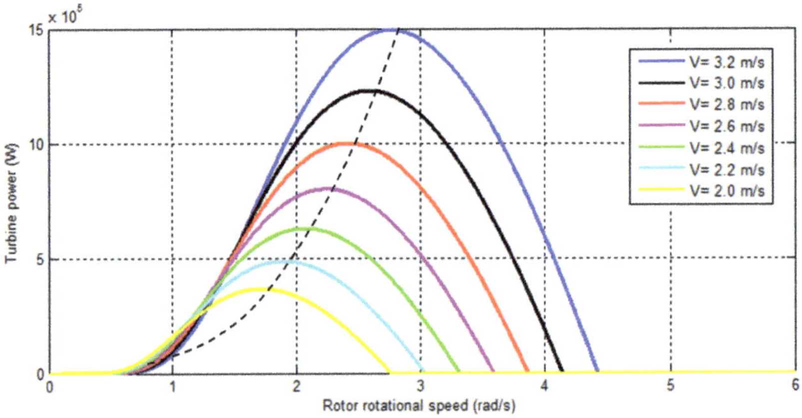

Figure 13. Extracted power as a function of the rotor speed for different tidal speeds.

In this sense, the MPPT will generate the optimum rotational speed depending on the tidal current speed. Using the developed MPPT for TSG, ω_{ref} is defined as the rotational speed control for which a reference signal is set to the rotor current q-axis i_{qr}^*. The current control loops calculate the reference signal of the rotor voltage defined in d–q synchronous frame. The expressions of the rotor voltages and currents are given by the following equations as defined in [45]:

$$\begin{cases} U_{dr} = R_r i_{dr} + \sigma L_r \frac{di_{dr}}{dt} \\ U_{qr} = R_r i_{qr} + \sigma L_r \frac{di_{qr}}{dt} \end{cases} \tag{22}$$

where σ is the leakage factor.

Also, the parameters of decoupling are added to the equations of the direct and quadrature component of the rotor voltages so as to improve the response of the system [46]. Therefore, the voltage references are given as follows:

$$\begin{cases} U_{dr}^* = -\omega_{slip}\sigma L_r i_{qr} + (K_{Pi}e_d + K_{Ii} \int e_d \, dt) \\ U_{qr}^* = \omega_{slip}(L_m i_m + \sigma L_r i_{dr}) + (K_{Pi}e_d + K_{Ii} \int e_d \, dt) \end{cases} \tag{23}$$

where ω_{slip} is the angular frequency of the slip given in (rad/s) and i_m is the current of stator magnetizing kept constant. K_{Pi} and K_{Ii} are the Proportional Integral (PI) controller parameters.

The PI controllers blocks are designed using the well-known Ziegler-Nichols method [47]. Also, a modification of the tuning on the first value of the parameters of the controller has been applied by means of the method robust response time algorithm [48]. The voltage references of the rotor are converted to the *abc* frame which will affect the RSC component through the Pulse Width Modulation (PWM) block.

4.2.2. GSC Control Design

The control scheme design of the GSC component is illustrated in Figure 14. The used method is the voltage oriented control. This strategy consists of two PI controllers for the current and one PI controller for the voltage. The investigated block design controls the voltage U_{dc} and the reactive power Q_g. In order to extract the phase of the input signal θ_g, the Phase Locked Loop (PLL) method is used in this study. The direct and quadrature components of the currents and voltages are obtained using Park's transformation method.

The expressions of the grid voltages given in the $d - q$ synchronous frame as:

$$\begin{cases} U_{gd} = i_{ds}R_g + L_g \frac{di_{ds}}{dt} - \omega_s L_g i_{qs} + U_{gd1} \\ U_{gq} = i_{qs}R_g + L_g \frac{di_{qs}}{dt} - \omega_s L_g i_{ds} + U_{gq1} \end{cases} \tag{24}$$

where R_g is the resistance of the grid given in (Ω), L_g is the inductance of the grid in (H), U_{gd1} and U_{gq1} are the two phases of the terminal voltages.

The active and reactive powers are controlled via the currents synchronous frame dq. The controllers of the currents are identical and give the grid reference voltages U_{ds}^* and U_{qs}^*. In order to enhance the system response, the compensator parameters and feed-forward voltages are added to the control signals [49]:

$$\begin{cases} U_{gd}^* = U_{gd} + \Omega_g L_g i_q - (K_{Pi}e_d + K_{Ii} \int e_d \, dt) \\ U_{gq}^* = U_{gq} - \Omega_g L_g i_d - (K_{Pi}e_q + K_{Ii} \int e_q \, dt) \end{cases} \tag{25}$$

The voltage controller is conceived to control the DC-link voltage in the way to maintain it at its reference. The i_{qs} current is intended to regulate the reactive power. The reference signal of the current in q-axis is considered zero. As the case of the RSC component, the PI controller parameters are calculated by means of the Ziegler-Nichols technique. Furthermore, the reference signals of the voltages are transformed to the *abc* frame and will give the PWM signals for the GSC component.

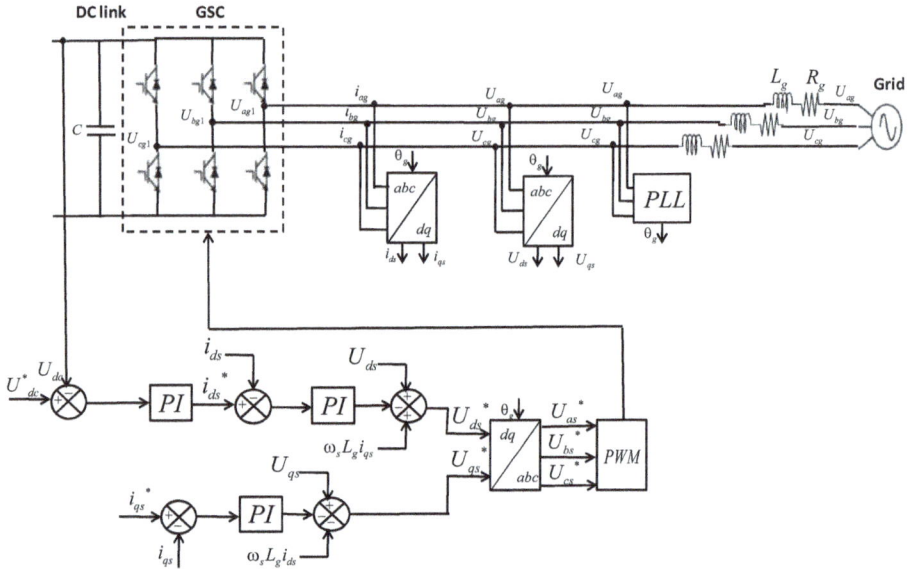

Figure 14. GSC control scheme design.

5. Validation Results and Discussion

In this section, based on the realistic tidal site Alderney Race profile two study cases were used to test the robustness and the effectiveness of the investigated control methods. The adaptive FGS-PID based control was analyzed regarding the disturbance in the tidal speed under regular and irregular profiles. The numerical implementation of the TSG in a digital environment including the hydrodynamic, mechanical and electrical parts of the power plant is shown in Figure 15 using the model parameters listed in Table 4.

Table 4. Tidal Stream Generator (TSG) system parameters.

Turbine	Drive-Train	DFIG	Converter
$\rho = 1027 \text{ kg/m}^3$	$H_t = 3 \text{ s}$	$P_n = 1.5 \text{ MW}$	$V_{dc} = 1150 \text{ V}$
$R = 8 \text{ m}$	$H_g = 0.5 \text{ s}$	$U_{rms} = 690 \text{ V}$	$C = 0.01 \text{ F}$
$C_{p\,max} = 0.44$	$K_{sh} = 2 \times 10^6 \text{ Nm/rad}$	$f_{req} = 50 \text{ Hz}$	
$\lambda_{opt} = 6.96$	$D_{sh} = 3.5 \times 10^5 \text{ Nms/rad}$	$R_s = 2.63 \text{ m}\Omega$	
$V_n = 3.2 \text{ m/s}$		$R_r = 2.63 \text{ m}\Omega$	**Choke**
		$L_s = 0.168 \text{ mH}$	$R_g = 0.595 \text{ m}\Omega$
		$L_r = 0.133 \text{ mH}$	$L_g = 0.157 \text{ mH}$
		$L_m = 5.474 \text{ mH}$	
		$p = 2$	

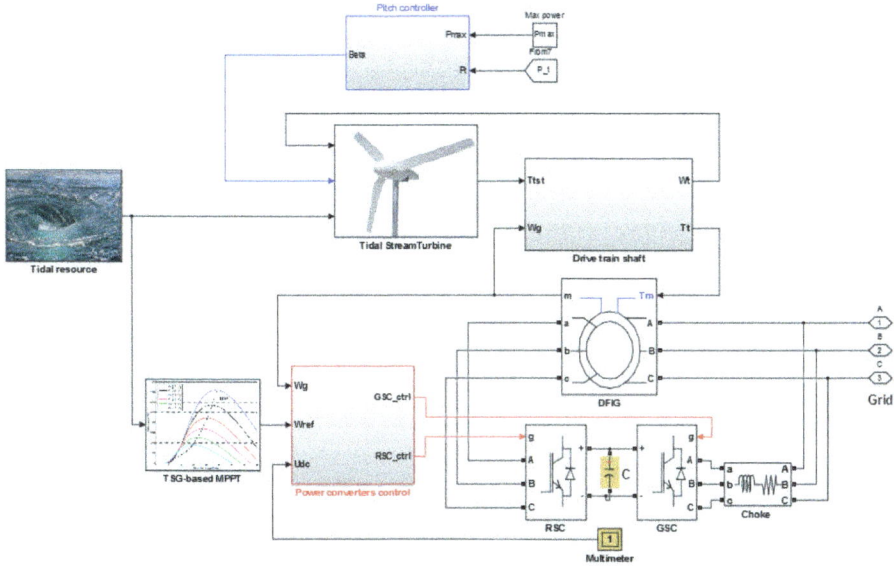

Figure 15. Model implementation of the TSG power plant.

In this first case, the sensibility of the proposed FGS-PID based pitch angle control was tested under a long time fluctuation of the tidal resource in the case of turbulence as depicted in Figure 16. The input considered has the shape of a regular neap and spring tides with a pic values of about 4 m/s and 4.5 m/s, respectively.

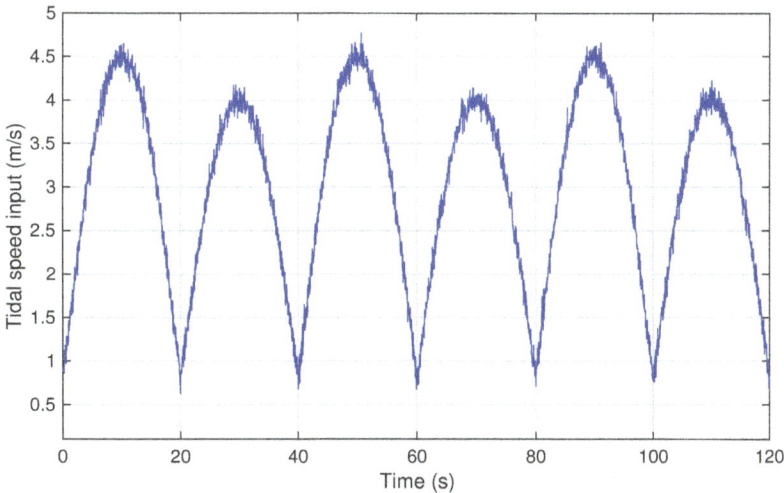

Figure 16. Case 1: Regular turbulent tidal resource speed.

The TSG control performances are illustrated in Figure 17. The power coefficient and the blade pitch angle curves are time varying for compensating to input disturbance. The FGS-PID based control provides the adaptive parameters of the pitch controller to respond to the behavior of the input change.

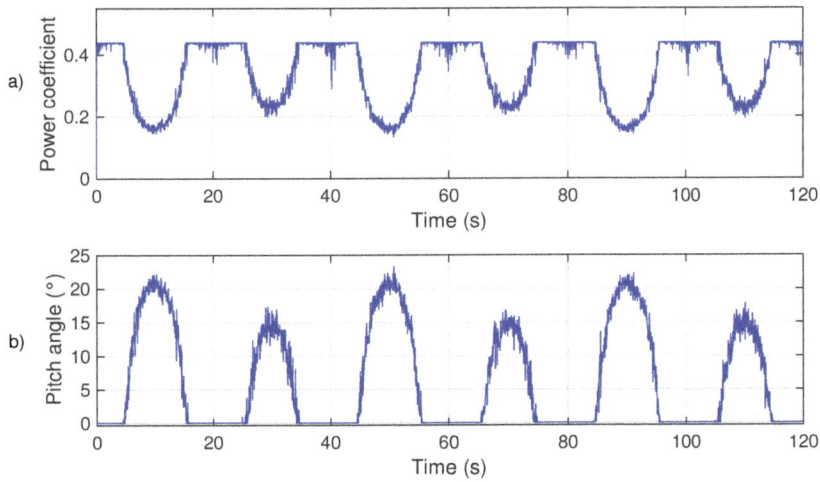

Figure 17. Control performances of case 1: (**a**) power coefficient variation; (**b**) pitch angle variation.

The generator speed response and the reference signal following the MPPT block are given in Figure 18. A zoom into the response within 1.2 s shows that the investigated control approach is robust regarding the speed tracking.

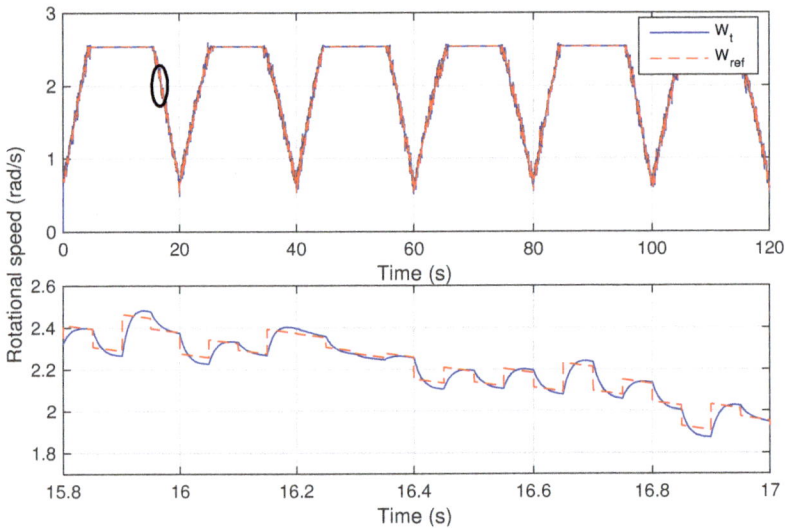

Figure 18. Case 1: Rotational speed and its reference curves.

The generated power variation is illustrated in Figure 19. The resulting power changes according to the variation of the marine velocity. It can be noted that the control schemes are able to limit the extracted power within a specific limit of about 1.497 MW.

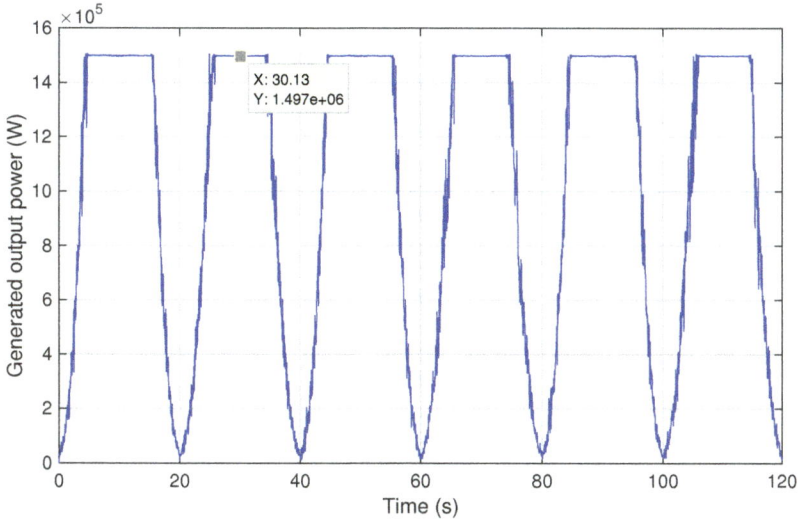

Figure 19. Case 1: Output power variation.

In the second case, the investigated control approach was analyzed regarding the swell effect disturbance which represents a short time fluctuation regarding the current speed input. The turbulent resource characteristic is shown in Figure 20. The average value taken is approximately about 3.7 m/s. The fluctuated tidal input admits a minimum value of 2.312 m/s and a maximum value of 5.022 m/s.

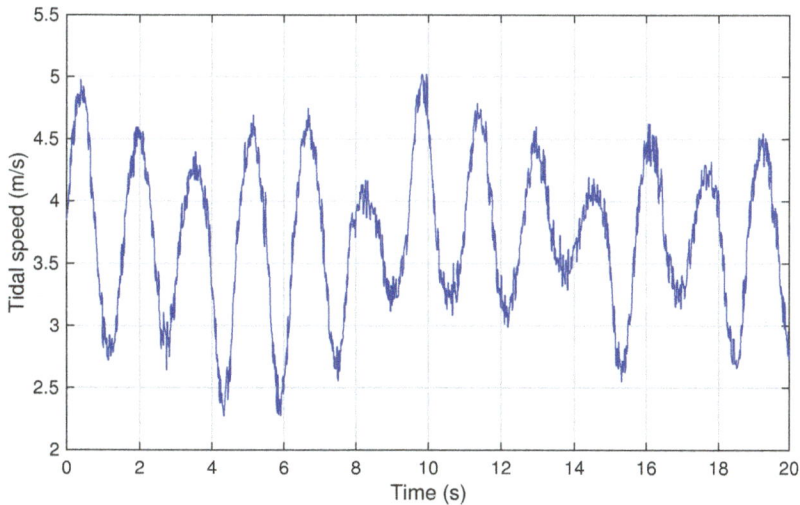

Figure 20. Case 2: Irregular disturbed tidal speed input.

Figure 21 shows the power coefficient and the pitch angle variations. It is obvious that the system adapts well to the short-time fluctuations. At high tidal speed reached, the power coefficient decreases and consequently the pitch angle signal increases.

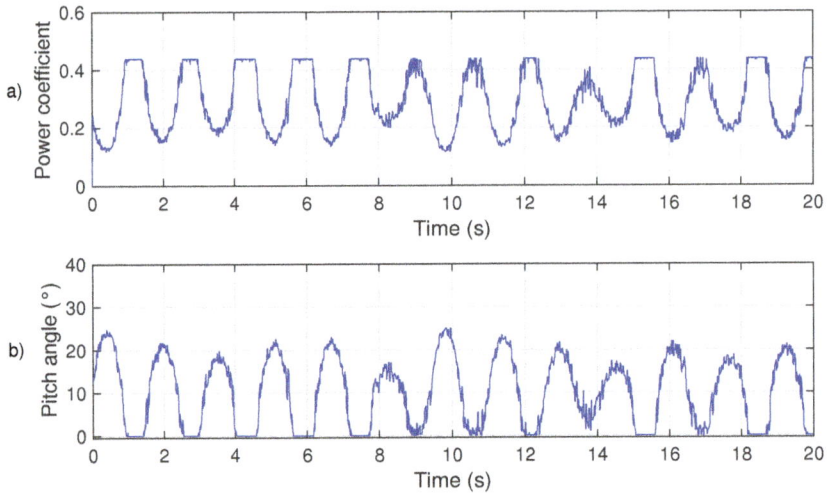

Figure 21. Control performances of case 2: (**a**) power coefficient variation; (**b**) pitch angle variation.

The response of the rotor speed and the reference gathered from the developed MPPT method are given in Figure 22. The controller shows a good tracking performance of the reference signal. This demonstrates that the FGS-PID based control has a reduced steady-state error due to the fact that the integral action is adequately changing regarding the variation of the tidal input.

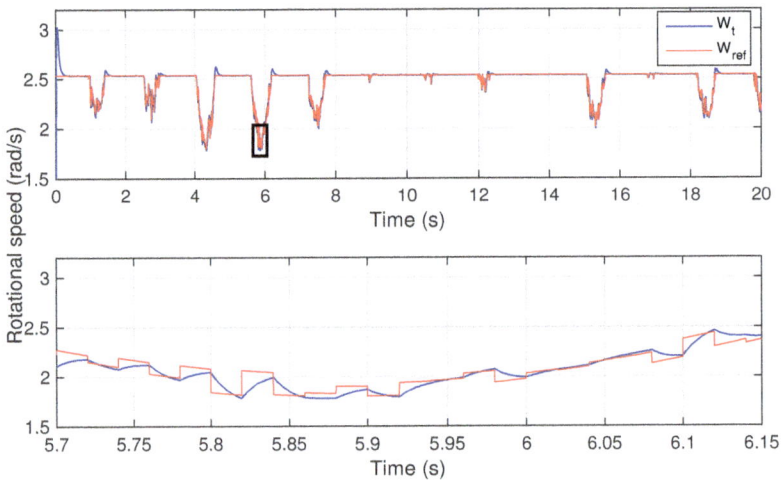

Figure 22. Case 2: Rotational speed and its reference curves.

The response of the generated power is illustrated in Figure 23. It can be seen that the power is limited to 1.496 MW. So, the system is able to optimize the extracted power in the case of the disturbed input under the swell effect phenomenon.

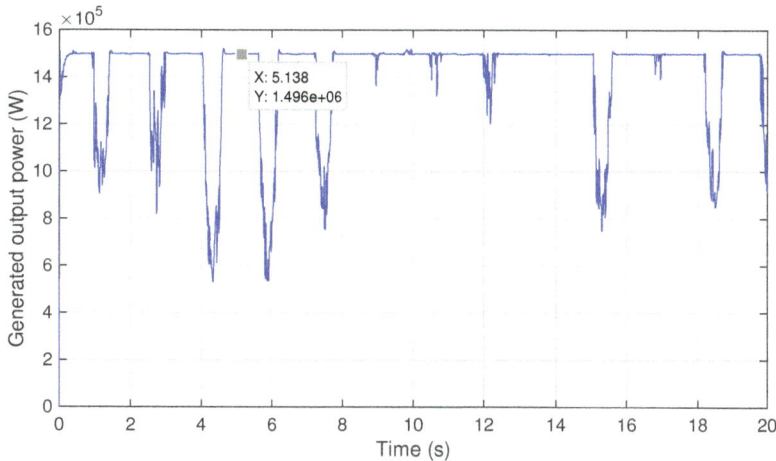

Figure 23. Case 2: Output power variation.

6. Conclusions

In this paper, a TSG system has been modeled and controlled. A fuzzy supervision has been conceived to the pitch controller in order to properly modify the gains of the PID in accordance with the variation of the tidal input. The MPPT strategy has been used to give the adequate rotational speed for the RSC control.

To test the robustness of the novel FGS-PID-based control the realistic tidal site Alderney Race site was investigated. The first experiment was performed using regular tidal speed under disturbance conditions. The results demonstrate that the control strategies successfully deal with these fluctuations which enable the plant to optimize the generated power.

A second case of study was used which considers a turbulent tidal profile under the swell effect disturbance. Simulation results show that the proposed control strategies are effective in terms of speed tracking and power regulation. Moreover, the sensitivity of the proposed fuzzy-based control strategy has been analyzed regarding the swell effect. The investigated control schemes seem to be a good solution when the resource is not well-known and even if the resource is heavily disturbed.

The dynamic performances of the tidal stream generator system have been evaluated versus intelligence control technique. The proposed fuzzy supervisor ensures the regulation of the blade pitch angle for the high marine currents. The sensitivity of the proposed control strategy has been analyzed regarding the swell effect. Indeed, any variation of the fluid speed consequently induces a variation of the rotor speed reference which is deduced from the MPPT strategy.

Author Contributions: Conceptualization, K.G., I.G., and A.J.G.; Formal Analysis, K.G.; Investigation, K.G.; Methodology, K.G.; Project administration, E.R.; Software, K.G.; Supervision, I.G., S.B., J.H. and A.J.G.; Writing, Review and Editing, K.G.

Funding: This research received no external funding.

Acknowledgments: This work was supported by the MINECO through the Research Project DPI2015-70075-R (MINECO/FEDER, UE) and in part by the University of the Basque Country (UPV/EHU) through PPG17/33. The authors would like to thank the collaboration of the Basque Energy Agency (EVE) through Agreement UPV/EHUEVE23/6/2011, the Spanish National Fusion Laboratory (EURATOM-CIEMAT) through Agreement UPV/EHUCIEMAT08/190 and EUSKAMPUS - Campus of international Excellence.

Conflicts of Interest: The authors declare no conflict of interest.

References

1. International Energy Outlook 2016. *International Energy Outlook 2016 With Projections to 2040*; Tech. Report; U.S. Energy Information Administration, Office of Energy Analysis U.S. Department of Energy: Washington, DC, USA, 2016.

2. Panwar, N.L.; Kaushik, S.C.; Kothari, S. Role of renewable energy sources in environmental protection: A review. *Renew. Sustain. Energy Rev.* **2011**, *15*, 1513–1524. [CrossRef]

3. Ellabban, O.; Abu-Rub, H.; Blaabjerg, F. Renewable energy resources: Current status, future prospects and their enabling technology. *Renew. Sustain. Energy Rev.* **2014**, *39*, 748–764. [CrossRef]

4. Frondel, M.; Ritter, N.; Schmidt, C.M.; Vance, C. Economic impacts from the promotion of renewable energy technologies: The German experience. *Energy Policy* **2010**, *38*, 4048–4056. [CrossRef]

5. Grino Colom, M. *Power Generation From Tidal Currents*; Application to Ria de Vigo; Escola de Comins, Departament d'Enginyeria Hidraulica, Maritima i Ambiental (DEHMA): Barcelone, Spain, 2015.

6. Shapiro, G.I. Effect of tidal stream power generation on the region-wide circulation in a shallow sea. *Ocean Sci.* **2011**, *7*, 165–174. [CrossRef]

7. Zhu, G.; Lin, Z.H.; Jing, Q.; Bai, P.; Pan, C.; Yang, Y.; Zhou, Y.; Wang, Z.L. Toward large-scale energy harvesting by a nanoparticle-enhanced triboelectric nanogenerator. *Nano Lett.* **2013**, *13*, 847–853. [CrossRef] [PubMed]

8. Wang, S.; Lin, L.; Wang, Z.L. Nanoscale triboelectric-effect-enabled energy conversion for sustainably powering portable electronics. *Nano Lett.* **2012**, *12*, 6339–6346. [CrossRef] [PubMed]

9. Uihlein, A.; Magagna, D. Wave and tidal current energy—A review of the current state of research beyond technology. *Renew. Sustain. Energy Rev.* **2016**, *58*, 1070–1081. [CrossRef]

10. Esteban, M.; Leary, D. Current developments and future prospects of offshore wind and ocean energy. *Appl. Energy* **2012**, *90*, 128–136. [CrossRef]

11. APEC Energy Working Group. *Marine and Ocean Energy Development An Introduction for Practitioners in APEC Economies*; Technical Report; Institute of Lifelong Education: Moscow, Russia, 2013.

12. Collin, A.J.; Nambiar, A.J.; Bould, D.; Whitby, B.; Moonem, M.A.; Schenkman, B.; Kiprakis, A.E. Electrical Components for Marine Renewable Energy Arrays: A Techno-Economic Review. *Energies* **2017**, *10*, 1973. [CrossRef]

13. Inger, R.; Attrill, M.J.; Bearhop, S.; Broderick, A.C.; Grecian, W.J.; Hodgson, D.J.; Godley, B.J. Marine renewable energy: Potential benefits to biodiversity? An urgent call for research. *J. Appl. Ecol.* **2009**, *46*, 1145–1153. [CrossRef]

14. Blackmore, T.; Myers, L.E.; Bahaj, A.S. Effects of turbulence on tidal turbines: Implications to performance, blade loads, and condition monitoring. *Int. J. Mar. Energy* **2016**, *14*, 1–26. [CrossRef]

15. Walker, S.; Cappietti, L. Experimental Studies of Turbulent Intensity around a Tidal Turbine Support Structure. *Energies* **2017**, *10*, 497. [CrossRef]

16. Wright, J.; Colling, A.; Park, D. (Eds.) *Waves, Tides, and Shallow-Water Processes*; Gulf Professional Publishing: Houston, TX, USA, 1999; Volume 4.

17. Zhou, Z.; Benbouzid, M.; Charpentier, J.F.; Sciuller, F.; Tang, T. A review of energy storage technologies for marine current energy systems. *Renew. Sustain. Energy Rev.* **2013**, *18*, 390–400. [CrossRef]

18. Ghefiri, K.; Bouallègue, S.; Haggège, J.; Garrido, I.; Garrido, A.J. Firefly algorithm based-pitch angle control of a tidal stream generator for power limitation mode. In Proceedings of the 2018 International Conference on Advanced Systems and Electric Technologies (IC ASET), Hammamet, Tunisia, 22–25 March 2018; pp. 387–392.

19. Kirke, B.K.; Lazauskas, L. Limitations of fixed pitch Darrieus hydrokinetic turbines and the challenge of variable pitch. *Renew. Energy* **2011**, *36*, 893–897. [CrossRef]

20. Whitby, B.; Ugalde-Loo, C.E. Performance of pitch and stall regulated tidal stream turbines. *IEEE Trans. Sustain. Energy* **2014**, *5*, 64–72. [CrossRef]

21. Zhou, Z.; Sciuller, F.; Charpentier, J.F.; Benbouzid, M.; Tang, T. Power limitation control for a PMSG-based marine current turbine at high tidal speed and strong sea state. In Proceedings of the 2013 IEEE International Electric Machines & Drives Conference (IEMDC), Chicago, IL, USA, 12–15 May 2013.

22. Kalogirou, S.A. Artificial neural networks in renewable energy systems applications: A review. *Renew. Sustain. Energy Rev.* **2001**, *5*, 373–401. [CrossRef]

23. Manas, M.; Kumari, A.; Das, S. An Artificial Neural Network based Maximum Power Point Tracking method for photovoltaic system. In Proceedings of the 2016 International Conference on Recent Advances and Innovations in Engineering (ICRAIE), Jaipur, India, 23–25 December 2016; pp. 1–6.

24. Ouammi, A.; Zejli, D.; Dagdougui, H.; Benchrifa, R. Artificial neural network analysis of Moroccan solar potential. *Renew. Sustain. Energy Rev.* **2012**, *16*, 4876–4889. [CrossRef]

25. Ghefiri, K.; Bouallègue, S.; Garrido, I.; Garrido, A.J.; Haggège, J. Multi-Layer Artificial Neural Networks Based MPPT-Pitch Angle Control of a Tidal Stream Generator. *Sensors* **2018**, *18*, 1317. [CrossRef] [PubMed]

26. Chang, C.S.; Fu, W. Area load frequency control using fuzzy gain scheduling of PI controllers. *Electr. Power Syst. Res.* **1997**, *42*, 145–152. [CrossRef]

27. Thièbot, J.; du Bois, P.B.; Guillou, S. Numerical modeling of the effect of tidal stream turbines on the hydrodynamics and the sediment transport-Application to the Alderney Race (Raz Blanchard), France. *Renew. Energy* **2015**, *75*, 356–365. [CrossRef]

28. SHOM. The Portal of Maritime and Coastal Geographic Information. Available online: http://www.shom.fr (accessed on 11 September 2018).

29. Lewis, M.J.; Neill, S.P.; Hashemi, M.R.; Reza, M. Realistic wave conditions and their influence on quantifying the tidal stream energy resource. *Appl. Energy* **2014**, *136*, 495–508. [CrossRef]

30. Ghefiri, K.; Garrido, I.; Bouallègue, S.; Haggège, J.; Garrido, A. Hybrid Neural Fuzzy Design-Based Rotational Speed Control of a Tidal Stream Generator Plant. *Sustainability* **2018**, *10*, 3746. [CrossRef]

31. Ghefiri, K.; Bouallègue, S.; Garrido, I.; Garrido, A.J.; Haggège, J. Complementary Power Control for Doubly Fed Induction Generator-Based Tidal Stream Turbine Generation Plants. *Energies* **2017**, *10*, 862. [CrossRef]

32. Ghefiri, K.; Bouallègue, S.; Haggège, J. Modeling and SIL simulation of a Tidal Stream device for marine energy conversion. In Proceedings of the 2015 6th International Renewable Energy Congress (IREC), Sousse, Tunisia, 24–26 March 2015; pp. 1–6.

33. Elghali, S.E.B.; Balme, R.; Le Saux, K.; Benbouzid, M.E.H.; Charpentier, J.F.; Hauville, F. A simulation model for the evaluation of the electrical power potential harnessed by a marine current turbine. *IEEE J. Ocean. Eng.* **2007**, *32*, 786–797. [CrossRef]

34. Fernandez, L.M.; Jurado, F.; Saenz, J.R. Aggregated dynamic model for wind farms with doubly fed induction generator wind turbines. *Renew. Energy* **2008**, *33*, 129–140. [CrossRef]

35. Amundarain, M.; Alberdi, M.; Garrido, A.J.; Garrido, I. Modeling and simulation of wave energy generation plants: Output power control. *IEEE Trans. Ind. Electron.* **2011**, *58*, 105–117. [CrossRef]

36. Alberdi, M.; Amundarain, M.; Garrido, A.J.; Garrido, I.; Maseda, F.J. Fault-ride-through capability of oscillating-water-column-based wave-power-generation plants equipped with doubly fed induction generator and airflow control. *IEEE Trans. Ind. Electron.* **2011**, *58*, 1501–1517. [CrossRef]

37. Baroudi, J.A.; Dinavahi, V.; Knight, A.M. A review of power converter topologies for wind generators. *Renew. Energy* **2007**, *32*, 2369–2385. [CrossRef]

38. Hu, J.; Nian, H.; Xu, H.; He, Y. Dynamic modeling and improved control of DFIG under distorted grid voltage conditions. *IEEE Trans. Energy Convers.* **2011**, *26*, 163–175. [CrossRef]

39. Zhao, Z.Y.; Tomizuka, M.; Isaka, S. Fuzzy gain scheduling of PID controllers. *IEEE Trans. Syst. Man Cybern.* **1993**, *23*, 1392–1398. [CrossRef]

40. Bouallègue, S.; Haggège, J.; Ayadi, M.; Benrejeb, M. PID-type fuzzy logic controller tuning based on particle swarm optimization. *Eng. Appl. Artif. Intell.* **2012**, *25*, 484–493. [CrossRef]

41. Tursini, M.; Parasiliti, F.; Zhang, D. Real-time gain tuning of PI controllers for high-performance PMSM drives. *IEEE Trans. Ind. Appl.* **2002**, *38*, 1018–1026. [CrossRef]

42. Bedoud, K.; Ali-rachedi, M.; Bahi, T.; Lakel, R. Adaptive fuzzy gain scheduling of PI controller for control of the wind energy conversion systems. *Energy Procedia* **2015**, *74*, 211–225. [CrossRef]

43. Dounis, A.I.; Kofinas, P.; Alafodimos, C.; Tseles, D. Adaptive fuzzy gain scheduling PID controller for maximum power point tracking of photovoltaic system. *Renew. Energy* **2013**, *60*, 202–214. [CrossRef]

44. Ghefiri, K.; Bouallègue, S.; Haggège, J.; Garrido, I.; Garrido, A.J. Modeling and MPPT control of a Tidal Stream Generator. In Proceedings of the 2017 4th International Conference on Control, Decision and Information Technologies (CoDIT), Barcelona, Spain, 5–7 April 2017; pp. 1003–1008.

45. Pena, R.; Clare, J.C.; Asher, G.M. Doubly fed induction generator using back-to-back PWM converters and its application to variable-speed wind-energy generation. *IEE Proc.* **1996**, *143*, 231–241. [CrossRef]

46. Twining, E.; Holmes, D.G. Grid current regulation of a three-phase voltage source inverter with an LCL input filter. *IEEE Trans. Power Electron.* **2003**, *18*, 888–895. [CrossRef]

47. Astrom, K.J.; Hagglund, T. *Advanced PID Control*; ISA—The Instrumentation, Systems, and Automation Society, Research Triangle: Park, NC, USA, 2006.

48. Vilanova, R.; Visioli, A. *PID Control in the Third Millennium*; Springer: London, UK, 2012.

49. Blaabjerg, F.; Teodorescu, R.; Liserre, M.; Timbus, A.V. Overview of control and grid synchronization for distributed power generation systems. *IEEE Trans. Ind. Electron.* **2006**, *53*, 1398–1409. [CrossRef]

energies

MDPI

Article

Validation of a Coupled Electrical and Hydrodynamic Simulation Model for a Vertical Axis Marine Current Energy Converter

Johan Forslund *[iD], Anders Goude and Karin Thomas

Department of Engineering Sciences, Uppsala University, P.O. Box 534, 751 21 Uppsala, Sweden;
anders.goude@angstrom.uu.se (A.G.); karin.thomas@angstrom.uu.se (K.T.)
* Correspondence: johan.forslund@angstrom.uu.se; Tel.: +46-(0)18-471-30-17

Received: 28 September 2018; Accepted: 30 October 2018; Published: 7 November 2018

check for updates

Abstract: This paper validates a simulation model that couples an electrical model in Simulink with a hydrodynamic vortex-model by comparing with experimental data. The simulated system is a vertical axis current turbine connected to a permanent magnet synchronous generator in a direct drive configuration. Experiments of load and no load operation were conducted to calibrate the losses of the turbine, generator and electrical system. The power capture curve of the turbine has been simulated as well as the behaviour of a step response for a change in tip speed ratio. The simulated results agree well with experimental data except at low rotational speed where the accuracy of the calibration of the drag losses is reduced.

Keywords: marine current energy converter; control system; vertical axis turbine; permanent magnet synchronous generator; load control; vortex model; coupled model

1. Introduction

Ocean energy is a field of growing interest when it comes to renewable energy thanks to its high density of energy per unit area, and to the high predictability. Waves and ocean currents conversion is being investigated through different concepts [1]. Hydrokinetic energy conversion implies use of the energy in free-flowing water for conversion to electric energy. There are many types of hydrokinetic conversion systems being investigated, from different turbine configurations to non-turbine systems [2–4]. Resource characterization, developing the power conversion technology and the design of arrays are some of the biggest areas to develop if marine current energy will become economically viable [5,6]. Numerical modeling is often used to study marine current conversion. A full Computational Fluid Dynamics (CFD) model is computationally demanding for simulating the flow in and around turbines. Instead, other approaches to simplify model of the flow has been developed, and many of these models originate from wind turbine research that has been modified for water environment. The most commonly used models are the double multiple streamtube (DMST) [7], vortex [8,9] and Actuator Line Model (ALM) [10]. Since the goal of the energy conversion is to generate electricity, it could be a big advantage to have a simulation model that can simulate both the hydrodynamic behaviour as well as the electrical output. This paper presents a coupled model of an electrical system with an hydrodynamic free vortex model. The model is validated with experimental data.

The Marine Current Power project at Uppsala University is investigating the possibilities of using a Vertical Axis Current Turbine (VACT) connected to a Permanent Magnet Synchronous Generator (PMSG) in a direct drive configuration, see Figure 1. The system has many similarities with a wind power system but one of the main differences is that the converter is rotating slower, subjecting the turbine and generator to a higher torque. A prototype turbine and generator has been deployed in the river Dal (Dalälven) in Söderfors, Sweden [11]. Water speeds in the river are usually in the interval of 0.4–1.5 m/s [12]. The output voltage and rotational speed of the generator can be controlled using an electrical system comprising a load control using a passive diode-bridge rectifier connected via a DC bus to a resistive dump load trough a switch, described in more detail in [13]. Using the coupled model, different discharge scenarios in the river can be simulated that cannot be tested at the experimental site. Control strategies and electrical systems can be optimized for maximum power capture and safe operation.

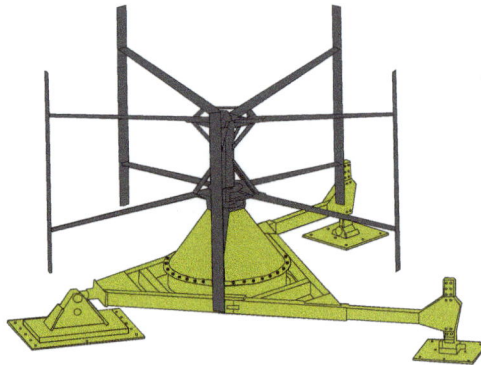

Figure 1. The Marine Current Energy converter rated at 7.5 kW, that can be placed on the river or sea bed. A five-bladed fixed pitch Vertical Axis Current Turbine is connected directly (no gearbox) to a Permanent Magnet Synchronous Generator. The radius, *r*, is 3 m and the height is 3.5 m to give a projected cross sectional area of 21 m^2.

2. The Söderfors Experimental Station

The experimental station has two acoustic doppler current profilers (ADCP), a turbine, a generator and a measurement cabin (housing control and measurement systems). The turbine and generator are placed approximately 800 m downstream of a conventional hydro power plant, at a depth of 7 m. The experimental station is fully described in [12,14].

2.1. The Turbine, the Generator and Load Control

The power in free-flowing water that reaches a turbine with cross-sectional area A is described by $P_{water} = \frac{1}{2} A \rho v^3$ where v is the water speed and ρ is the density of water. For a vertical axis turbine the projected area is the diameter of the turbine times the height. The fraction of power absorbed by the turbine is called power coefficient, C_P, defined as $C_P = \frac{P_{turbine}}{P_{water}}$. C_P is a function of the tip speed ratio (TSR or λ), i.e., the ratio of blade tip speed to undisturbed water speed, $\lambda = \frac{\omega r}{v}$, where ω is the angular speed of the turbine in rad/s and r the turbine radius in meters. The $C_P(\lambda)$-curve for the turbine has been experimentally verified in [15] to have a maximum power coefficient at tip speed ratio 3.1 for a power coefficient of 0.26. The turbine parameters can be found in Table 1.

Power for a rotating body can be described as $P = \omega T$ where T is torque. When the generator and turbine are rotating, the difference between torque delivered by the turbine, T_t, and the electrocmagnetic torque, T_e will determine the acceleration of the rotor, $d\omega/dt$, written as

$$\frac{d\omega}{dt} J = T_t - T_e \tag{1}$$

where J is the inertia of the rotating body. The electrical torque can be influenced by changing the magnitude of the resistive load. For a given water speed and resistive load, the rotational speed of the turbine and generator will settle at some value, resulting in a T_t and T_e. This is called load control because the magnitude of the resistive load is used for changing λ and thus controlling the power capture of the turbine.

For a Permanent Magnet Generator, the voltages of the generator are proportional to the rate of change of the magnetic flux established by the magnets, flux linkage, and to the rotational speed. For a slow-turning generator such as this one, the iron losses cannot be neglected as they can for the more common fast-turning generators. The efficiency of the generator was measured in [16] to be at least 80% in the range of 5–17 RPM. The parameters of the generator can be found in Table 1. After assembly of the turbine and generator, the iron losses and the losses in the seals and the bearings were estimated to be 350 ± 10 Nm times the rotational speed, presented in [17].

Table 1. Turbine and generator specifications at rated operation.

Turbine and generator rating	7.5 kW
Estimated iron, seal and frictional losses	350 Nm
The Vertical Axis Current Turbine	
$C_{P_{max}}$	0.26 at $\lambda = 3.1$
Rated water speed	1.35 m/s
Rated rotational speed	15 RPM
Number of blades	5
Blade pitch	Fixed at 0°
Blade profile	NACA0021
Rotor Radius	3 m
Rotor Height	3.5 m
Chord length	0.18 m
The Permanent Magnet Synchronous Generator	
Minimum efficiency	80 %
Nominal electrical frequency	14 Hz
Poles	112
Rated Line-to-line rms voltage	138 V
Rated stator rms current	31 A
Stator phase resistance	0.335 Ω
Armature inductance	3.5 mH

2.2. Electrical Layout and Control System

An enclosure containing all electrical components is located in the measurement cabin, see Figure 2. The entire system is controlled from LabVIEW using a CompactRIO and a FPGA module. The DC load operates at a switching frequency of 500 Hz and consists of a resistive load, a rectifier with a capacitor bank and an IGBT with a snubber circuit in parallel. The generator is connected to the measurement cabin by a three phase AC-power cable ∼200 m long with a resistance of 0.08 Ω/phase.

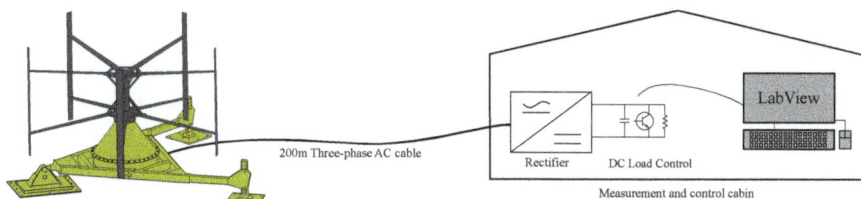

Figure 2. The turbine and generator are connected to the on-shore measurement and control cabin where the rectifier and DC load control is placed.

The DC load components are explained in more detail in [13]. This section will give a summary of the load control. The Target DC voltage aims to keep the DC bus voltage within a user defined range using a P-regulator loop that uses the error of measured DC bus voltage minus set DC bus voltage as input. The loop has three states depending on the size of the error. If the error is negative, the duty cycle is set to 0%, which means it is operating without load, to accelerate the turbine. If the error is more than 5 V, the duty cycle is set to 100%, full load operation, to decelerate the turbine. In between 0 V and 5 V the loop enforces a linear relationship between error and duty cycle. Experimental results presented in [13] shows that the rotational speed can be set to operate with a variance of 4%. Further details of the control and measurement system can be found in [18].

2.3. Water Speed Measurements

The ADCPs are placed about 15 m upstream and 15 m downstream of the turbine. The ADCP devices are Workhorse Sentinel 1200 kHz with an accuracy of 0.3% of the water speed. Measurements are taken every 3.6 s and give a velocity profile from one meter above the bottom of the river to one meter below the surface. Since the upstream ADCP is placed 15 m upstream from the turbine there will be a 10–15 s delay, depending on the water speed, between the measurement of the water speed and when water reaches the turbine. For this paper the speed of the water at the ADCP is assumed to remain constant until reaching the turbine. In reality, since the cross sectional area of the river increases downstream of the first ADCP, the water speed at the turbine will be a few percent lower. For all water speed measurements in the paper, the average water speed will be given. The variance in water speed was less than 1% of the mean for all measurements unless otherwise stated.

3. Coupling of the Electrical and the Hydrodynamic Vortex Model

The model of the electrical system is made in Simulink and the vortex model is imported to Simulink as a function with rotational speed and water speed as inputs, and gives the turbine torque as output.

3.1. Electrical Model in Simulink

The electrical system consists of Power Electronics components, which includes fast switching of devices to control voltages and currents. Such fast switching puts demands on the simulation to be able to compute continuous states at small time steps. When the measured voltage is far from the target voltage it leads to a long state of transition for many steps in the simulation. In this transitional phase the rotational speed and the voltages of the generator may be changing rapidly, to then change slower as they settle around the respective target values. The switching frequency will be much higher than that of any simulated big physical change in the system which leads to many consecutive steps of little change until the next switching state. This type of system is called a stiff system. The setup was modelled using Matlab SIMULINK™ because of its *powergui* blocks that have stiff solvers. Since the vortex code and the rest of the system updates at different time steps a variable step solver is best suited in order to maximize simulation speed and retain solver accuracy.

The Simulink model can be seen in Figure 3. The rectifier is modelled as a three phase passive diode bridge and the switch is an IGBT with a snubber circuit, see parameters in Table 2. The PWM duty cycle is determined by the size of the error described in Section 2.2.

The generator is modelled as a Permanent Magnet Synchronous Generator with a round rotor. It is a three phase machine with a sinusodial back electromotive force. The block is set to use torque as input and the outputs are three phase voltages and currents as well as rotational speed. The generator model does not account for the iron losses in the generator, and is instead included in the estimation of the losses related to the rotational speed, as discussed in Section 2.1.

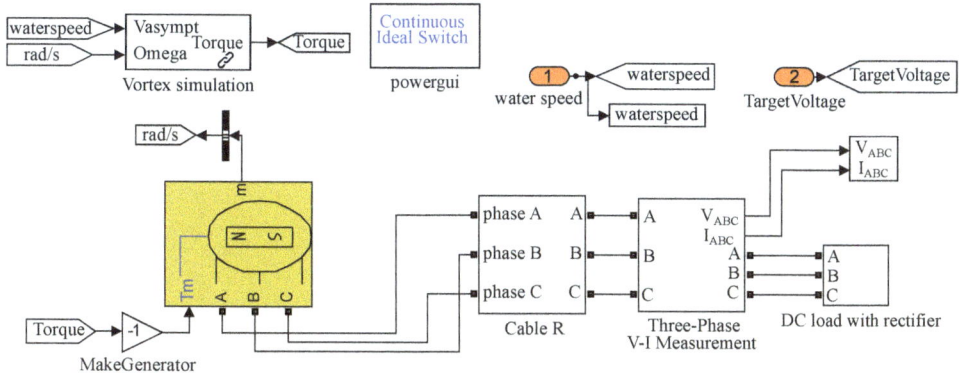

Figure 3. The Simulink model with the block *Vortex simulation* that imports the vortex code as a function. The *DC load with rectifier* block has been replaced with three resistors for the AC-load simulations.

Table 2. Rectifier and DC load parameters in Simulink.

The PMSG Generator block	
Flux linkage	1.28 Vs
Estimated moment of inertia	3000 kgm^2
DC load parameters	
Rectifier on-resistance	1 mΩ
Rectifier forward voltage drop	0 V
IGBT on-resistance	0.1 mΩ
IGBT forward voltage drop	1 V
Snubber resistance	47 kΩ
Snubber capacitance	470 nF

3.2. Hydrodynamic Vortex Model for Vertixal Axis Turbines

The hydrodynamic part of the simulation model is implemented using a two-dimensional free vortex method. The vortex method is a time dependent mesh-free method where the vorticity generated from the blades are used as the discretization variable. The method is designed for infinite domains with no external boundaries, which considering the width of the river at the current site should be considered a reasonable approximation. As the method already is well described in literature, only a brief summary of the method will be given here. For more general information regarding the vortex method, see e.g., reference [19], and for a detailed description of the model implemented in the current work, see e.g., references [20–22].

The vortex method is combined with a force model for the hydrodynamic forces to avoid having to solve the boundary layer flow of the blades, which is computationally demanding. Instead, the flow velocities are calculated at the blade positions. This gives the local Reynolds number and the local angle of attack for each blade, which can be used to calculate the forces. Due to the unsteady nature of the flow, as the angle of attack is constantly changing for vertical axis turbines, experimental data for lift and drag coefficients [23] are combined with a Leishman Beddoes type dynamic stall model for the force calculations, see references [21,24]. The forces are used to determine the vorticity that is released from the blades, and with the current method, one vortex is released from each blade at each time step. The vortex propagation can be evaluated using the fast multipole method (the current implementation uses the CPU version of the code described in reference [25]), which makes the evaluation time approximately linear with the number of vortices in the simulation.

The current vortex method has been validated for wind turbine applications in references [21,22]. It can be noted that the accuracy of the force calculation model decreases as the angle of attack increases, which means that the accuracy of the simulation model can be expected to decrease for low tip speed ratios of the turbine, which correspond to high angles of attack.

As support arms and their attachments to the blades are not properly modelled in the two-dimensional vortex model, a correction model has been applied to account for these losses. By assuming that the drag force generated by these parts can be given by

$$F_D = \frac{1}{2}C_D\rho A V_{rel}^2 \approx \frac{1}{2}C_D\rho A \left(r\Omega\right)^2 \tag{2}$$

where C_D is the drag coefficient, V_{rel} is the relative water speed to the blade and Ω is the rotational speed, one can approximate the torque as

$$T = rF_D \approx C\Omega^2 \tag{3}$$

where C is a constant. This constant will be experimentally determined by allowing the turbine to rotate without any load to determine its freespin velocity. The turbine is then simulated using this rotational velocity, and the constant has been adapted to make the simulation model give zero torque at the freespin velocity.

With the current implementation of the vortex method, it is possible to take much larger time steps for the hydrodynamics than for the electrical system, too small time steps should be avoided for the hydrodynamics to maintain reasonable computational speeds. To account for the difference in required time steps, the vortex method will use Heuns method for the time stepping, and for force evaluations between two time steps, the values are linearly interpolated between the value from the start of the time step, and the intermediate value in Heuns method, which is an approximation of the value at the end of the time step.

For each simulation the turbine will rotate at least 60 revolutions, around 500 s, in order for the vortex code to establish a wake. The hydrodynamic model is imported to Simulink as a function in a block. It receives the water speed and rotational speed of the turbine and returns the computed torque.

4. Calibrating the Simulation Model

Data from operation of the turbine and generator at the experimental site will be used to separately calibrate the generator and turbine models.

4.1. Calibration of Generator and Electrical System Losses

The generator model was implemented using the flux linkage, generator stator resistance and armature inductance from Table 1. The output voltage of the generator was calibrated by comparing the simulated generator voltage at no load operation with the measured output voltage of the generator. The RMS Line-to-Line voltage from six 30 m experiments during 2014 at different water speeds was recorded. The experimental and simulation results can be seen in Figure 4. The error between the measured and the simulated voltage was less than 1 %.

The model of the generator losses was calibrated using data from AC-load operation at the experimental site. Six load cases with resistive loads varying from 2.54 Ω to 13.0 Ω were carried out for 30 m each on the 20 and 21 January 2014 at around 1.3 m/s. The generator is given a fixed torque in the simulation that results in the same rotational speed for the generator as in the experiment. The simulated and measured power in the load is plotted in Figure 5a and the voltage over the load in Figure 5b. The simulations show good agreement except at the lowest rotational speed where the simulated power in the load is 2.2% lower and the simulated generator voltage 2.6% higher, showing that most probably the iron losses are overestimated at low rotational speed.

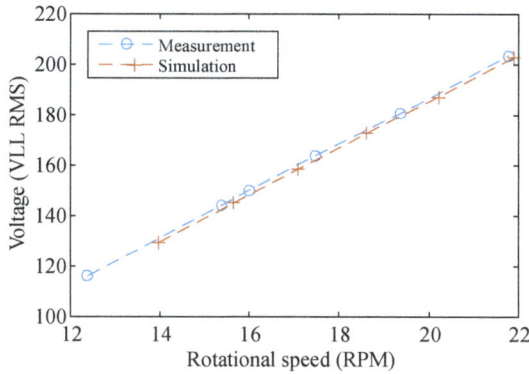

Figure 4. The generator no load line-to-line RMS-voltage.

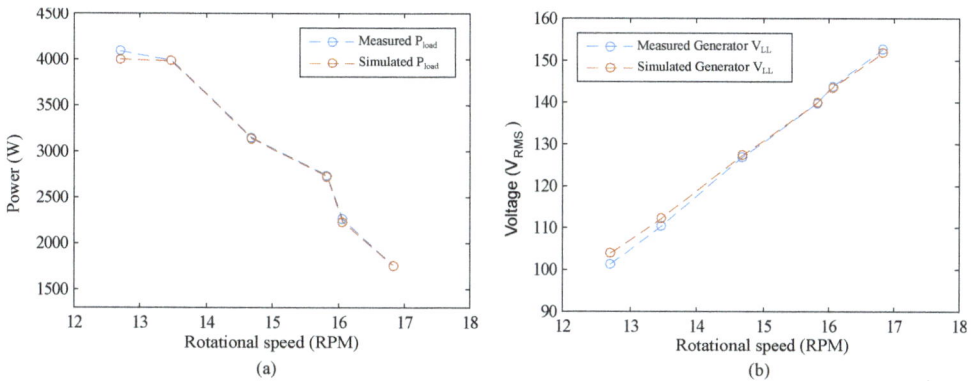

Figure 5. Calibration of the generator model using AC-load operation; (**a**) Power in the load; (**b**) Generator line-to-line RMS-voltage.

4.2. Calibration of Drag Losses

Freespin operation data at the experimental site will be used to calibrate the drag losses in Equation (2) of the turbine using a one parameter study. As discussed in Section 2.1, the iron losses and the losses in the seals and the bearings are set to 350 Nm. The turbine at the experimental site was operated without load for 30 m on 4 March 2014 at 1.42 m/s that resulted in a rotational speed of 20.5 RPM. This water speed and rotational speed was used as input to the vortex simulation where drag losses of 1000 Nms^2 was shown to give zero torque from the turbine. The hydrodynamic, electrical and mechanical torque giving losses dependent on the rotational speed are therefore estimated to be $350 + 1000\omega^2$ Nm. The accuracy of the estimation of the drag losses will decrease the further away from this calibration point. The calibrated model of the turbine and generator was now simulated at free spin operation for a range of water speeds between 1 m/s and 1.5 m/s. The simulation is compared with experimental data recorded on 8 seperate occasions where the turbine operated for 30 m without load. The results are shown in Figure 6a,b. The simulation is able to predict the rotational speed of the turbine at free-spin around the rated water speed of the turbine (1.35 m/s). The rotational speed is 0.8% lower than the experiment around 1.35 m/s and 15.9% higher at low water speed. λ is 0.9% lower around 1.35 m/s and 15.9% higher at low water speed. The difference in simulated and measured rotational speed at low water speed is explained by the loss of accuracy of the turbine calibration since these points are far away from the calibration point. At high water speed, there is a sudden drop in the measured rotational speed. The authors have no explanation for this behaviour other than something exterior affecting the performance of the turbine. Before each measurement at the experimental site, there is no possibility of a visual inspection of the status of the turbine. On some occasions, the turbine was showing some unexpected behaviour. For instance, it could be unusually difficult to start the turbine and not be able to free spin at water speeds where it can usually do so, for only the problems to disappear the next day at the same water speed. These occasions have been written off as something exterior temporarily affecting the turbine, but could not be verified since there is no visual inspection equipment available on site. This could explain the discrepancy of the two rotational speeds measured at just below 1.0 m/s.

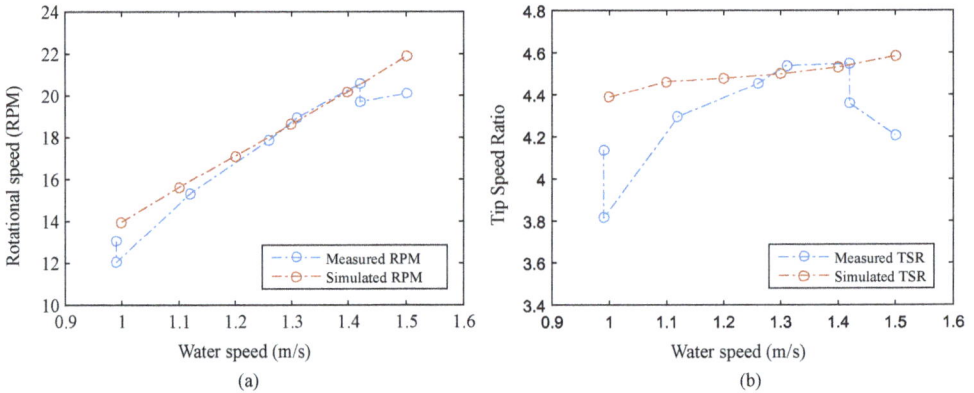

Figure 6. Simulated and experimentally measured free-spin operation of the turbine: (**a**) rotational speed vs water speed; (**b**) Tip-Speed-Ratio vs water speed.

5. Validating the Simulation Model

Using the calibrated generator and turbine models from Section 4, the C_P-curve for the turbine and step responses of DC-loads control can be simulated and compared with experimental results from the test site.

5.1. Simulations of the Power Capture of the Turbine

Experimental data at a range of water speeds and resistive AC-loads are in [15] used to investigate the performance of the turbine. In the study, the power produced by the turbine is estimated using the rotational speed of the turbine and by assuming that the power from the iron losses and mechanical losses in the generator are 180*ω plus the electrical power dissipated in the load. Hence the performance does not display only the hydrodynamic performance of the turbine. The results from the paper will be used to compare with the simulated C_P-curve and referred to as the $C_{P_{turbine}}$. The power capture curve of the turbine for water speeds 1.1 m/s to 1.5 m/s and AC loads from 1 Ω to 9 Ω at steps of 1 Ω has been simulated. The simulation has been plotted together with the experimentally obtained fitted curve presented in [15] in Figure 7a. The experimentally measured efficiency of the total system, including all losses in the turbine and generator, is plotted in Figure 7b. The simulated $C_{P_{turbine}}$-curve shows good agreement with the experimental results. The power capture of the turbine increases as the water speed increases, because the Reynolds number increases that in turn reduces the drag losses.

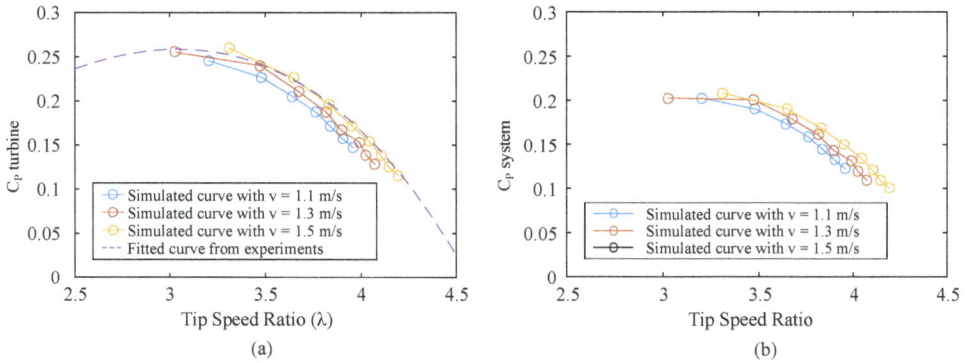

Figure 7. Simulated and experimentally measured C_P-curve: (**a**) $C_{P_{turbine}}$ (including generator iron losses); (**b**) C_P for the total system (including all mechanical and electrical losses).

5.2. Step Response of Change in Target DC Bus Voltage

The simulation model will also be evaluated on how well it can emulate the step response of the dc bus voltage and rotational speed for a change in λ. The torque output of the turbine will depend on the hydrodynamic model and the generator voltage and current will depend on the electrical model. The rise time and the overshoot will reveal how well the estimation of 3000 kgm^2 for the moment of inertia fits.

The experiment was carried out during 826 seconds of operation on 20 January 2014. The target value was changed with discrete steps and kept for a time period of at least one minute. The water speed interval of 1.1–1.25 m/s, seen in Figure 8a, and a DC bus voltage range of 75 V up to 180 V, seen in Figure 8b, covers operation in high and low λ and close to λ_{opt}. At the lowest DC voltage setting, in the experiment, the turbine reached a too low TSR for the turbine to absorb power, so it stopped. That water speed is far away from the calibration point used to estimate the drag losses for the simulation, see Figure 6a, where the rotational speed is overestimated by the simulation so the TSR is high enough to absorb power.

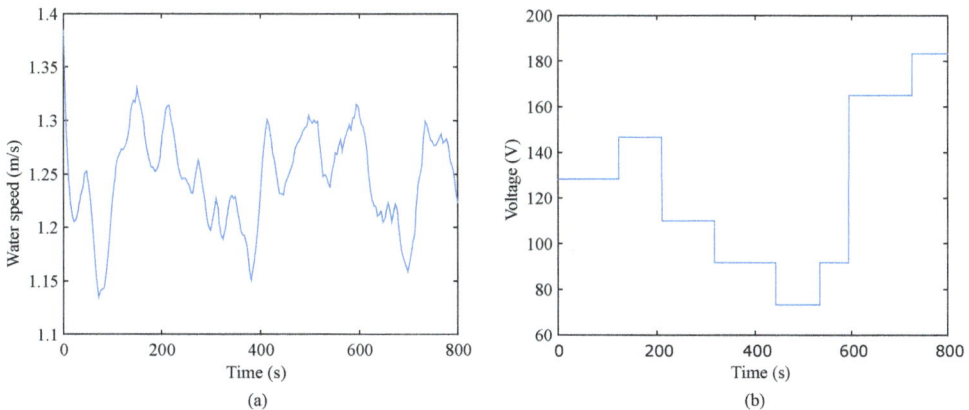

Figure 8. Experimental data that was used as input for the step response simulation: (**a**) Measured water speed in the river; (**b**) Set target DC bus voltage.

The first step response of the Target DC voltage and the rotational speed can be seen in Figure 9. λ is increased from 3.3 to 3.6 and the simulation and the experiment show good agreement. The second step response can be seen in Figure 10 where λ is decreased from 3.5 to 2.7. The simulated target DC voltage and the experiment shows good agreement, but the simulated rotational speed is lower. This is probably because the generator model overestimates the iron losses at low rotational speeds, causing more electrical power to be extracted from the generator.

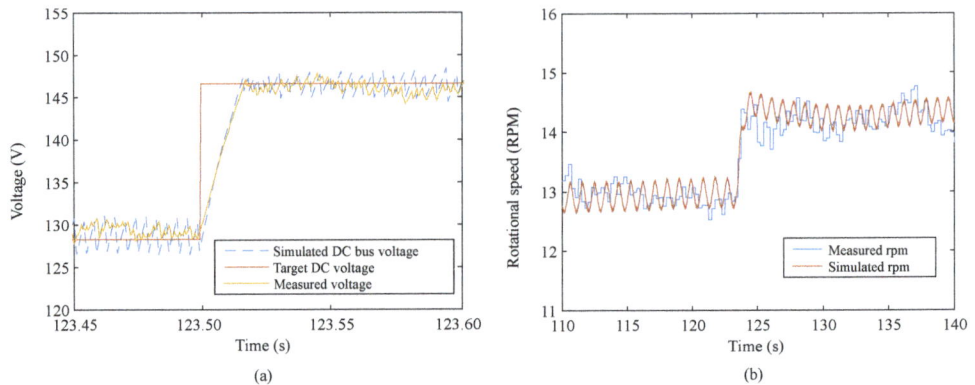

Figure 9. Simulated and experimental step response of turbine operation close to λ_{opt}, with a step corresponding to an increase of λ: (**a**) DC bus voltage; (**b**) Rotational speed of the turbine.

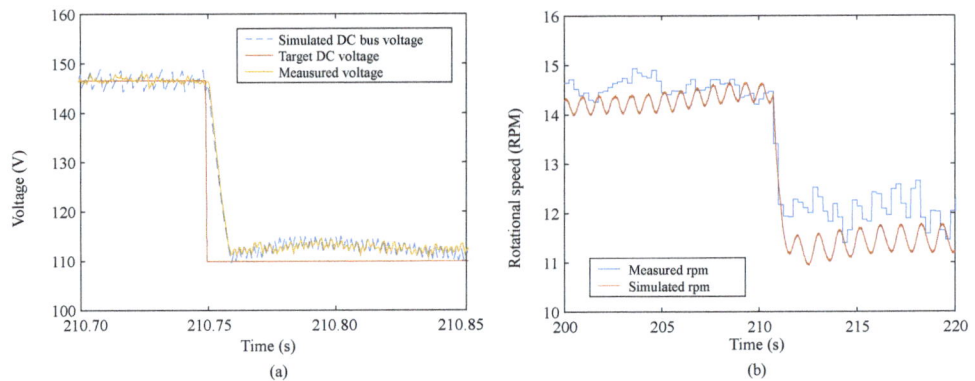

Figure 10. Simulated and experimental step response of turbine operation close to λ_{opt}, with a step corresponding to a decrease of λ: (**a**) DC bus voltage; (**b**) Rotational speed of the turbine.

Taking a closer look at the rotational speed and the DC voltage during the two steps, in Figure 11, it can be seen that the voltage reaches the set point value much faster than the rotational speed settles at the new operating point. When a higher target voltage is set, the DC control will disconnect the load causing the generator to accelerate. During the time the load is disconnected, there is no voltage drop over the transmission cable and the full generator voltage will reach the DC load control quickly charging the capacitor. When a lower target voltage is set, the voltage drop over the transmission line will be increased.

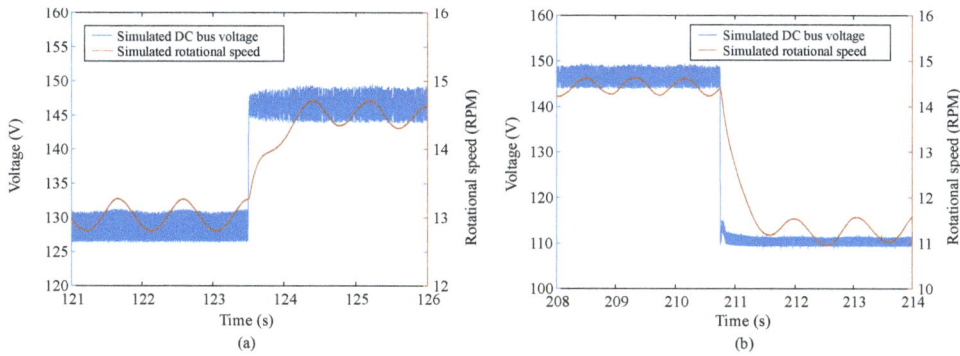

Figure 11. Simulated DC voltage and turbine rotational speed during (**a**) step one and (**b**) step two.

The sixth step is a big change in tip speed ratio from $\lambda = 2.2$, passing λ_{opt}, up to $\lambda = 3.9$, see Figure 12. Since the no-load voltage of the generator is lower than the target DC voltage set point, the DC control has to wait for the generator to accelerate in order to produce a higher voltage. Once the desired target voltage has been reached, the control system needs to brake the accelerating generator. Both in the simulation and in the experiment, there is an overshoot in rotational speed. The capability of the control system to brake the generator once it reaches the set point depends on the maximum power the load can extract. In the experiment it took less time to reach the target voltage, seen in Figure 12a, which is probably a result of the simulation predicting a lower rotational speed at the start of the step, see Figure 12b. At a lower rotational speed the turbine is operating at a lower power capture, so it needs more time to absorb the energy needed to reach the set point. Moreover, since the turbine starts at a lower rotational speed and accelerates freely, it will have a higher $d\omega/dt$ and requires more power to brake. This results in a bigger overshoot in the simulation. It has been shown in [26] that the forces on the turbine blades during runaway (overshoot related to lost control of the turbine) can be up to 2.7 times the forces during nominal operation. It is, therefore, of great importance that the control system can brake the turbine at high rotational speeds. By increasing the magnitude of the DC load the control system has a bigger load to brake the turbine with, and the overshoot could be reduced faster. However, controlling the rotational speed directly with a PI or PID regulator is a safer choice, since it can significantly reduce or completely remove the overshoot.

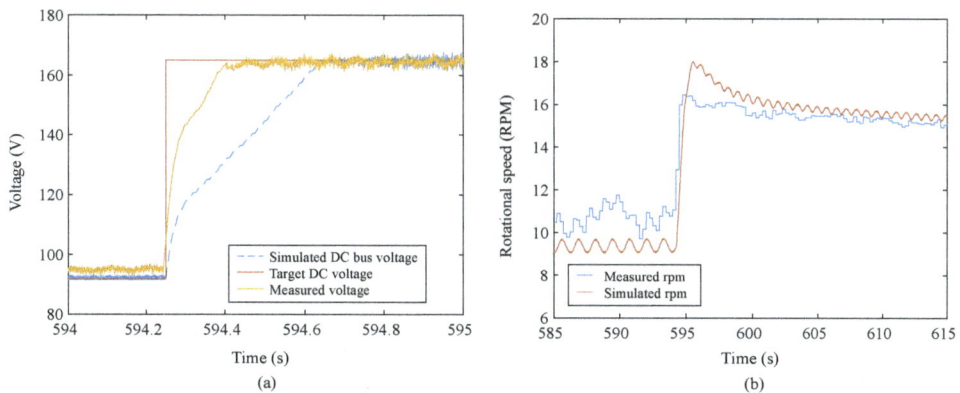

Figure 12. Simulated and experimental step response of turbine operation close from low to high λ: (**a**) DC bus voltage; (**b**) Rotational speed of the turbine.

6. Conclusions

A simulation model that couples the electric and hydrodynamic parts of a vertical axis marine current energy converter has been validated. The hydrodynamic model is calibrated using a one parameter study of the drag losses at the rated water speed of the turbine. Compared to experimental data, the simulation predicts a higher rotational speed at low water speeds. The electrical model is calibrated by comparing the efficiency of the generator to experimental data. The model overestimated the losses at low rotational speeds. The simulated power coefficient curve of the turbine agrees well with experimental data. The model has been shown to describe the behaviour of the turbine and generator for different water flow conditions by predicting the step response of the DC bus voltage and rotational speed for a change in λ. The simulation agrees overall well with experimental data except for a big change from low to high λ where the predicted rotational speed at the start of the step is lower causing the simulation to overestimate the rise time and overshoot. The ability of the DC voltage control system to brake the turbine depends on the size of the load available. Control of the turbine at high rotational speed is of great importance to ensure a safe operation of the turbine. Controlling the rotational speed directly with a PI or PID regulator instead of the DC voltage is a safer choice since it can remove the overshoot.

Author Contributions: The main author, J.F., created the electrical simulation model and conducted the experiments. A.G. provided the vortex simulation. The work was supervised by K.T. All authors contributed to editing and reviewing of the paper.

Funding: The work was carried out using grants from StandUp, Åforsk, Vattenfall, J Gust Richert and VR.

Acknowledgments: The authors would like to thank Senad Apelfröjd and Martin Fregelius for design and construction of the DC load control and Staffan Lundin for the joint effort with the experimental work. The authors would also like to thank the previous and current members of the marine current power group for a joint effort in construction and deployment of the experimental station in Söderfors.

Conflicts of Interest: The authors declare no conflict of interest.

References

1. Day, A.; Babarit, A.; Fontaine, A.; He, Y.P.; Kraskowski, M.; Murai, M.; Penesis, I.; Salvatore, F.; Shin, H.K. Hydrodynamic modelling of marine renewable energy devices: A state of the art review. *Ocean Eng.* **2015**, *108*, 46–69. [CrossRef]
2. Khan, M.J.; Bhuyan, G.; Iqbal, M.T.; Quaicoe, J.E. Hydrokinetic energy conversion systems and assessment of horizontal and vertical axis turbines for river and tidal applications: A technology status review. *Appl. Energy* **2009**, *86*, 1823–1835. [CrossRef]
3. Uihlein, A.; Magagna, D. Wave and tidal current energy—A review of the current state of research beyond technology. *Renew. Sustain. Energy Rev.* **2016**, *58*, 1070–1081. [CrossRef]
4. Zhou, Z.; Scuiller, F.; Charpentier, J.F.; Benbouzid, M.; Tang, T. An up-to-date review of large marine tidal current turbine technologies. In Proceedings of the 2014 International Power Electronics and Application Conference and Exposition, Shanghai, China, 5–8 November 2014; pp. 480–484.
5. Bahaj, A.S. Marine current energy conversion: The dawn of a new era in electricity production. *Philos. Trans. A Math. Phys. Eng. Sci.* **2013**, *371*. [CrossRef] [PubMed]
6. Domenech, J.; Eveleigh, T.; Tanju, B. Marine hydrokinetic (MHK) systems: Using systems thinking in resource characterization and estimating costs for the practical harvest of electricity from tidal currents. *Renew. Sustain. Energy Rev.* **2018**, *81*, 723–730. [CrossRef]
7. Paraschivoiu, I.; Allet, A. Aerodynamic analysis of the darrieus wind turbines including dynamic-stall effects. *J. Propuls. Power* **1988**, *4*, 472–477. [CrossRef]
8. Strickland, J.H.; Webster, B.T.; Nguyen, T. A vortex model of the darrieus turbine: An analytical and experimental study. *J. Fluids Eng.* **1979**, *101*, 500–505. [CrossRef]
9. Murray, J.; Barone, M. The development of CACTUS, a wind and marine turbine performance simulation code. In Proceedings of the 49th AIAA Aerospace Sciences Meeting including the New Horizons Forum and Aerospace Exposition, Orlando, FL, USA, 4–7 January 2011.

10. Sorensen, J.; Shen, W.Z. Numerical modeling of wind turbine wakes. *J. Fluids Eng.* **2002**, *124*, 393–399. [CrossRef]

11. Lundin, S.; Forslund, J.; Carpman, N.; Grabbe, M.; Yuen, K.; Apelfröjd, S.; Goude, A.; Leijon, M. The söderfors project: Experimental hydrokinetic power station deployment and first results. In Proceedings of the 10th European Wave and Tidal Energy Conference (EWTEC13), Aalborg, Denmark, 2–5 December 2013.

12. Yuen, K.; Lundin, S.; Grabbe, M.; Lalander, E.; Goude, A.; Leijon, M. The Söderfors Project: Construction of an Experimental Hydrokinetic Power Station. In Proceedings of the 9th European Wave and Tidal Energy Conference, EWTEC11, Southampton, UK, 2011, pp. 1–5.

13. Forslund, J.; Lundin, S.; Thomas, K.; Leijon, M. Experimental results of a DC bus voltage level control for a load controlled Marine Current Energy Converter. *Energies* **2015**, *8*, 4572–4586. [CrossRef]

14. Grabbe, M.; Yuen, K.; Goude, A.; Lalander, E.; Leijon, M. Design of an experimental setup for hydro-kinetic energy conversion. *Int. J. Hydropower Dams* **2009**, *15*, 112–116.

15. Lundin, S.; Forslund, J.; Goude, A.; Grabbe, M.; Yuen, K.; Leijon, M. Experimental demonstration of performance of a vertical axis marine current turbine in a river. *J. Renew. Sustain. Energy* **2016**, *8*, 064501. [CrossRef]

16. Grabbe, M.; Yuen, K.; Apelfröjd, S.; Leijon, M. Efficiency of a directly driven generator for hydrokinetic energy conversion. *Adv. Mech. Eng.* **2013**, *2013*, 1–8. [CrossRef]

17. Lundin, S.; Goude, A.; Leijon, M. One-dimensional modelling of marine current turbine runaway behaviour. *Energies* **2016**, *9*, 309. [CrossRef]

18. Yuen, K.; Apelfröjd, S.; Leijon, M. Implementation of control system for hydro-kinetic energy converter. *J. Control Sci. Eng.* **2013**, *2013*, 10. [CrossRef]

19. Cottet, G.H.; Koumoutsakos, P.D. *Vortex Methods: Theory and Practice*; Cambridge University Press: Cambridge, UK, 2008.

20. Goude, A. Fluid Mechanics of Vertical Axis Turbines: Simulations and Model Development. Ph.D. Thesis, Uppsala University, Uppsala, Sweden, 2012.

21. Dyachuk, E. Aerodynamics of Vertical Axis Wind Turbines. Development of Simulation Tools and Experiments. Ph.D. Thesis, Uppsala University, Uppsala, Sweden, 2015.

22. Dyachuk, E.; Goude, A. Numerical validation of a vortex model against experimental data on a straight-bladed vertical axis wind turbine. *Energies* **2015**, *8*, 11800–11820. [CrossRef]

23. Sheldahl, R.E.; Klimas, P.C. *Aerodynamic Characteristics of Seven Symmetrical Airfoil Sections through 180-Degree Angle of Attack for Use in Aerodynamic Analysis of Vertical Axis Wind Turbines*; Sandia National Laboratories: Albuquerque, New Mexico, 1981.

24. Dyachuk, E.; Goude, A.; Bernhoff, H. Dynamic stall modeling for the conditions of vertical axis wind turbines. *AIAA J.* **2014**, *52*, 72–81. [CrossRef]

25. Goude, A.; Bülow, F. Robust VAWT control system evaluation by coupled aerodynamic and electrical simulations. *Renew. Energy* **2013**, *59*, 193–201. [CrossRef]

26. Goude, A.; Lundin, S. Forces on a marine current turbine during runaway. *Int. J. Mar. Energy* **2017**, *19*, 345–356. [CrossRef]

energies

MDPI

Article

Estimation of the Near Future Wind Power Potential in the Black Sea

Daniel Ganea[ID], Elena Mereuta and Liliana Rusu *[ID]

Department of Mechanical Engineering, 'Dunarea de Jos' University of Galati, 47 Domneasca Street,
800 008 Galati, Romania; daniel.ganea@ugal.ro (D.G.); elena.mereuta@ugal.ro (E.M.)
* Correspondence: lrusu@ugal.ro; Tel.: +40-745-399-426

Received: 15 October 2018; Accepted: 15 November 2018; Published: 18 November 2018

Abstract: The main objective of the present study is to quantify the recent past and explore the near future wind power potential in the Black Sea basin, evaluating the possible changes. Furthermore, an analysis of the wind climate in the target area was also performed. The wind resources have been assessed using the wind fields provided by various databases. Thus, the wind power potential from the recent past was assessed based two different sources covering each one the 30-year period (1981–2010). The first source is the ERA-Interim atmospheric reanalysis provided by the European Centre for Medium-Range Weather Forecasts (ECMWF), while the second source represents the hindcast wind fields simulated by a Regional Climate Model (RCM) and provided by EURO-CORDEX databases. The estimation of the near future wind power potential was made based on wind fields simulated by the same RCM under future climate projections, considering two Representative Concentration Pathways (RCPs) scenarios (RCP4.5 and RCP8.5) and they cover also a 30-year time interval (2021–2050). Information in various reference points were analyzed in detail. Several conclusions resulted from the present work. Thus, as regards the mean wind power potential in winter season, in 51% of the locations a significant increase is projected in the near future (both scenarios). Besides providing a detailed description of the wind conditions from the recent past over the Black Sea basin considering two major sources, the novelty of the present work consists in the fact that it gives an estimation of the expected wind climate in the target area for the near future period and at the same time an evaluation of the climate change impacts on the wind speed and wind power potential.

Keywords: wind speed; wind power; Black Sea; EURO-CORDEX; ERA-Interim

1. Introduction

The global context generated by the rapid population growth and by the development of new technologies implies an increase in the energy needs. By combining this aspect with the actual energy producing capability, which uses mainly conventional methods (natural gas, lignite, petroleum, etc.) and has a negative impact to the environment, it results that it is urgently required to find effective methods of counteraction. From this perspective, the topic of energy remains an extremely important one [1]. Globally, many policies aiming to put into practice, on medium and long-term, methods for the energy sector development were adopted. In this way, the energy sector will become more efficient and sustainable. Developing and implementing energy policies is also an important condition in order to achieve strategic goals, which will propagate further in the economic development.

Various studies show that the feed-in tariff policy mechanism is really effective in fostering the sustainability transition of the energy sector and also to promote the investments in renewable energy [2,3]. Thus, the growth of renewable resource extraction (wind, wave, geothermal, photovoltaic, and hydro, etc.) for generating electricity is pursued in the detriment of the conventional methods.

For example, the European Union sets very specific targets for 2020 and 2030, as part of its long-term energy strategy, which covers the improved energy efficiency, emission reductions, and an increased share of renewables. An energy roadmap for the year 2050 has been also devolved. This aims to reduce the greenhouse gas emissions by 80–95% until 2050 when compared to the 1990 level. Through these energy policies, the European Union desires to ensure its citizens that they can access secure, affordable and sustainable energy supplies [4,5].

An example of good practice is followed by the Romanian Ministry of Energy, thought the energy policies. The Romanians energy strategy, according to the Ministry of Energy [6], covers a 15-year interval (2016–2030), with an outlook to 2050. According to this document, the energy policy covers the strategic goals, principles, main areas of state intervention and new directions for development. This trend is followed not only by the members of the European Union as Bulgaria [7] but also by countries like Georgia, Russia, Tukey, Ukraine, etc. The principles that all countries consider are: energy security, competitive market, consumer interests of first priority, transparency, smart grids and energy storage, smart buildings with energy self-sustainability and the most important is represented by the clean energy [8–12].

One of the most permanent and sustainable renewable resources is the wind energy. According to the European Wind Energy Association (EWEA), the wind energy potential extraction has gained more and more ground. By analyzing the EWEA annual reports it can be observed a substantial growth of the energy volume extracted. Since 2014, when the installed wind power capacity was 142 GW (about 92.2% onshore and only 7.8% offshore), to 2017 the onshore and offshore cumulative wind power installation grew by 18.8% (about 16.8% onshore and only 43.6% offshore). According to this statistic, Germany (56.1 GW), Spain (23.2 GW) and the United Kingdom (18.9 GW) together represent 58% of all the cumulative installed capacity of the European Union. In the middle of the rank are countries as: Romania (3 GW), Belgium (2.8 GW), Austria (2.8 GW), Greece (2.7 GW), Finland (2.1 GW) and Bulgaria (0.7 GW) [13].

However, a deep discrepancy between the EU countries can be observed, and this also regards the onshore versus offshore capabilities. In order to grow the offshore, wind energy exploitation it is required that researchers should find first the best new locations to exploit this green energy. The amount of land still available for the wind energy exploitation is becoming limited and there are also significant environmental issues. On the other hand, the offshore locations present some advantages, especially brought by the existence of large marine areas suitable for the wind farm development. The increase in wind speed with the distance from the coastline, together with the existence of less turbulence, allow for the turbines more energy extraction than the similar operating onshore [14,15].

The results of various previous researches indicate the fact that the Black Sea wind power potential cannot be neglected, especially for the countries located in the proximity of the sea [16]. From this perspective, the objective of this paper is to present a more complete picture of the wind energy potential of the Black Sea during the present and near future periods, by using two different data sources. The novelty of the present study also arises from the fact that such a detailed analysis has not yet been carried out for this area. A significant amount of researchers studied the green energy potential of the enclosed and semi-enclosed seas by using various techniques as reanalysis data, satellite data and climate models [14,17–22].

Davy et al. [23] carried out an analysis of the climate change impacts on wind energy potential in the European domain. The study was focused on the Black Sea and conducted by using a single-model-ensemble. The authors show that in the near future the wind intensity pattern in the Black Sea basin will not suffer relevant negative impact due to climate change. This feature would make the offshore wind-farms in the Black Sea to be a viable source of energy for the neighboring countries. Another important work that illustrates the wind power potential over the Mediterranean and the Black seas is performed by Koletsis et al. [24]. The authors analyzed by exploring six regional climate model simulations for the present period and two for the future periods. These models were produced in the framework of the ENSEMBLES project. The results for the Black Sea show that this basin is a suitable environment for the green energy extraction (average wind power being estimated in the range 500–900 W/m², with a deviation of ±50 W/m² during the future periods).

Onat et al. [25] conducted an analysis of the wind climate and of the wind energy potential for several regions in Turkey. The authors analyzed also a small region of the Black Sea, located in the west (Amasra) by using a five-layer Sugeno-type ANFIS model developed with MATLAB-Simulink software. The relationship between the wind speed and other climate variables was also determined and the resulted data confirm that the Amasra region is a location with a good potential for the wind energy extraction. According to this study, the average power density at 10 m height is 232 W/m² (nearly good), at 50 m height is 603 W/m² (good) and at 80 m height is 1300 W/m² (very good).

The wind pattern of the Black Sea was also evaluated by Onea and Rusu [17], considering 12 years of data from the U.S. National Centers for Environmental Prediction (NCEP). In that study, the authors analyzed the wind power distribution taking into account the diurnal versus nocturnal variations. According to the above mentioned study, the northwestern and northeastern sectors of the Black Sea are the most suitable for wind energy extraction. The northwestern sector of the Black Sea was also analyzed by Lin-Ye et al. [26]. Their approach considered a hybrid methodology involving the Simulating WAve Nearshore (SWAN) spectral wave-model to produce wave-climate projections. The wave model was forced with wind-fields corresponding to the two climate change scenarios.

From this perspective, the present study aims to characterize the offshore wind power potential of the Black Sea during the present and the near future, by analyzing four databases. Following this objective, the structure of the proposed work includes first a presentation of the materials and methods considered, focused on the description of the target area and of the databases taken into account. The next section presents the results providing in some reference points the wind speed and also the wind power. These relate both the 30-year period considered from the past (1981–2010) and that estimated for the near future period analyzed (2021–2050). Finally, a discussion of the results was also carried out. Thus, the novelty of this paper can be summarized as follows:

1. A detailed description of the wind conditions over the Black Sea basin from two major sources (Era-Interim and Euro-Cordex) covering the recent past (1981–2010).
2. An estimation of the expected wind climate in the near future (2021–2050) under two different RCP scenarios (RCP4.5 and RCP8.5).
3. Evaluation of the climate change impacts on the wind speed and wind energy potential by performing comparisons between the past and the future projections.

2. Materials and Methods

2.1. Target Area

In this study, the target area is the Black Sea. The Black Sea basin is under the influence of the NAO (North Atlantic Oscillations) mechanism, which causes, due to the influence of the cold air arriving from the northern regions, significant storm events during the winter [27].

Figure 1 illustrates the Black Sea basin and the geographical locations of the reference points further considered. In this study, the wind speed at 100 m height was evaluated in twenty-four points. The reference points are divided into two different categories. The first category includes the shallow water locations. More precisely, these points are in the range 29–39 m water depth (average depth being 34.9 m) and the distance to shore in the range 1.5 to 56 km (average distance to shore being 11.8 km). The second category refers to the deep water locations. These points are relatively close to shore in the range 3.6 to 109 km (the average distance to shore being 35.7 km) and a depth in the range 114 to 140 m, with an average value for the depth of 125.8 m.

Taking into account some previous studies that showed the existence of various areas in the Black Sea presenting different wind conditions, the target area was divided into five geographical zones with similar characteristics/patterns, labeled from A to E. Zone A contains seven points related to the coastlines of Romania and Ukraine (shallow water: A.n.1 to A.n.4 and deep water: A.o.1 to A.o.3), zone B contains six points (shallow water: B.n.1 to B.n.4 and deep water: B.o.1 and B.o.2) related to the coasts of Bulgaria's and the northwest of Turkey. Zone C contains two shallow water points C.n.1 and C.n.2 and only one deep water point C.o.1 located in the north and northeast of Turkey, zone D contains five points associated to the coastal environment of Georgia and Russia (shallow water: D.n.1 to D.n.3, deep water points: D.o.1 and D.o.2) and zone E with three points close to the Crimea Peninsula (shallow water: E.n.1 and E.n.2, deep water: E.o.1) (see Tables 1 and 2).

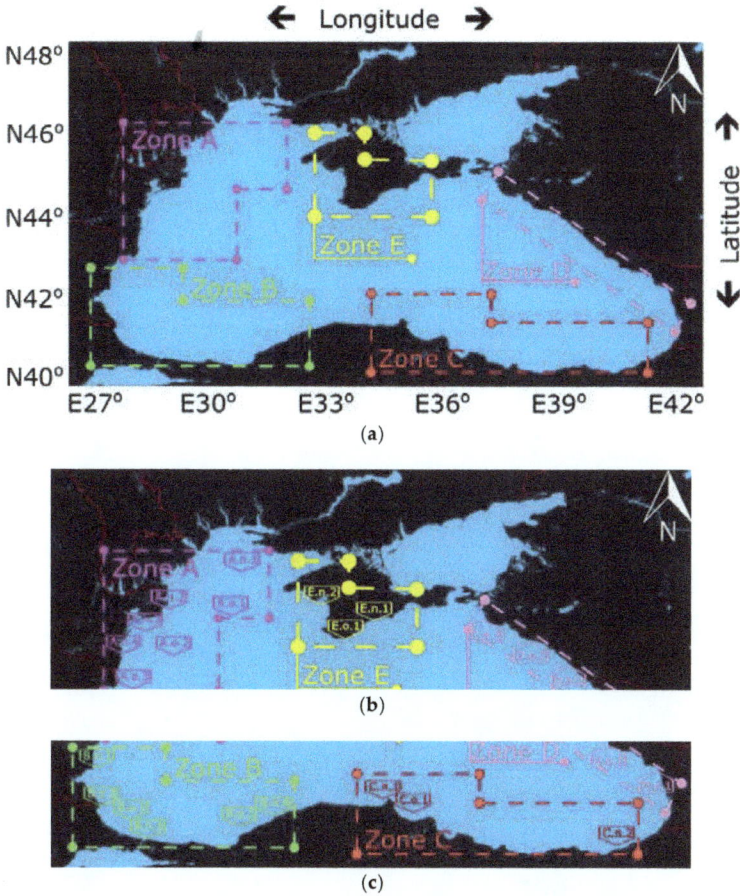

Figure 1. The geographical locations of the reference points corresponding to the 5 zones considered: zone A (shallow water points: A.n.1 to A.n.4 and deep water points: A.o.1 to A.o.3), zone B (shallow water points: B.n.1 to B.n.4 and deep water points: B.o.1 and B.o.2), zone C (shallow water points C.n.1 and C.n.2 and deep water point C.o.1), zone D (shallow water points: D.n.1 to D.n.3, deep water points: D.o.1 and D.o.2) and zone E (shallow water points: E.n.1 and E.n.2, deep water point: E.o.1); (**a**) Overview of the target zones; (**b**). The location of the points corresponding to zones A, E and a section of zone D; (**c**) The location of the points corresponding to zones B, C and a section of zone D.

Table 1. The geographical locations and the characteristics of the shallow water points.

Zone	Shallow Water Points	Latitude	Longitude	Sea Depth [m]	Distance to Shore [km]
A	A.n.1	45.75	31.50	38	56
	A.n.2	45.00	30.00	36	28.5
	A.n.3	44.50	29.40	39	33
	A.n.4	44.00	28.75	35	6.6
B	B.n.1	42.75	28.00	34	8.4
	B.n.2	42.06	28.06	36	5.6
	B.n.3	41.25	29.30	32	3.2
	B.n.4	41.65	32.15	39	2.3
C	C.n.1	42.11	35.00	39	2.9
	C.n.2	41.03	40.38	37	1.6
D	D.n.1	42.60	41.45	30	4.6
	D.n.2	43.85	39.35	37	3.2
	D.n.3	44.75	37.35	33	2
E	E.n.1	44.75	34.62	30	2.7
	E.n.2	45.00	33.35	29	16

Table 2. The geographical locations and the characteristics of the deep water points.

Zone	Deep Water Points	Latitude	Longitude	Sea Depth [m]	Distance to Shore [km]
A	A.o.1	44.75	31.50	128	101
	A.o.2	44.10	30.50	120	109
	A.o.3	43.35	29.15	114	47
B	B.o.1	41.70	28.75	123	13
	B.o.2	41.40	31.50	140	4.6
C	C.o.1	41.82	35.65	125	19
D	D.o.1	43.20	40.23	132	3.6
	D.o.2	44.34	38.30	125	5.6
E	E.o.1	44.25	34.10	126	18.3

2.2. ECMWF Dataset

One set of the data considered was obtained from the ERA-Interim database. This is a reanalysis project conducted by the European Centre for Medium-Range Weather Forecasts (ECMWF) [28]. ECMWF uses forecast models and data assimilation techniques that include a 4D analysis that has 12 h analysis windows to describe the atmosphere and oceans [29,30].

ERA-Interim is an ongoing project that comprises datasets with a various number of marine and atmospheric parameters from 1979 to the present. Among many data, it contains information about the wind speed components (u—zonal velocity and v—meridional velocity) at 10 m height. The wind speed at 10 m will be denoted as U_{10}. The data considered have a spatial resolution of $0.75° \times 0.75°$, cover a period of 30-years (1 January 1981 to 31 December 2010) and are dived in four hourly intervals (corresponding to 00:00:00, 06:00:00, 12:00:00, 18:00:00) for each day.

2.3. EURO-CORDEX Database

The Coordinated Regional Downscaling Experiment (CORDEX) initiated by the World Climate Research Program (WCRP), produces high-resolution 'downscaled' climate data through a global partnerships. EURO-CORDEX is the European branch of this project, where results from several RCMs were jointed to cover the European continent [31,32]. The data available have the spatial resolutions of 0.11° and 0.4°, respectively.

The EURO-CORDEX wind fields used in this study have a maximum range of 27° N ÷ 72° N and 22° W ÷ 45° E, with a spatial resolution of 0.11°. These data contain information about the wind speed components at 10 m height and are dived in four hourly intervals (corresponding to 00:00:00, 06:00:00, 12:00:00, 18:00:00) for each day. From this database, three types of wind fields were used in the present study. First, the evaluation (hindcast) wind fields that cover a period of 30-years (1 January

1981 to 31 December 2010) are analyzed. The wind fields are simulated by the Rossby Centre regional climate model—RCA4 model at the Swedish Meteorological and Hydrologic Institute (SMHI) and forced with initial and lateral boundary conditions provided by ECMWF ERA-Interim reanalysis data. The second and third datasets cover the 30-year interval 1 January 2021 to 31 December 2050, and there are wind fields simulated by the RCA4 model under future climate projections, considering two Representative Concentration Pathways (RCPs) emission scenarios, RCP4.5 and RCP8.5. More detailed information regarding the CORDEX scenarios for the European areas, as provided by the Rossby Centre regional climate model RCA4, are given in [33,34]. These two scenarios were simulated with the same RCM model as in the case of evaluation data, but forced by a Global Climate Model (GCM), namely EC-EARTH. In the global climate models, the related temporal evolution of atmospheric greenhouse gas and aerosol concentrations are prescribed, which then simulate the response of the climate system to the forcing. Thus, a range of potential future climate evolutions can be projected.

2.4. Data Evaluation

First, the data used in this study were evaluated in terms of daily, seasonal and yearly average. Both ERA-Interim and EURO-CORDEX databases provide the wind speed at 10 m height in terms of its components. For all the sites available and for all time scales various statistical analyses of the wind speed are conducted. Also, for each time series of the wind speeds the 5th and 95th percentiles were computed. The statistical parameters considered are the root mean square error, bias and the Pearson correlation coefficient. The root mean square error and bias are computed with the following relationships:

$$RMSE = \sqrt{\frac{\sum_{i=1}^{n}(x_i - y_i)^2}{n}}, \tag{1}$$

$$Bias = \frac{\sum_{i=1}^{n}(x_i - y_i)}{n} \tag{2}$$

where n is the total number of data pairs, x is the wind speed provided by ERA-Interim, while y is the evaluation wind speed value.

Maintaining the same notation of the data, the Pearson correlation coefficient is computed with the next equation:

$$r = \frac{\sum_{i=1}^{n}(x_i - \overline{x})(y_i - \overline{y})}{\sqrt{\sum_{i=1}^{n}(x_i - \overline{x}) \cdot \sum_{i=1}^{n}(y_i - \overline{y})}}, \tag{3}$$

where $\overline{x}$ and $\overline{y}$ are the mean values of the analyzed datasets.

The linear trend of the wind speeds was also estimated from a linear regression whose y dependent variable is represented by the wind speed time series, while the independent one is the time (x variable). The linear regression line has the following equation:

$$y = a + bx \tag{4}$$

where b is the slope of this line, and a is the intercept. The slope indicates the linear rate of wind speed change and its values can be positive or negative. A positive value corresponds to increasing trends while the negative value indicates a decreasing trend.

Taking into account that consistent wind speed measurements over the Black Sea basin are not available, a comparison between the present wind climate simulated by the RCM model with that resulted from a widely used reanalysis database is performed. In this way, the skill of the RCM wind fields to represent the present wind climate in the Black Sea basin can be determined.

Thus, the evaluation (hindcast) wind fields from EURO-CORDEX are compared with the wind fields from ERA-Interim reanalysis. The evaluation wind fields being provided by a RCM model, only statistical comparisons can be performed [35,36]. Near the Romanian coast, the wind speeds are recorded at an offshore platform, and the evaluation wind speeds were compared with these measurements by Rusu et al. [37].

3. Results

3.1. Wind Speed Analysis

This section provides a detailed analysis of the wind speed at 10 m height. First, comparisons between the present wind climate simulated by the RCM model with that resulted from a widely used reanalysis databases ERA-Interim were performed to evaluate the skill of the RCM model. Also, some comparisons between the wind fields simulated under RCP4.5 and RCP8.5 scenarios were performed in order to identify the differences induced by these two different scenarios. Actually, these are the most studied greenhouse gas concentration scenarios. Thus, RCP4.5 describes an intermediate concentration scenario with radiative forcing stabilized at around 4.5 W/m^2, while RCP8.5 describes a high concentration scenario under which the radiative forcing is expected to be higher than 8.5 W/m^2 by the end of the year 2100 [38,39]. On the other hand, the comparisons between the evaluation data and those obtained under various scenarios are indicating the possible further evolution of the wind conditions.

3.1.1. Hindcast Data

Figure 2 illustrates a comparative analysis in terms of maximum values, averages, 5th and 95th percentiles (lower than 5% and 95%) for the total time interval. In Figure 2a the values computed in each point for the present period are compared, considering the evaluation and ERA-Interim wind speeds. Some differences are observed between the magnitudes of the wind speeds. With the exception of the point B.n.3, located near the Bosporus Strait, the EURO-CORDEX Evaluation wind speeds are higher than those from ERA-Interim. The resulted values show that the difference is in average about 17% regarding the maximum computed values. At the same time, significant differences can be noticed regarding the total interval averages (12%) and for 95th (lower 95%) percentile analysis (14%).

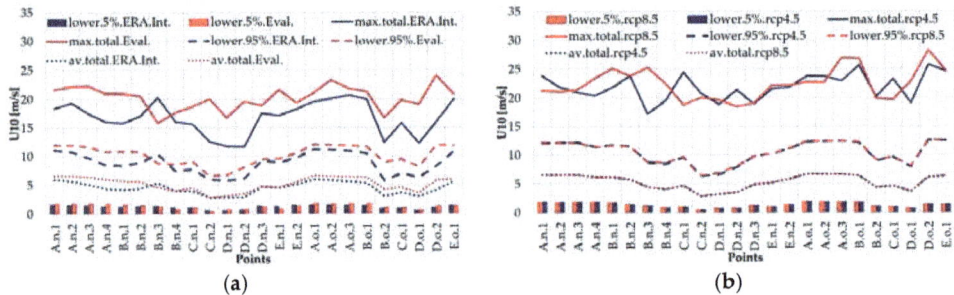

(a)   (b)

Figure 2. Analysis of the wind speed at 10 m (U10) height in terms of maximum values (max.total), average values (av.total), 5th percentile (lower.5%) and 95th percentile (lower.95%) values, corresponding to the reference points. (**a**) Comparison between ECMWF ERA-Interim (ERA.Int.) and EURO-CORDEX Evaluation (Eval.) wind speeds, for the 30-year period (1981–2010); (**b**) Comparison between EURO-CORDEX wind speeds corresponding to RCP4.5 and RCP8.5 scenarios for a 30-year period (2021–2050).

The analysis of the annual average values illustrated in Figures 3–6 (see also Figures A1 and A2) shows the fact that wind speed at 10 m height had and it is expected to have a different evolution in the areas considered. For instance, the data resulted in zone A (shallow water data, Figure 3) show that in the time interval 1981 to 2010, the wind speed has values varying in the range 4 to 7 m/s.

It can be noticed that the evaluation wind speed average values from RCM are higher than those from ERA-Interim. The difference is in a range of about 9% (point A.n.1) to 26% (point A.n.4).

Figure 4 illustrates two of the most representative shallow water points, close to the coastline of Bulgaria. By analyzing the data resulted for this area, it can be observed that in two points, the evaluation data are higher than ERA-Interim (point B.n.1: $\cong$ 26% and point B.n.2, not shown here $\cong$21%). In the case of the point B.n.3, the Evaluation data are lower with $\cong$ 16%. Regarding the point B.n.4, not shown here, both models (ECMWF ERA-Interim and EURO-CORDEX Evaluation) present similar behavior of the wind speed annual pattern evolution. According to the linear tendency, the annual means of the wind speeds show a different evolution for the period 1981 to 2010. As in zone A, in this zone the near future scenarios assess differently the wind speed pattern. Comparing the data from 1981–2010 with the near future interval for point B.n.1 it can be noticed a substantial increase of the wind speed. It is expected that the mean wind will reach values in the range of 5.7–6.6 m/s.

The wind speed analysis of two of the most representative deep water points located in zones A and B (A.o.1) and B (B.o.1) is presented in Figure 5. Regarding the deep water locations for zone A, the Evaluation wind speeds for the interval 1981–2010 are higher with approximately 9–12% than those from ERA-Interim. If zone B is analyzed, the difference between both data grows at about 15% to 29%. The data corresponding to the points associated with the coastline of Turkey are presented in Figure 6. By analyzing the data for the interval 1981–2010, in comparison with the data previously presented, it can be noticed that the wind is less intense in this area. Figure 6 also shows that according to both ERA-Interim and the Evaluation data, the mean wind speed has the tendency to increase from 1981 to 2010. No great variability of the annual averages along the entire 30-year interval is observed.

The wind pattern for the points D.n.3 and D.o.2 located in shallow water and deep water coastline of Russia, for the time interval 1981 to 2010 and 2021 to 2050 are illustrated in Figure A1. In this zone, there were analyzed initially five points (three in shallow water and two in deep water). In contrast with the previous cases, for the point located in shallow water (D.n.3), both ERA-Interim and Evaluation determined the same pattern for annual average values. This has not occurred in the case of the other points (D.n.1, D.n.2, D.o.1, and D.o.2). In these cases, the Evaluation data estimate higher wind speeds with about 8% (D.n.1), 17% (D.n.2), 47% (D.o.1) and 28% (D.o.2). By comparing the wind intensity for zone D with zone A, B, and C, here the lowest intervals in which the annual average wind speeds vary were encountered (D.n.1: 2.5–3.6 m/s, D.n.2: 2.5–4.1 m/s and D.o.2: 2.7–4.1). The wind speed at 10 m height was more intense in points D.n.3 (4.3–5.2 m/s) and D.o.2 (3.7–6.2 m/s).

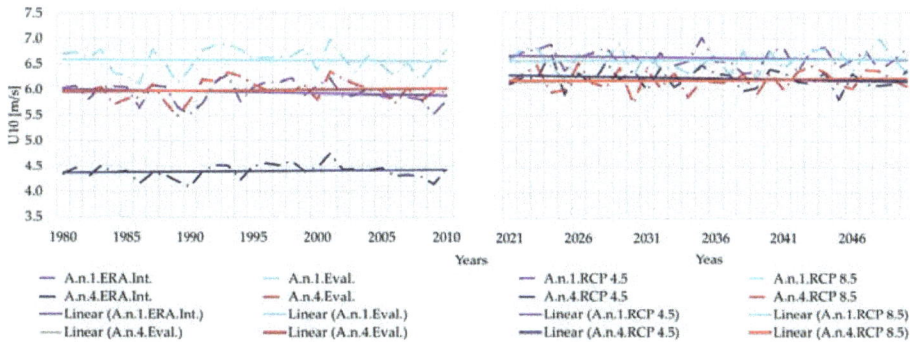

Figure 3. Annual averages of the wind speed at 10 m height for two of the most representative points located in the Romanian nearshore of. ERA.Int and Eval data, time interval 1981–2010, RCP4.5 and RCP8.5 data, time interval 2021–2050.

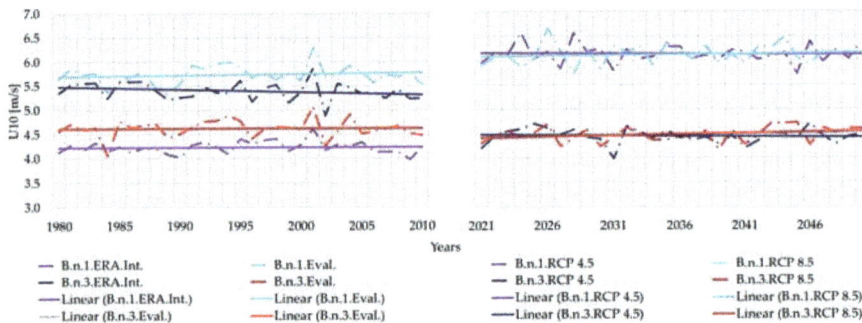

Figure 4. Annual averages of the wind speed at 10 m height for two of the most representative points located in the Bulgarian nearshore. ERA.Int and Eval data, time interval 1981–2010, RCP4.5 and RCP8.5 data, time interval 2021–2050.

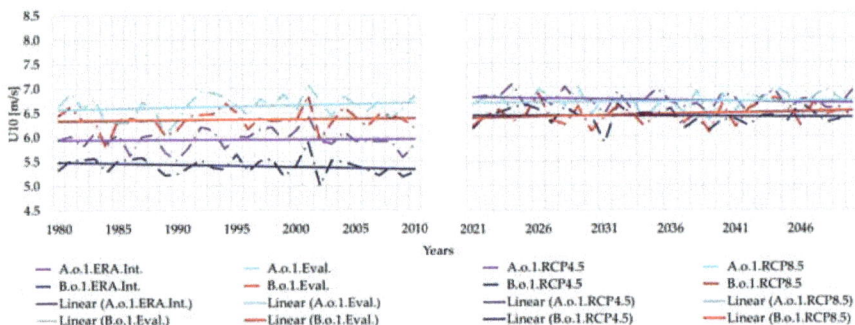

Figure 5. Annual averages of the wind speed at 10 m height annual averages for two of the most representative points located in deep water Bulgaria (B.o.1) and Romania (A.o.1). ERA.Int and Eval data, time interval 1981–2010, RCP4.5 and RCP8.5 data, time interval 2021–2050.

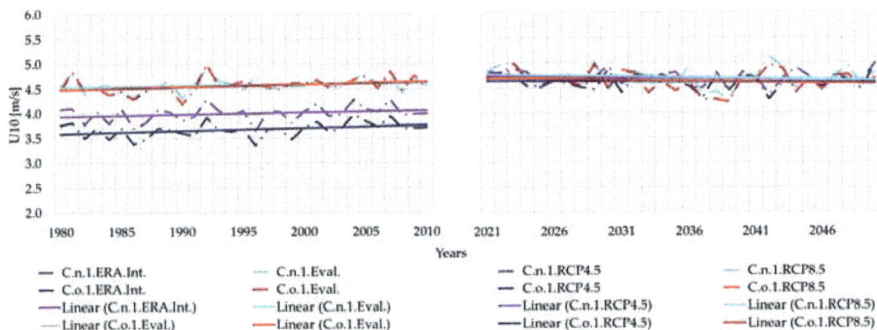

Figure 6. Annual averages of the wind speed at 10 m height for two of the most representative points located in shallow water (C.n.1) and in deep water (C.o.1) of Turkey. ERA.Int and Eval data, time interval 1981–2010, RCP4.5 and RCP8.5 data, time interval 2021–2050.

High wind speed values are encountered in the nearshore and offshore of the Crimea Peninsula. In this area, the wind speed at 10 m height was assessed for three locations, but here only the most representative results are presented. Thus, Figure A2 illustrates the annual averages of the wind speeds at 10 m height of two points located in shallow water (E.n.2) and deep water (E.o.1) of the Crimea Peninsula. By analyzing the time interval 1981–2010, it can be noticed that the wind has an uptrend evolution. Small differences are observed between both data. In the case of the point E.n.1, the Evaluation wind is higher with about 2.8%, for point E.n.2 with 7.1%, while for point E.o.1 with 7.3%.

The seasonal distributions of the wind intensity (total, winter, spring, summer, and autumn) are presented in Figures 7 and 8. By looking at these data, it can be noticed that during the winter the magnitude of the wind speed average is higher. In almost 90% of the cases presented, the wind speeds during the summer interval have the lowest values. For example, the points C.n.1 and C.o.1, which are located north of Turkey, have during summer wind speed averages almost as high as in the winter interval. However, it can be noticed that in these regions the difference between the seasonal averages is small.

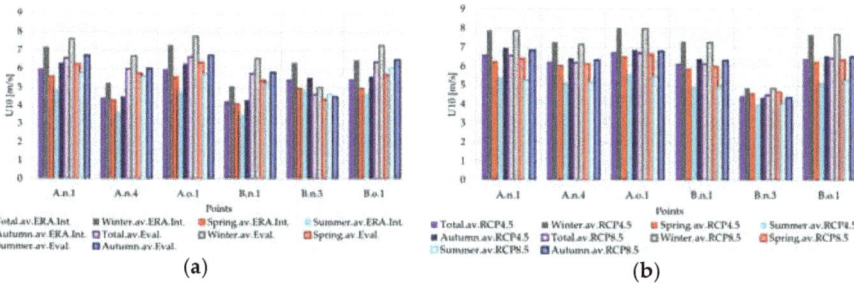

Figure 7. Seasonal wind speed averages at 10 m height compared with the total time averages for some representative points located in shallow water and deep water of Bulgaria and Romania coasts; (**a**) ERA.Int and Eval data, time interval 1980–2010; (**b**) RCP4.5 and RCP8.5 data, time interval 2021–2050.

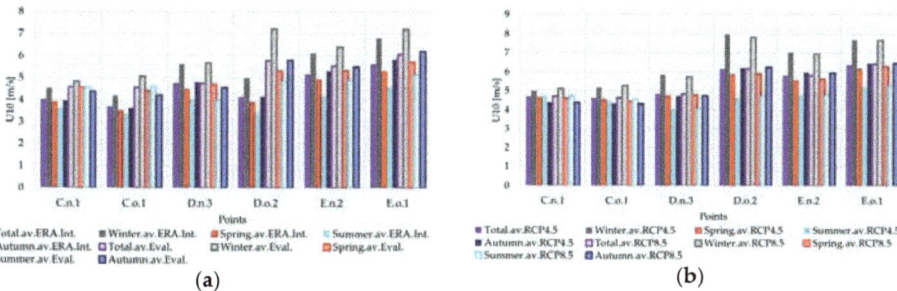

Figure 8. Seasonal wind speed averages at 10 m height compared with the total time averages for some representative points located in shallow water and deep water of Crimea Peninsula, Russia and Turkey coasts; (**a**) ERA.Int and Eval data, time interval 1981–2010; (**b**) RCP4.5 and RCP8.5 data, time interval 2021–2050.

The correlation level between the data provided by the ECWMF ERA-Interim databases and the EURO-CORDEX Evaluation climate model results is presented in Table 3 through the Pearson correlation coefficient. This analysis was performed considering the daily averages. By analyzing the data, it can be observed that the Pearson coefficient varies in the range 0.51–0.92. In order to evaluate the differences between the daily values, the root mean square error (RMSE) and the mean error (Bias) were also computed. The data presented in Table 3 show that RMSE is in the range 0.25–0.5 m/s. As regards the bias values, in all points they are negative (ranging from −0.25 to −0.06 m/s) indicating that the mean Evaluation wind speeds are higher than those computed for ERA-Interim.

According to the correlation levels defined in Table 4, it can be noticed that the correlation between both data ranges from a reasonable correlation to a very high correlation. More precisely, 71% of the data are in the interval of very high correlation, 25% are in the interval of high correlation and only 4% indicate a reasonable correlation. Thus, the lower RMSE and Bias values, together with the higher values of the correlation indexes, show that between the data provided by ERA-Interim and those from the RCM model there is a good agreement.

Table 3. Statistical evaluation in terms of the correlation coefficient (r) and RMSE for all 24 points.

Point	A.n.1	A.n.2	A.n.3	A.n.4	A.o.1	A.o.2	A.o.3	B.n.1	B.n.2	B.n.3	B.n.4	B.o.1
r	0.88	0.88	0.88	0.80	0.89	0.90	0.89	0.83	0.82	0.64	0.77	0.81
RMSE	0.43	0.43	0.42	0.38	0.45	0.44	0.43	0.40	0.40	0.37	0.27	0.50
Bias	−0.19	−0.19	−0.18	−0.14	−0.20	−0.19	−0.18	−0.16	−0.16	−0.14	−0.07	−0.23
Point	B.o.2	C.n.1.	C.n.2	C.o.1	D.n.1	D.n.2	D.n.3	D.o.1	D.o.2	E.n.1	E.n.2	E.o.1
r	0.76	0.51	0.88	0.65	0.80	0.92	0.88	0.91	0.90	0.68	0.87	0.90
RMSE	0.37	0.36	0.29	0.40	0.25	0.31	0.39	0.29	0.50	0.46	0.38	0.44
Bias	−0.14	−0.13	−0.09	−0.16	−0.06	−0.10	−0.15	−0.08	−0.25	−0.21	−0.14	−0.20

Table 4. Levels of interpretation of the Pearson correlation coefficient (r).

$r \in [0.8;1]$		$r \in [0.6;0.8)$		$r \in [0.4;0.6)$		$r \in [0.2;0.4)$		$r \in [0;0.2)$	
1		0.79		0.59		0.39		0.19	
	very high		high		reasonable		poor		very poor
0.9	correlation	0.7	correlation	0.5	correlation	0.3	correlation	0.1	correlation
0.8		0.6		0.4		0.2		0	

Based on the annual averages of the wind speeds, the linear trend in each point was computed and presented in Table 5. The ERA-Interim data shows a slight increasing (ranging from 0 to 0.132 m/s per decade) or decreasing (ranging from −0.054 to −0.002 m/s per decade) trends. The decreasing trends are found only in the western part of the Black Sea basin. These values of the linear trend are in line with those computed by Torralba et al. [40]. As regards the Evaluation data, in 23 points the linear trend has positive values from 0.004 to 0.119 m/s per decade.

Table 5. Linear trend (m/s per decade) values computed for all 24 points for recent past data.

Point	A.n.1	A.n.2	A.n.3	A.n.4	B.n.1	B.n.2	B.n.3	B.n.4	C.n.1	C.n.2	D.n.1	D.n.2
ERA-Interim	−0.041	−0.014	0.000	0.020	0.007	−0.021	−0.054	0.023	0.044	0.065	0.081	0.090
Evaluation	−0.012	0.024	0.050	0.024	0.029	0.019	0.004	0.007	0.034	0.065	0.086	0.107
Point	D.n.3	E.n.1	E.n.2	A.o.1	A.o.2	A.o.3	B.o.1	B.o.2	C.o.1	D.o.1	D.o.2	E.o.1
ERA−Interim	0.039	0.053	0.031	0.006	0.005	−0.002	−0.047	0.024	0.056	0.132	0.064	0.042
Evaluation	0.065	0.088	0.095	0.047	0.065	0.050	0.014	0.032	0.055	0.119	0.144	0.047

3.1.2. Future Projections

Another comparative analysis between the wind speeds projected by two of the future scenarios available in the framework of the EURO−CORDEX project, RCP4.5 and RCP8.5, is presented in Figure 2b. By analyzing Figure 2b it can be noticed that between both data no significant differences appear in terms of average values, 5th and 95th percentiles. On the other hand, major differences in the evaluation of the maximum wind speed values in the shallow water areas of Bulgaria and Romania are observed.

According to the near future data, both scenarios present approximately the same values in terms of annual averages, but with a lag of several years between the peaks. As previously presented, a slight difference occurs between the data from RCP4.5 and RCP8.5. In the case of zone A (deep water locations), the data for RCP4.5 scenarios are on average with 1% higher, while for zone B with almost 1% smaller than the 8.5 data. By conducting a comparative analysis of the shallow water and deep water locations for both zones A and B it can be concluded that there is no notable difference in terms of wind intensity.

The near future projections (time interval 2021–2050) near to the Turkish coast (see Figure 6) present similar average values of the wind speeds as those computed from the EURO-CORDEX Evaluation data for the present climate. The near future annual wind speed analysis does not highlight any trend. Moreover, the differences between the annual averages are lower than the previous cases.

Although the data for the interval 1981–2010 show an increase of the wind speed near to the Russian coastline (see Figure A1), both estimates an approximate constant pattern for the 30-year interval of the near future. Both RCP4.5 and RCP8.5 scenarios predict for the Crimea Peninsula region an increase of the wind at 10-m height (Figure A2).

Regarding the linear tendency, for the 2021–2050 time period, the RCP4.5 scenario shows that the wind speed will have a low decrease while the RCP8.5 a small increase. The values of the linear trends computed for near future data are presented in Table 6.

Table 6. Linear trend (m/s per decade) values computed for all 24 points for near future data.

Point	A.n.1	A.n.2	A.n.3	A.n.4	B.n.1	B.n.2	B.n.3	B.n.4	C.n.1	C.n.2	D.n.1	D.n.2
RCP4.5	−0.026	−0.026	−0.040	−0.039	−0.013	−0.016	−0.022	−0.028	−0.021	−0.013	−0.013	−0.011
RCP8.5	0.016	0.023	0.016	0.015	0.025	0.018	0.044	0.016	−0.037	−0.027	0.008	0.010

Point	D.n.3	E.n.1	E.n.2	A.o.1	A.o.2	A.o.3	B.o.1	B.o.2	C.o.1	D.o.1	D.o.2	E.o.1
RCP4.5	−0.009	0.006	−0.023	−0.044	−0.042	−0.040	−0.018	−0.042	−0.012	−0.018	−0.011	0.002
RCP8.5	0.034	0.044	0.031	0.009	0.011	0.023	0.044	0.022	−0.041	−0.013	0.023	0.024

3.1.3. Directional Analysis

The four datasets analyzed contain information about the zonal and meridional velocities needed for analyzing the wind main directions from which the winds are blowing. From the wind roses presented in Figure 9, it can be observed the dominant direction and also the predominant wind speed range for all five zones studied. At a first sight, the data show that over the Black Sea there are two main wind patterns. The first pattern is more spread and it is observed in the northwest and west of the Black Sea basin. According to all the models considered, the main directions for these zones are north, northeast and southwest. However, it can be noticed that all three data from EURO-CORDEX show also that the wind blows quite often from the west. Moreover, the most frequent values of the wind speeds are in the interval 0 to 12 m/s.

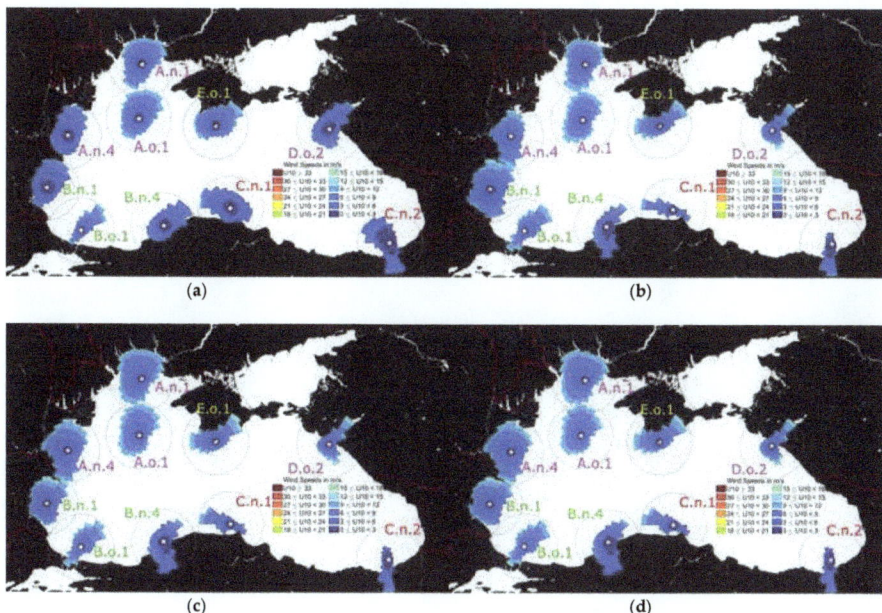

Figure 9. Wind roses corresponding to 30-year of data (1981–2010) coming from: (**a**) ECMWF ERA-Interim and (**b**) EURO-CORDEX Evaluation; 30-year interval (2021–2050) coming from: (**c**) EURO-CORDEX RCP4.5 scenario and (**d**) EURO-CORDEX RCP8.5 scenario.

The second dominant wind pattern is narrower and occurs in the north, east, south, and southwest. However, in these zones, the wind has different directions. It can also be observed that the ERA-Interim datasets show a different wind direction pattern in comparison with the other three models for the points located north and northeast of Turkey, east of Russia and south of Crimea Peninsula. Nevertheless, the four datasets show an identical pattern for the point located in deep water, near the Bosporus Strait.

3.2. Wind Power Analysis

The near future projections of the wind power potential of the Black Sea for the time interval 2021–2050 were also computed and compared with those corresponding to the time interval 1981–2010. Nowadays, the typical hub heights for the offshore wind turbines are ranging about 100 m [41], and for this reason, the wind data at 10 m height need to be recompiled. In order to adjust the wind speed to a level of 100 m for assessing the wind energy potential and output energy generated by a certain turbine, a logarithmic method is used. This method assumes neutral stability conditions [42,43]:

$$U_{100} = U_{10} \cdot \frac{\ln \frac{z_{100}}{z_0}}{\ln \frac{z_{10}}{z_0}}, \tag{3}$$

where U_{100} represents the wind speed at 100 m height (in m/s), U_{10} is the wind speed at 10 m height, $z_0 = 0.0002$ m is the surface roughness of the sea surface [44,45], while z_{100} and z_{10} are the reference heights at 100 m and 10 m, respectively.

Assessing the wind energy potential implies using several key parameters as wind power density generated by a certain wind turbine [46]. The wind power density [W/m^2] potential per unit of swept area can be determined as follows:

$$P_{wind100} = \frac{\rho \cdot U_{100}^3}{2}, \tag{4}$$

where ρ is the air density (1.22 kg/m^3).

Figures 10 and A3 illustrates the annual wind power and the total time averages [W/m^2] for 12 of the most representative points considered in the Black Sea basin. As expected, for the present period (1981–2010), the wind power values computed considering the EURO-CORDEX Evaluation data are higher than those based on the ERA-Interim data, the difference ranging between 12 W/m^2 (D.n.3) to 285 W/m^2 (D.o.2). The graphs presented in Figures 10 and A3 show that, in general, the evolution of the annual averages for both emission scenarios considered presents a similar pattern, while the wind power average values computed for the entire interval are very close.

In order to evaluate the near future evolution of the wind power averages, a comparison between the values computed based on data from RCP4.5 and RCP8.5 scenarios and those that represent the present data (Evaluation data) is performed. An increase of the averages is observed in the near future, except in the case of the point B.n.3. At this point, besides the fact that the average will decrease, the variability of the annual averages along the entire 30-year interval is lower by comparison with the 1981–2010 data. In Figures 10 and A3, the wind power potential computed in 4 points (A.n.4, B.n.1, D.o.2, and E.o.1) shows a significant increase for the near future, while in the case of 7 points (A.n.1, A.o.1, B.o.1, C.n.1, C.o.1, D.n.3, and E.n.2) a slight increase is noticed.

According to the wind power averages computed for the near future, it results that the most energetic zones of the Black Sea seem to be located in the northwest (points A.n.1 and A.o.1), southwest (point B.o.1), east (point D.o.2) and the Crimea Peninsula (point E.o.1). More precisely, in these points, the averages computed for the 30-year period are higher than 500 W/m^2. Regarding zone A, the higher increase of the average wind power in the near future is observed in point A.n.4 (Figure 10b), where this will pass the level of 400 W/m^2, while in the present it is below this threshold.

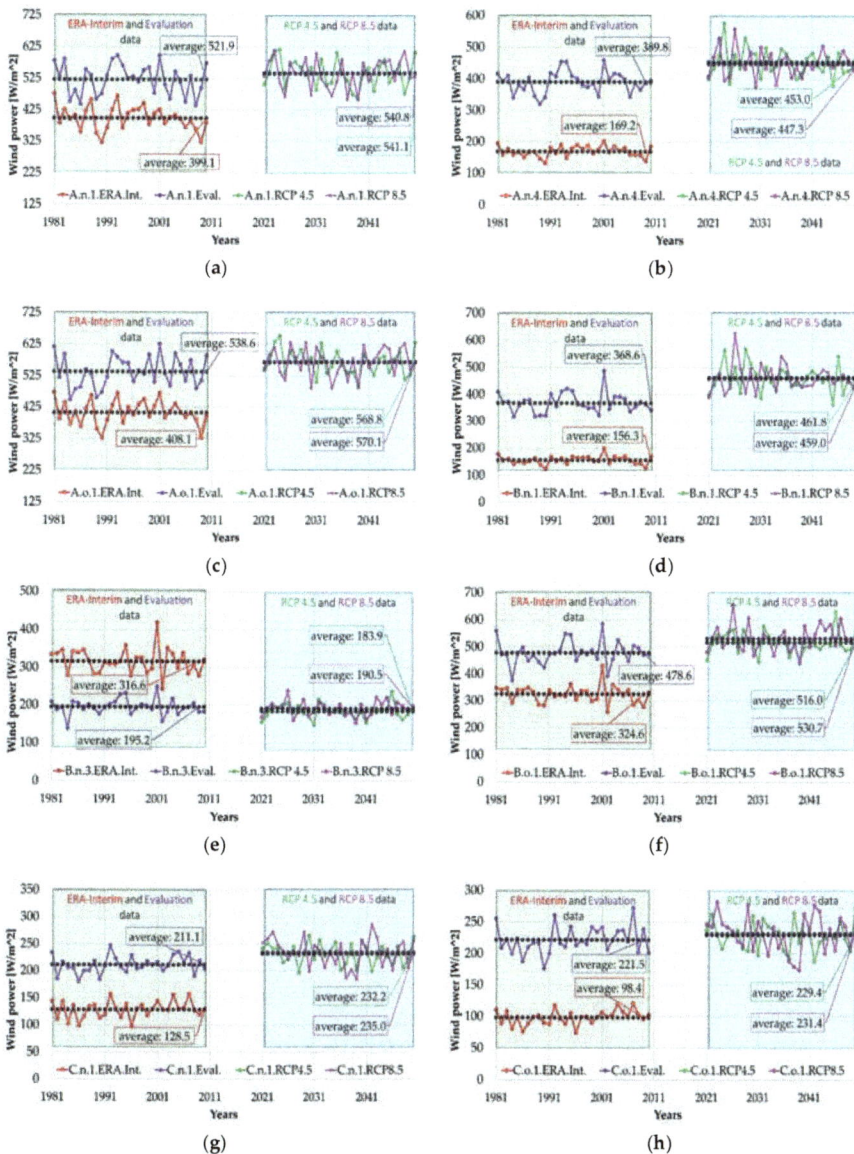

Figure 10. Annual wind power averages at 100 m height for 8 representative points: (**a**) A.n.1; (**b**) A.n.4; (**c**) A.o.1; (**d**) B.n.1; (**e**) B.n.3; (**f**) B.o.1; (**g**) C.n.1; (**h**) C.o.1.ERA.Int and Eval data, time interval 1981–2010; RCP4.5 and RCP8.5 data, time interval 2021–2050. For each dataset, the average of the wind power corresponding to the entire time interval is indicated in the box.

Regarding the zone B, two opposite cases are noticed. The averages for the points B.n.1 and B.o.1 present a significant increase. Thus, the wind power average for B.n.1 (Figure 10d) exceeds the value of 400 W/m^2, while in the case of the point B.o.1 (Figure 10f) this will be higher than 500 W/m^2. As above mentioned, an opposite behavior is found in the point B.n.3 (Figure 10e), which is located near Bosporus Strait. In the case of zone C, points C.n.1 (Figure 10g) and C.o.1 (Figure 10h), no major changes seem to be encountered in terms of the wind power averages. They will be maintained near the threshold of 200 W/m^2. As regards the reference points from the zone D, the higher increase in the average wind power is observed at the point located in deep water, D.o.2 (Figure A3b). This point represents in fact one of the most energetic locations in the Black Sea basin.

Another zone that can have a significant impact on the wind energy exploitation is zone E, situated near the Crimea Peninsula. Here, the data in the points E.n.2 (Figure A3c) and E.o.1 (Figure A3d) were studied in more detail. By analyzing these data, it can be observed that even though these two points are relatively close, the growth of the average value is higher in the deep water point E.o.1 (about 100 W/m^2), than in the shallow water point E.n.2.

The wind power averages at 100 m computed for the entire 30-year period and corresponding to the 12 reference points are now compared with the seasonal values, the results being presented in Figure 11. In this analysis, the four datasets available are also considered. The results clearly show that the winter averages are the highest, while the least energetic season is the summer. The other two seasons present almost equal values of the wind power averages.

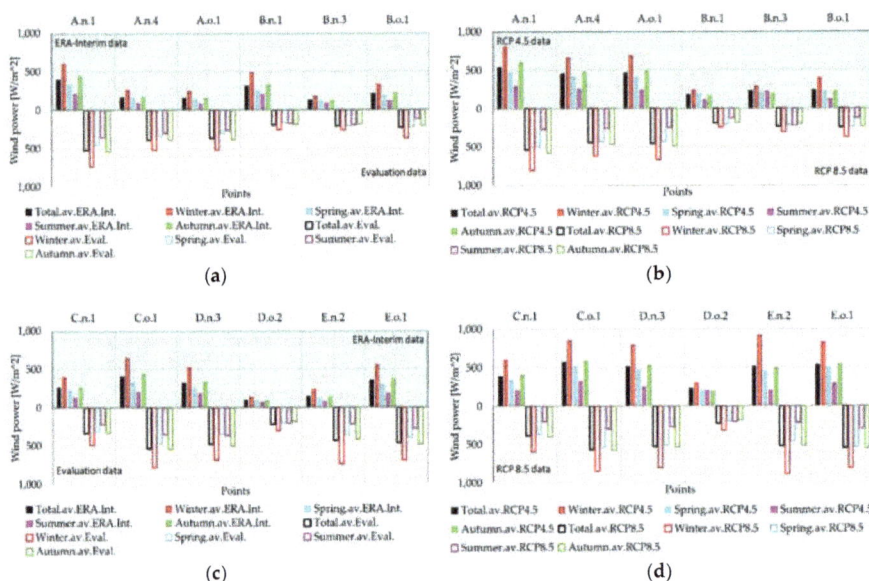

Figure 11. Seasonal wind power averages at 100 m height compared with the total time averages for 12 points considered the most representative. ERA.Int and Eval data, time interval 1981–2010; RCP4.5 and RCP8.5 data, time interval 2021–2050. Points A.n.1, A.n.4, A.o.1, B.n.1, B.n.3 and B.o.1: (a) ERA.Int and Eval data, (b) RCP4.5 and RCP8.5 data; Points C.n.1, C.o.1, D.n.3, D.o.2, E.n.2 and E.o.1: (c) ERA.Int and Eval data, (d) RCP4.5 and RCP8.5 data.

4. Conclusions

From the comparative analysis of the wind speed at 10 m over the sea level simulated by RCM from EURO-CORDEX, the RCA4 model respectively, against the data existent in ERA-Interim database, a good correlation was noticed between these two datasets in the Black Sea area. On the other hand, probably due to the fact that the spatial resolution of the RCM model is higher than that from ERA-Interim, the wind speeds provided by RCA4 in the reference points considered in the Black Sea basin are slightly higher than those indicated by ERA-Interim.

The analysis of the wind speed annual averages shows that the wind speeds at 10 m height have a different evolution in the areas considered in the Black Sea. In general, for the present period, it was observed that the Evaluation wind speed averages are higher than those from ERA-Interim with values ranging from 0.06 to 0.25 m/s. Through the linear trend, it was observed that in 75% of the cases both data have the same trend (upward trend—the maximum value is 0.144 m/s per decade).

The daily averages of the wind speeds provided by both databases present a very good correlation. The Pearson correlation coefficient varies in the range 0.51 to 0.92. More precisely 71% of the data are in the interval of very high correlation, 25% are in the interval of high correlation and only 4% have a reasonable correlation. The differences between the daily values were assessed by the analysis of the root mean square error and Bias. The results of the analysis show that RMSE is in the range 0.25 to 0.50 m/s, while the bias values, in all points they are negative (ranging from -0.25 to -0.06 m/s). Thus, the low RMSE and Bias values together with the higher values of the correlation indexes show that there is a good agreement between the data provided by ERA-Interim and those from the RCM model (Evaluation data).

As the results of the present work show, in most of the cases there are no relevant differences in the average wind speeds (smaller than 0.4 m/s) simulated under both scenarios. This is probably due to the fact that until the mid-century the differences between the two RCPs considered are not very high. According to the near future data, both scenarios present approximately the same values in terms of annual averages. Regarding the linear tendency for the period 2021–2050, the RCP4.5 scenario shows that the wind speed will have a low decrease in 92% of the points (values ranging from -0.044 to -0.009 m/s per decade) while the RCP8.5 a small increase in 83% of the cases (values ranging from 0.008 to 0.044 m/s per decade).

As regards the average wind power potential in winter season, for 51% of the reference points a significant increase was observed for the near future (both scenarios) compared with the present values, while for 41% only a slight growth. For both scenarios, a small decrease was noticed only in the point B.n.1. The most energetic zones of the Black Sea are the western part of the basin (northwest and southwest areas) and also in the east and south of the Crimea Peninsula. More precisely, in these points, the total wind power averages for the 30-year time interval are higher than 500 W/m^2. As expected, in the winter time the higher wind speeds and wind powers are encountered, while the lower values are found in the summer.

The fact that it was noticed an increase of the mean values of the wind resources in 95.6% of the reference points considered in the Black Sea, either under the RCP4.5 or RCP8.5 scenarios, can be considered beneficial from the perspective of the wind projects and this can give momentum to the installation of the wind farms in the areas already identified as having a good potential, as for example the western side of the basin. From this perspective, it is well known that some European areas are affected by a decrease of the wind resources (see for example [35,47,48]).

Author Contributions: D.G. processed and analyzed the data. E.M. performed the literature review and also gathered the data. L.R. has guided this research, discussed the data and drawn the main conclusions. The final manuscript has been approved by all authors.

Funding: This research was funded by a grant of Ministry of Research and Innovation, CNCS—UEFISCDI, project number PN-III-P4-ID-PCE-2016-0028 (Assessment of the Climate Change effects on the WAve conditions in the Black Sea—ACCWA), within PNCDI III.

Acknowledgments: The ECMWF data used in this study have been obtained from the ECMWF data servers. The EURO-CORDEX scenarios RCP4.5 and RCP8.5 and also the EURO-CORDEX Evaluation and Historical data used in this study have been obtained from the EURO-CORDEX data servers. The authors would like also to express their gratitude to the reviewers for their constructive suggestions and observations that helped in improving the present work.

Conflicts of Interest: The authors declare no conflict of interest.

Energies **2018**, *11*, 3198

Nomenclature

av.total	average value
CORDEX	Coordinated Regional Downscaling Experiment
ECMWF	European Centre for Medium-Range Weather Forecasts
ERA.Int.	ERA-Interim wind fields from ECMWF
EURO-CORDEX	European branch of CORDEX
Eval.	EURO-CORDEX Evaluation wind fields
EWEA	European Wind Energy Association
GCM	Global Climate Model
lower 5%	5th percentiles analysis
lower 95%	95th percentiles analysis
max. total	maximum values
NAO	North Atlantic Oscillations
NCEP	US National Centers for Environmental Prediction
$P_{wind100}$	wind power at 100 m height
r	Pearson correlation coefficient
RCA4	Rossby Centre regional climate model
RCM	Regional Climate Model
RCP	Representative Concentration Pathways
RMSE	root mean square error
u	zonal wind velocity
U_{10}	wind speed at 10 m height
U_{100}	wind speed at 100 m height
v	meridional wind velocity
z_{10}	reference height at 10 m
z_{100}	reference height at 100 m
WCRP	World Climate Research Program
WMO	World Meteorological Organization

Appendix A

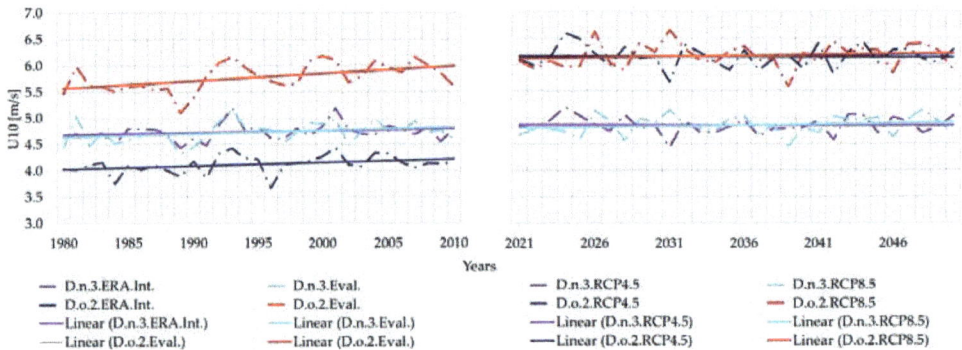

Figure A1. Annual averages of the wind speed at 10 m height for two of the most representative points located in shallow water (D.n.3) and in deep water (D.o.2) of Russia. ERA.Int and Eval data, time interval 1981–2010, RCP4.5 and RCP8.5 data, time interval 2021–2050.

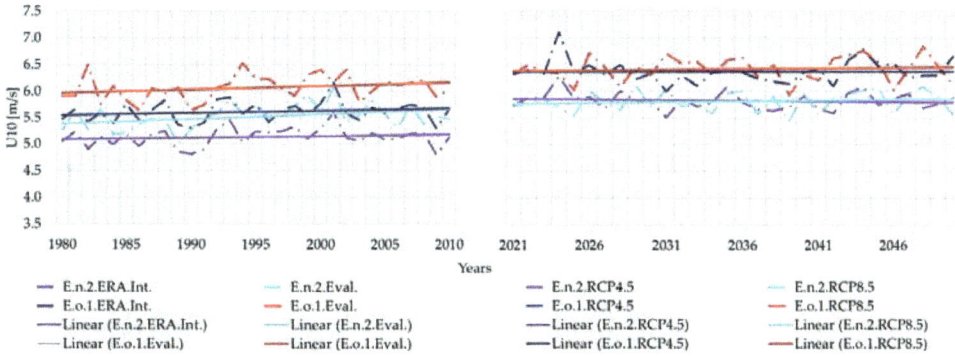

Figure A2. Annual averages of the wind speed at 10 m height for two of the most representative points located in shallow water (E.n.2) and in deep water (E.o.1) of Crimea Peninsula. ERA.Int and Eval data, time interval 1981–2010, RCP4.5 and RCP8.5 data, time interval 2021–2050.

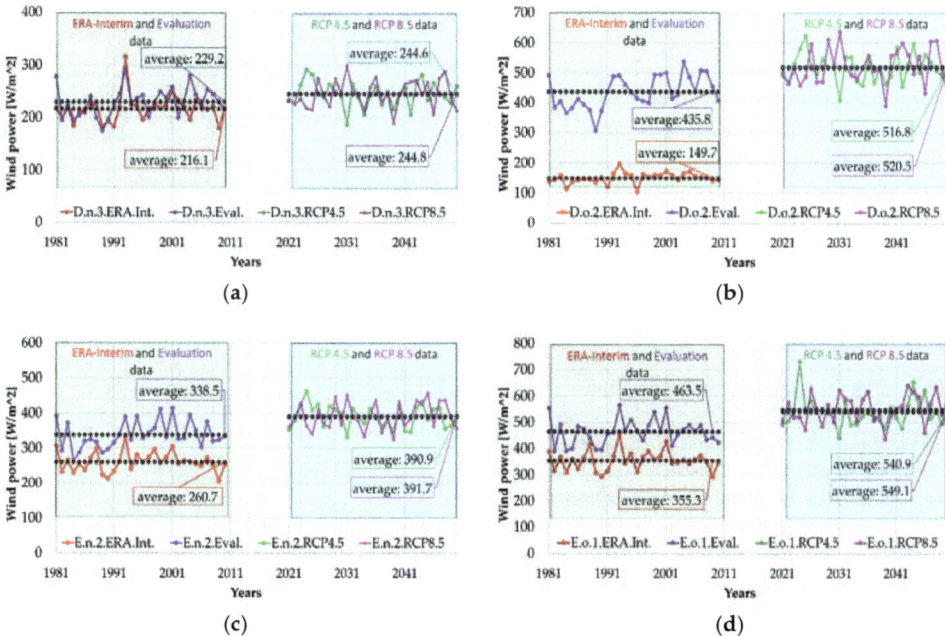

Figure A3. Annual wind power averages at 100 m height for the points: (**a**) D.n.3, (**b**) D.o.2, (**c**) E.n.2 and (**d**) E.o.1. ERA.Int and Eval data, time interval 1981–2010; RCP4.5 and RCP8.5 data, time interval 2021–2050. For each dataset, the average of the wind power corresponding to the entire time interval is indicated in the box.

References

1. Ganea, D.; Amortila, V.; Mereuta, E.; Rusu, E. A joint evaluation of the wind and wave energy resources close to the greek islands. *Sustainability* **2017**, *9*, 1025. [CrossRef]
2. Ponta, L.; Raberto, M.; Teglio, A.; Cincotti, S. An agent-based stock-flow consistent model of the sustainable transition in the energy sector. *Ecol. Econ.* **2018**, *145*, 274–300. [CrossRef]
3. De Filippo, A.; Lombardi, M.; Milano, M. User-aware electricity price optimization for the competitive market. *Energies* **2017**, *10*, 1378. [CrossRef]

4.   Rusu, E. Numerical modeling of the wave energy propagation in the Iberian nearshore. *Energies* **2018**, *11*, 980. [CrossRef]

5.   European Commission. Available online: https://ec.europa.eu/energy/en/topics/energy-strategy-and-energy-union (accessed on 14 November 2017).

6.   Romanian Energy Strategy, Ministry of Energy. Available online: http://energie.gov.ro/wp-content/uploads/2016/12/Strategia-Energetica-a-Romaniei-2016-2030_FINAL_19-decembrie-2.pdf (accessed on 2 December 2017).

7.   Nikolaev, A.; Konidari, P. Development and assessment of renewable energy policy scenarios by 2030 for Bulgaria. *Renew. Energy* **2017**, *111*, 792–802. [CrossRef]

8.   Chomakhidze, D.; Tskhakaia, K.; Shamaevi, D. Twenty years' experience of the regulation of energy in Georgia. *Energy Procedia* **2017**, *128*, 130–135. [CrossRef]

9.   Smeets, N. The green menace: Unraveling Russia's elite discourse on enabling and constraining factors of renewable energy policies. *Energy Res. Soc. Sci.* **2018**, *40*, 244–256. [CrossRef]

10.   Kaplan, Y.A. Overview of wind energy in the world and assessment of current wind energy policies in Turkey. *Renew. Sustain. Energy Rev.* **2015**, *43*, 562–568. [CrossRef]

11.   Child, M.; Breyer, C.; Bogdanov, D.; Fell, H.-J. The role of storage technologies for the transition to a 100% renewable energy system in Ukraine. *Energy Procedia* **2017**, *135*, 410–423. [CrossRef]

12.   Kurbatova, T.; Sotnyk, I.; Khlyap, H. Economical mechanisms for renewable energy stimulation in Ukraine. *Renew. Sustain. Energy Rev.* **2014**, *31*, 486–491. [CrossRef]

13.   WindEurope. Wind in Power 2017. Annual Combined Onshore and Offshore Wind Energy Statistics. Available online: https://windeurope.org/wp-content/uploads/files/about-wind/statistics/WindEurope-Annual-Statistics-2017.pdf (accessed on 28 March 2018).

14.   Onea, F.; Rusu, E. Wind energy assessments along the Black Sea basin. *Meteorol. Appl.* **2014**, *21*, 316–329. [CrossRef]

15.   Rosa, A.V.D. *Fundamentals of Renewable Energy Processes*; Academic Press: Cambridge, MA, USA, 2012; ISBN 978-0-12-397219-4.

16.   Rusu, L. Assessment of the wave energy in the Black Sea based on a 15-year hindcast with data assimilation. *Energies* **2015**, *8*, 10370–10388. [CrossRef]

17.   Onea, F.; Rusu, E. Efficiency assessments for some state of the art wind turbines in the coastal environments of the Black and the Caspian seas. *Energy Explor. Exploit.* **2016**, *34*, 217–234. [CrossRef]

18.   Valchev, N.; Davidan, I.; Belberov, Z.; Palazov, A.; Valcheva, N. Hindcasting and assessment of the western Black Sea wind and wave climate. *J. Environ. Prot. Ecol.* **2010**, *11*, 1001–1012.

19.   Rusu, L.; Bernardino, M.; Guedes Soares, C. Wind and wave modelling in the Black Sea. *J. Oper. Oceanogr.* **2014**, *7*, 5–20. [CrossRef]

20.   Santos, J.A.; Rochinha, C.; Liberato, M.L.R.; Reyers, M.; Pinto, J.G. Projected changes in wind energy potentials over Iberia. *Renew. Energy* **2015**, *75*, 68–80. [CrossRef]

21.   Luong, N.D. A critical review on potential and current status of wind energy in Vietnam. *Renew. Sustain. Energy Rev.* **2015**, *43*, 440–448. [CrossRef]

22.   Mostafaeipour, A.; Jadidi, M.; Mohammadi, K.; Sedaghat, A. An analysis of wind energy potential and economic evaluation in Zahedan, Iran. *Renew. Sustain. Energy Rev.* **2014**, *30*, 641–650. [CrossRef]

23.   Davy, R.; Gnatiuk, N.; Pettersson, L.; Bobylev, L. Climate change impacts on wind energy potential in the European domain with a focus on the Black Sea. *Renew. Sustain. Energy Rev.* **2018**, *81*, 1652–1659. [CrossRef]

24.   Koletsis, I.; Kotroni, V.; Lagouvardos, K.; Soukissian, T. Assessment of offshore wind speed and power potential over the Mediterranean and the Black Seas under future climate changes. *Renew. Sustain. Energy Rev.* **2016**, *60*, 234–245. [CrossRef]

25.   Onat, N.; Ersoz, S. Analysis of wind climate and wind energy potential of regions in Turkey. *Energy* **2011**, *36*, 148–156. [CrossRef]

26.   Lin-Ye, J.; García-León, M.; Gràcia, V.; Ortego, M.I.; Stanica, A.; Sánchez-Arcilla, A. Multivariate hybrid modelling of future wave-storms at the northwestern Black Sea. *Water* **2018**, *10*, 221. [CrossRef]

27.   Onea, F.; Raileanu, A.; Rusu, E. Evaluation of the wind energy potential in the coastal environment of two enclosed seas. *Adv. Meteorol.* **2015**, *2015*, 808617. [CrossRef]

28. Dee, D.P.; Uppala, S.M.; Simmons, A.J.; Berrisford, P.; Poli, P.; Kobayashi, S.; Andrae, U.; Balmaseda, M.A.; Balsamo, G.; Bauer, P.; et al. The ERA-interim reanalysis: Configuration and performance of the data assimilation system. *Q. J. R. Meteorol. Soc.* **2011**, *137*, 553–597. [CrossRef]

29. Cardinali, C. *Data Assimilation. Observation Impact on the Short Range Forecast*; ECMWF Lecture Notes; European Centre for Medium-Range Weather Forecasts (ECMWF): Reading, UK, 2013.

30. Shanas, P.R.; Sanil Kumar, V. Temporal variations in the wind and wave climate at a location in the eastern Arabian Sea based on ERA-interim reanalysis data. *Nat. Hazards Earth Syst. Sci.* **2014**, *14*, 1371–1381. [CrossRef]

31. EURO—CORDEX Guidelines. Available online: http://www.euro-cordex.net/imperia/md/content/csc/cordex/euro-cordex-guidelines-version1.0-2017.08.pdf (accessed on 12 December 2017).

32. Jacob, D.; Petersen, J.; Eggert, B.; Alias, A.; Christensen, O.B.; Bouwer, L.M.; Braun, A.; Colette, A.; Déqué, M.; Georgievski, G.; et al. EURO-CORDEX: New high-resolution climate change projections for European impact research. *Reg. Environ. Chang.* **2014**, *14*, 563–578. [CrossRef]

33. Strandberg, G.; Bärring, L.; Hansson, U.; Jansson, C.; Jones, C.; Kjellström, E.; Kupiainen, M.; Nikulin, G.; Samuelsson, P.; Ullerstig, A. *CORDEX Scenarios for Europe from the Rossby Centre Regional Climate Model RCA4*; Report Meteorology and Climatology No. 16; SMHI: Norrköping, Sweden, 2015. Available online: https://www.smhi.se/polopoly_fs/1.90273!/Menu/general/extGroup/attachmentColHold/mainCol1/file/RMK_116.pdf (accessed on 14 December 2017).

34. Kjellström, E.; Bärring, L.; Nikulin, G.; Nilsson, C.; Persson, G.; Strandberg, G. Production and use of regional climate model projections – A Swedish perspective on building climate services. *Clim. Serv.* **2016**, *2*, 15–29. [CrossRef] [PubMed]

35. Soares, P.M.; Lima, D.C.; Cardoso, R.M.; Nascimento, M.L.; Semedo, A. Western Iberian offshore wind resources: More or less in a global warming climate? *Appl. Energy* **2017**, *203*, 72–90. [CrossRef]

36. Hemer, M.A.; Trenham, C.E. Evaluation of a CMIP5 derived dynamical global wind wave climate model ensemble. *Ocean Model.* **2016**, *103*, 190–203. [CrossRef]

37. Rusu, L.; Raileanu, A.; Onea, F. A Comparative analysis of the wind and wave climate in the Black Sea along the shipping routes. *Water* **2018**, *10*, 924. [CrossRef]

38. Moss, R.H.; Edmonds, J.A.; Hibbard, K.A.; Manning, M.R.; Rose, S.K.; van Vuuren, D.P.; Carter, T.R.; Emori, S.; Kainuma, M.; Kram, T.; et al. The next generation of scenarios for climate change research and assessment. *Nature* **2010**, *463*, 747–756. [CrossRef] [PubMed]

39. Van Vuuren, D.P.; Edmonds, J.; Kainuma, M.; Riahi, K.; Thomson, A.; Hibbard, K.; Hurtt, G.C.; Kram, T.; Volker, K.; Lamarqu, J.-F.; et al. The representative concentration pathways: An overview. *Clim. Chang.* **2011**, *109*, 5. [CrossRef]

40. Torralba, V.; Doblas-Reyes, F.J.; Gonzalez-Reviriego, N. Uncertainty in recent near-surface wind speed trends: A global reanalysis intercomparison. *Environ. Res. Lett.* **2017**, *12*, 114019. [CrossRef]

41. Letcher, T. *Wind Energy Engineering: A Handbook for Onshore and Offshore Wind Turbines*, 1st ed.; Academic Press/Elsevier: Amsterdam, The Netherlands, 2017; 622p, ISBN 9780128094518.

42. Kubik, M.L.; Coker, P.J.; Hunt, C. Using meteorological wind data to estimate turbine generation output: A sensitivity analysis. In Proceeding of the World Renewable Energy Congress, Linköping, Sweden, 8–13 May 2011; Linköping University Electronic Press: Linköping, Sweeden, 2011; No. 057; pp. 4074–4081.

43. Yamada, T.; Mellor, G. A Simulation of the Wangara atmospheric boundary layer data. *J. Atmos. Sci.* **1975**, *32*, 2309–2329. [CrossRef]

44. WAsP. WAsP 9 Documentation: The Roughness of a Terrain. 2014. Available online: http://www.wasp.dk (accessed on 13 October 2017).

45. Onea, F.; Deleanu, L.; Rusu, L.; Georgescu, C. Evaluation of the wind energy potential along the Mediterranean Sea coasts. *Energy Explor. Exploit.* **2016**, *34*, 766–792. [CrossRef]

46. Wind Turbines Theory—The Betz Equation and Optimal Rotor Tip Speed Ratio IntechOpen. Available online: https://www.intechopen.com/books/fundamental-and-advanced-topics-in-wind-power/wind-turbines-theory-the-betz-equation-and-optimal-rotor-tip-speed-ratio (accessed on 15 March 2018).

47. Moemken, J.; Reyers, M.; Feldmann, H.; Pinto, J.G. Future changes of wind speed and wind energy potentials in EURO-CORDEX ensemble simulations. *J. Geophys. Res. Atmos.* **2018**, *123*, 6373–6389. [CrossRef]
48. Weber, J.; Wohland, J.; Reyers, M.; Moemken, J.; Hoppe, C.; Pinto, J.G.; Witthaut, D. Impact of climate change on backup energy and storage needs in wind-dominated power systems in Europe. *PLoS ONE* **2018**, *13*, e0201457. [CrossRef] [PubMed]

MDPI

St. Alban-Anlage 66

4052 Basel

Switzerland

Tel. +41 61 683 77 34

Fax +41 61 302 89 18

www.mdpi.com

Energies Editorial Office

E-mail: energies@mdpi.com

www.mdpi.com/journal/energies

www.ingramcontent.com/pod-product-compliance
Lightning Source LLC
Chambersburg PA
CBHW051710210326
41597CB00032B/5429